Lecture Notes in Mathematics

Edited by A. Dold, B. Eckmann and F. Takens

1454

F. Baldassarri S. Bosch
B. Dwork (Eds.)

p-adic Analysis

Proceedings of the International Conference
held in Trento, Italy, May 29–June 2, 1989

Springer-Verlag
Berlin Heidelberg New York London
Paris Tokyo Hong Kong Barcelona

Editors

Francesco Baldassarri
Dipartimento di Matematica Pura e Applicata
Università di Padova
Via Belzoni 7, 35131 Padova, Italy

Siegfried Bosch
Mathematisches Institut der Universität
Einsteinstr. 62, 4400 Münster, Federal Republic of Germany

Bernard Dwork
Department of Mathematics, Princeton University
Princeton, N.J. 08544, USA

Mathematics Subject Classification (1980): 12Jxx, 14Fxx, 12Hxx, 46PO5

ISBN 3-540-53477-6 Springer-Verlag Berlin Heidelberg New York
ISBN 0-387-53477-6 Springer-Verlag New York Berlin Heidelberg

Printing and binding: Druckhaus Beltz, Hemsbach/Bergstr.
2146/3140-543210 – Printed on acid-free paper

INTRODUCTION

The present volume contains the Proceedings of the Congress on «p-adic Analysis» held at Trento from May 28 to June 3, 1989.

The idea of organizing a meeting on this subject in Italy was first promoted by Philippe Robba, whose visits to Italy were always welcomed by his Italian colleagues for both the warmth and the illumination which he brought with him. He died prematurely on October 12, 1988, leaving a profound sense of loss in the world of p-adic analysis. We believe we have expressed the feelings of that whole community by dedicating this Meeting to him. At the opening of the Conference, Elhanan Motzkin commemorated Robba's exceptional character in a touching reminiscence, that will appear in the Seminars of the Groupe d'Etude d'Analyse Ultramétrique, of which Robba was one of the founders.

The conference was organized by the Centro Internazionale per la Ricerca Matematica (CIRM), of Trento, and was also sponsored by the Dipartimento di Matematica Pura e Applicata of the University of Padova. We are grateful to both these institutions.

We wish to express our gratitude in particular to Mr. Augusto Micheletti for his indefatigable efforts on behalf of the conference.

F. Baldassarri, S. Bosch, B. Dwork

Philippe Robba

March 18, 1941 - October 12, 1988

CONTENTS

Work of Philippe Robba

B. Dwork
Department of Mathematics
Princeton University

Among the subjects studied by Philippe Robba were:

1. domains of analyticity
2. p–adic Mittag–Leffler
3. index of differential operators
4. factorization of differential operators corresponding to radii of convergence and to order of logarithmic growth
5. effective estimates for logarithmic growth
6. weak frobenius (dimension one, precursor of work of Christol)
7. L–functions and exponential sums
8. application of p–adic methods to questions of irrationality and transcendence.

Robba's work was so involved with the p–adic theory of ordinary differential equations that it may be useful in an article devoted to his work to give a survey of the present status of this subject.

Let K be a field of characteristic zero, complete under a rank one valuation extending the ordinary p–adic valuation of $\mathbb{Q}$.

Let E be the completion of $K(X)$ under the gauss norm. Elements of E are admissible (resp: superadmissible) if they are analytic elements on the complement of a finite set of residue classes (resp: a finite set of disks of local radius strictly less than unity).

A p–adic Liouville number is an element $\alpha \in \mathbb{Z}_p$ (necessarily transcendental over $\mathbb{Q}$) such that either

$$\liminf_{m \to +\infty} |\alpha - m|^{1/m} < 1$$

or

$$\liminf_{m \to \infty} |\alpha + m|^{1/m} < 1.$$

These conditions are not equivalent and for the operator $L = x\frac{d}{dx} - \alpha$ the first condition gives difficulties at $x = 0$ while the second gives difficulties at $x = \infty$.

We recall the notion of a generic point t in a universal domain Ω which is algebraically closed and complete relative to a valuation extending that of K. We insist that $|t| = 1$ and the residue class of t be transcendental over the residue class field of K. The disks and annuli appearing in our theory involve subsets of Ω.

Let

$$\mathcal{A}_0 = \{\xi \in K[[x]] \mid \xi \text{ converges in } D(0,1^-)\}$$
$$\mathcal{B}_0 = \{\xi \in \mathcal{A}_0 \mid \xi \text{ is bounded on } D(0,1^-)\}$$
$$\mathcal{A}_t(r) = \{\xi \in \overline{K(t)}[[x-t]] \mid \xi \text{ converges in } D(0,r^-)\}$$
$$W_a^{r,\beta} = \{\xi = \Sigma A_j(x-a)^j \in \overline{K(a)}[[x-a]] \mid \operatorname*{Sup}_j A_j r^j/(1+j)^\beta < \infty\}.$$

The theory started in 1937 with Lutz's solution of the Cauchy problem:

Let $f(x,\vec{y})$ be an element of $K[[X,Y_1,\ldots,Y_n]]^n$ converging on a polydisk in $n+1$ space. Then the equation

$$\frac{d\vec{y}}{dx} = f(x,\vec{y})$$
$$\vec{y}(0) = 0$$

has a unique solution in $(xK[[x]])^n$ converging on a non–trivial disk about the origin.

Lutz estimated the radius of convergence and applied it to the study of rational points on elliptic curves.

Our own interest in ${}_2F_1(\frac{1}{2},\frac{1}{2},1,x)$ dates to the late 1950's and involved the (long unpublished) calculation of Tate's constant (cf. Dwork 1987). Our interest in the general theory of linear equations goes back to our study of the variation of cohomology of hypersurfaces (Dwork 1964, 66). Clark's work on linear equations at a singular point appeared in 1966. It was here that the question of p–adic Liouville exponents was first discussed. Adolphson [1976a, 1976b] investigated symmetric powers of ${}_2F_1(\frac{1}{2},\frac{1}{2},1,x)$ and studied index in the early 1970's. Robba's work started in 1974.

We will restrict our attention to linear equations but I cannot refrain from mentioning the splendid result of Sibuya and Sperber [1981].

THEOREM. *Let $y_0 \in K[[X]]$ be a formal solution of a non–linear polynomial differential equation, $P(x,y,y',\ldots,y^{(n)}) = 0$, where P is a polynomial in $n+2$ variables with coefficients in K. Substituting $y = y_0 + u$ we obtain the tangent linear operator*

$$L(u) = \frac{\partial P}{\partial y}(x,\vec{y}_0)u + \frac{\partial P}{\partial y'}(x,\vec{y}_0)Du + \cdots + \frac{\partial P}{\partial y^{(n)}}(x,\vec{y}_0)D^n u,$$

defined over $K[[X]]$. If the exponents of L at $x=0$ are p–adically non–Liouville then y_0 has a nontrivial p–adic radius of convergence.

The theory of ordinary p–adic differential equations addresses such questions as:

1. What are the radii of formal local solutions?
2. How do solutions grow as the boundary of the circle of convergence is reached?
3. What are the filtrations of the solution spaces relative to the growth and radii of convergence?
4. Index.

I. Order of growth

The main result of Robba on this question does not directly refer to differential equations.

THEOREM. (Robba 1980b) Let $u_1, \dots u_n \in \mathcal{A}_0$ and let the wronskian

$$w = \begin{vmatrix} u_1 & \dots & u_n \\ u_1' & \dots & u_n' \\ u_1^{(n-1)} & \dots & u_n^{(n-1)} \end{vmatrix}$$

never vanish in $D(0,1^-)$. Then each element $v = \Sigma a_s x^s$ in the K space spanned by $u_1, \dots u_n$ satisfies the condition

$$|a_s| \leq \operatorname*{Sup}_{0 \leq i < n-1} |a_i| \cdot \{s, n-1\}$$

where

$$\{s, n-1\} = 1/Inf\,|z_1 z_2 \cdots z_{n-1}| \quad (\leq s^{n-1}),$$

the inf being over all $1 \leq z_1 < z_2 < \cdots < z_{n-1} \leq s$.

This type of logarithmic estimate first appeared in the study of eigenvectors in our dual theory [Dwork 1964].

This type of estimate is a natural consequence of a strong Frobenius structure [Dwork, 1969]. For example if $\vec{y} = \Sigma \vec{a}_s x^s \in K[[X]]^n$ and if for some $\lambda \in K^\times$, we have $A\vec{y}(x^p) = \lambda \vec{y}(x)$ where $A \in \mathcal{M}_n(B_0)$ and is bounded by unity on $D(0,1^-)$ then $\vec{y}$ must converge in $D(0,1^-)$ and for $s \geq 1$

$$|\vec{a}_s| \leq |\vec{a}_0| |\lambda|^{-[\log s/\log p]}.$$

More recently effective bounds for solutions at a regular singular point have been found by Adolphson, et al. 1982 and by Christol,Dwork 1990, the former if the local monodromy is nilpotent of maximal rank, the second without the restriction of maximality of rank.

II. Filtration by growth and radius of convergence

For the second type of filtration we have

THEOREM. (Robba 1977a). Let L be a differential operator of order n with coefficients in $K(X)$ (or more generally with superadmissible coefficients). For $r \in (0,1]$, Ker$(L, \mathcal{A}_t(r))$ defines a monic factor, L_r, of L in $E[D]$ whose coefficients are indeed superadmisible.

In the case of a Frobenius structure, filtration by growth should correspond to filtration by magnitude of eigenvalue.

THEOREM. (Robba 1975a) *Let $L \in E[D]$, $\dim \operatorname{Ker}(L, \mathcal{A}_t) = order L = n$. Then this kernel lies in $W_t^{1,n-1}$ and for each $\beta \in [0, n-1]$, Ker$(L, W_t^{1,\beta})$ determines a monic factor L_β of L in $E[D]$.*

The great contribution of Robba to these questions was to view $\mathcal{R} = E[D]$ as a subspace of the Banach space $\mathcal{L}(W_t^{r,\beta}, W_t^{r,\beta})$. Thus he considered $\overline{\mathcal{R}L}$, the completion in $\mathcal{R}$ of the ideal

$\mathcal{R}L$ under the Banach space norm. This ideal has a generator shown by Robba to be the factor of L corresponding to $\text{Ker}(L, W_t^{r,\beta})$.

There are a number of unanswered questions.

1. If the coefficients of L (in this last theorem) are admissible (or even superadmissible) then the coefficients of L_β need not be superadmissible. But are they admissible? Conjecture: Yes.

2. Let L have coefficients which are analytic elements on $D(0,1^-)$. We may construct a Newton polygon for L at t whose slopes are the exceptional values β such that $\text{Ker}(L, W_t^{1,\beta-\varepsilon}) \subsetneq \text{Ker}(L, W_t^{1,\beta+\varepsilon})$ for an infinite sequence of $\varepsilon \to 0$, and whose vertices have abscissas given by $\lim_{\varepsilon\to 0} \dim \text{Ker}(L, W_t^{1,\beta-\varepsilon})$ for β exceptional. We may construct a similar polygon at $x = 0$.

CONJECTURE. *The polygon at $x = 0$ lies above the polygon at $x = t$.*

It is known (Dwork 1973, Robba 1975a)

$$\dim \text{Ker}(L, W_t^{1,0)}) \geq \dim \text{Ker}(L, W_0^{1,0})$$
$$\dim \text{Ker}(L, \mathcal{A}_t(1)) \geq 1 \ \text{ implies} \dim \ker(L, W_t^{1,0}) \geq 1.$$

A geometric example of the filtration by growth is given by ${}_2F_1(\frac{1}{2}, \frac{1}{2}, 1, x)$. This was analyzed (Dwork 1969, 1971) in two ways:

(a) by directly demonstrating the admissibility of $F(X)/F(X)^b$ via congruences associated with the p-adic gamma function

(b) by constructing a unit root crystal from the given superadmissible two dimensional crystal.

For ${}_2F_1(a, b, 1, x)$ Robba [1976(b)] gave a treatment based on a weak form of the Hahn Banach theorem. He avoided all references to Frobenius structure.

Dwork (1983) discussed ${}_2F_1(a, b, c, x)$ on the basis of Frobenius structure.

The nature of the factorization subject to geometric type hypotheses have been investigated for hypersurfaces, (Dwork 1973) for kloosterman sums (Adolphson, Sperber 1984) and hyperkloosterman sums (Sperber 1980, Sibuya, Sperber 1985).

Subject to geometric type hypotheses, Sperber and the author [Dwork, Sperber 1990] have found the coefficients of the factor corresponding to the bounded solutions to have mittag–leffler decompositions in which the components are of the form $\Sigma\, A_j/(x-\alpha)^j$ with $\text{ord}\, A_j > k \log(1+j)$ for some $k > 0$. This has played a role in investigating the unit root zeta function.

III. Index

This question had great interest for Robba. At least four of his articles mention index in the title while others are devoted to applications of index. In his early work (1975, 76) there were no indications of applications but these appeared subsequently (1982c). His 1984 Asterisque article was dominated by the application to one dimensional cohomology and by 1986 he began studying symmetric powers of the Bessel differential equation.

Both Robba and Adolphson used patching arguments to reduce the question of index to the case of $L \in K[X][D]$ and to the calculation of either $\mathcal{A}_0/L\mathcal{A}_0$ or $\mathcal{B}_0/L\mathcal{B}_0$. For the applications it made no difference which one was finite. We consider only this elementary form.

For his application Adolphson was able to reduce to the case of order one and more explicitly to $X\frac{d}{dx} - a$, $a \in \mathbf{Q}$.

Robba [1975a] showed

If $\ker(L, \mathcal{A}_t(1)) = 0$ then $\chi(L, \mathcal{A}_0(1)) = \chi(L, \mathcal{B}_0(1))$.

This result is of interest as it seems to capture the essential point of dagger type cohomology involving over convergent series. Unfortunately this has not been extended to the case of several variables.

Of course if order $L = \dim \mathrm{Ker}(L, \mathcal{A}_0(1))$ then L has index as operator on $\mathcal{A}_0$ (but not on $\mathcal{B}_0$). In particular, by the transfer principle:

If L has no singularity on $D(0, 1^-)$ and if order $L = \dim(\mathrm{Ker} L, \mathcal{A}_t(1))$ then L has index on $\mathcal{A}_0(1)$.

By means of Christol's transfer theorem [Christol 1984] we may extend this last result to the case in which L has just one regular singularity in $D(0, 1^-)$ with non–Liouville exponents.

The operator (Robba 1977a) $L = p(1 - x)D^2 - xD - a$ where $\liminf |a - m|^{1/m} = +1$, $\liminf |a + m|^{1/m} < 1$ is an example of an operator with no singularity in $D(0, 1^-)$ but without index in $\mathcal{A}_0$.

For operators of the first order $L = aD + b$, $a, b \in K[X]$, Robba (1985a) gave a beautiful formula

$$\left(\frac{d}{d\log r} \log \rho(L, r)\right)^- = \chi^-(L, r) + \mathrm{ord}^-(a, r)$$

where

$$\begin{aligned}
\chi^- &= \dim \mathrm{Ker} - \dim \mathrm{cokernel} \text{ for } L \text{ as operator on } H(D(0, 1^-)) \\
\rho(L, r) &= \text{radius of convergence of the solution at } t_r, \\
\mathrm{ord}^-(a, r) &= \text{abcissa of point of contact of the Newton polygon of a with the} \\
&\qquad \text{line of support of slope } -\log r/\log p,
\end{aligned}$$

i.e. if $a = \Sigma A_n X^n$, ν minimal such that $|a|_0(r) = |A_\nu| r^\nu$ then $\nu = \mathrm{ord}^-(a, r)$.

This formula showed how the Turritin form may be used to compute the index if the origin is an irregular singular point. It gives further motivation for extending Christol's transfer theorem to the case of irregular singular points. This index need not be equal to the algebraic index.

In view of the failure of crystalline cohomology to provide a proof finiteness of cohomology, the question of finiteness of index in the sense of this section must still be viewed as pertinent. It is our opinion that the critical case is that in which $D(0, 1^-)$ contains more than one regular singularity and order $L = \dim(\mathrm{Ker}\, L, \mathcal{A}_t(1))$.

We mention a few aspects of Robba's mathematical personality.

He was a very clear expositor and did much to popularize p–adic analysis. Together with Amice and Escassut he organized the GEAU. He gave a total of 24 written exposès, three in the first year, five in the second year.

He had many beautiful ideas. One was his abstract construction of the generic disk (1977c), a second was his explanation of Turrittin's theorem by means of the valuation polygon (1980a), a third was his method for removal of apparent singularities (cf. Christol 1981, Theorem 8.3), a fourth was his construction of a transcendental π, $\mathrm{ord}\pi = 1/(p-1)$ which had the property that ord $\pi(x - x^p)$ converges for $\mathrm{ord} x > -1/p$.

Bibliography of Ph. Robba

[1966] "Dérivée en moyenne de fonction moyennable", Communication an Congrés des Mathematiciens, Moscow, 2p.

[1967] "Solutions presque-periodiques d'équations differentielles aux différences", Seminaire Lions–Schwartz, 25p.

[1969a] (with E. Motzkin), "Ensembles satisfaisant au principe du prolongement analytique en analyse p–adique" C.R. Acad. Sc. Paris **269**, p. 126–129.

[1969b] ———, "Ensembles d'analyticité en analyse p–adique", C.R. Acad. Sc. Paris **269**, p. 450–453.

[1970] "Decomposition en élements simple d'une fonction analytique sur un corps valué ultramétrique. Application au prolongement analytique", Sem. Delange–Pisot–Poitou, (1969/70), No. 22, 14p.

[1971] "Prolongement analytique pour les fonction de plusieurs variables sur un corps valué complet", Sem. Delange–Pisot–Poitou, (1971/72).

[1972a] "Decomposition en élements simple d'un elément analytique sur un corps ultramétrique", C.R. Acad. Sc. Paris **274**, p. 532–535.

[1972b] "Fonctions analytiques sur un corps ultramétrique", C.R. Acad. Sc. Paris **274**, p. 721–723.

[1972c] "Prolongement analytique en analyse ultramétrique", C.R. Acad. Sci. Paris **274**, p. 830–833.

[1973a] "Fonctions analytiques sur les corps valués ultramétriques complets", *Asterisque*, **10**, p. 109–220.

[1973b] "Prolongement analytique pour les fonctions de plusieurs variables sur un corps valué complet", *Bull. Soc. Math. France* **101**, pp. 193–217.

[1974a] "Croissance des solutions d'une équation differentille homogéne", GEAU 73/74, exp. 1, 15p.

[1974b] "Indice d'un opérateur différentiel", GEAU 73/74, exp. 2, 14p.

[1974c] "Prolongement des solutions d'un équation différentielle p–adique", GEAU 73/74 exp 11, 3p. (also C.R. Acad. Sc. **279**, p. 153–154).

[1974d] "Quelques propriétes des éléments et fonctions analytiques au sens de Krasner", *Bull. Soc. Math. France* Mémoire 39–40, p. 341–349.

[1974e] "Un classe d'ensembles analytiques á plusieurs dimensions", *Bull. Soc. Math France*, Mémoire 39–40, p. 351–357.

[1975a] "On the index of p–adic differential operators I", *Annals of Math.* **101** p. 280–316.

[1975b] "Equations différentielles p–adiques", Sem. Delange–Pisot–Pitou, 74/75, 15p.

[1975c] "Factorisation d'un opérateur différentiel", GEAU 74/75 exp. 2, 16p.

[1975d] "Factorisation d'un opérateur diffèrentiel", Applications, GEAU 74/75 exp. 10, 6p.

[1975e] "Caracterisation des dérivées logarithmiques", GEAU 74/75 exp. 12, 6p.

[1975f] "Lemma de Hensel pour les opérateurs différentielles" GEAU 74/75 exp. 16, 11p.

[1975g] "Structure de Frobenius faible pour les équations différentielles de premier ordre", GEAU 74/75 exp. 20, 11p.

[1976a] "On the index of p–adic differential operators II", *Duke Math. J.* **43**, p. 19–31.

[1976b] "Solutions bornées des systems différentiels linéaires, Application aux fonctions hypergéométriques", GEAU 75/76 exp. 5, 16p.

[1976c] "Factorisation d'un opérateur différentiel II", GEAU 75/76 exp. 6, 6p.

[1976d] "Prolongement algébrique et equations différentielles", Proc. Conference on p–adic analysis, Nijmegen (1978), p. 172–184. Math. Inst. Katholieke Univ. Toernooiveld.

[1976e] "Théorie de Galois p–adique", GEAU 75/76 exp. 14, 8p.

[1976f] "Lemme de Schwarz p–adique pour plusieurs variables" GEAU 75/76 exp. J9, 12p.

[1977a] (with B. Dwork) "On ordinary linear p–adic differential equations", *Trans. Amer. Math. Soc.* **231**, p. 1–46.

[1977b] "Nouveau point de vue sur le prolongement algébrique", GEAU 76/77 exp. 5, 14p.

[1977c] "Disque générique et équations différentielles", GEAU 76/77 exp. 8, 7p.

[1977d] "Nombre de zéros des fonctions exponentielles–polynômes", GEAU 76/77 exp. 9, 3p.

[1978a] "Lemmes de Schwarz of lemmes d'approximation p–adiques en plusieurs variables", *Invent. Math.* **48**, 245–277.

[1978b] "Zéros de suites récurrentes linéaires", GEAU 77/78 exp. 13, 5p.

[1979a] (with B. Dwork) "On natural radii of p–adic convergence", *Trans. Amer. Math. Soc.* **256**, p. 199–213.

[1979b] "Points singuliers d'equations différentielles, d'apres F. Baldassarri", GEAU 78/79 exp. 5, 2p.

[1979c] (with B. Dwork) "Majorations effective", GEAU 78/79 exp. 18, 2p.

[1979d] "Clôture algébrique relative de certain anneaux de séries entières", GEAU 78/79 exp. 21, 2p.

[1979e] "Lemmes de Hensel pour les opérateurs differéntielles", GEAU 78/79 exp. 23, 8p.

[1980a] "Lemmes de Hensel pour les opérateurs différentielles, Application a la reduction formelle des equations différentielles", *Ens. Math.* **26**, p. 279–311.

[1980b] (with B. Dwork) "Effective p–adic bounds for solutions of homogeneous linear differential equations", *Trans. Amer. Math. Soc.* **259**, p. 559–577.

[1980c] (with S. Bosch and B. Dwork) "Un theorem de prolongement pour les fonctions analytiques", *Math. Ann.* **242**, p. 165–173.

[1980d] "Factorisation d'opérateurs différentiel á coefficients rationnels", GEAU 79/81 exp. 12, 3p.

[1981] "Une propriété de specialisation continue, d'après Artin, Bosch, Lang, Van den Dries", GEAU 79/81 exp. 26, 11p.

[1982a] "Sur les equations différentielles linéaires p–adiques II", *Pacific J. Math.* **98**, p. 393–418.

[1982b] "Calcul des residus en analyse p–adique, d'apres Gerritzen et Van der Put", GEAU 81/82 exp. 28, 8p.

[1982c] "Indice d'un opérateur différentiel linéaire p–adique d'ordre 1 et cohomologies p–adiques", GEAU 81/82 exp. J15, 10p.

[1983] "Cohomologie de Dwork, I", GEAU 82/83 exp. 3, 10p.

[1984a] "Une Introduction naive aux cohomologies de Dwork", *Bull. Soc. Math. France* **114** (Supp) Memoire **23**, p. 61–105.

[1984b] "Index of p–adic differential operators III, Application to twisted exponential sums", *Astérisque*, **119–120**, p. 191–266.

[1984c] "Quelques remarques sur les opérateurs différentiels d'order 1", GEAU 83/84 exp. 12, 8p.

[1985a] "Indice d'un opérateur différentiel p–adique IV, Cas des systèms. Mesure d'irregularité dans un disque", *Ann. Inst. Fourier* **35**, p. 13–35.

[1985b] "Conjecture sur les equations différentielles p–adiques linéaires", GEAU exp. 2, 8p.

[1985c] "Product symmetrique de l'equation de Bessel", GEAU exp. 7, 3p.

[1986a] "Symmetric powers of the p–adic Bessel Function", *Crelle* **366**, p. 194–220.

[1986b] (with J. P. Bezivin) "Rational solutions of linear differential equations", Proc. of conference on p–adic analysis, Hengelhoef, Belg. p. 11–18.

[1987] "Propriete d'approximation pour les elements algébriques", *Comp. Math.* **63**, p. 3–14.

[1989a] (with J. P. Bezivin) "A new p–adic method for proving irrationality and transcendence results", *Ann. of Math.* **129**, p. 151–160.

[1989b] "Rational solutions of linear differential equations", *J. Australian Math. Soc.* **46**, p. 184–196.

General Bibliography

[1976a] Adolphson, A., "An index theorem for p– adic differential operators", *Trans. Amer. Math. Soc.* **216**, 279–293.

[1976b] ——————, "A p–adic theory of Hecke polynomials", *Duke Math. J.* **43**, 115–145.

[1982] ——————, Dwork, B., Sperber, S., "Growth of solutions of linear differential equations at a logarithmic singularity", *Trans. Amer. Math. Soc.* **271**, 245–252.

[1984] ——————, Sperber, S., "Twisted kloosterman sums and p–adic Bessel functions", *Amer. J. Math.* **105**, 549–591.

[1981] Christol, "Systémes différentiels linéaires p–adiques", Structure de Frobenius Faible, *Bull. Soc. Math. France* **109**, 83–122.

[1984] ________, "Un theorem de transfert pour les disques singuliers réguliers", *Astérisque* **119–120**, 151–168.

[1990 ________, Dwork, B. "Effective p–adic bounds at regular singular points", *Duke Math. J.* (submitted).

[1966] Clark, D. "A note on the p–adic convergence of solutions of linear differential equations", *Proc. Amer. Math. Soc.* **17**, 262–269.

[1964] Dwork, B. "On the zeta function of a hypersurface II", *Ann. Math.* **80**, 227–299.

[1966] ________, "On the zeta function of a hypersurface III", *Ann. Math.* **83**, 457–519.

[1969] ________, "p–adic cycles", *Pub. Math. IHES* **37**, 27–116.

[1971] ________, "Normalized period matrices I", *Ann. Math.* **93**, 337–388.

[1973] ________, "Normalized period matrices II", *Ann. Math.* **98**, 1–57.

[1982] ________, *Lectures on p–adic differential equations*, Springer–Verlag.

[1987] ________, "On the Tate constant", *Comp. Math.* **61**, 43–59.

[1990] ________, Sperber, S., "Logarithmic decay and over convergence of the unit root and associated zeta functions", *Ann. Sc. ENS*, to appear.

[1973] Katz, N.,*Travaux de Dwork*, Sem. Bourbuki 71/72 exp. 409 (Lecture Notes in Math., Springer–Verlag).

[1937] Lutz, E., "Sur l'equation $y^2 = x^3 - Ax - B$ dans les corps p–adiques", *J. Reine Angew. Math.* **177**, 238–247.

[1981] Sibuya, Y., Sperber, S., "Arithmetic properties of power series solutions of algebraic differential equations", *Ann. Math.* **113**, 111–157.

[1985] ________, On the p–adic continuation of the logarithmic derivative of certain hypergeometric functions, GEAU 84/85 exp. 4, 5p.

[1980] Sperber, S., "Congruence properties of the hyperkloosterman sum", *Comp. Math.* **40**, 3–32.

p-Adic estimates for exponential sums

Alan Adolphson*
Department of Mathematics
Oklahoma State University
Stillwater, Oklahoma 74078
USA

Steven Sperber†
School of Mathematics
University of Minnesota
Minneapolis, Minnesota 55455
USA

1 Introduction

Let k be the finite field of $q = p^a$ elements, $f = \sum_{j \in J} a_j x^j \in k[x_1, \ldots, x_n, (x_1 \cdots x_r)^{-1}]$ where $0 \le r \le n$, $\Psi : k \to \mathbb{Q}(\zeta_p)^\times$ be a nontrivial additive character, and $\chi_1, \ldots, \chi_r : k^\times \to \mathbb{Q}(\zeta_{q-1})^\times$ be multiplicative characters. Define

$$S_1(\chi_1, \ldots, \chi_r, f) = \sum_{x \in (k^\times)^r \times k^{n-r}} \chi_1(x_1) \cdots \chi_r(x_r) \Psi(f(x)). \tag{1}$$

The first problem we consider in this article is:

<u>Problem 1:</u> Find a p-adic estimate for S_1.

Let k_m be the extension of k of degree m. We can define for each m an exponential sum related to (1):

$$S_m(\chi_1, \ldots, \chi_r, f) = \sum_{x \in (k^\times)^r \times k^{n-r}} \left(\prod_{i=1}^{r} \chi_i(N_m(x_i)) \right) \Psi(Tr_m(f(x))), \tag{2}$$

where $Tr_m : k_m \to k$ is the trace map and $N_m : k_m \to k$ is the norm map. This data can be encapsulated in an L-function:

$$L(\chi_1, \ldots, \chi_r, f; t) = \exp\left(\sum_{m=1}^{\infty} S_m(\chi_1, \ldots, \chi_r, f) \frac{t^m}{m} \right). \tag{3}$$

The following result is well-known:

Theorem 1 *(Dwork, Grothendieck) $L(\chi_1, \ldots, \chi_r, f; t)$ is a rational function, i. e.,*

$$L(\chi_1, \ldots, \chi_r, f; t) = \frac{\prod_{\text{finite}}(1 - \alpha_i t)}{\prod_{\text{finite}}(1 - \beta_j t)}. \tag{4}$$

---

*Partially supported by NSF Grant No. DMS-8803085.
†Partially supported by NSF Grant No. DMS-8601461.

By logarithmic differentiation of (4),

$$S_m(\chi_1,\ldots,\chi_r,f)=\sum_j \beta_j^m-\sum_i \alpha_i^m. \tag{5}$$

Thus a more precise version of Problem 1 is:

Problem 2: Find p-adic estimates for the α_i and β_j.

For Problem 1, we use an elementary approach due to Ax[6], namely, reduction to Stickelberger's calculation of the p-ordinals of Gauss sums. Ax's theorem (an improvement of the theorem of Chevalley-Warning[13]) was generalized by Katz[9], Sperber[11], and Adolphson-Sperber[3] using Dwork's trace formula and the theory of completely continuous operators on p-adic Banach spaces. Recently, however, Wan[12] showed that Katz's improvement can be obtained using Ax's original method. We then realized that the results of [11] and [3] can also be obtained by Ax's method. Our theorem is a generalization of the main result of [3], where only an additive character was considered.

Problem 2 is much harder, and good results are known only in those cases where the L-function or its reciprocal is a polynomial. In such cases, Problem 2 can be interpreted as asking for the p-adic Newton polygon of that polynomial. In [4] we dealt with the case where all the multiplicative characters are trivial. We gave conditions on f that insure that $L(f;t)^{(-1)^{n-1}}$ is a polynomial and found a lower bound for its p-adic Newton polygon. And we conjectured that this lower bound is attained generically for primes p in certain residue classes.

We are now able to handle the case of nontrivial multiplicative characters using similar arguments. These results are described in section 3. Since the proofs are like those in [4], we only sketch them here.

2 Problem 1

We assume throughout this section that f is not a polynomial in some proper subset of the variables $x_1,\ldots,x_n$. This involves no loss of generality, since it is trivial to reduce the sum (1) to one of this type.

First we embed $S_1(\chi_1,\ldots,\chi_r,f)$ into a p-adic field. Let K be the unramified extension of Q_p of degree a and let T be the set of $(q-1)$-st roots of unity in K together with 0. We regard our characters as taking values in $K(\zeta_p)$. Our sum may then be written as

$$S(d_1,\ldots,d_r,f)=\sum_{t\in(T^\times)^r\times T^{n-r}} t_1^{q-1-d_1}\cdots t_r^{q-1-d_r}\Psi(f(\bar{t})), \tag{6}$$

where $d_1,\ldots,d_r\in\{0,1,\ldots,q-2\}$ and $\bar{t}\in k$ is the reduction of t mod p. For later convenience we set $d_i=q-1$ for $i=r+1,\ldots,n$.

For each $d\in\{0,1,\ldots,q-1\}$, define

$$d'=\begin{cases}0 & \text{if } d=0\\ \text{least \underline{positive} residue of } pd \text{ modulo } q-1 & \text{if } d\neq 0.\end{cases} \tag{7}$$

We denote the i-fold iteration of this operation by $d^{(i)}$. Note that since $q=p^a$, we have $d^{(a)}=d$. Put $\mathbf{d}=(d_1,\ldots,d_n)\in\mathrm{R}^n$ and $\mathbf{d}'=(d_1',\ldots,d_n')\in\mathrm{R}^n$. Define $\Delta(f)$, the *Newton polyhedron* of f, to be the convex hull of J, the exponents of the monomials appearing in

f considered as points in $\mathbf{R}^n$, together with the origin. For $i = 0, \ldots, a-1$, let $\omega(\mathbf{d}^{(i)}, f)$ be the least nonnegative real number ω such that $\omega\Delta(f) \stackrel{\text{def}}{=} \{\omega x \mid x \in \Delta(f)\}$ contains a point of $(q-1)^{-1}\mathbf{d}^{(i)} + (\mathbf{Z}^r \times \mathbf{N}^{n-r})$, where $\mathbf{N}$ denotes the nonnegative integers. (The condition that f involve all n variables ensures that the set of such ω's is nonempty.) Let ord_q be the additive p-adic valuation on $K(\zeta_p)$ normalized by $\mathrm{ord}_q\, q = 1$.

Theorem 2 $\mathrm{ord}_q\, S(\mathbf{d}, f) \geq \dfrac{1}{a} \sum_{i=0}^{a-1} \omega(\mathbf{d}^{(i)}, f)$

Example. For $0 \leq d \leq q-2$, consider the Gauss sum $g_d = \sum_{t \in T^\times} t^{q-1-d} \Psi(\bar{t})$. Then $\Delta(f) = [0,1] \subseteq \mathbf{R}^1$ and $\omega(d^{(i)}, f) = (q-1)^{-1} d^{(i)}$. The theorem implies (after an easy calculation) that

$$\mathrm{ord}_q\, g_d \geq \sigma(d)/a(p-1), \tag{8}$$

where $\sigma(d)$ is the sum of the digits in the p-adic expansion of d. (It is well-known by Stickelberger that the two sides of (8) are equal for $1 \leq d \leq q-2$.) It will be seen that the estimate (8) is all that is needed to prove the theorem, so the theorem is, in fact, equivalent to (8).

Proof of the theorem. Let $F = \sum_{j \in J} A_j x^j \in K[x_1, \ldots, x_n, (x_1 \cdots x_r)^{-1}]$ be the Teichmüller lifting of f, i. e., $A_j \in T$ and $A_j \equiv a_j \pmod{p}$. Define a polynomial $P = \sum_{m=0}^{q-1} c_m u^m \in K(\zeta_p)[u]$ by the condition

$$P(t) = \Psi(\bar{t}) \in K(\zeta_p)$$

for all $t \in T$. One checks that $c_0 = 1$, $c_{q-1} = -q/(q-1)$, and that for $1 \leq m \leq q-2$, $c_m = g_m/(q-1)$, where g_m is the Gauss sum defined above. Then Stickelberger implies that for $0 \leq m \leq q-1$,

$$\mathrm{ord}_q\, c_m = \sigma(m)/a(p-1). \tag{9}$$

Let $\mathbf{e} = (1, \ldots, 1) \in \mathbf{R}^n$. We have

$$\begin{aligned}
S(\mathbf{d}, f) &= \sum_{t \in (T^\times)^r \times T^{n-r}} t^{(q-1)\mathbf{e}-\mathbf{d}} \Psi(f(\bar{t})) \\
&= \sum_{t \in (T^\times)^r \times T^{n-r}} t^{(q-1)\mathbf{e}-\mathbf{d}} \prod_{j \in J} \Psi(a_j \bar{t}^j) \\
&= \sum_{t \in (T^\times)^r \times T^{n-r}} t^{(q-1)\mathbf{e}-\mathbf{d}} \prod_{j \in J} P(A_j t^j) \\
&= \sum_{t \in (T^\times)^r \times T^{n-r}} t^{(q-1)\mathbf{e}-\mathbf{d}} \prod_{j \in J} \left(\sum_{m=0}^{q-1} c_m A_j^m t^{mj} \right).
\end{aligned}$$

Now expand the product: Let Φ be the set of all functions from J to $\{0, 1, \ldots, q-1\}$. Then

$$\begin{aligned}
S(\mathbf{d}, f) &= \sum_{t \in (T^\times)^r \times T^{n-r}} t^{(q-1)\mathbf{e}-\mathbf{d}} \sum_{\phi \in \Phi} \left(\prod_{j \in J} c_{\phi(j)} A_j^{\phi(j)} \right) t^{\sum_{j \in J} \phi(j) j} \\
&= \sum_{\phi \in \Phi} \left(\prod_{j \in J} c_{\phi(j)} A_j^{\phi(j)} \right) \left(\sum_{t \in (T^\times)^r \times T^{n-r}} t^{(q-1)\mathbf{e}-\mathbf{d}+\sum_{j \in J} \phi(j) j} \right).
\end{aligned}$$

Since the A_j's are roots of unity,

$$\operatorname{ord}_q S(\mathbf{d}, f) \geq \min_{\phi\in\Phi}\Big\{\operatorname{ord}_q \Big(\prod_{j\in J} c_{\phi(j)}\Big)\Big(\sum_{t\in(T^\times)^r\times T^{n-r}} t^{(q-1)\mathbf{e}-\mathbf{d}+\sum_{j\in J}\phi(j)j}\Big)\Big\}. \tag{10}$$

The second factor on the right-hand side of (10) can be evaluated explicitly. For $i \in \mathbf{Z}$

$$\sum_{t\in T^\times} t^i = \begin{cases} 0 & \text{if } (q-1)\not|\, i \\ q-1 & \text{if } (q-1) \mid i, \end{cases}$$

and for $i \in \mathbf{N}$

$$\sum_{t\in T} t^i = \begin{cases} 0 & \text{if } (q-1)\not|\, i \\ q & \text{if } i = 0 \\ q-1 & \text{if } i > 0 \text{ and } (q-1) \mid i. \end{cases}$$

Hence

$$\sum_{t\in(T^\times)^r\times T^{n-r}} t^{(q-1)\mathbf{e}-\mathbf{d}+\sum_{j\in J}\phi(j)j} =$$
$$\begin{cases} 0 & \text{if } (q-1)\mathbf{e}-\mathbf{d}+\sum_{j\in J}\phi(j)j \notin (q-1)(\mathbf{Z}^r\times\mathbf{N}^{n-r}) \\ q^{n-r-m(\phi)}(q-1)^{r+m(\phi)} & \text{if } (q-1)\mathbf{e}-\mathbf{d}+\sum_{j\in J}\phi(j)j \in (q-1)(\mathbf{Z}^r\times\mathbf{N}^{n-r}), \end{cases} \tag{11}$$

where $m(\phi)$ is the number of nonzero entries in the last $n-r$ coordinates of the vector $(q-1)\mathbf{e}-\mathbf{d}+\sum_{j\in J}\phi(j)j$.

We show that the minimum on the right-hand side of (10) occurs when $m(\phi) = n-r$. Say the $(r+1)$-st coordinate of $(q-1)\mathbf{e}-\mathbf{d}+\sum_{j\in J}\phi(j)j$ is zero. Since f is assumed to involve all the variables, there exists $j_0 \in J$ whose $(r+1)$-st coordinate is nonzero, so we must have $\phi(j_0) = 0$. Define $\hat{\phi} \in \Phi$ by

$$\hat{\phi}(j) = \begin{cases} \phi(j) & \text{if } j \neq j_0 \\ q-1 & \text{if } j = j_0. \end{cases}$$

Then $(q-1)\mathbf{e}-\mathbf{d}+\sum_{j\in J}\hat{\phi}(j)j$ is in $(q-1)(\mathbf{Z}^r\times\mathbf{N}^{n-r})$. In the last $n-r$ coordinates, it has a nonzero entry wherever $(q-1)\mathbf{e}-\mathbf{d}+\sum_{j\in J}\phi(j)j$ does and, in addition, it is nonzero in the $(r+1)$-st coordinate. Hence $m(\hat{\phi}) \geq m(\phi)+1$. It follows from (9) and (11) that

$$\operatorname{ord}_q \Big(\prod_{j\in J} c_{\phi(j)}\Big)\Big(\sum_{t\in(T^\times)^r\times T^{n-r}} t^{(q-1)\mathbf{e}-\mathbf{d}+\sum_{j\in J}\phi(j)j}\Big) \geq$$
$$\operatorname{ord}_q \Big(\prod_{j\in J} c_{\hat{\phi}(j)}\Big)\Big(\sum_{t\in(T^\times)^r\times T^{n-r}} t^{(q-1)\mathbf{e}-\mathbf{d}+\sum_{j\in J}\hat{\phi}(j)j}\Big),$$

hence the minimum on the right-hand side of (10) occurs when the last $n-r$ coordinates of $(q-1)\mathbf{e}-\mathbf{d}+\sum_{j\in J}\phi(j)j$ are nonzero. This implies by (10) and (11) that

$$\operatorname{ord}_q S(\mathbf{d}, f) \geq \min_{\phi\in\Phi_0}\Big\{\sum_{j\in J}\operatorname{ord}_q c_{\phi(j)}\Big\}, \tag{12}$$

where $\Phi_0 = \{\phi \in \Phi \mid \sum_{j\in J}\phi(j)j \in \mathbf{d} + (q-1)(\mathbf{Z}^r\times\mathbf{N}^{n-r})\}$.

For $\phi \in \Phi$, define $\phi' \in \Phi$ by

$$\phi'(j) = \begin{cases} 0 & \text{if } \phi(j) = 0 \\ \text{least } \underline{\text{positive}} \text{ residue of } p\phi(j) \text{ modulo } q-1 & \text{if } \phi(j) \neq 0. \end{cases}$$

We denote the i-fold iteration of this operation by $\phi^{(i)}$. In terms of p-adic expansions, if $\phi(j) = m_0 + m_1 p + \cdots + m_{a-1} p^{a-1}$, where $0 \leq m_i \leq p-1$, then $\phi'(j) = m_{a-1} + m_0 p + m_1 p^2 + \cdots + m_{a-2} p^{a-1}$. Hence $\sum_{i=0}^{a-1} \phi^{(i)}(j) = \sigma(\phi(j))(1+p+\cdots+p^{a-1}) = \sigma(\phi(j))(q-1)/(p-1)$. It then follows from (9) and (12) that

$$\operatorname{ord}_q S(\mathbf{d}, f) \geq \min_{\phi \in \Phi_0} \left\{ \frac{1}{a} \sum_{i=0}^{a-1} \sum_{j \in J} \frac{\phi^{(i)}(j)}{q-1} \right\}. \tag{13}$$

Let Ψ_i be the set of all functions ψ from J to the nonnegative real numbers such that $\sum_{j \in J} \psi(j) j \in (q-1)^{-1} \mathbf{d}^{(i)} + (\mathbf{Z}^r \times \mathbf{N}^{n-r})$. If $\phi \in \Phi_0$, then $(q-1)^{-1} \phi^{(i)} \in \Psi_i$ for $i = 0, 1, \ldots, a-1$. So if we put

$$\rho(\mathbf{d}^{(i)}, f) = \inf_{\psi \in \Psi_i} \left\{ \sum_{j \in J} \psi(j) \right\},$$

then (13) implies

$$\operatorname{ord}_q S(\mathbf{d}, f) \geq \frac{1}{a} \sum_{i=0}^{a-1} \rho(\mathbf{d}^{(i)}, f).$$

The theorem is an immediate consequence of the following lemma.

Lemma 1 *Let J be a finite subset of $\mathbf{R}^n$, $\Delta(J)$ its convex hull, and L an arbitrary subset of $\mathbf{R}^n$. Define*

$$\begin{aligned} \omega(J, L) &= \inf \left\{ \omega \in \mathbf{R}_{\geq 0} \,\middle|\, \omega \Delta(J) \cap L \neq \emptyset \right\}, \\ \rho(J, L) &= \inf \left\{ \sum_{j \in J} \psi(j) \,\middle|\, \psi : J \to \mathbf{R}_{\geq 0} \text{ and } \sum_{j \in J} \psi(j) j \in L \right\}. \end{aligned}$$

Then $\omega(J, L) = \rho(J, L)$.

Proof. If $x \in \omega \Delta(J)$, then by properties of the convex hull,

$$x = \omega \sum_{j \in J} \phi(j) j,$$

where $\phi : J \to \mathbf{R}_{\geq 0}$ and $\sum_{j \in J} \phi(j) = 1$. Put $\psi = \omega \phi$. If $x \in L$ also, then $\sum_{j \in J} \psi(j) j \in L$ and $\sum_{j \in J} \psi(j) = \omega$. This argument is reversible, so the two sets defined in the statement of the lemma are identical.

3 Problem 2

In this section, we restrict attention to sums where all variables are assumed to be nonzero: Let

$$f = \sum_{j \in J} a_j x^j \in k[x_1, \ldots, x_n, (x_1 \cdots x_n)^{-1}], \tag{14}$$

$$S_m(\chi_1, \ldots, \chi_n, f) = \sum_{(x_1,\ldots,x_n)\in(k_m^\times)^n} \Big(\prod_{i=1}^n \chi_i(N_m(x_i))\Big) \Psi(Tr_m(f(x))). \tag{15}$$

Let $\mathbf{d} = (d_1, \ldots, d_n) \in \mathbf{R}^n$ be the vector associated to the multiplicative characters as in (6), and let $L(\mathbf{d}, f; t)$ denote the L-function associated to the sums (15). Dwork's theory provides us with a p-adic Banach space $B(\mathbf{d}, f)$ and a completely continuous operator α on $B(\mathbf{d}, f)$ such that

$$L(\mathbf{d}, f; t)^{(-1)^{n-1}} = \det(I - t\alpha \mid B(\mathbf{d}, f))^{\delta^n}, \tag{16}$$

where δ is the operator on formal power series with constant term 1 defined by $g(t)^\delta = g(t)/g(qt)$. Futhermore, $\det(I - t\alpha \mid B(\mathbf{d}, f))$ is a p-adic entire function of the variable t.

Equation (16) shows that there is a close connection between the zeros and poles of $L(\mathbf{d}, f; t)$ and the zeros of $\det(I - t\alpha \mid B(\mathbf{d}, f))$. We recall the lower bound we gave in [5] for the Newton polygon of $\det(I - t\alpha \mid B(\mathbf{d}, f))$.

Let $C(f)$ be the *cone over* $\Delta(f)$, i. e., $C(f)$ is the union of all rays emanating from the origin and passing through $\Delta(f)$. For $r \in C(f)$, let $w(r)$ be the least nonnegative real number w such that $r \in w\Delta(f)$. It is clear that there exists a positive integer D such that for all $r \in (q-1)^{-1}\mathbf{Z}^n$, $Dw(r)$ is a nonnegative integer. For $i = 0, \ldots, a-1$, put $R^{(i)} = (q-1)^{-1}\mathbf{d}^{(i)} + \mathbf{Z}^n$, and for k a positive integer let

$$W^{(i)}(k) = \text{card}\,\{r \in R^{(i)} \cap C(f) \mid w(r) = k/D\}. \tag{17}$$

Order the elements of $R^{(i)} \cap C(f) = \{r_1^{(i)}, r_2^{(i)}, \ldots\}$ so that $w(r_j^{(i)}) \le w(r_{j'}^{(i)})$ for $j \le j'$. We define

$$W(k) = \text{card}\left\{j \,\middle|\, \sum_{i=0}^{a-1} w(r_j^{(i)}) = k/D\right\}. \tag{18}$$

The main technical result of [5] is:

Theorem 3 *The Newton polygon of* $\det(I - t\alpha \mid B(\mathbf{d}, f))$ *with respect to the valuation* ord_q *lies on or above the polygon with vertices* $(0,0)$ *and*

$$\left(\sum_{k=0}^M W(k), \frac{1}{aD}\sum_{k=0}^M kW(k)\right), \qquad M = 0, 1, \ldots.$$

In other words, it is bounded below by the Newton polygon of $\prod_{k=0}^\infty (1 - q^{k/aD}t)^{W(k)}$.

Theorem 2 can be deduced from this theorem just as [3, Theorem 1.2] was deduced from [2, Proposition 3.13], although, of course, this proof is less elementary than the one in section 2 since it relies on the theory of p-adic Banach spaces. Roughly speaking, Theorem 2 comes from looking at the slope of the first side of the lower bound of Theorem 3. Further information about $L(\mathbf{d}, f; t)$ can be derived from Theorem 3. See [5] for the proofs of the following two theorems.

Theorem 4 *Let $V(f)$ be the volume of $\Delta(f)$ with respect to Lebesgue measure on $\mathbf{R}^n$. Then*

$$0 \leq \text{degree } L(\mathbf{d}, f; t)^{(-1)^{n-1}} \leq n!\, V(f),$$

where the degree *of a rational function is the degree of its numerator minus the degree of its denominator.*

Suppose that $\dim \Delta(f) = n$. If σ is an $(n-1)$-dimensional face of $\Delta(f)$ containing the origin and $E(\sigma)$ is the $(n-1)$-dimensional subspace of $\mathbf{R}^n$ generated by σ, let $S(\sigma)$ be the $(n-1)$-dimensional volume of σ *relative to* $\mathbf{d}$, i. e., $S(\sigma) = 0$ if $E(\sigma) \cap R^{(i)} = \emptyset$ for some (and hence for all) $i = 0, 1, \ldots, a-1$; otherwise, $S(\sigma)$ is the volume of σ with respect to Haar measure on $E(\sigma)$ normalized so that a fundamental domain for the lattice $E(\sigma) \cap \mathbf{Z}^n$ has volume 1. Let $S(f) = \sum_\sigma S(\sigma)$, where the sum is over all $(n-1)$-dimensional faces σ of $\Delta(f)$ that contain the origin.

Theorem 5 *Suppose that* $\dim \Delta(f) = n$ *and that* $L(\mathbf{d}, f; t)^{(-1)^{n-1}}$ *and its dual* $L(-\mathbf{d}, -f; t)^{(-1)^{n-1}}$ *are polynomials of degree* $n!\, V(f)$. *Let* $\{\rho_i\}_{i=1}^{n!\,V(f)}$ *(resp.* $\{\bar{\rho}_i\}_{i=1}^{n!\,V(f)}$*) be the reciprocal roots of* $L(\mathbf{d}, f; t)^{(-1)^{n-1}}$ *(resp.* $L(-\mathbf{d}, -f; t)^{(-1)^{n-1}}$*). Then*

$$\sum_{i=1}^{n!\,V(f)} \operatorname{ord}_q \rho_i + \sum_{i=1}^{n!\,V(f)} \operatorname{ord}_q \bar{\rho}_i \geq n \cdot n!\, V(f) - (n-1)!\, S(f).$$

In particular, if the origin is an interior point of $\Delta(f)$ *or if* $E(\sigma) \cap R^{(i)} = \emptyset$ *for every* $(n-1)$*-dimensional face* σ *of* $\Delta(f)$ *that contains the origin, then*

$$\sum_{i=1}^{n!\,V(f)} \operatorname{ord}_q \rho_i + \sum_{i=1}^{n!\,V(f)} \operatorname{ord}_q \bar{\rho}_i \geq n \cdot n!\, V(f).$$

At this point, we are unable to say anything more precise about Problem 2 unless we make assumptions about f that ensure that $L(\mathbf{d}, f; t)^{(-1)^{n-1}}$ is a polynomial. For f as in (14) and σ a face of $\Delta(f)$, let $f_\sigma = \sum_{j \in \sigma \cap J} a_j x^j$. Recall that f is said to be *nondegenerate with respect to* $\Delta(f)$ if for every face σ of $\Delta(f)$ that does not contain the origin, the Laurent polynomials $\partial f_\sigma / \partial x_1, \ldots, \partial f_\sigma / \partial x_n$ have no common zero in $(\bar{k}^\times)^n$, where $\bar{k}$ denotes the algebraic closure of k. The proof of the following theorem will be sketched later.

Theorem 6 *Suppose f is nondegenerate and* $\dim \Delta(f) = n$. *Then* $L(\mathbf{d}, f; t)^{(-1)^{n-1}}$ *is a polynomial of degree* $n!\, V(f)$.

Suppose that $\dim \Delta(f) = n$ and consider the generating series $\sum_{k=0}^\infty W^{(i)}(k) t^{k/D}$, where the $W^{(i)}(k)$ are as defined in (17). The proof of the following lemma will be sketched later.

Lemma 2 *For $i = 0, \ldots, a-1$ there exists polynomials $P^{(i)}(t) \in \mathbf{Z}[t]$ of degree $\leq nD$ with nonnegative coefficients satisfying* $P^{(i)}(1) = n!\, V(f)$ *and*

$$\sum_{k=0}^{\infty} W^{(i)}(k) t^{k/D} = \frac{P^{(i)}(t^{1/D})}{(1-t)^n}. \tag{19}$$

Since the $P^{(i)}(t)$ have nonnegative integral coefficients they may be written as

$$P^{(i)}(t) = \sum_{j=1}^{n!\,V(f)} t^{a_j^{(i)}}, \tag{20}$$

where the $a_j^{(i)}$ are nonnegative integers ordered so that

$$a_1^{(i)} \leq a_2^{(i)} \leq \cdots \leq a_{n!\,V(f)}^{(i)}.$$

We use this data to define a new polynomial $P(t) \in \mathbb{Z}[t]$ of degree $\leq anD$ with nonnegative coefficients satisfying $P(1) = n!\,V(f)$:

$$P(t) = \sum_{j=1}^{n!\,V(f)} t^{\sum_{i=0}^{a-1} a_j^{(i)}}. \tag{21}$$

Rewrite $P(t)$ as

$$P(t) = \sum_{j=0}^{anD} b_j t^j, \tag{22}$$

where the b_j are nonnegative integers.

Theorem 7 *Suppose f is nondegenerate and $\dim \Delta(f) = n$. Then the Newton polygon of $L(\mathbf{d}, f; t)^{(-1)^{n-1}}$ (which is a polynomial of degree $n!\,V(f)$ by Theorem 6) with respect to the valuation ord_q lies on or above the Newton polygon of the polynomial $\prod_{j=0}^{anD}(1 - q^{j/aD}t)^{b_j}$ and their endpoints coincide.*

Remark. D. Wan has pointed out that under the hypothesis of Theorem 7, the inequality of Theorem 5 is actually an equality. Since the endpoints of the Newton polygons coincide, this amounts to checking that

$$\frac{1}{aD} \sum_{j=0}^{anD} j(b_j + \bar{b}_j) = n \cdot n!\,V(f) - (n-1)!\,S(f),$$

where the $\bar{b}_j$ are associated to $-\mathbf{d}$ in the same manner the b_j are associated to $\mathbf{d}$.

Outline of proof. Consider the ring $k[\{x^u \mid u \in \mathbb{Z}^n \cap C(f)\}]$ generated over k by all monomials x^u with $u \in \mathbb{Z}^n \cap C(f)$. The weight function w defines a filtration on this ring, whose associated graded ring we denote by A. These two rings are identical as vector spaces over k, but multiplication in A satisfies the rule:

$$x^u \cdot x^{u'} = \begin{cases} x^{u+u'} & \text{if } w(u)^{-1}u \text{ and } w(u')^{-1}u' \text{ lie on the same face of } \Delta(f) \\ 0 & \text{otherwise.} \end{cases} \tag{23}$$

Let M_i be the vector space over k generated by all x^u with $u \in R^{(i)} \cap C(f)$. Equation (23) defines a structure of graded A-module on M_i. The generating series $\sum_{k=0}^{\infty} W^{(i)}(k) t^{k/D}$ is then the Poincaré series of the graded module M_i. For $j = 1, \ldots, n$, let f_j be the element of A of degree 1 that is the image of $\partial f/\partial x_j$ when one passes to the associated graded. The method of Kouchnirenko[10] shows that the Koszul complex on M_i determined by $\{f_j\}_{j=1}^n$ is acyclic in positive dimension, i. e.,

$$0 \to (M_i)^{\binom{n}{n}} \xrightarrow{\partial_n} \cdots \xrightarrow{\partial_2} (M_i)^{\binom{n}{1}} \xrightarrow{\partial_1} M_i \tag{24}$$

is exact, and $\dim_k M_i/\partial_1(M_i)^{\binom{n}{i}} = n!\,V(f)$. Since the boundary maps in (24) have degree 1, this establishes Lemma 2.

The sequence (24) can be lifted to a sequence of p-adic Banach spaces with action of Frobenius, still acyclic in positive dimension and with H^0-term of dimension $n!\,V(f)$. (It takes some work to fill in the details of this step: see [4, section 2].). Theorem 6 then follows from equation (16) by passing to homology. Finally, the lower bound of Theorem 7 follows by the method of [8, section 7].

We conjecture that, except in finitely many characteristics, if $p \equiv 1 \pmod D$ then the lower bound of Theorem 7 is equal to the Newton polygon of $L(\mathrm{d}, f; t)^{(-1)^{n-1}}$ for all f in a nonempty Zariski-open subset of the set of all Laurent polynomials having the same Newton polyhedron $\Delta(f)$. In fact, we believe that it would suffice to choose D so that $Dw(r)$ is a nonnegative integer for all $r \in e^{-1}\mathbf{Z}^n \cap C(f)$, where e is the least common multiple of the orders of the multiplicative characters $\chi_1, \ldots, \chi_n$.

4 Examples

For $x \in k^\times$, consider the Kloosterman sum

$$K(d, x) = \sum_{t \in T^\times} t^{q-1-d}\Psi\left(\bar{t} + \frac{x}{\bar{t}}\right).$$

In [1], we determined precisely the generic Newton polygon for the associated L-function. It can lie strictly above the lower bound of Theorem 7. However, if all the digits in the p-adic expansion of d are $\geq (p-1)/2$ (or if they are all $\leq (p-1)/2$), then [1, Theorem 1] and a short calculation show that the lower bound of Theorem 7 equals the actual Newton polygon for generic x. In particular, the conjecture in the last paragraph of section 3 is true for this example because the condition $p \equiv 1 \pmod D$ forces all digits in the p-adic expansion of d to be equal. See Carpentier[7] for the determination of the Newton polygons of some hyperkloosterman sums with multiplicative characters.

The next example illustrates a simpler situation where everything can be worked out explicitly using Gauss sums. Assume that $3 \mid (q-1)$, so that k has multiplicative characters of order 3. In this section we illustrate the above results by calculating the Newton polygons of the sums

$$\sum_{x,y \in k^\times} \chi_1(x)\chi_2(y)\Psi(x + xy^4), \tag{25}$$

where $\chi_1^3 = \chi_2^3 = 1$. We assume that p is a prime ≥ 5, which implies $p \equiv 1, 5, 7, 11 \pmod{12}$. (The answer will turn out to depend on $p \pmod{12}$.) Referring to equations (4) and (5), we see that we may always extend scalars when calculating the Newton polygon since $\mathrm{ord}_q\, \alpha_i = \mathrm{ord}_{q^m}\, \alpha_i^m$. To evaluate (25) explicitly, it will be necessary to work over a field with multiplicative characters of order 12; we therefore take $q = p^2$, which ensures that $12 \mid (q-1)$.

The results are summarized in Table 1, where as usual ω denotes the Teichmüller character. We carry out the calculation for $\chi_1 = \chi_2 = \omega^{-(q-1)/3}$, $p \equiv 7 \pmod{12}$, the other cases being similar. In all cases, the point is that the number of terms appearing in the additive character equals the number of variables, hence the given sum

χ_1	χ_2	p (mod 12)	Lower bound	Actual slopes
1	1	1,5,7,11	$0,1,1,1$	$0,1,1,1$
1	$\omega^{-(q-1)/3}$	1,5,7,11	$1,1,1,1$	$1,1,1,1$
1	$\omega^{-2(q-1)/3}$	1,5,7,11	$1,1,1,1$	$1,1,1,1$
$\omega^{-(q-1)/3}$	1	1	$\frac{1}{3},\frac{1}{3},\frac{4}{3},\frac{4}{3}$	$\frac{1}{3},\frac{1}{3},\frac{4}{3},\frac{4}{3}$
		5	$\frac{1}{2},\frac{1}{2},1,\frac{3}{2}$	$\frac{1}{2},\frac{1}{2},1,\frac{3}{2}$
		7	$\frac{1}{3},\frac{1}{3},\frac{4}{3},\frac{4}{3}$	$\frac{1}{3},\frac{5}{6},\frac{5}{6},\frac{4}{3}$
		11	$\frac{1}{2},\frac{1}{2},1,\frac{3}{2}$	$\frac{1}{2},1,1,1$
$\omega^{-(q-1)/3}$	$\omega^{-(q-1)/3}$	1	$\frac{1}{3},\frac{1}{3},\frac{4}{3},\frac{4}{3}$	$\frac{1}{3},\frac{1}{3},\frac{4}{3},\frac{4}{3}$
		5	$\frac{1}{2},\frac{1}{2},1,\frac{3}{2}$	$\frac{1}{2},\frac{1}{2},1,\frac{3}{2}$
		7	$\frac{1}{3},\frac{1}{3},\frac{4}{3},\frac{4}{3}$	$\frac{1}{3},\frac{5}{6},\frac{5}{6},\frac{4}{3}$
		11	$\frac{1}{2},\frac{1}{2},1,\frac{3}{2}$	$\frac{1}{2},1,1,1$
$\omega^{-(q-1)/3}$	$\omega^{-2(q-1)/3}$	1	$\frac{1}{3},\frac{4}{3},\frac{4}{3},\frac{4}{3}$	$\frac{1}{3},\frac{4}{3},\frac{4}{3},\frac{4}{3}$
		5	$\frac{1}{2},1,1,\frac{3}{2}$	$1,1,1,1$
		7	$\frac{1}{3},\frac{4}{3},\frac{4}{3},\frac{4}{3}$	$\frac{1}{3},\frac{4}{3},\frac{4}{3},\frac{4}{3}$
		11	$\frac{1}{2},1,1,\frac{3}{2}$	$1,1,1,1$
$\omega^{-2(q-1)/3}$	1	1	$\frac{2}{3},\frac{2}{3},\frac{2}{3},\frac{5}{3}$	$\frac{2}{3},\frac{2}{3},\frac{2}{3},\frac{5}{3}$
		5	$\frac{1}{2},\frac{1}{2},1,\frac{3}{2}$	$\frac{1}{2},\frac{1}{2},1,\frac{3}{2}$
		7	$\frac{2}{3},\frac{2}{3},\frac{2}{3},\frac{5}{3}$	$\frac{2}{3},\frac{2}{3},\frac{7}{6},\frac{7}{6}$
		11	$\frac{1}{2},\frac{1}{2},1,\frac{3}{2}$	$\frac{1}{2},1,1,1$
$\omega^{-2(q-1)/3}$	$\omega^{-(q-1)/3}$	1	$\frac{2}{3},\frac{2}{3},\frac{2}{3},\frac{5}{3}$	$\frac{2}{3},\frac{2}{3},\frac{2}{3},\frac{5}{3}$
		5	$\frac{1}{2},1,1,\frac{3}{2}$	$1,1,1,1$
		7	$\frac{2}{3},\frac{2}{3},\frac{2}{3},\frac{5}{3}$	$\frac{2}{3},\frac{2}{3},\frac{2}{3},\frac{5}{3}$
		11	$\frac{1}{2},1,1,\frac{3}{2}$	$1,1,1,1$
$\omega^{-2(q-1)/3}$	$\omega^{-2(q-1)/3}$	1	$\frac{2}{3},\frac{2}{3},\frac{2}{3},\frac{5}{3}$	$\frac{2}{3},\frac{2}{3},\frac{2}{3},\frac{5}{3}$
		5	$\frac{1}{2},\frac{1}{2},1,\frac{3}{2}$	$\frac{1}{2},\frac{1}{2},1,\frac{3}{2}$
		7	$\frac{2}{3},\frac{2}{3},\frac{2}{3},\frac{5}{3}$	$\frac{2}{3},\frac{2}{3},\frac{7}{6},\frac{7}{6}$
		11	$\frac{1}{2},\frac{1}{2},1,\frac{3}{2}$	$\frac{1}{2},1,1,1$

Table 1: Comparison of the lower bound of Theorem 7 with the actual slopes of the Newton polygon for $\sum_{x,y\in k^\times}\chi_1(x)\chi_2(y)\Psi(x+xy^4)$.

can be evaluated explicitly in terms of Gauss sums. One then uses Stickelberger to find the exact p-ordinal. Note that the cases $p \equiv 7 \pmod{12}$, $(\chi_1, \chi_2) = (\omega^{-(q-1)/3}, 1)$, $(\omega^{-(q-1)/3}, \omega^{-(q-1)/3})$, $(\omega^{-2(q-1)/3}, 1)$, $(\omega^{-2(q-1)/3}, \omega^{-2(q-1)/3})$, do not contradict the conjecture at the end of section 3. One has $D = 3$ and $p \equiv 1 \pmod 3$, but the sum (25) is not generic. The generic polynomial with this Newton polyhedron has the form $a_0x + a_1xy + a_2xy^2 + a_3xy^3 + a_4xy^4 + a_5$.

We have

$$\sum_{x,y\in k^\times} \omega^{-(q-1)/3}(x)\omega^{-(q-1)/3}(y)\Psi(x + xy^4) = \sum_{x,y\in k^\times} \omega^{-(q-1)/3}(x)\omega^{-(q-1)/12}(y^4)\Psi(x + xy^4)$$

$$= \sum_{x,y\in k^\times} \omega^{-(q-1)/3}(x)\omega^{-(q-1)/12}(y)(1 + \omega^{-(q-1)/4}(y) + \omega^{-(q-1)/2}(y) + \omega^{-3(q-1)/4}(y))\Psi(x + xy)$$

$$= g(\tfrac{q-1}{4})g(\tfrac{q-1}{12}) + g(\tfrac{q-1}{3}) + g(\tfrac{3(q-1)}{4})g(\tfrac{7(q-1)}{12}) + g(\tfrac{q-1}{2})g(\tfrac{5(q-1)}{6}), \tag{26}$$

where the last line is obtained by the change of variable $y \to y/x$ and where for typographical convenience we have written $g(d)$ for the Gauss sum denoted by g_d in the example following Theorem 2. We have for $p \equiv 7 \pmod{12}$ that $(q-1)/4 = (3p-1)/4 + p(p-3)/4$, so by Stickelberger $\operatorname{ord}_q g(\frac{q-1}{4}) = ((3p-1)/4 + (p-3)/4)/2(p-1) = 1/2$. The ordinals of the other Gauss sums are evaluated similarly. The terms appearing in (26) have ord_q equal to 5/6, 1/3, 5/6, and 4/3, respectively.

References

[1] A. Adolphson and S. Sperber: Twisted Kloosterman sums and p-adic Bessel functions, II: Newton polygons and analytic continuation. Amer. J. Math. **109**, 723–764 (1987)

[2] A. Adolphson and S. Sperber: Newton polyhedra and the degree of the L-function associated to an exponential sum. Invent. Math. **88**, 555–569 (1987)

[3] A. Adolphson and S. Sperber: p-Adic estimates for exponential sums and the theorem of Chevalley-Warning. Ann. Scient. E. N. S. **20**, 545–556 (1987)

[4] A. Adolphson and S. Sperber: Exponential sums and Newton polyhedra: Cohomology and estimates. Ann. of Math. **130**, 367–406 (1989)

[5] A. Adolphson and S. Sperber: On twisted exponential sums (preprint)

[6] J. Ax: Zeroes of polynomials over finite fields. Amer. J. Math. **86**, 255–261 (1964)

[7] M. Carpentier: Sommes exponentielles dont la géométrie est tres belle: p-adic estimates. Pac. J. Math. **141**, 229–277 (1990)

[8] B. Dwork: On the zeta function of a hypersurface, II. Ann. of Math. **80**, 227–299 (1964)

[9] N. Katz: On a theorem of Ax. Amer. J. Math. **93**, 485–499 (1971)

[10] A. G. Kouchnirenko: Polyèdres de Newton et nombres de Milnor. Invent. Math. **32**, 1–31 (1976)

[11] S. Sperber: On the p-adic theory of exponential sums. Amer. J. Math. **108**, 255–296 (1986)

[12] D. Wan: An elementary proof of a theorem of Katz. Amer. J. Math. **111**, 1–8 (1989)

[13] E. Warning: Bemerkung zur vorstehenden Arbeit von Herr Chevalley. Abh. Math. Sem. Univ. Hamburg **11**, 76–83 (1936)

p-ADIC BETTI LATTICES

Yves André
Intitut H. Poincaré, 11 rue P. et M. Curie,
F-75231 Paris 5, France

Under the label "p-adic Betti lattices", we shall discuss two kinds of objects.

The first type of lattices arises via Artin's embedding of integral Betti cohomology into p-adic étale cohomology for complex algebraic varieties; there are comparison theorems with algebraic De Rham cohomology both over the complex numbers (Grothendieck) and p-adically (Fontaine-Messing-Faltings). The second type of lattices, which we believe be new, arises in connection with p-adic tori. Although its definition is purely p-adic, it is closely tied to the classical Betti lattice of some related complex torus, and can be viewed as a bridge between the Dwork and Fontaine theories of p-adic periods; "half" of this lattice is provided by the cohomology of the rigid analytic constant sheaf $\mathbb{Z}$. In fact, both themes of this paper are motivated by a question of Fontaine about the p-adic analog of the Grothendieck period conjecture, as follows.

1. Let X be a proper smooth variety over the field of rational numbers $\mathbb{Q}$. The singular rational cohomology space $H^n_B := H^n(X_{\mathbb{C}},\mathbb{Q})$ carries a rational Hodge structure (for any n); this structure is defined by a complex one-parameter subgroup of $GL(H^n_B \otimes \mathbb{C})$, whose rational Zariski closure in $GL(H^n_B)$ is the so-called Mumford-Tate group of H^n_B.

Let H^n_{DR} denote the $n^{\underline{th}}$ algebraic De Rham cohomology group of X. There is a canonical isomorphism

$$\mathcal{P} : H^n_{DR} \otimes_{\mathbb{Q}} \mathbb{C} \xrightarrow{\sim} H^n_B \otimes_{\mathbb{Q}} \mathbb{C}$$

provided by the functor GAGA and the analytic Poincaré lemma. The entries in $\mathbb{C}$ of a matrix of $\mathcal{P}$ w.r.t. some bases of H^n_{DR}, H^n_B, are usually called periods.
One variant of the Grothendieck period conjecture [G1] predicts that the transcendence degree of the extension of $\mathbb{Q}$ generated by the periods is the dimension of the Mumford-Tate group.

2. On the other hand, let $H^n_{et} := H^n_{et}(X_{\overline{\mathbb{Q}}},\mathbb{Q}_p)$ denote the $n^{\underline{th}}$ p-adic étale cohomology group of

$X_{\overline{\mathbb{Q}}}$, where $\overline{\mathbb{Q}}$ stands for the complex algebraic closure of $\mathbb{Q}$.

Let us choose an embedding γ of $\overline{\mathbb{Q}}$ into the field $\mathbb{C}_p = \hat{\overline{\mathbb{Q}}}_p$. The successive works of Fontaine, Messing and Faltings [FM] [Fa] managed to construct a canonical isomorphism of filtered $\mathrm{Gal}(\overline{\mathbb{Q}}_p/\mathbb{Q}_p)$–modules:

$$H^n_{DR} \otimes_{\mathbb{Q}} B_{DR} \xrightarrow{\sim} H^n_{et}(X_{\mathbb{C}_p}, \mathbb{Q}_p) \otimes_{\mathbb{Q}_p} B_{DR},$$

where B_{DR} denotes the quotient field of the universal pro–infinitesimal thickening of $\mathbb{C}_p$. Via Artin's comparison theorem and the theorem of proper base change for étale cohomology (applied to γ) [SGA 4] III, this supplies us with an isomorphism

$$\mathcal{P}_\gamma : H^n_{DR} \otimes_{\mathbb{Q}} B_{DR} \xrightarrow{\sim} H^n_{et} \otimes_{\mathbb{Q}_p} B_{DR} \xrightarrow{\sim} H^n_B \otimes_{\mathbb{Q}} B_{DR}.$$

The entries in B_{DR} of a matrix of $\mathcal{P}_\gamma$ w.r.t. some bases of H^n_{DR}, H^n_B, are called (γ)–p–adic periods.

Fontaine asked whether the analog of Grothendieck's conjecture for p–adic periods holds true. The answer turns out to be negative; indeed, we shall prove:

Proposition 1. Let X be the elliptic modular curve $X_0(11)$, and $n = 1$, $p = 11$. There are two choices of γ for which the transcendence degree of the extension of $\mathbb{Q}$ generated by the respective p–adic periods differ.

Nevertheless, one can still ask in general whether the property holds true for "sufficiently general" γ. This would be a consequence of a standard conjecture on "geometric p–adic representations":

Proposition 2. Let G be the image of $\mathrm{Gal}(\overline{\mathbb{Q}}/\mathbb{Q}) \longrightarrow GL(H^n_{et}) \simeq GL(H^n_B) | \mathbb{Q}_p$. Assume that the rational Zariski closure of G in $GL(H^n_B)$ contains the Mumford–Tate group. Then for "sufficiently general" γ, the transcendence degree of the extension of $\mathbb{Q}$ generated by the p–adic periods is not smaller than the dimension of the Mumford–Tate group; if moreover $n = 1$, there is equality.

3. Let us next turn to p–adic Betti lattices of the second kind, the construction of which is modelled on the following pattern. Let us assume that over some finite extension E of $\mathbb{Q}$ in $\mathbb{C}_p$, X_E acquires semi–stable reduction, i.e. admits locally a model over the valuation ring of the p–adic completion K of E, which is smooth over the scheme defined by an equation $x_1x_2...x_n$ = some uniformizing parameter of K. In this situation Hyodo and Kato showed the

existence of a <u>semi-stable structure</u> on H^n_{DR} (as was conjectured by Jannsen and Fontaine): namely an isocrystal (H_0,φ) endowed with a nilpotent endomorphism N satisfying $N\varphi = p\varphi N$, together with an isomorphism $H^n_{DR} \otimes_E K \xrightarrow{\sim} H_0 \otimes_{K^0} K$ depending on the choice of a branch β of the p-adic logarithm on $K^\times$ (here K^0 denotes the maximal absolutely unramified subfield of K).

On the other side, one can sometimes use the combinatorics of the intersection graph of the reduction to provide lattices, well-behaved under φ, in suitable twited graded (w.r.t. the "p-adic monodromy" N) forms of H^n_{DR}, and then use φ in order to lift them to H^n_{DR}. For instance, this works pretty well when $X_E = A$ is an Abelian variety with multiplicative reduction at p.

4. Before we describe this situation, let us remind the classical situation $(E \subset \mathbb{C}) : A(\mathbb{C})$ is a complex torus $\mathbb{C}^g/L$, where L is a lattice of rank 2g ; furthermore $H^1_{DR} \otimes \mathbb{C} \simeq \mathrm{Hom}(L,\mathbb{C})$. Composition with a suitably normalized exponential map yields the Jacobi parametrization:

$A(\mathbb{C}) \simeq \mathbb{C}^{\times g}/M$ where M is a lattice of rank g ; thus L appears as an extension of M by $2i\pi M'^{\vee}$, where M' denotes the character group of $\mathbb{C}^{\times g}$. The bilinear map on M, say q, obtained by composing any "polarization" $M' \longrightarrow M$ with the bilinear map $M \times M' \longrightarrow \mathbb{G}_m$ (the multiplicative group) describing $M \longrightarrow \mathbb{C}^{\times g}$, enjoys the following property: $-\log|q|$ is a scalar product.

Similarly, at any place of multiplicative reduction above p, there is the Tate parametrization:

$A(\mathbb{C}_p) \simeq \mathbb{C}_p^{\times g}/M$ where M is again a lattice of rank g ; there is an analogous bilinear map q' on M such that $-\log|q|_p$ is a scalar product.

Using the semi-stable structure, we construct the "p-adic" lattice L_β of rank 2g, formed of φ-invariants and depending on β, which sits in an exact sequence like L (in this new context, $2i\pi$ has to be understood as a generator of $\mathbb{Z}_p(1)$ inside B_{DR}).

Setting $K_{HT} := K[2i\pi,(2i\pi)^{-1}]$, we have moreover a canonical isomorphism: $\mathcal{P}_\beta : H^1_{DR} \otimes_E K_{HT} \xrightarrow{\sim} \mathrm{Hom}_{\mathbb{Z}}(L_\beta, K_{HT})$.

5. We call the entries in K_{HT} of a matrix of $\mathcal{P}_\beta$ w.r.t. some basis of H^1_{DR}, L_β, "(β)-p-adic periods". We may now state a more rigid p-adic transcendence conjecture:

<u>Conjecture 1</u>: for suitable choice of β, the transcendence degree of the extension of E generated by the β- p-adic periods equals the dimension of the Mumford-Tate group of H^1_B.

This conjecture splits into two parts:
We first prove the inequality $\text{tr.deg}_E\, E[\mathcal{P}_\beta] \leq \dim M.T.$ under some extra hypothesis (*) (theorem 3); this amounts to showing the <u>rationality of Hodge classes</u> w.r.t. L_β. (The hypothesis (*) concerns the Shimura variety associated to A , but we think it is unnecessary, or even always satisfied). On the other side, we use G–function methods to prove inequalities of the type "boundary $\text{tr.deg}_E\, E[\mathcal{P}_\beta] \geq \dim M.T.$" refering to polynomial relations of bounded degree between periods (theorems 4 and 5).
Roughly speaking, this is made possible because, when A varies in a degenerating family defined over E , the β–p–adic periods involve the β–logarithm and p–adic evaluations of Taylor series with coefficients in E , whose complex evaluations give the usual periods (theorem 2).

6. The previous considerations suggest the possibility of a purely <u>p–adic definition of</u> (absolute) <u>Hodge classes</u> on A .

<u>Conjecture 2</u>: Let E' be any extension of E , and let ξ be a mixed tensor on $H^1_{DR} \otimes E'$ lying in the 0–step of the Hodge filtration. Then ξ is an absolute Hodge class (i.e. rational w.r.t. L for every $E' \hookrightarrow \mathbb{C}$) if and only if ξ is rational w.r.t. L_β for every $E' \hookrightarrow \mathbb{C}_p$ and every branch β of the p–adic logarithm.

<u>Convention</u>: In this text, a smooth separated commutative group scheme will be called semi–abelian if each fiber is an extension of an abelian variety by a torus.

<u>Acknowledgements</u>: I first thank the organizers of the Trento conference for inviting me to such a nice congress. This article was written out during a stay at the Max–Planck–Institut at Bonn, partially supported by the Humboldt Stiftung; I thank both institutions heartily. The paper benefited very much from several conversations with W. Messing, J.M. Fontaine and especially J.P. Wintenberger.

I. The p–adic comparison isomorphism

1. The "Barsotti rings" B_{DR} and B_{cris}

Let K be a p–adic field, i.e. a finite extension of $\mathbb{Q}_p$ with valuation $v|p$. Let K^0, resp. $\bar{K}$, $\mathbb{C}_p$ denote the maximal nonramified extension of $\mathbb{Q}_p$ inside K, resp. an algebraic closure of K, and its completion. Let $R, R^0, \bar{R}, \mathbb{R}_p$ denote the respective rings of integers, and let $\mathcal{G}$ denote the Galois group $\mathrm{Gal}(\bar{K}/K)$. Fontaine has constructed a universal p–adic pro–infinitesimal thickening of $\mathbb{C}_p$, see e.g. [F1] [F2].

It is denoted by B^+_{DR} and can be constructed as follows.

Let us consider the Witt ring W of the perfection $\varprojlim_{x\mapsto x^p} \bar{R}/p\bar{R}$ of the residual ring $\bar{R}/p\bar{R}$. It sits in an exact sequence $0 \longrightarrow F^1 \longrightarrow W \longrightarrow \mathbb{R}_p \longrightarrow 0$, where the ring homomorphism is defined by the diagram:

$$\begin{array}{ccccccc}
 & & W & \longrightarrow & \varprojlim \bar{R}/p\bar{R} & & \\
 & & \uparrow\wr\wr & & \uparrow\wr\wr & & \\
{[v_n]} & \in & \mathrm{Witt}(\varprojlim \bar{R}) & \longrightarrow & \varprojlim \bar{R} & \ni & v_n = (v_n^0, v_n^1, \dots) \\
\text{(Teichmüller lifting of } v_n) & & \downarrow & \Sigma p^n [v_n] & & & \\
 & & & \downarrow & & & \\
 & & \mathbb{R}_p & \Sigma p^n v_n^0 & & &
\end{array}$$

This provides a continuous surjective homomorphism $B^+_{DR} \longrightarrow \mathbb{C}_p$, where B^+_{DR} denotes the F^1–adic completion of $W[\frac{1}{p}]$. The fraction field B_{DR} of B^+_{DR} is a $\bar{K}[\mathcal{G}]$–module, endowed with the F^1–adic (called Hodge) filtration $F^{\cdot}$, and $\mathrm{Gr}^{\cdot}_F B_{DR} \simeq \bigoplus_{r\in\mathbb{Z}} \mathbb{C}_p(r)$ (Tate twists).

On the other hand, there is a universal PD–thickening of $\mathbb{C}_p$, denoted by B^+_{cris}. It is obtained by inverting p in the p–adic completion of the subalgebra of $W[\frac{1}{p}]$ generated by the $\frac{p^n}{n!}$'s. For instance, if $\epsilon = (\epsilon_0, \epsilon_1, \dots)$ is a generator of $\mathbb{Z}_p(1) = \varprojlim \mu_{p^n}(\bar{K})$, $t_p := \log[\epsilon] = \sum \frac{(-)^{n-1}([\epsilon]-1)^n}{n} \in B^+_{cris}$. The Frobenius φ of W then extends to

$B_{cris} = B^{+}_{cris}[\frac{1}{t_p}]$ $(\varphi t_p = pt_p)$ and commutes with the $\mathcal{G}$–action. Moreover $B_{cris} \otimes_{K_0} K$ imbeds into B_{DR}.

2. The comparison theorem for Abelian varieties

Let $A = A_K$ be an Abelian variety over K. According to Fontaine–Messing [F1] [FM], there is a canonical isomorphism of filtered $\mathcal{G}$–modules:

$$\text{F.M.}: H^{*}_{DR}(A) \underset{K}{\otimes} B_{DR} \xrightarrow{\sim} H^{*}_{et}(A_{\overline{K}}, \mathbb{Q}_p) \otimes_{\mathbb{Q}_p} B_{DR} .$$

In particular H^{*}_{DR} can be recovered from H^{*}_{et} as the space of $\mathcal{G}$–invariants in the R.H.S. This isomorphism can be reformulated as a pairing:

$$H^{1}_{DR}(A) \otimes T_p(A_{\overline{K}}) \longrightarrow B_{DR} .$$

[Faltings and Wintenberger have generalized this pairing to the relative case [W]; the relative H^{*}_{DR} and B_{DR} are endowed with connections and the relative comparison isomorphism is horizontal.]

In order to describe part of this pairing in down–to–earth terms, let us assume that A has semi–stable reduction, i.e. extends to a semi–abelian scheme A_R over R. (By a fundamental result of Grothendieck this always happens after replacing K by a finite extension). Let $\hat{A}_R$ be the formal group attached to A_R; then $T_p(\hat{A}_R)(\overline{K})$ is the "fixed part" of $T_p(A_{\overline{K}})$ [G2]. Now the restricted pairing $H^{1}_{DR}(A) \otimes T_p(\hat{A}_R)(\overline{K}) \longrightarrow B_{cris} \otimes_{K_0} K$ is easily described as follows:

a) It factorizes through the quotient $H^{1}_{DR}(\hat{A}_R)_K \otimes T_p(\hat{A}_R)(\overline{K})$.

b) Using the formal Poincaré lemma, write any $\omega \in H^{1}_{DR}(\hat{A}_R)_K$ in the form $\omega = df$, $f \in \mathcal{O}_{\hat{A}_K}$.

c) For any $\gamma = (\gamma_0, \gamma_1, \ldots) \in T_p(\hat{A}_R)(\overline{R}) = T_p(\hat{A}_R)(\overline{K})$, lift every $\gamma_n \in \overline{R}$ to $\tilde{\gamma}_n \in B_{cris}$.

d) The coupling constant $<\omega,\gamma> \in B_{cris} \otimes_{K_0} K$ is then given by $\lim_n p^n f(\tilde{\gamma}_n)$. See [Co].

3. The crystalline and semi–stable structures

a) Let us first assume that A has good reduction, i.e. extends to an Abelian scheme A_R over

R, and let us denote the special fiber of A_R by $\bar{X}$. In this case $H^*_{DR}(A)$ carries a natural K^0–structure, namely $H^*_0 := H^*_{cris}(\bar{X}/R^0) \otimes_{R^0} K^0$; moreover this K^0–space is a crystal: it is canonically endowed with a semi–linear "Frobenius" isomorphism φ. The Fontaine–Messing isomorphism is then induced by an isomorphism of filtered φ– and $\mathcal{G}$– modules:

$$H^*_0 \otimes_{K^0} B_{cris} \xrightarrow{\sim} H^*_{et}(A_{\bar{K}}, \mathbb{Q}_p) \otimes_{\mathbb{Q}_p} B_{cris} .$$

In particular H^*_{et} can be recovered from H^*_0 as the space of φ–invariants in the F^0–subspace of the L.H.S.

b) If contrawise A has bad reduction, let us use Grothendieck's theorem to reduce to the case of semi–stable reduction. [Jannsen had the idea that there is still a fine structure on H^*_{DR}, involving some "monodromy operator", and such that H^*_{et} could be recovered in a similar way as in the good reduction case [J]. Fontaine then formulated a precise conjecture and proved it in the case of Abelian varieties]. The result is [F2]:

Choose a branch β of the v–adic logarithm. Then

b_1) there exists a canonical K^0–structure H^*_0 on $H^*_{DR}(A)$, endowed with a nilpotent endomorphism N; $N = 0$ iff A has good reduction.

b_2) H^*_0 is naturally endowed with a semi–linear "Frobenius" $\varphi = \varphi_\beta$, related to N by means of the formula: $N\varphi = p\varphi N$.

b_3) there exists $u_\beta \in B_{DR}$ such that $B_{ss} := B_{cris}[u_\beta]$ is $\mathcal{G}$–stable, and such that $N = d/du_\beta$ and the extension of φ to B_{ss} given by $\varphi u_\beta = pu_\beta$ commute with the $\mathcal{G}$–action.

b_4) the p–adic comparison isomorphism is induced by an isomorphism of filtered $K^0(\mathcal{G})$–modules compatible with φ and N:

$$H^*_0 \otimes_{K^0} B_{ss} \xrightarrow{\sim} H^*_{et}(A_{\bar{K}}) \otimes_{\mathbb{Q}_p} B_{ss} .$$

In particular H^*_{et} can be recovered from H^*_0 as the space $[F^0(H^*_{et} \otimes B_{ss})]^{\varphi=1,N=0}$.
For a concrete description of the semi–stable structure due to Raynaud, see below III 4, IV 1 and [R2].

4. Rigid 1–motives and Fontaine's LOG

In the study of the comparison isomorphism, it is useful to embed Abelian varieties into the bigger category of 1–motives [D1].

a) Recall that a smooth 1–motive $[\underline{M} \xrightarrow{\psi} \underline{G}]$ on a scheme S consists in

i) an étale sheaf $\underline{M}$ locally defined by a free abelian group of finite rank

ii) a semi–abelian scheme $\underline{G}$ over S

iii) a morphism $\psi : \underline{M} \longrightarrow \underline{G}$.

For each prime p, one attaches to $[\underline{M} \xrightarrow{\psi} \underline{G}]$ a (Barsotti–Tate) p–divisible group, and its étale cohomology (= étale realization of $[\underline{M} \xrightarrow{\psi} \underline{G}]$).

On the other hand, the universal vectorial extension $\underline{M} \longrightarrow \underline{G}^{\natural}$ of $\underline{M} \longrightarrow \underline{G}$ provides the De Rham realization $H^1_{DR}[\underline{M} \longrightarrow \underline{G}] := \underline{\text{Colie}}\ \underline{G}^{\natural}$, with its Hodge filtration $F^1H^1_{DR} = \underline{\text{Colie}}\ \underline{G}$.

b) There is a notion of duality for 1–motives. We shall only consider symmetrizable 1–motives, i.e. 1–motives isogeneous to their duals (the isogeny inducing a polarization of the Abelian quotient of $\underline{G}$). This amounts to giving

i) a polarized Abelian scheme $(\underline{A},\lambda)$ over S

ii) a morphism $\chi ; \underline{M} \longrightarrow \underline{A}$, where $\underline{M}$ is an étale sheaf of lattices; let $\chi^{\vee} = \lambda \circ \chi$

iii) a symmetric trivialization of the inverse image by $(\chi,\chi^{\vee})$ of the Poincaré biextension of $\underline{A} \times \underline{A}'$.

c) It is convenient to view 1–motives as complexes in degree $(-1,0) : \overset{-1}{\underline{M}} \longrightarrow \overset{0}{\underline{G}}$. When S = Spec K, K = p–adic field, it is more convenient, according to Raynaud [R2], to identify 1–motives which are quasi–isomorphic in the rigid analytic category; for instance, if A is isomorphic to the rigid quotient G/M, we consider A (or $[0 \longrightarrow A]$) and $[M \longrightarrow G]$ as two incarnations of the same rigid 1–motive.

Indeed, the associated p–divisible groups, resp. filtered De Rham realizations, are isomorphic; furthermore this isomorphism is compatible with the Fontaine–Messing comparison isomorphism, which extends to the case of 1–motives (its semi–stable refinement also extends to this case (Fontaine–Raynaud)).

d) Let us illustrate this in the simple case $[\mathbb{Z} \xrightarrow{\psi} \mathbb{G}_m]$, $1 \longmapsto q$ (when q is not a unit in K, this is the 1–motive attached to the Tate curve $K^{\times}/q^{\mathbb{Z}}$). The Tate module sits in an exact sequence

$$0 \longrightarrow \mathbb{Z}_p(1) \longrightarrow T_p \longrightarrow q^{\mathbb{Z}} \otimes_{\mathbb{Z}} \mathbb{Z}_p \longrightarrow 0$$

Let t_p be a generator of $\mathbb{Z}_p(1)$, and let $u_p \in T_p$ lift q. Let moreover μ be a generator of the character group $X(\mathbb{G}_m)$, so that $d\mu/1+\mu$ generates the K–space $\Omega^1_{\mathbb{G}_m}$. At last let us represent u_p by a sequence $(q,q_1,\ldots)$ with $q_{n+1}^p = q_n$, and let $\tilde{q}_n$ lift q_n in B_{DR}. The p–adic periods of $[\mathbb{Z} \xrightarrow{\psi} \mathbb{G}_m]$ are given by:

$$<t_p, d\mu/1+\mu> = \pm t_p \text{ in } B_{DR},$$

$$<u_p, d\mu/1+\mu> = \lim_n \log \tilde{q}_n^{p^n}/q .$$

By abuse language, one denotes this limit by LOG q ; its class mod $\mathbb{Z}_p(1)$ depends only on q . If one requires more rigidity, one may embed K into $\mathbb{C}$ somehow, and choose u_p in the $\mathbb{Z}$–lattice given by the Betti realization of the corresponding complex 1–motive; LOG q is then defined up to addition by $\mathbb{Z}t_p$, as in the classical case.

e) More generally, let us consider a 1–motive $[\underline{M} \xrightarrow{\psi} T]$, where T is a torus. In this case the universal extension splits canonically: $G^{\natural} = T \times \mathrm{Hom}(M,\mathbb{G}_a)^{\vee}$; this induces a canonical splitting of the Hodge filtration: $H^1_{DR}[\underline{M} \longrightarrow T] = F^1 \oplus \mathrm{Hom}(\underline{M},K)$. On the other hand, let M' denote the character group of T and $q: M \times M' \longrightarrow \mathbb{G}_m$ the bilinear form induced by ψ. Again the étale cohomology sits in an extension

$$0 \longrightarrow \mathrm{Hom}(M,\mathbb{Q}_p) \longrightarrow H^1_{et}[M \longrightarrow T] \longrightarrow M' \otimes_{\mathbb{Z}} \mathbb{Q}_p(-1) \longrightarrow 0 .$$

Now assume that M and M' are constant.

Let $(m_i^{\vee})$ denote a basis of $\mathrm{Hom}(M,\mathbb{Z})$ as well as its images in H^1_{et} and H^1_{DR} resp.; let (μ_j) denote a basis of M', let $d\mu_j/1+\mu_j$ be the corresponding basis in F^1, and let $\tilde{\mu}_j$ lift μ_j/t_p inside H^1_{et}. At last, let (m_i) denote the basis of M dual to $(m_i^{\vee})$, and set $q_{ij} = q(m_i,\mu_j)$. Then in the bases of H^1_{DR} (resp. H^1_{et}) given by $\{d\mu_j/1+\mu_j; m_j^{\vee}\}$ (resp. $\{\tilde{\mu}_i, m_i^{\vee}\}$), the matrix of the comparison isomorphism takes the shape:

$$\left[\begin{array}{c|c} t_p I & 0 \\ \hline (\mathrm{LOG}\ q_{ij}) & I \end{array}\right] .$$

This completes the description of this isomorphism for any Abelian variety with split multiplicative reduction.

II. Hodge classes.

1. The complex setting.

a) Let E be a field embeddable into $\mathbb{C}$, and let A_E be an Abelian variety over E. An element $\xi \in F^0\left[H^1_{DR}(A_E)^{\otimes n} \otimes H^1_{DR}(A_E)^{\vee\otimes n}\right] = F^0\left[\mathrm{End}\, H^1_{DR}(A_E)\right]^{\otimes n}$ (for any n) is called a Hodge class if its image in $\left[\mathrm{End}\, H^1_B(A_{\mathbb{C}},\mathbb{C})\right]^{\otimes n}$ lies in the rational subspace $\left[\mathrm{End}\, H^1_B(A_{\mathbb{C}},\mathbb{Q})\right]^{\otimes n}$. By Deligne's theorem on absolute Hodge cycles $[D_2]$, this definition does not depend on the chosen embedding $E \hookrightarrow \mathbb{C}$. Moreover, after a preliminary finite extension of E, one gets no more Hodge class by further extending E. It follows that the connected component of identity of the Hodge group of A_E (which is by definition the algebraic subgroup of $GL\left[H^1_{DR}(A_E)\right]$ which fixes the Hodge classes) is an E–form of the Mumford–Tate group of $H^1_B(A_{\mathbb{C}},\mathbb{Q})$. It is known that the Hodge group is a classical reductive group.

b) Let us fix an embedding $\iota : E \hookrightarrow \mathbb{C}$. For any E–algebra E′, the E′–linear bijections $H^1_{DR}(A_E) \otimes_E E' \xrightarrow{\sim} H^1_B(A_E \otimes_\iota \mathbb{C},\mathbb{Q}) \otimes_{\mathbb{Q}} E'$ which preserve Hodge classes form the set of E′–valued points of a E–torsor P_ι under the Hodge group; for $E' = \mathbb{C}$, one has a canonical point $\mathscr{P}_\iota$ given by "integration of differential forms of second kind".

Lemma 1: the torsor P_ι is irreducible.

Indeed, there exists a finite Galois extension E′ of E such that the Hodge group of $A_{E'}$ is connected; hence the associated torsor P'_ι is geometrically irreducible. But via the isomorphism $H^*_{DR}(A_E) \otimes_E E' = H^*_{DR}(A_{E'})$, a Hodge class on A_E is just a Hodge class on $A_{E'}$ which is fixed by Gal(E′/E). Therefore P_ι is the Zariski closure of P'_ι over E, and is irreducible.

Conjecture (Grothendieck): if E is algebraic over $\mathbb{Q}$, $\mathscr{P}_\iota$ is a (Weil) generic point of P_ι (over E).

Thanks to the irreducibility lemma, this amounts to say that the transcendence degree over $\mathbb{Q}$ of the periods equals the dimension of the Mumford–Tate group (here, "periods" means entries of a matrix of $\mathscr{P}_\iota$ w.r.t. bases of $H^1_{DR}(A_E)$, $H^1_B(A_{\mathbb{C}},\mathbb{Q})$). [This deep problem is solved only for Abelian varieties isogeneous to some power of an elliptic curve with complex multiplication (Chudnovsky).

The conjecture can also be formulated as follows: every polynomial relation between periods, with coefficients in E, comes from Hodge classes. A major result in transcendence theory establishes this for linear relations (Wüstholz); the only Hodge classes which appear in this context are classes of endomorphisms.]

2. Behaviour under the p–adic comparison isomorphism

Assume now that E is a number field; let $v|p$ be a finite place of E, and $K = E_v$ be the completion of E w.r.t. v; $\overline{E}$ denotes the algebraic closure of E in $\overline{K}$.

Let us choose an embedding $\gamma : \overline{K} \hookrightarrow \mathbb{C}$ and denote by ι its restriction to K.

At last, let $\mathcal{P}_\gamma : H^1_{DR}(A_E) \otimes_E B_{DR} \xrightarrow{\sim} H^1_B(A_E \otimes_\iota \mathbb{C}, \mathbb{Q}) \otimes_{\mathbb{Q}} B_{DR}$ denote the composed isomorphism:

$$H^1_{DR}(A_E) \otimes_E B_{DR} \xrightarrow{\sim} H^1_{DR}(A_K) \otimes_K B_{DR} \xrightarrow[\sim]{F.M.} H^1_{et}(A_{\overline{E}}) \otimes_{\mathbb{Q}_p} B_{DR}$$

$$\xrightarrow{\sim} H^1_{et}(A_{\overline{E}}) \otimes_{\mathbb{Q}_p} B_{DR} \xrightarrow{\sim} H^1_B(A_{\overline{E}} \otimes_\gamma \mathbb{C}, \mathbb{Q}) \otimes_{\mathbb{Q}} B_{DR}$$

$$\xrightarrow{\sim} H^1_B(A_E \otimes_\iota \mathbb{C}, \mathbb{Q}) \otimes_{\mathbb{Q}} B_{DR}$$

Blasius–Ogus [Bl] and independently Wintenberger have recently proved the following striking result:

Theorem 1. For every γ above ι, $\mathcal{P}_\gamma$ is a B_{DR}–valued point of P_ι.

[The Wintenberger proof uses the relative comparison isomorphism while the Blasius–Ogus proof uses Faltings's comparison theorem applied to smooth compactifications of "total spaces" of Abelian schemes]. With the notation of I 3, it follows formally that Hodge classes lie in $(\mathrm{End}\ H^1_0)^{\otimes n}$, are Frobenius–invariant and killed by N.

In view of this theorem, it is natural to ask whether the p–adic analog of Grothendieck's conjecture holds, namely whether $\mathcal{P}_\gamma$ is a (Weil) generic point of P_ι over K. [After I communicated the counterexample in prop. 3 to Fontaine, he suggested the following:]

Conjecture 4: for "sufficiently general" γ above ι, $\mathcal{P}_\gamma$ is a (Weil) generic point of P_ι over

See below, § 4.

3. Proof of proposition 1.

In this example $E = \mathbb{Q}$, and $A_{\mathbb{Q}}$ is the elliptic curve $X_0(11)$. For $p = 11$, $A_{\mathbb{Q}_p}$ is a Tate curve $\mathbb{Q}_p^{\times}/q^{\mathbb{Z}}$, $q \in p\mathbb{Z}_p$. With the notations of I 4b, consider the exact sequence $0 \longrightarrow \mathbb{Z}_p(1) \longrightarrow T_p(A_{\overline{\mathbb{Q}}}) \longrightarrow q^{\mathbb{Z}} \otimes_{\mathbb{Z}} \mathbb{Z}_p \longrightarrow 0$, and let t_p be a $\mathbb{Z}_p$-generator of $\mathbb{Z}_p(1)$ such that $t_p \wedge u_p$ is a $\mathbb{Z}$-generator of the image of $\overset{2}{\wedge} H_1(A_{\overline{\mathbb{Q}}} \otimes_{\gamma} \mathbb{C},\mathbb{Z})$ in $\overset{2}{\wedge} T_p(A_{\overline{\mathbb{Q}}})$ for some fixed $\gamma : \overline{\mathbb{Q}} \hookrightarrow \mathbb{C}$; this determines t_p up to sign. Let moreover ν be a unit in $\mathbb{Z}_p$ such that $\omega := \frac{1}{\nu} \frac{d\mu}{1+\mu}$ belongs to the rational subspace $\Omega^1_{A_{\mathbb{Q}}}$ of $\Omega^1_{A_{\mathbb{Q}_p}}$. According to I 4b, we then have:

$$< \nu t_p, \omega > = \pm t_p .$$

Now let $g \in \mathrm{Gal}(\overline{\mathbb{Q}}/\mathbb{Q})$; changing γ into $\gamma \circ g$ modifies the Betti lattice inside T_p via the formula:

$$T_p(A_{\overline{\mathbb{Q}}}) \xrightarrow[g^*]{\sim} T_p(A_{\overline{\mathbb{Q}}}) \simeq H_1(A_{\overline{\mathbb{Q}}} \otimes_{\gamma} \mathbb{C},\mathbb{Z}) \otimes \mathbb{Z}_p ,$$

where g^* denotes the image of g under the group homomorphism

$$\mathrm{Gal}(\overline{\mathbb{Q}}/\mathbb{Q}) \longrightarrow GL(T_p) .$$

But in our case, this homomorphism is surjective, according to Serre [S 1]. In particular, there exists some $g \in \mathrm{Gal}(\overline{\mathbb{Q}}/\mathbb{Q})$, with $\det g^* = 1$, and such that $\nu t_p \in T_p$ lies in the Betti lattice $H_1(A_{\overline{\mathbb{Q}}} \otimes_{\gamma \circ g} \mathbb{C},\mathbb{Z})$; since $\det g^* = 1$, changing γ to $\gamma \circ g$ preserves t_p.

It then follows from the relation $< \nu t_p, \omega > = \pm t_p$ that the Zariski closure of $\mathcal{P}_{\gamma \circ g}$ over $\mathbb{Q}$ is contained in a hypersurface of P. On the other hand, it follows from Serre's result and the next lemma that for some other $\gamma' : \overline{\mathbb{Q}} \hookrightarrow \mathbb{C}$, the Zariski closure of $\mathcal{P}_{\gamma'}$ over $\mathbb{Q}$ is the full torsor P.

4. Proof of proposition 2 (Abelian case).

We prove the following variant for an Abelian variety A_E over a number field E [Proposition 2 itself is proved in the same way with only minor modifications involving simple general facts about absolute Hodge cycles contained in the beginning of [D$_3$]] .

Let us fix $\gamma_0 : E \hookrightarrow \mathbb{C}$ and denote by $H^1_{\gamma_0}$ the rational structure $H^1_B(A \otimes_{E\ \gamma_0} \mathbb{C},\mathbb{Q})$ inside $H^1_{et}(A_{\bar{E}},\mathbb{Q}_p) = H^1_{et}(A_{\bar{K}},\mathbb{Q}_p)$ (for $\bar{E}$ = algebraic closure of E in $\bar{K}$, where $K = E_v$, $v|p$). The Galois representation $H^1_{et}(A_{\bar{E}},\mathbb{Q}_p)$ is described by a homomorphism :
$\mathrm{Gal}(\bar{E}/E) \longrightarrow GL(H^1_{\gamma_0})(\mathbb{Q}_p)$.

Let us denote by G_{γ_0} the Zariski closure of the image of $\mathrm{Gal}(\bar{E}/E)$ over $\mathbb{Q}$, which is the smallest algebraic subgroup of $GL(H^1_{\gamma_0})$ whose group of p–adic points contains the image of $\mathrm{Gal}(\bar{E}/E)$.

Conjecture 5: the Mumford–Tate group of $H^1_{\gamma_0}$ is the connected component of identity in G_{γ_0} .

[One easily checks that the truth of this conjecture does not depend on the choice of γ_0 ; on the other side, the fact that the Mumford–Tate group contains $G^0_{\gamma_0}$ is a theorem of Borovoi [Bo]]. This conjecture is a weak form of the well–known conjecture of Mumford–Serre–Tate (replace $\mathbb{Q}$ by $\mathbb{Q}_p$ in the statement).

Proposition 2′: Conjecture 5 implies conjecture 4.

Proof: let $\mathcal{P}^E_{\gamma_0}$ denote the Zariski closure of $\mathcal{P}_{\gamma_0}$ over E , inside the torsor $P = P_\iota(\iota = \gamma_{0|E})$; let $G^\alpha_{\gamma_0}$ denote any connected component of G_{γ_0} . For any $g_\alpha \in G^\alpha_{\gamma_0}(\mathbb{Q}_p)$, let

$$\psi_{g_\alpha} : \mathrm{Spec}\, B_{DR} \longrightarrow \mathrm{Spec}\, E[\mathcal{P}_{\gamma_0}] \times \mathrm{Spec}\, \mathbb{Q}_p \longrightarrow \mathcal{P}^E_\gamma \times G^\alpha_{\gamma_0|E}$$

be the composed morphism of affine schemes given by $(\mathcal{P}_{\gamma_0}, g_\alpha)$.

From lemma 1 and conjecture 5, it follows that $G_{\gamma_0|E} = \cup\, G^\alpha_{\gamma_0|E}$ acts transitively on P , and that $Q \cdot G^\alpha_{\gamma_0|E} = P$ for any non–empty E–subscheme Q of P . We can now make the expression "sufficiently general γ" (in conjecture 4) precise: it means "any γ of the form $\gamma = \gamma_0 \circ g_\alpha$ where $g_\alpha \in \mathrm{Im\, Gal}(\bar{E}/E)$ is such that ψ_{g_α} maps to the generic point"; indeed for these embeddings γ,

$$\overline{\mathcal{P}_\gamma}^{E} = \overline{\mathcal{P}_{\gamma_0} \cdot g_\alpha}^{E} = \overline{\mathcal{P}_{\gamma_0}}^{E} \cdot (\overline{g_\alpha}^{\mathbb{Q}})_{|E} = \overline{\mathcal{P}_{\gamma_0}}^{E} \cdot G^{\alpha}_{\gamma_0 | E} = P .$$

It remains to prove the existence of (uncountably many) such g_α. To this aim, let us remark that there are only countably many subvarieties of $G^{\alpha}_{\gamma_0 | E(\mathcal{P}_{\gamma_0})}$; we denote them by Q_n, $n \in \mathbb{N}$. Hence there exist linear subspaces $\prod$ of $\operatorname{End} H^1_{\gamma_0} \otimes \mathbb{Q}_p$, of codimension $\dim P - 1$, such that $\prod \cap G^{\alpha}_{\gamma_0}(\mathbb{Q}_p) \cap Q_n \neq \prod \cap G^{\alpha}_{\gamma_0}(\mathbb{Q}_p)$ for every n. Any $g_\alpha \in \prod \cap G^{\alpha}_{\gamma_0}(\mathbb{Q}_p)$ being outside the countable subset $\bigcup_n \prod \cap G^{\alpha}_{\gamma_0}(\mathbb{Q}_p) \cap Q_n$ then satisfies the required property $\overline{g_\alpha}^{E(\mathcal{P}_{\gamma_0})} = G^{\alpha}_{\gamma_0 | E(\mathcal{P}_{\gamma_0})}$.

III. Covanishing cycles and the monodromy filtration.

1. Covanishing cycles.

a) Let again A be an Abelian variety of dimension g over the p–adic field K, with semi–stable reduction. For any finite extension K' of K, let $A_{K'}^{rig}$ denote the associated rigid analytic variety ("Abeloid variety") over K'.

The (Čech) cohomology $H^1(A_{K'}^{rig},\mathbb{Z})$ of the constant sheaf $\mathbb{Z}$ on $A_{K'}^{rig}$ can be interpreted as the group of Galois covers of $A_{K'}^{rig}$ with group $\mathbb{Z}$ [R$_1$] [U].

For reasons which will soon be clear, we denote this group by $\underline{M}^\vee(K')$. One defines this way an etale sheaf $\underline{M}$ on Spec K, described by the $\mathcal{G}$– module $M^\vee := \underline{M}^\vee(\overline{K})$; points of the lattice $M^\vee$ will be called (integral) covanishing cycles.

b) In order to understand the geometrical meaning of $M^\vee$, let us consider the Raynaud extension G (resp. G') of A (resp. of the dual Abelian variety A') :

$$\begin{array}{ccccccccccc} & & \underline{M} & & & & & & \underline{M}' & & \\ & & \downarrow & & & & & & \downarrow & & \\ T & \to & G & \to & B & \qquad & T' & \to & G' & \to & B' \\ & & \downarrow & & & & & & \downarrow & & \\ & & A & & & & & & A' & & \end{array}$$

G is an extension of an Abelian variety B by an unramified torus T of dimension $r \le g$ (lifting the torus part of the semi–stable reduction of A), and A (resp. $A_{\mathbb{C}_p}$) is the rigid analytic quotient of G (resp. $G_{\mathbb{C}_p}$) by the lattice $\underline{M}(K)$ of K–characters (resp. the lattice $M := \underline{M}(\overline{K})$ of characters) of T' ; and symmetrically for G' ...

This description of $A_{\mathbb{C}_p}$ shows that $M^\vee$ is the dual of M ; in particular the (finite) $\mathcal{G}$– action is unramified (since T' is). [In Berkovich's astonishing theory of analytic spaces, one associates with A some pathwise connected locally simply connected topological space A^{an} ; $\underline{M}(K)$ should then appear as its fundamental group in the ordinary topological sense [Be]].

c) Composing the morphisms

$$M^{\vee} \otimes_{\mathbb{Z}} \mathbb{Z}_p \longrightarrow \varprojlim_n H^1(A^{rig}_{\mathbb{C}_p}, \mathbb{Z}/p^n\mathbb{Z}) \xrightarrow{\text{GAGR}} \varprojlim_n H^1_{et}(A_{\mathbb{C}_p}, \mathbb{Z}/p^n\mathbb{Z})$$

[where GAGR denotes the functor studied by Kiehl [K]], yields a natural injection of $\mathbb{Z}_p[\mathcal{G}]$–modules:

$$\iota_{et} : M^{\vee} \otimes_{\mathbb{Z}} \mathbb{Z}_p \hookrightarrow H^1_{et}(A_{\bar{K}}, \mathbb{Z}_p) .$$

d) On the other side, the lattice $\underline{M}^{\vee}(K)$ is naturally isomorphic to the group of rigid analytic homomorphisms from $\mathbb{G}_m$ to A' [R_1], see also [BL] for the variant over $\mathbb{C}_p$.

Composing the morphisms

$$\mathrm{Hom}_{rig}(\mathbb{G}_m, A') \xrightarrow{\text{pull–back}} \mathrm{Hom}(H^1_{DR}(A'^{rig}), H^1_{DR}(\mathbb{G}^{rig}_m))$$

$$\xrightarrow{\text{duality}} H^1_{DR}(A''^{rig}) \xrightarrow{\text{GAGR}} H^1_{DR}(A'')$$

yields a natural embedding:

$$\iota_{DR} : \underline{M}^{\vee}(K) \otimes_{\mathbb{Z}} K \hookrightarrow H^1_{DR}(A) .$$

[Le Stum [lS] interprets the image of ι_{DR} as follows. By means of some compactification $\bar{A}$ of the semi–abelian group scheme A_R over R extending A, there is the notion of strict neighborhood in $\bar{A}^{rig}_K$ of the formal completion $\hat{A}$. For any $\mathcal{O}_A$rig–module $\mathcal{F}$, set $j^+\mathcal{F} = \varinjlim_{\lambda} j_{\lambda *} j_{\lambda}^* \mathcal{F}$, where j_λ runs over all embeddings of strict neighborhoods of $\hat{A}$ inside A^{rig}; j^+ is an exact functor, and there is a canonical epimorphism $\mathcal{F} \twoheadrightarrow j^+ F$ [B]. Define the <u>covanishing complex</u> by $\phi := \mathrm{Ker}(\Omega^{\cdot}_{A\ rig} \longrightarrow j^+\Omega^{\cdot}_{A\ rig})$, which gives rise to a long exact sequence

$$\longrightarrow H^n(A^{rig}, \phi) \longrightarrow H^n_{DR}(A) \longrightarrow H^n_{rig}(\tilde{A}) \longrightarrow$$

involving Berthelot's rigid cohomology of the special fiber $\tilde{A}$. The group $H^1(A^{rig}, \phi)$ can then be identified with $\mathrm{Im}\ \iota_{DR}$; this justifies the label "covanishing cycles" by analogy with the complex case.]

e) It turns out that the maps ι_{et} and ι_{DR} are compatible with the Fontaine–Messing isomorphism; More precisely:

Proposition 3: the following triangle is commutative:

$$\begin{array}{ccc} & \underline{M}^{\vee}(K) \underset{\mathbb{Z}}{\otimes} \mathbb{Q}_p & \\ \swarrow \iota_{DR} & & \searrow \iota_{et} \\ H^1_{DR}(A) \otimes_K B_{DR} & \xrightarrow{\text{F.M.}} & H^1_{et}(A_{\bar{K}}) \otimes_{\mathbb{Q}_p} B_{DR} \end{array}$$

Proof: let us introduce the Raynaud realization $[\underline{M} \longrightarrow G]$ of the (rigid) 1–motive A .

The map ι_{et} can be identified with the natural injection of $\mathcal{G}$–modules : $\mathrm{Hom}(\underline{M}(K),\mathbb{Q}_p) \hookrightarrow H^1_{et}[\underline{M} \longrightarrow G]$.

On the other side, getting rid of double duality, one easily sees that ι_{DR} can be identified with the natural embedding $\mathrm{Hom}(\underline{M}(K),K) \hookrightarrow H^1_{DR}[\underline{M} \longrightarrow G]$, see also [IS] 6.7. The required commutativity then follows from the fact that F.M. is tautological for the quotient 1–motive $[\underline{M}(K) \longrightarrow 1]$ (whose associated p–divisible group is $\cong (\mathbb{Q}_p/\mathbb{Z}_p)^n$) .

g) An orientation of $\mathbb{C}_p$ is an embedding of $\mu_{p^\infty}(\mathbb{C}_p) = \mathbb{Z}_p(1)$ into $\mathbb{C}^\times$; this amounts to the choice of a generator t_p of the $\mathbb{Z}_p$–module $\mathbb{Z}_p(1)$ up to sign, [a further orientation of $\mathbb{C}$ itself would fix the sign], or else to the choice of an embedding of Abelian groups $X_*(\mathbb{G}_m) \longrightarrow T_p(\mathbb{G}_m)(= \mathbb{Z}_p(1))$.

By using an orientation of $\mathbb{C}_p$ and duality, we get from c) an injection:

$$j_{et} : M'^{\vee}(1) := M'^{\vee} \otimes X_*(\mathbb{G}_m) \longrightarrow H^1_{et}(A'_{\bar{K}},\mathbb{Z}_p) \otimes T_p(\mathbb{G}_m) \simeq T_p(A_{\bar{K}}) .$$

Using the Raynaud 1–motive $[M \longrightarrow G]$ over $\mathbb{K}$, it is then clear that the Fontaine–Messing pairing between H^1_{DR} and $M'^{\vee}(1)$ takes its values in $K't_p$ for some finite unramified extension K' of K (even in Kt_p if the torus part of the semi–stable reduction $\mathcal{A}$ splits).

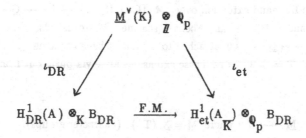

2. Raynaud extensions and the q–matrix.

Let $f : \underline{A} \longrightarrow S$ be a semi–abelian scheme with proper generic fiber, S being an affine normal connected noetherian scheme; we put $S = \mathrm{Spec}\ \mathcal{R}$, $\mathcal{K} = \mathrm{Frac}\ \mathcal{R}$..

a) Let us first assume that $\mathcal{R}$ is complete w.r.t. some ideal I (we set $S_0 := \mathrm{Spec}\ \mathcal{R}/I$), and that the rank r of the toric part T_0 of $A_0 = \underline{A} \times_S S_0$ is constant.

One constructs the Raynaud extension over $\mathcal{R}$ [CF] II, $0 \longrightarrow T \longrightarrow \underline{G} \longrightarrow \underline{B} \longrightarrow 0$, where T lifts T_0 and $\underline{B}$ is an Abelian scheme. There is also the Raynaud extension $0 \longrightarrow T' \longrightarrow \underline{G}' \longrightarrow \underline{B}' \longrightarrow 0$ attached to the dual Abelian scheme $\underline{A}'$, and $\underline{B}'$ is the dual of $\underline{B}$; moreover $\mathrm{rk}\ T = \mathrm{rk}\ T' = r$. These extensions arise via push–out from morphisms of fppf sheaves

$$\underline{M} \longrightarrow \underline{B}\,, \text{ where } \underline{M} = \underline{X}^*(T') \text{ (character groups).}$$

$$\underline{M}' \longrightarrow \underline{B}' \qquad \underline{M}' = \underline{X}^*(T)$$

The objects $\underline{G}$, T , $\underline{M}$,$\underline{B}$ (resp. $\underline{G}'$, ...) are functorial in $\underline{A}$ (resp. $\underline{A}'$) .

b) Replacing S by some open dense subset U , the Faltings construction (using an auxiliary ample line bundle $\mathcal{L}$ on $G_{\mathcal{K}}$ [CF] II 5.1), or methods of rigid analytic geometry ($[BL_1]$ with less generality), provide a trivialization q (independent of $\mathcal{L}$ [CF] III 7.2) of the $\mathbb{G}_m$–biextension of $\underline{M} \times \underline{M}'$ obtained as inverse image of the Poincaré biextension of $\underline{B} \times \underline{B}'$; this amounts to giving a lifting $\underline{M}_U \longrightarrow \underline{G}_U$ of $\underline{M} \longrightarrow \underline{B}$ (whence a smooth 1–motive $[\underline{M} \xrightarrow{\psi} \underline{G}]$ on U). When T_0 splits, so that $\underline{M} = M$ and $\underline{M}' = M'$ are constant, one can use some basis $\{(m_i,\mu_j)\}$ of $M \times M'$ in order to express the bilinear form $q : M \times M' \longrightarrow \mathbb{G}_{m,U}$ by a matrix with entries $q_{ij} \in \mathcal{K}^\times$. [If moreover $\underline{A}$ is principally polarizable, such a polarization induces an isomorphism $M \cong M'$, and then $q : M \otimes M \longrightarrow \mathcal{K}^\times$ is symmetric. In the literature on Abeloid varieties, the associated q–matrix is often referred to as the "period matrix"; however this terminology conflicts with the Fontaine–Messing theory, but some precise relation will be exhibited in IV] .

c) In order to understand the complex counterpart, we replace S by Δ^n , where Δ denotes the unit disk in $\mathbb{C}$. Assume that the restriction of f to the inverse image of $S^* = \Delta^{*n}$ is proper, where Δ^* stands for the punctured Δ .

The kernel $\underline{\Lambda}$ of the exponential map $\exp\colon \underline{\mathrm{Lie}}\ \underline{A}/S \longrightarrow \underline{A}$ is a sheaf of lattices extending the local system $\{H_1(A_s,\mathbb{Z})\}_{s\in S^*}$. The (unique) extension in $\underline{A}$ of the fiber of $\underline{A}$ over 0 is a local system $\underline{N}$ of rank $2g-r$. Via exp (which factorizes through $\underline{N}$), $\underline{A}$ becomes a quotient of the semi–abelian family $\underline{G} = (\underline{\mathrm{Lie}}\ \underline{A}/S)/\underline{N} : \underline{A} = \underline{G}/\underline{M}$, where $\underline{M}$ denotes the sheaf of lattices $\underline{A}/\underline{N}$ (which degenerates at 0).

This supplies us with a (complex analytic) smooth symmetrizable 1–motive $[\underline{M} \longrightarrow \underline{G}]$ over S^*. Both the Betti realizations H^1_B and the De Rham realizations H^1_{DR} (endowed with the Hodge filtration) of $\underline{A}$ and $[\underline{M} \longrightarrow \underline{G}]$ are canonically isomorphic. However, one may not identify these "1–motives" because the weight filtrations differ, see below § 4.

d) We now start with the following global situation:

S_1 is an affine variety over a field E of characteristic 0 ; 0 is a smooth rational point of S_1, and $x_1, \ldots, x_n$ are local coordinates around 0 ;

$f_1 : \underline{A}_1 \longrightarrow S_1$ is a semi–Abelian scheme, proper outside the divisor $x_1x_2 \ldots x_n = 0$, and the toric rank is constant on this divisor.

Because f_1 is of finite presentation, it arises by base change from a semi–abelian scheme $\tilde{f}_1 : \tilde{\underline{A}}_1 \longrightarrow \tilde{S}_1$ (where $\tilde{E}$ is a sub–$\mathbb{Z}$–algebra of E of finite presentation), with the same properties as f_1. If we put $\mathcal{R} = \tilde{E}[[x_1, \ldots, x_n]]$, $S = \mathrm{Spec}\ \mathcal{R}$ (the completion of $\tilde{S}_1$ at 0), $I = (x_1x_2 \ldots x_n)$, $f = \tilde{f}_{1/S}$, we are in the situation a) b). Moreover, the open subscheme U may be defined by the condition $x_1x_2 \ldots x_n \neq 0$. It follows that the entries q_{ij} of the q–matrix belong to $\tilde{E}[[x_1, \ldots, x_n]]\left[\dfrac{1}{x_1x_2 \ldots x_n}\right]$.

e) Assume moreover that E is a number field, with ring of integers $\mathcal{O}_E$. Then $\tilde{E}$ can be chosen in the form $\mathcal{O}_E\left[\frac{1}{\nu}\right]$, where ν is a product of distinct prime numbers. Thus for every finite place v of E not dividing ν, the q_{ij} entries are meromorphic functions on Δ_v^n, analytic on Δ_v^{*n} (Δ_v, resp. Δ_v^* denotes the v–adic "open" unit disk, resp. punctured unit disk), and bounded away from 0. On the other hand, one can also see (using construction c)) that the q_{ij}'s define meromorphic functions on some complex polydisk centered at 0.

[Remark: following [C], an element y of $E[[x_1, \ldots, x_n]]$ is said to be <u>globally bounded</u> if $y \in \mathcal{O}_E\left[\frac{1}{\nu}\right][[x_1, \ldots, x_n]]$ for some ν, and if y has non–zero radius of convergence at every place of E. (Such series form a regular noetherian ring with residue field E, and the filtered union of these rings over all finite extensions of E , is strictly henselian). One can show that the $(x_1 \dots x_n)^m q_{ij}$'s are globally bounded series (for suitable m) . The problem is to show that the v–adic radius of converge is not 0 for any $v | \nu$. Using the compactification of Siegel modular stacks over $\mathbb{Z}$, one can find a semi–abelian extension of $\tilde{\mathcal{Y}}_1$ over an $\mathcal{O}_E$–model of some covering of $\tilde{S}_1$, and afterwards, one has to use the 2–step construction of [CF] III 10 to keep track of the possible variation of the torus rank of the reduction, after replacing the divisor $x_1 \dots x_n = 0$ by $\nu x_1 \dots x_n = 0]$.

b) Lemma 2. If $v \nmid \nu$, then the entries of the q–matrix are units w.r.t. the v–adic Gauss norm.

Proof (sketch): let $\mathcal{E}$ denote the completion of the quotient field of $\mathcal{R} = \tilde{E}[[x_1, \dots x_n]]$ w.r.t. the v–adic Gauss norm $| \ |_{Gauss}$ (= "sup" norm on $\mathcal{R}$) . Because v is discrete, so is $| \ |_{Gauss}$ by Gauss' lemma, hence $\mathcal{E}$ is a complete discretely valued field of unequal characteristics.

By construction of the Raynaud extension, the Barsotti–Tate groups associated to $\underline{A}/\mathcal{E}$ resp. to the 1–motive $[\underline{M}/\mathcal{E} \xrightarrow{\psi} \underline{G}/\mathcal{E}]$ coincide. It follows that Grothendieck's monodromy pairing associated to $\underline{A}/\mathcal{E}$ is induced by the pairing $M \times M' \longrightarrow \mathcal{E}^\times \longrightarrow \mathbb{Z}$ given by the valuation of the q–matrix w.r.t. $| \ |_{Gauss}$. Since $\underline{A}/\mathcal{E}$ has good reduction modulo the valuation ideal of $\mathcal{E}$ (indeed its reduction is the generic fiber of the reduction of $\underline{A}$ modulo v , which is proper when $v \nmid \nu$) , this pairing has to be trivial:

$$|q_{ij}|_{Gauss} = 1 .$$

g) An example: let us consider the Legendre elliptic pencil with parameter $x = \lambda$, given by the affine equation

$$v^2 = u(u-1)(u-x) .$$

Here one can choose $\tilde{E} = \mathbb{Z}\left[\frac{1}{2}\right]$, and one has the explicit formulae:

$$16q = x(1-x)^{-1} e^{-G/F}$$

$$x = 16q\left(\prod_{m=1}^{\infty} (1+q^{2m})(1+q^{2m-1})^{-1}\right)^8 ,$$

$$\text{where } F = \sum_{m=0}^{\infty} ((\tfrac{1}{2})_m/m!)^2 x^m$$

$$G = 2\sum_{m=1}^{\infty} ((\tfrac{1}{2})_m/m!)^2 (\sum_{\ell=1}^{\infty} \tfrac{1}{\ell}) x^m .$$

This example is studied thoroughly in [Dw].

3. Vanishing periods.

a) Let us take up the situation 2d again, and assume that E is contained in the p–adic field K, with $\breve{E} \subset R$. Assume also that the torus part of the semi–stable reduction splits.

As before, we then have our constant sheaves of lattices $\underline{M} = M$, $\underline{M}' = M$ on the v–adic unit polydisk Δ_v^n; let $\{\mu_i\}$ be a basis of M', and let $\{\mu'_i\}$ be the image of the dual basis of $M'^{\vee}(1)$ under j_{et} (defined up to sign, see III 1g).

On the other hand, we have the relative De Rham cohomology sheaf $H^1_{DR}(\underline{A}/S^*)$ which admits a canonical locally free extension to S (where the Gauss–Manin connection acquires a logarithmic singularity with nilpotent residue); in fact this extension is free because S is local, and we denote by $\{\omega_j\}$ a basis of global sections. We are aiming to give some analytic recipe to compute the Fontaine–Messing "vanishing periods" $\frac{1}{t_p} < \mu'_i, \omega_j(s) >$ of the fiber $\underline{A}_1(s)$, $s \in \Delta_v^{*n}$, see III 1f.

b) Let us express the composed morphism

$$H^1_{DR}(\underline{A}/S^*)^{can} \longrightarrow H^1_{DR}(\hat{\underline{A}}/\hat{S}) \longrightarrow H^1_{DR}(\hat{T})_{\hat{S}} \simeq M' \otimes \mathcal{O}_{\hat{S}}$$

(roof = formal completion) in terms of the bases ω_j, $d\mu_i/1+\mu_i$. We get a (2g,r)–matrix (ω_{ij}) with entries in $\mathcal{R} = \breve{E}[[x_1, \ldots, x_n]]$.

Lemma 3. For any $s \in \Delta_v^{*n}$, one has the relation $\omega_{ij}(s) = \pm \frac{1}{t_p} < \mu'_i, \omega_j(s) >$. Moreover ω_{ij} is a bounded solution of the Gauss–Manin partial differential equations on Δ^n.

Proof: the first assertion is easily proved by considering Raynaud's incarnation $[M(s) \longrightarrow \underline{G}(s)]$ of the rigid 1–motive associated to $\underline{A}_1(s)$, together with the trivial computation of Fontaine–Messing periods of the split torus $T = \underline{T}(s) : < \mu'_i, d\mu_j/1+\mu_j > = \pm \delta_{ij} t_p$. The second assertion follows from the horizontality of the map $H^1_{DR}(\underline{A}/S^*)^{can} \longrightarrow H^1_{DR}(\hat{A}/\hat{S})$ w.r.t. the Gauss–Manin connections ∇ , and the fact that M' is formed of horizontal sections of $H^1_{DR}(\hat{T})_{\hat{S}}$ (see also [vM]).

c)   Let $\tilde{\omega}$ denote a uniformizing parameter of R . We modify slightly the setting of 2. d) by assuming that f_1 extends to a semi–abelian scheme $\tilde{\Upsilon} : \underline{A}^{\sim}_{Spec\ R \cap E} \longrightarrow \tilde{S}$, proper outside the divisor $\tilde{\omega}\, x_1 \dots x_n = 0$, and with constant split toral part on this divisor. Again, the ω_{ij}'s converge on Δ^n_v , and for every point $s \in S^*_1(E) \cap \Delta^{*n}_v$, the v–adic evaluation of ω_{ij} at s may be interpreted as in lemma 3 (if furthermore E is a number field, the ω_{ij}'s are in fact globally bounded series). We next look for complex interpretation.

d)   Let $\iota : E \hookrightarrow \mathbb{C}$ be a complex embedding. We now assume that $s \in S^*_1(E)$ satisfies the following property: $\tilde{S}(\mathbb{C})$ should contain the polydisk of radius $|x_i(s)|$ (to insure the convergence of the analytic solutions of Gauss–Manin in this polydisk).

By specializing to s , construction 2c provides an embedding: $\iota_B : M^\vee \hookrightarrow H^1_B(A_s \otimes_\iota \mathbb{C}, \mathbb{Z})$, where $A_s := \underline{A}_1(s)$. Dually, we also have an embedding:

$$j_B : M'^\vee(1) = 2i\pi M'^\vee \hookrightarrow H_{1B}(A_s, \mathbb{C}, \mathbb{Z}) .$$

In addition to the orientation of $\mathbb{C}_p$, we choose an orientation of $\mathbb{C}$; this eliminates all ambiguities of signs, and allows to identify $j_B(\mu^\vee_j(1))$ with μ'_j .

<u>Proposition 4.</u> The following diagram is commutative:

$$
\begin{array}{ccccc}
& & M^\vee & & \\
& \overset{\iota_{et}}{\swarrow} & & \overset{\iota_B}{\searrow} & \\
H^1_{et}(A_s, \overline{K}, \mathbb{Q}_p) & & & & H^1_B(A_s, \mathbb{C}, \mathbb{Z}) \\
\Big\downarrow {\scriptstyle F.M.^{-1}} & & & & \Big\downarrow {\scriptstyle \mathscr{F}^1_\iota} \\
H^1_{DR}(A_s) \otimes_E B_{DR} & & & & H^1_{DR}(A_s) \otimes_E \mathbb{C} \\
\underset{\substack{v-adic \\ evaluation}}{} & \nwarrow & H^1_{DR}(\underline{A}/S^*)^\vee & \nearrow & complex\ evaluation
\end{array}
$$

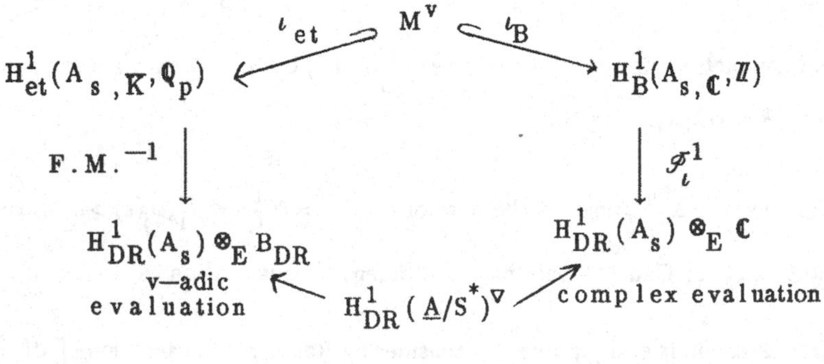

In particular (by duality), the complex evaluation of ω_{ij} at s gives the "usual" period $\frac{1}{2i\pi} < \mu'_i, \omega_j(s) >$.

Proof: let us draw a middle vertical arrow $\begin{matrix} M^{\vee} \\ \downarrow \\ H^1_{DR}(\underline{A}/S^*)^{\nabla} \end{matrix}$, defined by the obvious embedding

$M^{\vee} = \Gamma\underline{M}^{\vee} \hookrightarrow \Gamma H^1_{DR}[\underline{M} \xrightarrow{\psi} \underline{G}]_{/S^*} = \Gamma H^1_{DR}(\underline{A}/S^*)$ (or equivalently, when $n = 1$, by the analog of ι_{DR} in the rigid analytic category over the discretely valued field $E((x))$.

Then the commutativity on the L.H.S. is essentially the content of prop. 3; the commutativity on the R.H.S. follows immediately from the definition of ι_B (details are left to the reader).

This proposition suggests the following open question: assume that E is a number field, and denote by $\bar{E}$ its algebraic closure of E inside $\mathbb{C}_p$. Does there exist $\gamma : \bar{E} \hookrightarrow \mathbb{C}$ above ι such that the following diagrams commute?

$$\begin{array}{ccc} M^{\vee} & \xrightarrow{\iota_B} & H^1_B(A_{\mathbb{C}}, \mathbb{Z}) \\ {\scriptstyle \iota_{et}}\downarrow & & \downarrow \\ H^1_{et}(A_{\bar{E}}, \mathbb{Q}_p) & \xrightarrow{\gamma^*} & H^1_{et}(A_{\mathbb{C}}, \mathbb{Q}_p) \end{array} \qquad \begin{array}{ccc} M'^{\vee}(1) & \xrightarrow{j_B} & H_{1B}(A_{\mathbb{C}}, \mathbb{Z}) \\ \downarrow{\scriptstyle j_{et}} & & \downarrow \\ T_p(A_{\bar{E}}) & \xleftarrow{\gamma^*_{\sim}} & T_p(A_{\mathbb{C}}) \end{array}$$

(We leave it as an exercise to answer positively, when A_E is an elliptic curve, with help of $[S_2]$).

4. The monodromy filtration.

a) In $[G_2]$, Grothendieck constructs and studies thoroughly a 3–step filtration on $T_p(A_{\bar{K}})$, the "monodromy filtration" (here, we turn back to the setting of section 1)). By duality, we get a filtration W_{et} on H^1_{et} ; it turns out that this filtration is the natural weight filtration on the H^1_{et} of Raynaud's incarnation of the associated rigid 1–motive, loc. cit. § 14.

b) According to the semi–stable philosophy (motivated by higher dimensional motives), it should be natural to handle the monodromy business on the De Rham realization. The monodromy filtration $W_{-1} = 0$, $W_0 \simeq \underline{M}^{\vee}(K) \otimes_{\mathbb{Z}} K$, $W_2 = H^1_{DR}$, $Gr^W_2 \simeq \underline{M}'(K) \otimes_{\mathbb{Z}} K$, is the canonical filtration associated with the nilpotent operator of level 2 defined by:

$$\begin{array}{ccc}
N: \quad H^1_{DR}(A) & \dashrightarrow & H^1_{DR}(A) \\
\| & & \| \\
H^1_{DR}[\underline{M} \to G] & & H^1_{DR}[\underline{M} \to G] \\
\downarrow & & \uparrow \\
H^1_{DR}[0 \to T] & & H^1_{DR}[\underline{M} \to 0] \\
\| & & \| \\
\underline{M}'(K) \otimes_{\mathbb{Z}} K & \longrightarrow & \underline{M}^{\vee}(K) \otimes_{\mathbb{Z}} K
\end{array}$$

where the arrow at the bottom $\underline{M}'(K) \longrightarrow \underline{M}^{\vee}(K)$ is the map induced by opposite of Grothendieck's monodromy pairing: $v(q) : M \otimes M' \xrightarrow{q} \mathbb{G}_m\big|_{K^{nr}} \xrightarrow{v} \mathbb{Z}$ (v = valuation), ibid (we change the sign because we work on H^1_{DR}, not on the covariant H_{1DR}). Assume moreover that $M = \underline{M}(K)$. Then the cokernel of the map $M' \longrightarrow M^{\vee}$ inducing μ is canonically isomorphic to the group of connected components of the special fiber of the Néron model A, see [CF] III 8.1. The weight filtrations W and W_{et} are related via F.–M. :

Lemma 4 (for $M = \underline{M}(K)$) : $Gr_0^W \oplus Gr_2^W(1) \xrightarrow{\sim} (Gr_0^{W_{et}} \oplus Gr_2^{W_{et}}) \otimes_{\mathbb{Q}_p} K$.

In case A is a Jacobian variety, there is moreover a Picard–Lefschetz formula (loc. cit. § 12), where ${}^{\ell}et(M^{\vee})$ appears once again as the module of covanishing cycles.

c) Like the Raynaud extension, the operator N admits a complex analog (which is well–known). In the situation 2 c), let $D_j = \Delta^{j-1} \times \{0\} \times \Delta^{n-j} \subset \Delta^n$ be the jth divisor "at infinity". For any $s \in \Delta^{*n}$, there is a monodromy action "around $D_j^{"}$: $M_j^{\infty} \in GL(H_1(A_s,\mathbb{Z}))$, which is unipotent of level 2. Set $N_j^{\infty} := \frac{1}{2i\pi} \log^t (M_j^{\infty})^{-1} \in \mathrm{End}\, H^1(A_s,\mathbb{C})$. These nilpotent operators are constant on Δ^{*n}, and can be computed on the limit fiber by: $N_j^{\infty} = -\mathrm{Res}_{D_j}\nabla$ (the opposite of the residue at D_j of the Gauss–Manin connection).

Under the identification $H^1(A_s,\mathbb{Q}) \simeq H^1_B[\underline{M}(s) \longrightarrow \underline{G}(s)] \otimes \mathbb{Q}$, the "monodromy" filtration on the L.H.S. associated with N_j^{∞} is just the standard <u>weight filtration</u> on the R.H.S. [D1].

d) One can mimic the construction a) over any complete discretely valued ring instead of K, e.g. over $\mathcal{R} = \tilde{\mathbb{E}}[[x]]$, $I = (x)$, in the situation 3 d), with $n = 1$; We denote by $N^{for} \in \mathrm{End}\, H^1_{DR}\left[\underline{A}/\mathcal{R}\left[\frac{1}{x}\right]\right]$ the nilpotent endomorphism obtained this way.

Next, we wish to compare N , N^{for} and N^{∞} .

Let us consider a double embedding $\tilde{E} \begin{smallmatrix} \nearrow R \\ \searrow \mathbb{C} \end{smallmatrix}$ and let $s \in S_1(E)$. Assume that $|x(s)|_v < 1$ and that $S_1(\mathbb{C})$ contains the disk of radius $|x(s)|_2$.

At last, set $A_s = \underline{A}_{(s)}$.

Proposition 5. In this situation, the complex evaluation of N^{for} at s is $N^{\infty} \in \mathrm{End}\, H^1_{DR}(A_s \otimes \mathbb{C}) \simeq \mathrm{End}\, H^1(A_{s,\mathbb{C}},\mathbb{C})$; the v–adic evaluation of N^{for} at s is $v(x(s))N \in \mathrm{End}\, H^1_{DR}(A_s \otimes K)$.

Proof: the complex fact is well–known. The v–adic assertion relies on the equality $v(q_{ij}(s)) = (\mathrm{val}_x q_{ij}) \cdot v(x(s))$, which follows immediately from lemma 2.

[Remarks: d_1) if we only assume that $\underline{A}^{\sim} \longrightarrow \tilde{S}_1$ is proper outside $\tilde{\omega}\, x = 0$ (instead of $x = 0$) , the monodromy filtrations corresponding to $N^{for}_{(s)}$ and $N = N_s$ still coincide at the limit.

d_2) A quite general definition of N is given in [CF] III 10.]

IV. Frobenius and the p-adic Betti lattice.

1. Semi-stable Frobenius.

We take up again the situation I 3b), and explain a construction of the Frobenius semi-linear endomorphism φ_β (due to Raynaud $[R_2]$).

a) Let β denote a branch of the logarithm on $K^{\times}$. This amounts to the choice of some uniformizing parameter of R, say $\tilde{\omega}$, characterized (up to a root of unity) by the fact that $\beta : K^{\times} \simeq \tilde{\omega}^{\mathbb{Z}} \times (R/\tilde{\omega}R)^{\times} \times (1 + \tilde{\omega}R) \longrightarrow K$ factorizes through $1 + \tilde{\omega}R$.

b) Let A be an Abelian variety over K with semi-stable reduction, and let $[\underline{M} \xrightarrow{\psi} G]$ the Raynaud realization of the associated rigid 1-motive (G sits in an extension $0 \longrightarrow T \longrightarrow G \longrightarrow B \longrightarrow 0$, and ψ is described by $q : \underline{M} \times \underline{M}' \longrightarrow \mathbb{G}_m$).

Let us factorize $q = \tilde{\omega}^{v(q)} \cdot q^0$, so that $q^0 : \underline{M} \times \underline{M}' \longrightarrow \mathbb{G}_m$ extends over R. This amounts to a factorization $\psi = \chi_{\tilde{\omega}} \cdot \psi^0$, where $\chi_{\tilde{\omega}} : \underline{M} \longrightarrow T = \underline{\mathrm{Hom}}(M', \mathbb{G}_m)$ is induced by $\tilde{\omega}^{v(q)}$ and $\psi^0 : \underline{M} \longrightarrow G$ extends over R (we use the same notation ψ^0 for this extension). Because T is a torus, the universal $(\underline{M}, \chi_{\tilde{\omega}})$-equivariant vectorial extension of T splits canonically, which yields a canonical isomorphism of (Hodge) filtered K-vector spaces:

$$\Delta_\beta : H^1_{DR}[\underline{M} \xrightarrow{\psi^0} G]_{/R} \otimes_R K \xrightarrow{\sim} H^1_{DR}[\underline{M} \xrightarrow{\psi} G] = H^1_{DR}(A).$$

For two uniformizing parameters $\tilde{\omega}_1$, $\tilde{\omega}_2$, the map Δ_{β_1}, Δ_{β_2} are related by:

(i) $\Delta_{\beta_2} \Delta_{\beta_1}^{-1} = \exp(-\log \tilde{\omega}_2/\tilde{\omega}_1 \cdot N)$, where N is the operator defined in the previous section.

[Note the similarity with the definition of the canonical extension in the theory of regular connections, and also with [CF] III 9].

c) Let BT denote the Barsotti–Tate group attached to the reduction mod. $\tilde{\omega}$ of $[\underline{M} \xrightarrow{\psi^0} G]_{/R}$ (1), and let H^1_{crys/K^0} denote the K^0-space obtained by inverting p in its

---

(1) Remember that the Barsotti–Tate group attached to $[\underline{M} \xrightarrow{\psi^0} G]_{/R}$ is given by the image of ψ^0 under the connecting homomorphism $\mathrm{Hom}(\underline{M}, {}_{p^n}G) \longrightarrow \mathrm{Ext}(\underline{M}, {}_{p^n}G)$ associated with the exact sequence $0 \longrightarrow {}_{p^n}G \longrightarrow G \xrightarrow{p^n} G \longrightarrow 0$.

first crystalline cohomology group with coefficients in R^0. Up to isogeny, BT splits into the sum of two Barsotti–Tate groups: the constant one $\underline{M}(K) \otimes_{\mathbb{Z}} \mathbb{Q}_p/\mathbb{Z}_p$, and $\varinjlim_{p^n} G \otimes_R R/\tilde{\omega}R$. [It follows that H^1_{crys/K^0} does not depend on $\tilde{\omega}$; in fact, it depends only on $A_R \otimes R/\tilde{\omega}^2R$, which determines $G \underset{R}{\otimes} R/\tilde{\omega}^2R$.]

The K^0–structure H^1_0 mentioned in I 3 b_1) is just the image of H^1_{crys/K^0} under Δ_β inside $H^1_{DR}(A)$; the element u_β is $u_\beta := -\,\mathrm{LOG}\,\tilde{\omega}$ (defined up to translation by $\mathbb{Z}_p(1) \subset B^+_{cris}$).

By transport of structure, the σ–semi–linear Frobenius on H^1_{crys}/K^0 provides the σ–semi–linear endomorphism $\varphi = \varphi_\beta$ on H^1_0 (σ = Frobenius on K^0). Using (i), one gets the following relation:

(ii) $$\varphi_{\beta_2} \circ \varphi_{\beta_1}^{-1} = \exp(-\tfrac{1}{p}\log(\tilde{\omega}_2/\tilde{\omega}_1)^{p-\sigma}\cdot N).$$

From the functoriality of Raynaud extensions G and of the rigid analytic isomorphisms $G^{rig}/M = A^{rig}$, it follows that the semi–stable structure (H^1_0,φ,N) is functorial in A.

e) That construction of Raynaud may be extended to the relative situation III 2, i.e. over $\mathcal{R} = \tilde{\omega}$ – adically complete noetherian normal R^0–algebra.

Let $U \subset \mathrm{Spec}\,\mathcal{R}$ be as in loc. cit., and let us choose a lifting $\sigma \in \mathrm{End}\,U$ of the char. p Frobenius. By analogy with step c), we can construct, locally for the "loose" topology on U, a horizontal morphism $\phi_\beta(\sigma) : \sigma^* H^1_{DR}(\underline{A}/U) \longrightarrow H^1_{DR}(\underline{A}/U)$; furthermore, this morphism "stabilizes" $\underline{M}^\vee_U$, and it can be globally defined there. [This is the "stability of vanishing cycles" mentioned in [Dw]; indeed, when say $\mathcal{R} = R[\widehat{\tilde{\omega}x}]$, $\sigma : x \longmapsto x^p$, ϕ_β is nothing but the analytic Dwork–Frobenius mapping].

If A is the fiber $\underline{A}(s)$ of $\underline{A}$ at some point $s \in U$ fixed under σ, we recover $\phi_\beta(\sigma) = \varphi_\beta$.

2. <u>Construction of $H^1_\beta(\underline{A})$</u>.

From now onwards, we shall assume that A <u>has multiplicative reduction</u>.

a) With our previous notations, we then obtain the following consequences:

a_1): $G = T$, $r = g$,

a_2): the Hodge filtration splits canonically:

$$H^1_{DR} = (\underline{M}^{\vee}(K) \otimes_{\mathbb{Z}} K) \oplus F^1$$

a_3): the monodromy filtration consists of only two steps:

$$Gr^{W_{et}} H^1_{et} \simeq (M^{\vee} \otimes_{\mathbb{Z}} \mathbb{Q}_p) \oplus (M' \otimes_{\mathbb{Z}} \mathbb{Q}_p(-1)) \text{ (via } \iota_{et} \text{ and } j_{et}) ,$$

$$Gr^W H^1_{DR} \simeq (M^{\vee}(K) \otimes_{\mathbb{Z}} K) \oplus (\underline{M}'(K) \otimes_{\mathbb{Z}} K) ,$$

(these isomorphisms being compatible via F.M., by prop. 3 and its dual) $\underline{M}^{\vee}(K) \otimes_{\mathbb{Z}} K = \operatorname{Ker} N$, and F^1 projects onto $\underline{M}'(K) \otimes_{\mathbb{Z}} K$ (this isomorphic projection being given by $F^1 = \operatorname{Colie} A^{rig} \simeq \operatorname{Colie} T^{rig} \xrightarrow{\sim} \underline{M}'(K) \otimes_{\mathbb{Z}} K)$.

a_4): the Fontaine–Messing isomorphism F.M. is described in I 4 c).

b) The splitting of BT (up to isogeny) reflects on H^1_0 , and translates into an isomorphism:

$$\Sigma_\beta : Gr^W H^1_0 \xrightarrow{\sim} H^1_0$$

(φ acts trivially on $Gr_0 = \underline{M}^{\vee}(K) \otimes_{\mathbb{Z}} K^0$, and by multiplication by p on the image of $Gr_1 = \underline{M}'(K) \otimes_{\mathbb{Z}} K^0)$.

Let us now choose an orientation of $\mathbb{C}_p$ (see III 1f): $\mathbb{Z}(-1) := X^*(\mathbb{G}_m) \hookrightarrow \mathbb{Z}t_p^{-1} \subset B_{DR}$,and let us consider the etale lattice $\underline{\Lambda} := \underline{M}^{\vee} \oplus \underline{M}'(-1)$, and let $\Lambda := \underline{\Lambda}(\overline{K}) = \underline{\Lambda}(\overline{R}) = \underline{\Lambda}(K^{nr})$, where K^{nr} denotes the maximal subfield of $\overline{K}$ non ramified over K .

Using Σ_β and the orientation, we can embed Λ into $H^1_{DR} \otimes_K K^{nr}\left[\frac{1}{t_p}\right] \subset H^1_{DR} \otimes_K B_{DR}$, and we call <u>p–adic Betti lattice</u> its image, which we denote by H^1_β [This is the dual of the lattice L_β mentioned in the introduction. The introduction of t_p , the "p–adic $2i\pi$" , is motivated by the fact that the complex Betti lattice (in the setting III 4c) is stable under $2i\pi N_\infty$, not N_∞] .

We thus get a tautological isomorphism:

$$\mathscr{P}_\beta : H^1_{DR} \underset{K}{\otimes} K^{nr}[t_p] \xrightarrow{\sim} H^1_\beta \underset{\mathbb{Z}}{\otimes} K^{nr}[t_p]$$

where in fact K^{nr} could be replaced by some finite extension of K, or else by K itself if T is split.

From formulae (i) (ii), it follows:

(iii) $$H^1_{\beta_2} = \exp(-\log \tilde{\omega}_2/\tilde{\omega}_1 \cdot N) \cdot H^1_{\beta_1} .$$

From the very construction of H^1_β and the formula $\varphi t_p = p t_p$, we get:

Lemma 5: The lattice H^1_β spans the $\mathbb{Q}_p$-space of φ_β-invariants in $H^1_{DR} \underset{K}{\otimes} K^{nr}[t_p]$.

Remark: the image of $\mathscr{P}_\beta^{-1} H^1_\beta$ under F.M. does not lie in $H^1_{et}(A,\mathbb{Q}_p)$; compare with lemma 4.

c) Let us now describe the complex analog of $\Sigma_\beta : \Lambda \longrightarrow H^1_\beta$. So let $A_\mathbb{C}$ be a complex Abelian variety in Jacobi form $T_\mathbb{C}/M$ (the quotient being alternatively described by $q : M \otimes M' \longrightarrow \mathbb{C}^\times$, where $M' = X^*(T_\mathbb{C})$). Let us orient $\mathbb{C}$, and choose a branch β_∞ of the complex logarithm, and compose with $q : M \otimes M' \xrightarrow{\beta_\infty \circ q} \mathbb{C}$. We get an embedding $M \hookrightarrow M'^\vee \otimes_\mathbb{Z} \mathbb{C} \simeq \operatorname{Lie} T_\mathbb{C} \simeq H_{1B}(A_\mathbb{C},\mathbb{Z}) \otimes_\mathbb{Z} \mathbb{R}$ which factorizes through $H_{1B}(A_\mathbb{C},\mathbb{Z})$. This in turn provides an isomorphism $\Sigma_{\beta_\infty} : \Lambda = M^\vee \oplus M'(-1) = M^\vee \oplus \frac{1}{2i\pi} M' \xrightarrow{\sim} H^1_B(A_\mathbb{C},\mathbb{Z})$ (the injectivity is a consequence of the Riemann condition $\operatorname{Re} \beta_\infty(q) < 0$).

[d) One can imitate the construction of the p-adic lattice in the case of an Abelian variety B with ordinary good reduction over $K = K_0$. Over $\widehat{K^{nr}}$ indeed, the Barsotti–Tate group $B(p) = \varprojlim {}_{p^n}B$ becomes isomorphic to the B.–T. group associated to a 1-motive $[M \xrightarrow{\psi} T]$, where ψ is given by the Serre–Tate parameters [K]. However, in contrast to the multiplicative reduction case, the lattice $\simeq M^\vee \oplus M'(-1)$ obtained in this way is not functorial, as is easily seen from the case of complex multiplication $(\psi = 1)$.

e) The construction of Frobenius generalizes easily to the case of 1-motives. This allows to construct p-adic Betti lattices for 1-motives whose Abelian part has multiplicative reduction. We shall not pursue this generalization any further here.]

3. Computation of periods.

a) We shall compute the matrix of the restriction of $\mathcal{P}_\beta$ to $F^1H^1_{DR}$ w.r.t. the bases $\{d\mu_j/1+\mu_j\}_{j=1}^g$ in F^1, $\{\mu''_i = \Sigma_\beta(\mu_i(-1))\ ,\ m^v_i\}_{i=1}^g$ in H^1_β, assuming that T splits over K. In other words, we compute half of the (β)–p–adic period matrix.

Proposition 6. Let $q_{ij} = q(m_i,\mu_j)$, as in I 4 c). The following identity holds in $H^1_{DR}(A_K) \otimes_K K[t_p]$:

$$d\mu_j/1 + \mu_j = t_p\mu''_j + \sum_{i=1}^{g} \beta(q_{ij})m^v_i .$$

b) Proof: it relies on a deformation argument. First of all, one may replace M by a sublattice of finite index, such that $q \equiv \tilde{\omega}^{v(q)}q^0$ with $q^0 \equiv 1 \bmod \tilde{\omega}$ (in this situation BT splits actually, not only up to isogeny). Let us consider the analytic deformation $\left[M \xrightarrow{\underline{\Psi} = X_{\tilde{\omega}} \cdot \underline{\Psi}^0} T\right]$ of $\left[M \xrightarrow{\psi = X_{\tilde{\omega}} \cdot \psi^0} T\right]$ over $\mathcal{R} = R[[\xi_{ij} - \delta_{ij}]]_{i,j=1}^g$ δ_{ij} = Kronecker symbol, $\underline{\Psi}^0$ being given by the matrix ξ_{ij} (so that $[M \xrightarrow{\psi} T]$ arise as the fiber at $\xi_{ij} = q^0_{ij}$). For the fiber at $\xi_{ij} = \delta_{ij}$: $[M \xrightarrow{X_{\tilde{\omega}}} T]$, the $F^1H^1_{DR}$ coincides with $\Sigma_\beta(Gr^W_1 H^1_{DR})$; more precisely $d\mu_j/1 + \mu_j = t_p\mu''_j$, at $\xi_{ij} = \delta_{ij}$. By definition of the Kodaira–Spencer mapping K.S. (see e.g. [CF] III. 9), one deduces that

$$d\mu_j/1 + \mu_j = t_p\mu''_j + \left(\int_{\xi_{ij}=\delta_{ij}}^{q^0_{ij}} K.S.\right)m^v_i \ , \ \text{at } \xi_{ij} = q^0_{ij} .$$

But in our bases, K.S. is expressed by the matrix $d\xi_{ij}/\xi_{ij}$ (see [Ka], or [CF] ibid, where there is a minus sign because of a slightly different convention). One concludes by noticing that $\log q^0_{ij} = \beta(q_{ij})$.

c) One could also argue as follows, using F.M.: it follows from 2 a 3) that $d\mu_j/1 + \mu_j$ may be expressed in the form $t_p\mu''_j + \Sigma\beta_{ij}m^v_i$, $\beta_{ij} \in K$; furthermore, these coefficients β_{ij} are uniquely determined by the property that $d\mu_j/1 + \mu_j - \Sigma\beta_{ij}m^v_i$ lies in $H^1_0 \otimes_{K^0} B_{ss}$ and is multiplied by p under φ_p. Let us show that $\beta_{ij} = \beta(q_{ij})$ satisfies this property: by I 4 c), we have

$$d\mu_j/1+\mu_j = t_p FH^{-1}(\tilde{\mu}_j) + \Sigma\, \mathrm{LOG}(q_{ij}) m_i^{\vee}\text{, so that}$$

$$d\mu_j/1+\mu_j - \Sigma\, \beta_{ij} m_i^{\vee} = \Sigma(\mathrm{LOG}(q_{ij}) - \beta(q_{ij}))m_i^{\vee} + t_p FM^{-1}(\tilde{\mu}_j)\,.$$

Because $\tilde{\mu}_i \in H^1_{et}$, $t_p FM^{-1}(\tilde{\mu}_j) \in (H^1_0 \otimes B_{ss})^{\varphi = p}$, and we conclude by the following:

Lemma 6: let $c \in K^{x}$. Then "the" element $\mathrm{LOG}\, c - \beta c$ of B_{ss} is multiplied by p under the Frobenius φ_β.

Proof: let us write $c = \tilde{\omega}^{v(c)} c^0$, so that $\mathrm{LOG}\, c - \beta c = -v(c)u_\beta + \mathrm{LOG}\, c^0 - \log c^0$. Now $\mathrm{LOG}\, c^0 - \log c^0 = -\log \lim (\tilde{c}_n)^{p^n}$ in B^+_{cris}, where $\tilde{c}_n$ is any lifting of $c_n = (c^0)^{p^{-n}} \in \bar{R}$. Let $c'_n = (\dots c_{n+1}, c_n) \in \varprojlim_{x \mapsto x^p} \bar{R}$, and let $\tilde{c}_n$ be the Teichmüller representative $[c'_n] \in W(\varprojlim \bar{R})$. We have $[c'_n]^\varphi = [c'^{\,p}_n] = [c'_{n-1}] = \tilde{c}_{n-1}$, whence $(\lim \tilde{c}_n p^n)^\varphi = (\lim \tilde{c}_n^{p^n})^p$. It remains only to take logarithms and remind that $\varphi_\beta u_\beta = pu_\beta$.

d) Let us examine the complex counterpart, as in 2 c). The lattice $M \oplus M'^{\vee}(1) = \Lambda^{\vee} \xrightarrow[\sim]{({}^t\Sigma_{\beta_\infty})^{-1}} H_{1B}(A_{\mathbb{C}}, \mathbb{Z})$ embeds into Lie $T_{\mathbb{C}}$; the subspace $F^0H_{1DR}(A_{\mathbb{C}})$ of $H_1(A_{\mathbb{C}}, \mathbb{Z}) \otimes \mathbb{C} \simeq H_{1DR}(A_{\mathbb{C}})$ is just the kernel of the complexification of this embedding. It follows that the canonical lifting $\tilde{m}_i$ of m_i inside $F^0H_{1DR}(A_{\mathbb{C}})$ is given by $\tilde{m}_i = m_i - \frac{1}{2i\pi} \Sigma\beta_\infty(q_{ij})\mu'_i$ (we set $\mu'_i = ({}^t\Sigma_{\beta_\infty})^{-1}(\mu_i^{\vee}(1))$, and $\mu''_j = \Sigma_{\beta_\infty}(\mu_j(1))$). By orthogonality $(F^1H^1_{DR} = (F_0H_{1DR})^{\perp})$, we obtain:

Proposition 7: the following identity holds in $H^1_{DR}(A_{\mathbb{C}})$:

$$d\mu_j/1+\mu_j = 2i\pi\mu''_j + \Sigma\beta_\infty(q_{ij})m_i^{\vee}\,.$$

[The compatibility (resp. analogy) between prop. 6 and formula (iii) resp. prop. 7., is a good test for having got the right signs. Although μ''_j is defined quite differently in the p-adic, resp. complex case, the exterior derivative of the coefficients of $m_i^{\vee}$'s describes in both cases the Kodaira–Spencer mapping.]

4. Periods in the relative case, and Dwork's p–adic cycles.

a) Let us consider the relative situation as in 1. d with $r = g$; U being subject to be the complement of divisor with normal crossings $\tilde{\omega}x_1 \dots x_n = 0$. We set $\mathcal{R} = R[[x_1, \dots, x_n]]$, and we denote by $\mathcal{A}$ the K–algebra generated by $\mathcal{R}\left[\frac{1}{x_1 \cdots x_n}\right]$ and (β)–logarithms of non–zero elements of $\mathcal{R}\left[\frac{1}{x_1 \cdots x_n}\right]$. The construction of H^1_β can be transposed to this relative setting: We use "the" relative Frobenius $\phi_\beta(\sigma)$ to construct an embedding

$$\underline{\Lambda} \xrightarrow{\sim} \underline{H}^1_{\beta,\sigma} \subsetneq H^1_{DR}(\underline{A}/\mathcal{A}[t_p]),$$

such that $\phi_\beta(\sigma)\Big|_{\mathrm{Im}\,\underline{\Lambda}} = \sigma_*$. Of course, when $\sigma s = s$, we recover $\underline{H}^1_{\beta,\sigma}(s) = H^1_\beta$.

Because $\phi_\beta(\sigma)$ is horizontal, so is $\underline{H}^1_{\beta,\sigma}$ (it is locally constant w.r.t. the loose topology), and we get:

Lemma 7: $H^1_{DR}(\underline{A}/\mathcal{A}[t_p])^\nabla = \underline{H}^1_{\beta,\sigma} \otimes_{\mathbb{Z}} K[t_p]$.

b) In order to interpret the lattice $\underline{H}^1_{\beta,\sigma}$ (for $n = 1$, $\phi : x \longmapsto x^p$) in terms of Dwork's p–adic cycles [Dw], one forgets about t_p (or better, one specializes t_p to $1 : K[t_p] \longrightarrow K$, $\underline{H}^1_{\beta,\sigma} \simeq \underline{M}^\vee \oplus \underline{M}'$) . Let us for instance take back the example III 2g (Legendre). For $K = \mathbb{Q}_p(\sqrt{-1})$ $(p \neq 2)$, we have $M = \underline{M}(K)$, with base m . Setting $v = uw$, the period of the differential of the first kind $\omega = \frac{du}{2v}$ for the covanishing cycle $m^\vee$ at $x = 0$ is given by the residue of $\frac{du}{2uw} = \frac{dw}{w^2+1}$ at one of the two points above $u = 0$ on the rational curve $w^2 = u - 1$; namely, this is $\frac{\sqrt{-1}}{2}$.

Let μ be the basis of $M' = \underline{M}'(K)$ lifted to H^1_β , such that $q = q(m,\mu)$ is given by the formula displayed in III 2. g. Then after specializing t_p to 1 , the matrix of $\mathcal{P}_\beta$ in terms of the bases ω , $\omega' = \nabla(x\frac{d}{dx})\omega$ is

$$\frac{\sqrt{-1}}{2}\begin{bmatrix} F & x\,\dot{F} \\ F\,\log q & x(F\,\dot{\log}\,q) \\ = F\,\log x - \log 16 + \dots & = 1 + x\dot{F}\,\log x + \dots \end{bmatrix} \quad \text{(with determinant } (4x(x-1))^{-1}\text{)}.$$

Here "log" is standing for the branch β, and $\dot{F}$ for $\frac{d}{dx}F$.

In fact, Dwork prefers to get rid of the constants $\log 16$ and $\frac{\sqrt{-1}}{2}$, by changing the basis $\{\mu, m^\vee\}$ into $-2\sqrt{-1}\{\mu + (\log 16)m^\vee, m^\vee\}$. In this new basis, the entries of the period matrix lie in $\mathbb{Q}[[x]][[\log x]]$, and the matrix of $\phi_\beta(x \longmapsto x^p)$ becomes $(-1)^{\frac{p-1}{2}}\begin{bmatrix} p & 0 \\ \log 16^{1-p} & 1 \end{bmatrix}$ see [Dw] 8. 11.

c) In section 3, we computed periods of one–forms of the first kind. The "horizontality lemma" 7 then allows to obtain other periods by taking derivatives; still, we have to show that, in the multiplicative reduction case, any one–form of the second kind is the Gauss–Manin derivative of some one–form of the first kind. In other words:

Lemma 8. Let us consider a relative situation, as in III 2c or 2d. If $r = g$, then for any $k = 1, \dots, n$, the smallest $\mathcal{O}_s[\nabla(x_k \partial/\partial x^k)]$–submodule of $H^1_{DR}(\underline{A}/S^*)$ containing F^1 is $H^1_{DR}(\underline{A}/S^*)$.

Indeed, this amounts to the surjectivity of K.S., which follows from the invertibility of its residue at $x_k = 0$; this follows in turn from the fact that this residue $(F^1)^{can}_{x_k=0} \simeq \underline{M}'(\mathcal{K}) \otimes E \to (H^1_{DR}/F^1)^{can}_{x_k=0} \simeq \underline{M}^\vee(\mathcal{K}) \otimes E$ is induced by the non–degenerate pairing $\mathrm{val}(x_k) \circ q$. In the situation of III 3 a) b), we can now complete the analytic description of the period matrix: take a basis ω_j of the canonical extension of $H^1_{DR}(\underline{A}/S^*)$ in the form

$$\begin{cases} \omega_j \in F^1 \\ \omega_{j+g} = \nabla(x_k \partial/\partial x_k)\omega_j \end{cases} \quad j = 1, \dots g .$$

Lemma 9. The matrix of $\mathcal{P}_\beta$ w.r.t. the bases $\{\omega_j\}$, $\{\mu''_i, m_i\}$ has the form:

$$\begin{bmatrix} \pm t_p \omega_{ij}(s) & \pm t_p (x_k \partial/\partial x_k \omega_{ij})(s) \\ \omega_{ij}(s) \log q_{ij}(s) & (x_k \partial/\partial x_k(\omega_{ij} \log q_{ij}))(s) \end{bmatrix} \quad (\text{for } A = \underline{A}_1(s))$$

d) We are now in position to state the main result of this section IV, relating p–adic and complex Betti lattices.

Data: d_1): a field E , doubly embedded $E \begin{smallmatrix} \nearrow \mathbb{C} \\ \searrow K \end{smallmatrix}$; orientations of $\mathbb{C}$ and $\mathbb{C}_p$. A branch β (resp. β_∞) of the logarithm on $K^\times$ (resp. on $\mathbb{C}^\times$) ; a uniformizing parameter $\tilde{\omega}$ such that $\beta(\tilde{\omega}) = 0$.

d_2): an affine curve S_1 over E ; a smooth point $0 \in S_1(E)$, and a local parameter x around 0 ; a regular model $\tilde{S}_1$ of S_1 over $E \cap R$.

d_3): a semi–abelian scheme $f : \underline{A} \longrightarrow \tilde{S}_1$, proper outside the divisor $\tilde{\omega} x = 0$, and given by a split torus on this divisor. To f , one attaches as before the constant sheaf of lattices $\Lambda = M^{\vee} \oplus M'(-1)$ (outside $x = 0$) , and the bilinear form $q : M \otimes M' \longrightarrow \mathbb{G}_m$ (outside $\tilde{\omega} x = 0$) . Taking bases of M , resp. M' , one may expand the entries of a matrix of q into Laurent series: $q_{ij} = \eta_{ij} x^{n_{ij}} + \text{h.o.t.}$, and consider the double homomorphism from the E–algebra $E_1 := E[\log \eta_{ij}, t] \begin{smallmatrix} \nearrow K[t_p] \\ \searrow \mathbb{C} \end{smallmatrix}$ induced by β , $t \longmapsto t_p$ (resp. β_∞ , $t \longmapsto 2i\pi$) .

d_4): a simply connected open neighborhood of 0 in $S_{\mathbb{C}}$, say $\mathcal{U}$; over $\mathcal{U} \setminus 0$, Λ is identified with the graded form (w.r.t. the local monodromy N_∞) of $R^1 f_*^{an} \mathbb{Z}$.

d_5): a point $s \in S_1(E)$ such that $s \in \mathcal{U}$ and $|x(s)|_v < 1$ (from this last condition, it follows that the fiber $\underline{A}(s)$ has multiplicative reduction mod $\tilde{\omega}$) .

Combining the previous lemma with propositions 4, 6, and 7, we obtain:

Theorem 2. The following diagram is commutative:

$$\begin{array}{ccccc}
 & & \Lambda & & \\
 & \swarrow{\scriptstyle \Sigma\beta} & & \searrow{\scriptstyle \Sigma\beta_\infty} & \\
H^1_\beta(\underline{A}(s)) \otimes_{\mathbb{Z}} K[t_p] & & & & H^1_B(\underline{A}(s) \otimes \mathbb{C}, \mathbb{Z}) \otimes_{\mathbb{Z}} \mathbb{C} \\
 & \nwarrow & & \nearrow & \\
 & & H^1_{DR}(\underline{A}/E_1[[x]][\log x])^{\nabla} & &
\end{array}$$

left arrow: p–adic evaluation at s (via $\log x \longmapsto \beta(x(s))$ and $E_1 \longrightarrow K[t_p]$)

right arrow: complex evaluation at s (via $\log x \mapsto \beta_\infty(x(s))$ and $E_1 \rightarrow \mathbb{C}$)

(In the example III 2g, E_1 is just $\mathbb{Q}(\sqrt{-1})[t]$, and the parameter $x = \lambda$ should be replaced by $x = 16\lambda$) .

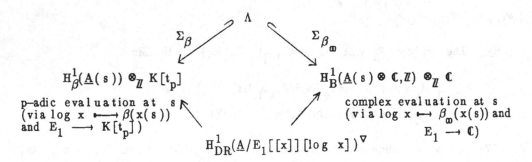

V. p–Adic lattice and Hodge classes.

1. Rationality of Hodge classes.

a) Let A_E be an Abelian variety over a number field E. Let v be a finite place of E where A_E has multiplicative reduction, and let $K = E_v$ denote the completion.

Conjecture 6. Let $\xi \in (\mathrm{End}\, H^1_{DR}(A_E))^{\otimes n}$ be some Hodge class [(1)]. Then for every branch β of the logarithm on $K^\times$, the image of ξ under $\mathcal{P}_\beta$ lies in the rational subspace $(\mathrm{End}\, H^1_{\beta,\mathbb{Q}})^{\otimes n}$, where $H^1_{\beta,\mathbb{Q}} := H^1_\beta \otimes_{\mathbb{Z}} \mathbb{Q}$. (For instance, this holds if $n = 1$ just by functoriality of H^1_β).

b) Let $\iota : E \hookrightarrow \mathbb{C}$ and let Sh be the connected Shimura variety associated to the Hodge structure $H^1_B(A_E \otimes_\iota \mathbb{C}, \mathbb{Z})$ and to some (odd prime–to–p) N–level–structure; Sh descends to an algebraic variety over some finite extension E' of E, and $A_{E'}$ is the fiber of an Abelian scheme $\underline{A} \longrightarrow \mathrm{Sh}$ at some point $s \in \mathrm{Sh}(E')$. In terms of Siegel's modular schemes $A_{g,N}$ [CF] IV, we have a commutative diagram

$$\begin{array}{ccc} \mathrm{Sh} & \hookrightarrow & A_{g,N} \otimes_{\mathbb{Z}\left[\frac{1}{N},\zeta_N\right]} E' \\ \cap & & \cap \\ \overline{\mathrm{Sh}} & \hookrightarrow & \overline{A}_{g,N} \otimes_{\mathbb{Z}\left[\frac{1}{N},\zeta_N\right]} E' \end{array},$$

where the superscript $-$ denotes suitable projective toroidal compactifications, see [H].

In fact $\underline{A} \longrightarrow \mathrm{Sh}$ extends to a semi–abelian scheme over a normal projective model $\widetilde{\mathrm{Sh}}$ of $\overline{\mathrm{Sh}}$ over $\mathcal{O}_{E'}$ (namely $\widetilde{\mathrm{Sh}}$ = normalization of the schematic adherence of $\overline{\mathrm{Sh}}$ in $\overline{A}_{g,N} \otimes_{\mathbb{Z}\left[\frac{1}{N},\zeta_N\right]} \mathcal{O}_{E'}$).

---

(1) Some authors prefer to look at Hodge classes in the more general twisted tensor spaces $(H^1_B)^{\otimes m_1} \otimes (H^{1\vee}_B)^{\otimes m_2}(m_3)$. However such spaces contain Hodge classes only if $m_1 + m_2$ is even (in fact if $m_1 - m_2 = 2m_3$), and any polarization then provides an isomorphism of rational Hodge structures $H^{1\,\otimes m_1}_B \otimes (H^{1\vee}_B)^{m_2}(m_3) \simeq (\mathrm{End}\, H^1_B)^{\otimes \frac{m_1+m_2}{2}}$. In particular, these extra Hodge classes do not change the Hodge group.

We consider the following condition:

(*) There exists a zero–dimensional cusp in $\overline{Sh}$, say 0, such that 0 and s have the same reduction mod. the maximal ideal of R'. In fancy terms, this means that any Abelian variety with multiplicative reduction in characteristic p should also degenerate multiplicatively (in characteristic 0) inside the family "of Hodge type" that it defines [M].

Remark: condition (*) should follow from Gerritzen classification [Ge] of endomorphism rings of rigid analytic tori (which is the same in equal or unequal characteristics), in the special case of Shimura families of p_{EL}–type [Sh] (i.e. characterized by endomorphisms).

Theorem 3. Conjecture 6 follows from (*).

Proof: by definition of the Shimura variety, and by the theory of absolute Hodge classes [D_2], $\xi = \underline{\xi}(s)$ is the fiber at s of a global horizontal section $\underline{\xi} \in \Gamma(\mathrm{End}\, H^1_{DR}(\underline{A}/Sh)^{\otimes n})^{\nabla}$.

Let S_1 be an algebraic curve on $\overline{Sh}$, joining 0 and s, and smooth at 0; let x be a local parameter around 0, with $|x(s)|_v < 1$. Then because 0 is a 0–dimensional cusp, $\underline{A}$ degenerates multiplicatively at 0 and we are in the situation where theorem 2 applies.

The β–periods of $\underline{\xi}$ admit an expansion in the form $\sum_{\ell=0}^{n} \alpha_\ell \log^\ell x$, with $\alpha_0 \in E'[[x]]$, $\alpha_\ell \in E'_1[[x]]$, whose complex evaluation (w.r.t $\iota : E' \hookrightarrow \mathbb{C}$) gives the corresponding complex period of $\underline{\xi}$, according to theorem 2. Since $\underline{\xi}$ is a global horizontal section and a Hodge class at s, the complex periods are rational constants: $\alpha_\ell = 0$ for $\ell > 1$, and $\alpha_0 \in \mathbb{Q}$. Thus the β–periods of $\xi = \underline{\xi}(s)$ are rational numbers.

Remark: it follows (inconditionally) from theorem 1 and Fontaine' semi–stable theorem that the image of ξ under $\mathcal{P}_\beta$ lies in $(\mathrm{End}\, H^1_\beta)^{\otimes n} \otimes_{\mathbb{Z}} \mathbb{Q}_p$.

2. p–Adic Hodge classes.

Let E' be some finitely generated extension of E. We define a p–adic Hodge class on $A_{E'}$ to be any element ξ of $F^0(\mathrm{end}\, H^1_{DR}(A_{E'})^{\otimes n})$ such that for every E–embedding of E' into any finite extension K' of K, and for every branch β of the logarithm on $K'^{\times}$, the image of ξ under $\mathcal{P}_\beta$ lies in the rational subspace $(\mathrm{End}\, H^1_{\beta,\mathbb{Q}})^{\otimes n}$. Conjecture 6 predicts that any Hodge class is a p–adic Hodge class, and conjecture 2 would identify the two notions.

Proposition 8: if E is algebraically closed in E', then any p–adic Hodge class ξ comes from $(\mathrm{End}\, H^1_{DR}(A_E))^{\otimes n}$, and is sent into $[(\mathrm{End}\, H^1_{et})^{\otimes n}]^{\mathcal{G}}$ by F.M.

Proof: the first assertion follows Deligne's proof in the complex case $[D_2]$. To prove the second one, we remark that $\xi \in F^0[(\mathrm{End}\, H^1_0)^{\otimes n}]^{\varphi=1}$; moreover, by changing β continuously, the lattice H^1_β is moved by $\exp(-\log u.N)$, $u \in R^x$. Since ξ has to remain rational w.r.t. all these lattices, we deduce that $N\xi = 0$, and we conclude by Fontaine semi–stable theorem.

Remark: it is essential to take all E–embedding $E' \hookrightarrow K$ into account; for instance, $m^v \in F^0 H^1_{DR}(A_{E'})$ for $E' = K$, and $m^v \in H^1_\beta$, $FM(m^v) \in (H^1_{et})^{\mathcal{G}}$, but it is highly probable that m^v is not defined over $\bar{E} \cap K$.

3. A p–adic period conjecture.

For any E–algebra E', the E'–linear bijections $H^1_{DR}(A_E) \otimes_E E' \xrightarrow{\sim} (H^1_{\beta,\mathbb{Q}}) \otimes_{\mathbb{Q}} E'$ which preserve p–adic Hodge classes form the set of E'–valued points of an irreducible E–torsor P_β under the "p–adic Hodge group" of A_E (which is by definition the algebraic subgroup of $GL\, H^1_{DR}(A_E)$ which fixes the p–adic Hodge classes; conjecture 2 would identify this group with the Hodge group). One has a canonical $K[t_p]$–valued point of P_β given by $\mathcal{P}_\beta$. A variant of conjecture 1 may be stated as follows:

Conjecture 1′: for sufficiently general β, $\mathcal{P}_\beta$ is a (Weil) generic point of P_β.

The next section will offer two partial positive answers.

4. Period relations of bounded degree.

a) We denote by $E[\mathcal{P}_{\beta_v}]_{\le\delta}$ the quotient of the polynomial ring in $4g^2$ indeterminates over E by the ideal generated by relations of degree $\le\delta$ among (β_v)–p–adic periods $(v|p)$. Hence for sufficiently large δ, there is a natural embedding $\mathrm{Spec}\, E[\mathcal{P}_{\beta_v}]_{\le\delta} \subset P_{\beta_v}$. The same construction works simultaneously at several places of multiplicative reduction: $E[(\mathcal{P}_{\beta_v})_{v\in V}]_{\le\delta} \subset \prod_v E_v[t_p]$, and we have projections $\mathrm{Spec}\, E[(\mathcal{P}_{\beta_v})_{v\in V}]_{\le\delta} \longrightarrow P_{\beta_v}$.

b) Assume that A_E is the fiber at $s \in S_1(E)$ of a semi–abelian scheme $\underline{A} \longrightarrow S_1$ over an affine curve $S_1/\mathrm{Spec}\, E$, proper outside some smooth point $0 \in S_1(E)$, and degenerating to a split torus at this point. Let x be a local parameter around 0, and let $\delta >> 0$.

We lay down an extra normalization hypothesis:

(**) the entries of the q–matrix expand $q_{ij} = \eta_{ij} x^{n_{ij}} + \ldots$ where η_{ij} are roots of unity (this is the case in example III 2g), if we set $x = 16\lambda$ and $E = \mathbb{Q}(\sqrt{-1})$.

In these circumstances, we have the following two results:

Theorem 4. Assume that $|x(s)|_v$ is sufficiently small – w.r. to δ – so that in particular $A_E = \underline{A}(s)$ has multiplicative reduction at v . Let us choose $\beta = \beta_v$ such that $\beta(x(s)) = 0$.

Then $\mathrm{Spec}\, E[\mathcal{P}_\beta]_{\leq\delta} = P_\beta$, and moreover any p–adic Hodge class on A_E is a Hodge class.

Theorem 5. Assume that $\underline{A} \longrightarrow S_1$ extends to a semi–abelian scheme over some regular model of S_1 over $\mathcal{O}_E$, proper outside the divisor $\nu x = 0$, $\nu \in \mathbb{N}$. Let $V(s)$ denote the finite set of finite places v of E where $|x(s)|_v < |\nu|_v$ (so that $\underline{A}(s)$ has multiplicative reduction at $v \in V$) . Let us choose β_v such that $\beta_v(x(s)) = 0$, $v \in V(s)$, and let $\varepsilon > 0$. If for every $\iota : E \hookrightarrow \mathbb{C}$, $|x(s)|_\iota \geq \varepsilon$, then the projections $\mathrm{Spec}\, E[(\mathcal{P}_{\beta_v})_{v\in V}]_{\leq\delta} \longrightarrow P_{\beta_v}$ are surjective, except possibly if s belong to a certain finite exceptional set (depending on δ,ε) .

c) In fact, the proof shows a little bit more: one can replace P_{β_v} in the statements by the specialization at s of the S_1–torsor formed of isomorphisms $H^1_{DR}(\underline{A}/S_1^*) \otimes ? \longrightarrow \underline{H}^1_\beta \otimes ?$ preserving global horizontal classes; this makes sense because any such class is automatically a $\mathcal{O}_{S_1^*}$–linear combination of relative Hodge classes, in virtue of:

Proposition 9 (Mustafin). On an Abelian scheme $\underline{A} \longrightarrow S_1^*$ degenerating to a torus at $0 \in S_1 \backslash S_1^*$, any element ξ of $\Gamma(\mathrm{End}\, H^1_{DR}(\underline{A}/S_1^*)^{\otimes n})^\nabla$ is a linear combination of relative Hodge cycles.

See e.g. [A] IX 3.2. The argument given in the course of proving theorem 3 then shows that ξ is also a linear combination of relative p–adic Hodge cycles.

d) We thus have to show that any relation (resp. "global relation" for theorem 5) of degree $\leq \delta$ with coefficients in E between (β)–periods of $\underline{A}(s)$ is the specialization at s of some relation of degree $\leq \delta$ with coefficients in $E[x]$ between the relative β_v–periods (which belong to $E[t_p, \log x][[x]]$ in virtue of (**) and lemma 9).

Because t_p is transcendental over E_v , and $\beta_v(\eta_{ij} x^{n_{ij}}(s)) = 0$, it suffices to replace in this

statement β_v–periods by the v–adic evaluations of the G–functions ω_{ij}, ω'_{ij}, $\omega_{ij}\log q^1_{ij}$, $(\omega_{ij}\log q^1_{ij})'$, where $q^1_{ij} = \frac{1}{\eta_{ij}} q_{ij} x^{-n_{ij}} = 1 + \ldots$

This can be now deduced from standard results in G–function theory [A] VII thm. 4.3, resp. 5.2. See also, ibid IX for more details about the proof of a (complex) analogous statement.

References

[A] André Y.; G–functions and Geometry, Aspects of Mathematics E13, Vieweg 1988.

[Be] Berkovich V.G.; Spectral theory and analytic geometry over non–archimedean fields II. Preprint IHES 1987.

[B] Berthelot P.; Cohomologie rigide, to appear in the "Astérisque" series.

[Bo] Borovoi, M.; On the action of the Galois group on rational cohomology classes of type (p.p) of Abelian varieties, Mat. Sbornik 94 (1974) n°4, 649–652 (Russian).

[BL] Bosch S., Lütkebohmert W.; stable reduction and Uniformization of Abelian varieties II, Inv. Math. 78, 257–297 (1984).

[Bl 1] Bosch S., Lütkebohmert W.; Degenerating Abelian Varieties, preprint Münster 1988.

[CF] Chai C.L., Faltings G.; Semiabelian degeneration and compactification of Siegel modular spaces. Forthcoming book (1990).

[C] Christol G.; Globally bounded solutions of differential equations, to appear in Proc. of the Tokyo conference on analytic number theory.

[Cn] Coleman R.; The universal vectorial bi–extension, preprint Berkeley 1989.

[Co] Colmez P.; Périodes des variétés abéliennes de type CM, preprint Bonn 1987.

[D_1] Deligne P.; Théorie de Hodge III, IHES Publ. Math. N° 44 (1974), 5–78.

[D_2] Deligne P.; Hodge cycles on Abelian varieties, in Lecture Notes in Math. 900 Springer–Verlag 1982 (notes by J. Milne).

[Dw] Dwork B.; p–adic cycles, IHES Publ. Math. N° 37 (1969), 27–116.

[Fa] Faltings G.; preprint Princeton on the "De Rham conjecture" 1989.

[F_1] Fontaine J.M.; Sur certains types de représentations p–adiques du groupe de Galois d'un corps local, construction d'un anneau de Barsotti–Tate, Ann. of Math. 115, (1982), 529–577.

[F_2] Fontaine J.M.; contribution to the proc. of IHES seminar 88 on "p–adic periods".

[F.M.] Fontaine J.M.; Messing W.; p–adic periods and p–adic etale cohomology, Contemporary mathematics, Vol. 67 AMS (1985), 179–207.

[Ge] Gerritzen L.; On multiplication algebras of Riemann matrices, Math. Ann. 194 (1971), 109–122.

[G_1] Grothendieck A.; On the De Rham cohomology of algebraic varieties, IHES Publ. Math. 29 (1966) (especially footnote 10).

[G_2] Grothendieck A.; Modèles de Néron et monodromie, exp. IX, SGA 7, in Lecture Notes in Math. 288 Springer–Verlag 1970.

[G_3] Grothendieck A.; Groupes de Barsotti–Tate et cristaux de Dieudonné, Séminaire de Mathématiques Supérieures, n° 45, Presses de l'Université de Montréal 1974.

[H] Harris J.; Functorial properties of toroidial compactifications of locally symmetric varieties. Proc. London Math. Soc. (3) 59 (1989), 1–22.

[Ka] Katz N.; Serre–Tate local moduli, in Surfaces Algébriques, exposé 5 bis, Springer LNM 868 (1981), 138–202.

[K] Kiehl R.; Die De Rham Kohomologie algebraischer Mannigfaltigkeiten über einem bewerteten Körper. IHES Publ. Math. 33 (1967), 5–20.

[lS] Le Stum B.; Cohomologie rigide et variétés abéliennes, thèse Univ. Rennes 1985.

[M] Mumford D.; Families of Abelian varieties, Algebraic groups and Discontinuous subgroup, Proc. Symp. Pure Math. vol. 9, AMS, Providence R.I. (1966), 347–351.

[Mu] Mustafin G.A.; Families of Algebraic varieties and invariant cycles, Math. USSR Izvestiya Vol 27 (1986) n° 2, 251–278.

[O] Ogus A.; Contribution to this conference.

[R_1] Raynaud M.; Variétés abéliennes et géométrie rigide. Actes du congrès international de Nice 1970, t.1, 473–477.

[R_2] Raynaud M.; Contribution to the proc. of IHES seminar 88 on "p–adic periods".

[S_1] Serre J.P.; Propriétés galoisiennes des points d'ordre fini des courbes elliptiques, Inv. Math., 15, 1972, 259–331.

[S_2] Serre J.P.; Abelian ℓ–adic representations and elliptic curves. Benjamin, New–York, 1968.

[SGA_4] Artin M., Grothendieck A., Verdier J.–L.; Théorie des topos et cohomologie étale des schémas, t. 3, LNM 269 Springer Verlag (1972).

[Sh] Shimura G.; On analytic families of polarized abelian varieties and automorphic functions, Annals of Math. 78 (1963), 149.

[St] Steenbrink J.; Limits of Hodge structures, Inv. Math. 31 (1975/76), 229–257.

[T] Tate J.; p–divisible groups. Proc. of a Conf. on local fields. Springer Verlag (1967).

[U] Ullrich P.; Rigid analytic covering maps. Proc. of the Conf. on p–adic analysis, Hengelhoef (1986), 159–171.

[vM] Van der Marel B.; thesis Rijksuniversiteit Groningen 1987.

[W] Wintenberger J.P.; p–adic Hodge theory for families of abelian varieties, preprint 1989.

THE NONARCHIMEDEAN BANACH-STONE THEOREM

Jesús ARAUJO, Universidad de Oviedo,
ETSII, Castiello de Bernueces, 33204 GIJON, Spain

J.MARTINEZ-MAURICA[1], Universidad de Cantabria,
Facultad de Ciencias, Avda. los Castros, 39071 SANTANDER, Spain

In 1987, E. Beckenstein and L. Narici ([BN]) introduced the Banach-Stone maps as those linear surjective isometries between two given spaces $C(X)$ and $C(Y)$ of K-valued continuous functions (X, Y are Hausdorff zerodimensional compact spaces and K is a nonarchimedean valued field) which derive in a natural way from some homeomorphism of Y onto X (see Definition 1 for details).

In that paper they proved that, unlike the case of real or complex ground field, one can find linear surjective isometries $T: C(X) \to C(X)$ which are not Banach-Stone maps whenever X is not rigid. In this paper, we extend this result to any X with more than one point (Theorem 2 and the closely related Theorems 3 and 7). Another evidence of the non existence of a nonarchimedean Banach-Stone theorem had been given by A. C. M. van Rooij ([vR], p.190).

Also, we characterize the Banach-Stone maps in several different ways (Theorems 4, 5, 8 and Corollary 4). As a result some topological properties of the space of all Banach-Stone maps are given (Theorem 7 and Corollary 5).

Throughout this paper K denotes a nonarchimedean commutative and complete valued field endowed with a nontrivial valuation. X, Y are always two separated zerodimensional compact spaces. By $C(X)$ we indicate the Banach space of all continuous functions from X into K endowed with the sup-norm. $C(X)'$ stands for the normed dual of $C(X)$.

A subset of a topological space is said to be clopen if it is both open and closed. We say that a clopen subset of X is proper if it is nonempty and different from X.

---

1 Research partially supported by the Spanish Dirección General de Investigación Científica y Técnica (DGICYT, PS87-0094).

1. BANACH-STONE MAPS.

DEFINITION 1 ([BN]) *A linear isometry T of $C(X)$ onto $C(Y)$ is a Banach-Stone map if there exists a homeomorphism h of Y onto X and an $a \in C(Y)$, $|a(y)| \equiv 1$, such that $(Tf)(y) = a(y)f(h(y))$, for every $y \in Y, f \in C(X)$.*

REMARKS

1 If the ground field is R or C and X, Y are general (this is, zerodimensional or not) separated compact spaces, the Banach-Stone theorem allows us to conclude that every linear isometry of $C(X)$ onto $C(Y)$ is a Banach-Stone map. So the above definition of Banach-Stone map is of no use in the case of real or complex field.

2 There are non-homeomorphic spaces X and Y for which it is possible to find a linear surjective isometry $T:C(X) \to C(Y)$. In fact, let X, Y be two non-homeomorphic compact ultrametrizable spaces (e.g. let Y be the space which consists in adding an isolated point to the Cantor space X). Then $C(X)$ and $C(Y)$ are both linearly isometric to c_0 ([vR], p.74).

For $f \in C(X)$, we define the cozero set of f to be

$$c(f) = \{x \in X : f(x) \neq 0\}.$$

For $U \subset X$, ξ_U stands for the characteristic function on U.

DEFINITION 2 ([BN]) *A map T of $C(X)$ into $C(Y)$ has the disjoint cozero set property if $c(f) \cap c(g) = \emptyset$ implies $c(Tf) \cap c(Tg) = \emptyset$ for $f, g \in C(X)$.*

REMARK We have

$$\begin{aligned} c(f) \cap c(g) &= \{x \in X : f(x) \neq 0\} \cap \{x \in X : g(x) \neq 0\} \\ &= \{x \in X : f(x)g(x) \neq 0\} \\ &= c(fg). \end{aligned}$$

Hence, the cozero set property means nothing else but:

$$fg = 0 \Rightarrow (Tf)(Tg) = 0.$$

If $x \in X$, we define $e_x : C(X) \to K$ by $e_x(f) = f(x)$ $(f \in C(X))$, and we shall say that e_x is the evaluation map at x. It is obvious that $e_x \in C(X)'$ and that $\| e_x \| = 1$ for every $x \in X$.

THEOREM 1 ([BN]) *For a linear isometry T mapping $C(X)$ onto $C(Y)$, the following statements are equivalent:*

(1) T has the disjoint cozero set property.

(2) Its adjoint T' maps each norm-one multiple of an evaluation map into a norm-one multiple of an evaluation map.

(3) T is a Banach-Stone map.

COROLLARY 1 *Let T be a linear isometry of $C(X)$ onto $C(Y)$. If $c(T\xi_U)\cap c(T\xi_V)=\emptyset$ whenever U and V are two disjoint clopen subsets of X, then T is a Banach-Stone map.*

Proof. Let $x \in X$. Let U, V be two disjoint clopen subsets of X. If $c(T\xi_U)\cap c(T\xi_V)=\emptyset$, then

$$(T'e_x)(\xi_U)=(T\xi_U)(x)=0 \text{ or}$$
$$(T'e_x)(\xi_V)=(T\xi_V)(x)=0.$$

Hence by ([BN], lemma 2), $T'e_x$ is a scalar multiple of an evaluation map. Also

$$\|T'e_x\| = \sup_{f\neq 0}\frac{|(T'e_x)(f)|}{\|f\|}=\sup_{f\neq 0}\frac{|e_x(Tf)|}{\|f\|}$$

$$=\sup_{g\neq 0}\frac{|e_x(g)|}{\|g\|}=\|e_x\| = 1.$$

Therefore, T is a Banach-Stone map because of Theorem 1.

THEOREM 2 *If $\#X\geq 2$, then there are linear surjective isometries $T:C(X)\to C(X)$ which are not Banach-Stone maps.*

Proof. Let $B\subset X$ be a nonempty proper clopen subset of X.

By ([vR], p.188), every orthonormal set of characteristic functions of clopen sets can be extended to an orthonormal base of $C(X)$ and by ([vR], p.167) all the orthonormal bases of a Banach space have the same cardinality.

Let $\mathcal{B}_1=\{\xi_X\}\cup\{g_i : i\in I\}$ and $\mathcal{B}_2=\{\xi_B\}\cup\{h_i : i\in I\}$ be two orthonormal bases of $C(X)$ extending $\{\xi_X\}$ and $\{\xi_B\}$ respectively. Now define $T(\xi_X)=\xi_B$ and $T(g_i)=h_i$ for every $i\in I$. We can extend T by linearity and continuity to a map $T:C(X)\to C(X)$. It is obvious that T is a surjective linear isometry.

Suppose now that T is a Banach-Stone map. Then there exists an $a\in C(X)$, $|a(x)|\equiv 1$, and a homeomorphism $h:X\to X$ such that for every f in $C(X)$ and x in X, $(Tf)(x)=a(x)f(h(x))$. If we take $x_0\in X-B$, we have $|(T\xi_X)(x_0)|=|a(x_0)|=1$ and $|(T\xi_X)(x_0)|=|\xi_B(x_0)|=0$, a contradiction.

Thus we can conclude that T is not a Banach-Stone map.

REMARKS

1 This result had been proved in ([BN]) when X is not rigid, this is, when there exists a homeomorphism $h:X \to X$ different from the identity map. (In [KR] and [L] it is shown that there exist rigid spaces of arbitrary large cardinality).

2 If X has only one point, every linear surjective T of $C(X)$ onto $C(Y)$ is a Banach-Stone map.

Banach-Stone maps can be also considered as algebra almost-morphisms in the following sense:

PROPOSITION 1 *Let T be a linear isometry from $C(X)$ onto $C(Y)$. Then, T is a Banach-Stone map if and only if $T\xi_X$ is an invertible element of $C(Y)$ and*

$$T(fg) = (T\xi_X)^{-1}(Tf)(Tg) \qquad (f, g \in C(X)).$$

Proof. If T is a Banach-Stone map, it is obvious that $T\xi_X$ is invertible in $C(Y)$. Also

$$(T(fg))(y) = a(y)f(h(y))g(h(y))$$
$$= ((T\xi_X)(y))^{-1}(Tf)(y)(Tg)(y)$$

for each $y \in Y$, where $a \in C(Y)$ and h are as in Definition 1.

Conversely, let $e_y \in C(Y)'$ be an evaluation map. Let U, V be two disjoint clopen subsets of X.

We have that

$$0 = e_y(T(\xi_U\xi_V))$$
$$= (T(\xi_U\xi_V))(y)$$
$$= ((T\xi_X)(y))^{-1}(T\xi_U)(y)(T\xi_V)(y).$$

Then, $(T\xi_U)(y) = 0$ or $(T\xi_V)(y) = 0$. So T is a Banach-Stone map because of Corollary 1.

The following result is a slight generalization of Theorem 2.

THEOREM 3 *If $\#X \geq 2$, then there are linear surjective isometries $T:C(X) \to C(X)$ such that $T(\xi_X) = \xi_X$ which are not algebra morphisms.*

Proof. We shall distinguish two cases:

i) If X is not rigid ([KR],[L]) there exist two subsets of X, U and V, proper, clopen and disjoint, which are homeomorphic. Let $\alpha \in K$, $0 < |\alpha| < 1$, and let us modify slightly the example given in ([BN], p.245), by defining:

$$(Tf)(x)=\begin{cases}\frac{1}{1+\alpha}(\alpha f(x)+f(h(x))) & \text{if } x\in U\\ \frac{1}{1+\alpha}(f(h^{-1}(x))+\alpha(f(x))) & \text{if } x\in V\\ f(x) & \text{otherwise}\end{cases}$$

By using the same arguments as in ([BN], p.245) and applying Proposition 1, we arrive at the required conclusion.

ii) If X is rigid, let B be a proper clopen subset of X. By ([vR], Corollary 5.23), we can find an orthonormal base of $C(X)$, $\mathcal{B}=\{\xi_B,\xi_{X-B}\}\cup\{g_i : i\in I\}$, for some adequate $g_i\in C(X)$, $i\in I$. Define now $T(\xi_B)=\xi_{X-B}$, $T(\xi_{X-B})=\xi_B$ and $T(g_i)=g_i$ $(i\in I)$. We arrive at the the result by proceeding as in Theorem 2 and taking in account that the identity map is the only homeomorphism from X onto X.

2. SOME CHARACTERIZATIONS.

For $f\in C(X)$, we denote $v(f)=\{|f(x)| : x\in X\}-\{0\}$.

LEMMA 1 *The following properties are equivalent for $f\in C(X)$:*

(1) $c(f)$ is clopen.

(2) $v(f)$ is finite.

(3) $\inf v(f)>0$.

Proof. (1) $\Rightarrow$ (2). Let, for each $r>0$, $A_r=\{x\in X : |f(x)|=r\}$. Obviously, A_r is clopen for every $r>0$, and $c(f)=\bigcup_{r>0}A_r$.

Also, since $c(f)$ is clopen, then it is compact and there exist $r_1,r_2,\ldots,r_n$ in R such that $c(f)=A_{r_1}\cup A_{r_2}\ldots\cup A_{r_n}$. Then $v(f)$ is contained in $\{r_1,r_2,\ldots,r_n\}$.

(2) $\Rightarrow$ (3) is obvious.

(3) $\Rightarrow$ (1) Let $\alpha\in K$ such that $0<|\alpha|<\inf v(f)$. Then, we have that

$$\{x\in X : f(x)=0\}=|f|^{-1}[0,|\alpha|)$$
$$=\{x\in X : |f(x)|<|\alpha|\}$$

is clopen. Hence $c(f)$ is also clopen.

LEMMA 2 *If $T: C(X) \to C(Y)$ is a linear surjective isometry, then the following are equivalent:*

(1) *T is a Banach-Stone map.*

(2) *If $c(f) = c(g)$, then $c(Tf) = c(Tg)$ $(f, g \in C(X))$*

(3) *If $c(f) = c(g)$ is a clopen subset of X, then $c(Tf) = c(Tg)$ $(f, g \in C(X))$.*

Proof. (1) $\Rightarrow$ (2), (2) $\Rightarrow$ (3) are obvious.

(3) $\Rightarrow$ (1) Let U, V be proper clopen subsets of X such that $U \cap V = \varnothing$. Let

$$f_\alpha = \xi_U + \alpha \xi_V \quad (\alpha \in K - \{0\}).$$

Clearly, we have that $c(f_\alpha) = U \cup V$ for every $\alpha \in K - \{0\}$. Then $c(Tf_\alpha) = c(T\xi_{U\cup V})$ because of (3).

We are going to prove that $c(T\xi_U) \cap c(T\xi_V) = \varnothing$. Otherwise, there exists $s_0 \in c(T\xi_U) \cap c(T\xi_V)$. Let

$$\alpha_0 = -\frac{(T\xi_U)(s_0)}{(T\xi_V)(s_0)}.$$

Then $s_0 \notin c(T\xi_U + \alpha_0 T\xi_V) = c(Tf_{\alpha_0})$ and hence $s_0 \notin c(T\xi_{U\cup V})$. Then if $\beta \in K - \{\alpha_0, 0\}$, $s_0 \in c(T\xi_U + \beta T\xi_V)$, which implies $s_0 \in c(T\xi_{U\cup V})$ and we arrive at a contradiction. So $c(T\xi_U) \cap c(T\xi_V) = \varnothing$.

Then, by Corollary 1, T is a Banach-Stone map.

LEMMA 3 *Let $f, g \in C(X)$. Then $f + \alpha g \in C(X)$ is invertible for every $\alpha \in K$, $\alpha \neq 0$, if and only if $c(f) \cap c(g) = \varnothing$ and $c(f) \cup c(g) = X$.*

Proof. First assume $f + \alpha g$ is invertible for every $\alpha \in K - \{0\}$. Since $X = c(f+g) \subset c(f) \cup c(g)$, then $X = c(f) \cup c(g)$.

Let us see that $c(f) \cap c(g) = \varnothing$. Otherwise, take $s_0 \in c(f) \cap c(g)$ and let

$$\alpha_0 = -\frac{f(s_0)}{g(s_0)} \in K - \{0\}.$$

Since $(f + \alpha_0 g)(s_0) = 0$, then $f + \alpha_0 g$ is not invertible. Hence $c(f) \cap c(g) = \varnothing$.

The converse is easy.

LEMMA 4 *Let $\alpha, \beta \in K - \{0\}$. Then, if $a \in R, a > 0$, there exists $\gamma \in K$ such that $|\gamma| = |\alpha| / |\beta|$ and $0 < |\alpha + \beta\gamma| < a$.*

Proof. Let $\delta \in K - \{0\}$ such that $|\delta| < \min\{|\alpha|, a\}$. Let $\gamma = (\delta - \alpha)/\beta$. It is obvious that γ satisfies the requirements.

THEOREM 4 *If $T: C(X) \to C(Y)$ is a linear surjective isometry, then the following properties are equivalent:*

(1) *T is a Banach-Stone map.*

(2) $\inf v(f) \le \inf v(Tf)$ *for every* $f \in C(X)$.

(3) *If* $v(f) = \{1\}$, *then* $v(Tf) = \{1\}$.

(4) $v(f) \subset v(Tf)$ *for every* $f \in C(X)$.

(5) *If* $f \in C(X)$ *is invertible, then so is* $Tf \in C(Y)$.

Proof. (1) $\Rightarrow$ (2), (2) $\Rightarrow$ (3) are trivial.

(3) $\Rightarrow$ (4) We infer from (3) that if $v(f) = \{r\}$, for some $r > 0$, then $v(Tf) = \{r\}$. Suppose that there exists f in $C(X)$ such that $a \in v(f)$ and $a \notin v(Tf)$, $a > 0$. We can assume without loss of generality that $\|f\| = 1$. Then $a < 1$, because $1 \in v(Tf)$. We can decompose f as $g + h$ where $g = f\xi_A$, $h = f\xi_{X-A} \in C(X)$ and $A = \{x \in X : |f(x)| \ge a\}$.

Obviously, $a \in v(g)$. Now, assume there exists $y \in Y$ such that $|(Tg)(y)| = a$. Since $\|h\| = \|Th\| < a$, then $|(Tf)(y)| = a$. This proves that $a \notin v(Tg)$.

On the other hand, $c(g)$ is clopen because $\inf v(g) > 0$ (see Lemma 1). So $v(g) = \{a = a_1 < a_2 < \ldots < a_n = 1\}$ and then $g = \sum_{i=1}^{n} \alpha_i$, where $\alpha_i = f\xi_{|f|^{-1}(a_i)}$ for each $i \in \{1, 2, \ldots, n\}$.

We obtain that $Tg = \sum_{n=1}^{n} T\alpha_i$.

Next, we are going to prove that $c(T\alpha_1) \subset \bigcup_{i=2}^{n} c(T\alpha_i)$. Suppose that there exists an $x_0 \in c(T\alpha_1)$ such that $(T\alpha_i)(x_0) = 0$ for $i = 2, 3, \ldots, n$. Then $|(Tg)(x_0)| = |(T\alpha_1)(x_0)| = a_1 = a$, which is impossible. Hence $c(T\alpha_1) \subset \bigcup_{i=2}^{n} c(T\alpha_i)$.

So we know that there exists $i_0 \in \{2, 3, \ldots, n\}$ such that $c(T\alpha_1) \cap c(T\alpha_{i_0})$ is not empty. Let x_0 be an element of $c(T\alpha_1) \cap c(T\alpha_{i_0})$. Then we have that $|(T\alpha_1)(x_0)| = a$ and $|(T\alpha_{i_0})(x_0)| = a_{i_0}$. Let $b_1, b_{i_0} \in K$ such that

$$|b_1| = \frac{1}{a}, \quad |b_{i_0}| = \frac{1}{a_{i_0}}.$$

Let us consider now $\beta_1 = b_1\alpha_1 \in C(X)$, $\beta_{i_0} = b_{i_0}\alpha_{i_0} \in C(X)$. Obviously, we have that $v(\beta_1) = v(\beta_{i_0}) = \{1\}$. By Lemma 4 there exists $b \in K$ such that $0 < r = |b(T\beta_1)(x_0) + (T\beta_{i_0})(x_0)| < 1$ and $|b| = 1$.

Let $f' = b\beta_1 + \beta_{i_0}$. Clearly we have that $v(f') = \{1\}$ because $c(\beta_1) \cap c(\beta_{i_0}) = \emptyset$. On the other hand $r \in v(Tf')$, with $r \ne 1$. Now, we have arrived at a contradiction with our assumption.

Then $v(f) \subset v(Tf)$ for every f in $C(X)$.

(4) $\Rightarrow$ (5) Suppose that there exists $f \in C(X)$ invertible such that $Tf \in C(Y)$ is not. Then $\inf v(f) > 0$. Define $A = \{y \in Y : |Tf(y)| < \inf v(f)\}$. Clearly A is clopen and nonempty because Tf is not invertible. Consider $g = T^{-1}\xi_A$. Since T^{-1} is an isometry, $\|g\| = 1$, and then there exists a point $x_0 \in X$ in which g attains its norm, this is, $|g(x_0)| = 1$. By Lemma 4, there exists $\alpha \in K$ such that

$$0 < r = |f(x_0) + \alpha g(x_0)| < \inf v(f),$$

where $|\alpha| = |f(x_0)| > 0$.

Then $r \in v(f + \alpha g)$ but $r \notin v(T(f + \alpha g))$, which is a contradiction. So T is invertible.

(5) $\Rightarrow$ (1) Let f, g in $C(X)$ such that $c(f) = c(g)$ is clopen and let $A = X - c(f)$. We have that $f + \alpha\xi_A$ is invertible in $C(X)$ for every $\alpha \in K - \{0\}$ because of Lemma 3. By (5), $Tf + \alpha T\xi_A$ is invertible for every $\alpha \neq 0$. One can infer from Lemma 3 that $c(Tf) \cap c(T\xi_A) = \varnothing$ and $c(Tf) \cup c(T\xi_A) = Y$. We can obtain analogous results if we replace f by g.

Hence $c(Tf) = Y - c(T\xi_A) = c(Tg)$ and T is a Banach-Stone map by Lemma 2.

REMARK It is easy to see that if $T: C(X) \to C(Y)$ is a Banach-Stone map, then its inverse $T^{-1}: C(Y) \to C(X)$ is also a Banach-Stone map. Indeed T^{-1} is a linear surjective isometry. Since T is a Banach-Stone map, then there exist $a \in C(Y)$, $|a(y)| \equiv 1$, and a homeomorphism $h: Y \to X$ such that $Tf = a(f \circ h)$ for every $f \in C(X)$. It is easy to see that

$$T^{-1}g = \frac{1}{a \circ h^{-1}}(g \circ h^{-1}) \quad (g \in C(Y))$$

which proves that T^{-1} is a Banach-Stone map.

COROLLARY 2 *If $T: C(X) \to C(Y)$ is a linear surjective isometry, then the following properties are equivalent:*

(1) *T is a Banach-Stone map.*

(2) $\inf v(Tf) \leq \inf v(f)$ *for every* $f \in C(X)$.

(3) $\inf v(f) = \inf v(Tf)$ *for every* $f \in C(X)$.

(4) *If* $v(Tf) = \{1\}$, *then* $v(f) = \{1\}$.

(5) $v(Tf) \subset v(f)$ *for every* $f \in C(X)$.

(6) $v(Tf) = v(f)$ *for every* $f \in C(X)$.

(7) *If $Tf \in C(Y)$ is invertible, then so is $f \in C(X)$.*

THEOREM 5 *Let $T: C(X) \to C(Y)$ be a linear surjective isometry. Then, the following are equivalent:*

i) If $f \in C(X)$ is a topological divisor of zero, then so is $Tf \in C(Y)$.

ii) T is a Banach-Stone map.

Proof. $i) \Rightarrow ii)$ Let us suppose that there is $f \in C(X)$ which is not invertible but such that Tf is invertible. Then there exists a point $x_0 \in X$ in which f vanishes, and

$$r = \inf_{y \in Y} |(Tf)(y)| > 0.$$

Let

$$h(x) = \begin{cases} f(x) & \text{if } x \notin U \\ 0 & \text{if } x \in U \end{cases}$$

where $U = \{x \in X : |f(x)| < r/2\}$.

We have that h is a topological divisor of zero. Then so is Th. This implies the existence of a sequence $(g_n)_{n \in N}$ in $C(Y)$ such that $\| g_n \| = 1$ for each $n \in N$ and $\lim_{n \to \infty} (Th)g_n = 0$.

Since $\|Tf - Th\| = \|f - h\| < r/2$ and $|(Tf)(y)| > r$ for each $y \in Y$, we deduce that $|(Th)(y)| = |(Tf)(y)| > r$ for every $y \in Y$.

For each $n \in N$, let $y_n \in Y$ such that $|g_n(y_n)| = 1$. Then, $\|(Th)g_n\| \geq |(Th)(y_n)| \, |g_n(y_n)| \geq r$ for each $n \in N$, and we arrive at a contradiction. So T is a Banach-Stone map.

$ii) \Rightarrow i)$ is easy.

REMARK With slight changes in the previous proof, it is possible to assure that the theorem also holds if we replace $i)$ by

$i')$ If $f \in C(X)$ is a divisor of zero, then so is $Tf \in C(Y)$.

COROLLARY 3 *Let $T: C(X) \to C(Y)$ be a linear surjective isometry. Then the following are equivalent:*

(1) *T is a Banach-Stone map.*

(2) *T maps maximal ideals of $C(X)$ onto maximal ideals of $C(Y)$.*

(3) *T maps closed ideals of $C(X)$ onto closed ideals of $C(Y)$.*

Proof. By ([vR], p. 209), we know that the maximal ideals of $C(X)$ can be expressed in the form

$$\mathcal{M} = \{f \in C(X) : f(x_{\mathcal{M}}) = 0\}$$

for some $x_{\mathcal{M}} \in X$ and the map

$$\mathcal{M} \to x_{\mathcal{M}}$$

is a bijection from the set of all maximal ideals of $C(X)$ onto X.

Also, $\mathcal{N}$ is a closed ideal in $C(X)$ if (and only if) there exists a subset $B_{\mathcal{N}}$ of X such that

$$\mathcal{N} = \{f \in C(X) : f(x) = 0 \;\; \forall x \in B_{\mathcal{N}}\}.$$

(1) $\Rightarrow$ (2) If T is a Banach-Stone map, then there exists a homeomorphism h of Y onto X and an $a \in C(Y)$, $|a(y)| \equiv 1$, such that for each f in $C(X)$ and y in Y,

$$(Tf)(y) = a(y)f(h(y)).$$

Let $\mathcal{M}$ be a maximal ideal of $C(X)$. It is easy to check that $T(\mathcal{M}) = \{g \in C(Y): g(h^{-1}(x_{\mathcal{M}})) = 0\}$ which shows that $T(\mathcal{M})$ is a maximal ideal of $C(Y)$.

(2) $\Rightarrow$ (3) Let $\mathcal{M}$ be a closed ideal in $C(X)$. Then $\mathcal{N}$ is the intersection of all the maximal ideals containing it:

$$\mathcal{N} = \bigcap_{\mathcal{N} \subset \mathcal{M}} \mathcal{M}.$$

Since T maps maximal ideals onto maximal ideals,

$$T(\mathcal{N}) = \bigcap_{\mathcal{N} \subset \mathcal{M}} T(\mathcal{M})$$

is closed.

(3) $\Rightarrow$ (1) Let $f \in C(X)$ such that $f(x_0) = 0$ for some $x_0 \in X$. Let $\mathcal{M}$ be a maximal (and hence closed) ideal such that $f \in \mathcal{M}$. We have that $T(\mathcal{M})$ is a closed ideal of $C(Y)$, and then there exists a point $y_{\mathcal{M}} \in Y$ in which Tf vanishes. Then, by Corollary 2 ((7) $\Rightarrow$ (1)), T is a Banach-Stone map.

COROLLARY 4 *If $T: C(X) \to C(Y)$ is a linear surjective isometry, then the following are equivalent:*

(1) *T is a Banach-Stone map.*

(2) *If $v(f) \subset v(g)$, then $v(Tf) \subset v(Tg)$* $(f, g \in C(X))$.

(3) *If $v(f) = v(g)$, then $v(Tf) = v(Tg)$* $(f, g \in C(X))$.

(4) *If* $\inf v(f) \leq \inf v(g)$, *then* $\inf v(Tf) \leq \inf v(Tg)$ $(f, g \in C(X))$.

(5) *If* $\inf v(f) = \inf v(g)$, *then* $\inf v(Tf) = \inf v(Tg)$ $(f, g \in C(X))$.

(6) *If $|f| \leq |g|$, then $|Tf| \leq |Tg|$* $(f, g \in C(X))$, *(where $|f| \leq |g|$ means $|f(x)| \leq |g(x)|$ for every $x \in X$).*

(7) *If $|f| = |g|$, then $|Tf| = |Tg|$* $(f, g \in C(X))$.

Proof.

(1) $\Rightarrow$ (2) $\Rightarrow$ (3), (1) $\Rightarrow$ (4) $\Rightarrow$ (5), (1) $\Rightarrow$ (6) $\Rightarrow$ (7) are obvious.

We shall prove simultaneously (3) $\Rightarrow$ (1) and (5) $\Rightarrow$ (1).

Let $f \in C(X)$ such that $v(f) = \{1\}$ and let $g = T^{-1}\xi_Y$. Let $A = \{x \in X : |g(x)| = 1\}$. It is obvious that $A \neq \emptyset$ because T^{-1} is an isometry and there exists some point in X where g attains its norm.

Let us put g in the way $g = g\xi_A + g\xi_{X-A}$. Clearly, if $x \in X$, $|g\xi_A(x)| \in \{0, 1\}$, this is , $v(g\xi_A) = \{1\}$, because there is at least one point $x_0 \in X$ with $|g\xi_A(x_0)| = 1$.

Also we have that for each $y \in Y$,

$$|(T(g\xi_A))(y)| = |(Tg)(y) - (T(g\xi_{X-A}))(y)|$$
$$= |1 - (T(g\xi_{X-A}))(y)|$$
$$= 1$$

because $\|T(g\xi_{X-A})\| = \|g\xi_{X-A}\| < 1$. Thus $v(T(g\xi_A)) = \{1\}$.

Now, by using the hypothesis (3), $v(Tf) = \{1\}$ and by using the hypothesis (5), $\inf v(Tf) = 1$, this is, $v(Tf) = \{1\}$.

By Theorem 4 we conclude that T is a Banach-Stone map.

(7) $\Rightarrow$ (1) Let $f, g \in C(X)$ such that $c(f) = c(g)$ is a clopen subset of X. Then, by Lemma 1, there exists $\alpha \in K$ such that $|f(x)| < |\alpha g(x)|$ for every $x \in c(f)$. Then $|f(x) + \alpha g(x)| = |\alpha g(x)|$ for every $x \in X$. By hypothesis,

$$|(Tf)(y) + \alpha(Tg)(y)| = |\alpha(Tg)(y)|$$

for every $y \in Y$. Then if $y \notin c(Tg)$, $(Tg)(y) = 0$ and then $(Tf)(y) = 0$. This implies $c(Tf) \subset c(Tg)$. If we interchange the roles of f and g we shall have the equality $c(Tf) = c(Tg)$ which is equivalent, by Lemma 2, to being T a Banach-Stone map.

REMARK We cannot replace (7) for the weaker form

(7') *If* $|f| = 1$, *then* $|Tf| = 1$, $f \in C(X)$.

For that, let B be a clopen proper subset of X. We know ([vR] p.167) that there exist $g_i \in C(X)$ $(i \in I)$ such that $\{\xi_B, \xi_X\} \cup \{g_i : i \in I\}$ is an orthonormal base of $C(X)$. Then, so is $\{\alpha\xi_B, \xi_X\} \cup \{g_i : i \in I\}$, where $\alpha \in K$, $|\alpha| = 1$ and $0 < |1 - \alpha| < 1$.

Define

$$T : C(X) \to C(X)$$
$$\xi_B \to \alpha\xi_B$$
$$\xi_X \to \xi_X$$
$$g_i \to g_i \qquad (i \in I)$$

and extend it by linearity and continuity to the whole space $C(X)$.

For every $f \in C(X)$, $|f| = 1$, there exist $\alpha_i \in K$ $(i \in I)$, $\nu \in K$ and $\beta \in K$ such that $|\beta| \le 1$ and

$$f = \sum_{i \in I} \alpha_i g_i + \nu\xi_X + \beta\xi_B.$$

Then, for every $x \in X$,

$$|(Tf)(x)| = \left|\left(\sum_{i\in I}\alpha_i g_i + \nu\xi_x + \beta\xi_B\right)(x) + (\alpha\beta\xi_B - \beta\xi_B)(x)\right|$$

$$= |f(x) + \beta(\alpha-1)\xi_B(x)|$$

$$= |f(x)|$$

$$= 1.$$

It is easy to see that T is a linear surjective isometry. However, T is not a Banach-Stone map. In order to see that , suppose there exist an $a \in C(X)$, $|a(x)| \equiv 1$, and a homeomorphism h of X onto X, such that for each f in $C(X)$ and x in X, $(Tf)(x) = a(x)f(h(x))$,. Then, for every $x \in X$, $a(x) = (T\xi_X)(x) = 1$.

Hence $\xi_B(h(x)) = \xi_B(h(x))a(x) = (T\xi_B)(x) = \alpha\xi_B(x)$which is impossible because $\alpha \neq 0, 1$. This completes the proof.

3. TOPOLOGICAL PROPERTIES.

Let us introduce the following notation:

$$I(C(X), C(Y)) = \{T : C(X) \to C(Y) : T \text{ is linear surjective isometry}\}.$$

$I(C(X), C(Y))$ will be considered as a topological subspace of $L(C(X), C(Y))$. We shall write $I(C(X))$ instead of $I(C(X), C(X))$.

From now on, $BS(C(X)), C(Y))$ stands for the topological subspace of $I(C(X), C(Y))$ consisting of all the Banach-Stone maps of $C(X)$ onto $C(Y)$. We shall write $BS(C(X))$ instead of $BS(C(X), C(X))$.

THEOREM 6 *$I(C(X), C(Y))$ is complete.*

Proof. Let $(T_n)_{n\in N}$ be a Cauchy sequence of elements of $I(C(X), C(Y))$. Then, $(T_n)_{n\in N}$ converges to $T \in L(C(X), C(Y))$. Obviously $\|Tf\| = \lim_{n\to\infty}\|T_n f\| = \|f\|$ for every $f \in C(X)$, and hence T is an isometry.

On the other hand, let us see that $(T_n^{-1})_{n\in N}$ is a Cauchy sequence in $I(C(Y), C(X))$. In order to see that, let $\varepsilon > 0$. Then there exists $n_0 \in N$ such that if $n, m \geq n_0$, $\|T_m - T_n\| < \varepsilon$. So, for $n, m \geq n_0$,

$$\begin{aligned}\|T_n^{-1}-T_m^{-1}\| &\le \|T_n^{-1}T_m - I\|\;\|T_m^{-1}\| \\ &\le \|T_n^{-1}\|\;\|T_m - T_n\|\;\|T_m^{-1}\| \\ &= \|T_m - T_n\| \\ &< \varepsilon.\end{aligned}$$

Then $(T_n^{-1})_{n\in N}$ is a Cauchy sequence in $I(C(Y),C(X))$ and, as above, it converges to an isometry $S\in L(C(Y),C(X))$.

It is left to verify that T is surjective. For that, let us see that $T\circ S$ is the identity map on $C(Y)$. Let $\varepsilon>0$, $g\in C(Y)$ and $n_1\in N$ such that $\|T_n^{-1}g - Sg\|<\varepsilon$ for every $n\ge n_1$. Also, let $n_0\in N$ $(n_0\ge n_1)$ such that $\|T_n(Sg)-T(Sg)\|<\varepsilon$ for every $n\ge n_0$.

Then, for $n\ge n_0$, we have

$$\begin{aligned}\|T(Sg)-g\| &\le \max\{\|T(Sg)-T_n(Sg)\|,\|T_n(Sg)-g\|\} \\ &\le \max\{\|T(Sg)-T_n(Sg)\|,\|T_n\|\;\|Sg-T_n^{-1}g\|\} \\ &< \varepsilon.\end{aligned}$$

Since this inequality holds for all $\varepsilon>0$, we conclude that $T(Sg)=g$ which implies that T is surjective.

COROLLARY 5 *$BS(C(X),C(Y))$ is a closed subset of $I(C(X),C(Y))$.*

Proof. Let $(T_n)_{n\in N}$ be a Cauchy sequence in $BS(C(X),C(Y))$. Let T be its limit in $L(C(X),C(Y))$. By Theorem 6, we know that T is a surjective isometry. Now, suppose that $f,g\in C(X)$ are such that $|f|=|g|$. Then $|T_nf|=|T_ng|$ for every $n\in N$ and then, $|(Tf)(x)|=|(Tg)(x)|$ for every $x\in X$. So T is a Banach-Stone map (Corollary 4).

THEOREM 7 *If $\#X\ge 2$, then $BS(C(X))$ has empty interior in $I(C(X))$ (this is, $\overline{I(C(X))-BS(C(X))}=I(C(X))$).*

Proof. Let $T\in BS(C(X))$. Let $\varepsilon>0$. Consider $\alpha\in K$ such that

$$0<|\alpha|<\min\left\{\frac{1}{2},\varepsilon\right\}.$$

Let U be a proper clopen subset of X. Since T is a Banach-Stone map, $c(T\xi_U)$ is clopen and proper (Lemma 1 and Corollary 2) and hence $\xi_{X-c(T\xi_U)}$ belongs to $C(X)$. Let $\{\xi_U\}\cup\{g_i : i\in I\}$ be an orthonormal base of $C(X)$ extending $\{\xi_U\}$. Define $S(\xi_U)=\xi_{X-c(T\xi_U)}$, $S(g_i)=0$ for every $i\in I$ and extend S by linearity and continuity to the whole space $C(X)$. It is obvious that $\|S\|\le 1$, and hence $T+\alpha S$ belongs to the ball of center T and radius ε, $B_\varepsilon(T)$. It is also easy to see that $T+\alpha S$ is a linear isometry.

Let us see that $T+\alpha S$ is surjective: Since $\{T\xi_U\}\cup\{Tg_i : i\in I\}$ is an orthonormal base of $C(X)$ and:

$$\|(Tg_i)-(T+\alpha S)(g_i)\| \leq \frac{1}{2}=\frac{1}{2}\|Tg_i\| \quad (i\in I),$$

$$\|(T\xi_U)-(T+\alpha S)(\xi_U)\| \leq \frac{1}{2}=\frac{1}{2}\|T\xi_U\|$$

then $\{(T+\alpha S)(\xi_U)\}\cup\{(T+\alpha S)(g_i): i\in I\}$ is another orthonormal base of $C(X)$ ([vR], p. 183).

Now it is easy to see that this last property implies that $T+\alpha S$ is surjective.

Also, ξ_U is not invertible in $C(X)$. However, $(T+\alpha S)(\xi_U)$ is invertible and we can deduce from Corollary 2 that $T+\alpha S$ is not a Banach-Stone map.

REMARKS

1 If $\#X=1$, then $BS(C(X))=I(C(X))$.

2 Also, $I(C(X))$ is clopen in $L(C(X))$. We have proved in Theorem 6 that it is closed. Let us see that it is open. Let $T\in I(C(X))$ and $S\in B_{\frac{1}{2}}(T)$. Then, for every $f\in C(X)$,

$$\|S(f)\| = \|S(f)-T(f)+T(f)\| = \|T(f)\| = \|f\|$$

which proves that S is a linear isometry. By using the same arguments as in Theorem 7, we can prove that S is surjective and hence $B_{\frac{1}{2}}(T)\subset I(C(X))$. Thus, $I(C(X))$ is clopen in $L(C(X))$.

4. A SPECIAL CASE.

THEOREM 8 *Suppose Y has not isolated points. Let $T: C(X)\to C(Y)$ be a linear surjective isometry. Then, T is a Banach-Stone map if and only if* $\inf v(Tf)>0$ *whenever* $\inf v(f)>0$ $(f\in C(X))$.

Proof. First, suppose that $\inf v(Tf)>0$ whenever $\inf v(f)>0$, $f\in C(X)$. Let us see that T maps invertible elements of $C(X)$ onto invertible elements of $C(Y)$ and then apply Theorem 4.

In order to prove that, suppose there exists an invertible map $f\in C(X)$ such that Tf is not invertible in $C(Y)$. Then, $Y-c(Tf)$ is a proper clopen subset of Y (Lemma 1). Since Y has not isolated points, there exists a map $h: Y-c(Tf)\to K$ which is not locally constant at some $x_0\in Y-c(Tf)$ by ([GH] Corollary 5.4.). Without loss of generality we can suppose that $h(x_0)=0$ and then we extend h to the whole space Y by defining $h(x)=0$ if $x\in c(Tf)$. Obviously $h\in C(Y)$ because $c(Tf)$ is clopen. If $\|h\|<\inf v(f)$, then $\inf v(f+T^{-1}h)=\inf v(f)>0$, because $c(f)=X$ and $\|T^{-1}h\| = \|h\| < \inf v(f)$.

However $\inf v(Tf+h)=0$, because $(Tf+h)(x_0)=0$ and $Tf+h$ is not locally constant at x_0. So we arrive at a contradiction.

The converse is trivial.

REMARK Compare this result with Theorem 4, where some stronger conditions on $v(f)$ appear (e.g. $\inf v(f)=\inf v(Tf)$ or $v(f)=v(Tf)$).

THEOREM 9 *If $\#X \geq 2$ and X has isolated points, then there exists a linear surjective isometry $T:C(X)\to C(X)$ such that $\inf v(Tf)>0$ whenever $\inf v(f)>0$ $(f\in C(X))$ which is not a Banach-Stone map.*

Proof. Let $a\in X$ be an isolated point. Choose $b\in X$, $b\neq a$, and set

$$Tf:=f+f(b)\xi_{\{a\}} \qquad (f\in C(X))$$

T is a linear continuous map from $C(X)$ onto $C(X)$ and $\|T\|\leq 1$. One easily infers that T has an inverse, viz.

$$f\to f-f(b)\xi_{\{a\}}.$$

So, T is a surjective linear isometry.

It is easily seen that $v(f)$ and $v(Tf)$ differ by at most one point, so $\inf v(f)>0 \Rightarrow \inf v(Tf)>0$.

Now assume T is a Banach-Stone map. Then, for every $f\in C(X)$, $Tf=c(f\circ h)$, where $c\in C(X)$, $|c|\equiv 1$ and $h:X\to X$ is a homeomorphism. So, in particular,

(*) $$c(a)f(h(a))=f(a)+f(b).$$

By taking $f=\xi_{\{a\}}$ we find $c(a)\xi_{\{a\}}(h(a))=1$, so that $\xi_{\{a\}}(h(a))\neq 0$, i. e. $h(a)=a$ and $c(a)=1$. Then, from (*),

$$f(a)=f(a)+f(b) \qquad (f\in C(X)).$$

Hence $f(b)=0$ for every $f\in C(X)$, a contradiction.

We are grateful to Dr. W. H. Schikhof for his valuable suggestions.

REFERENCES

[BN].- E. BECKENSTEIN and L. NARICI. *A nonarchimedean Stone-Banach theorem*, Proc. Amer. Math. Soc, **100** (1987) 242-246.

[GH].- L. GILLMAN, M. HENRIKSEN. *Concerning rings of continuous functions*, Trans. Amer. Math. Soc. **77** (1954) 340-362.

[KR].- V. KANNAN and M. RAJAGOLAPAN. *Rigid spaces.III*, Canad. J. Math, **30** (1978), 926-932.

[L].- F. W. LOZIER. *A class of compact rigid 0-dimensional spaces*, Canad. J. Math. **21** (1969), 817-821.

[vR].- A. C. M. VAN ROOIJ. *Non-archimedean functional analysis*. Marcel Dekker, New York, 1978.

Cohomologie rigide et théorie des $\mathscr{D}$-modules

Pierre BERTHELOT
(IRMAR, Université de Rennes)

Introduction

Soient k un corps parfait de caractéristique p, $W = W(k)$ l'anneau des vecteurs de Witt à coefficients dans k, K le corps des fractions de W. Si ℓ est un nombre premier différent de p, on sait associer à tout k-schéma X une catégorie de faisceaux pour la topologie étale, la catégorie des faisceaux ℓ-adiques constructibles, à partir de laquelle on peut construire une catégorie dérivée $D^b_c(X)$, la catégorie des complexes bornés de faisceaux ℓ-adiques à cohomologie constructible, satisfaisant au "formalisme des six opérations" dégagé par Grothendieck (notamment dans [11] pour les faisceaux cohérents et [SGA4] pour les faisceaux étales, et pour lequel nous renvoyons aussi à l'ouvrage de Mebkhout [17, II § 9]) : produit tensoriel, faisceau d'homomorphismes, image directe et image directe à support propre, image inverse et image inverse exceptionnelle.

Dans le cas p-adique, la situation actuelle est très loin d'être aussi satisfaisante. On peut associer à tout k-schéma séparé de type fini X des groupes de cohomologie rigide $H^i_{\mathrm{rig}}(X/K)$, coïncidant avec la cohomologie cristalline (tensorisée par K) si X est propre et lisse, et avec la cohomologie de Monsky et Washnitzer si X est affine et lisse (cf. [2], [4]). On peut également également associer à X une catégorie de coefficients, la catégorie des F-isocristaux surconvergents, permettant notamment d'interpréter les espaces de cohomologie de Dwork comme groupes de cohomologie rigide à coefficients dans un isocristal [3]. Mais, outre le fait qu'on ne sait en général démontrer certaines propriétés fondamentales de ces groupes de cohomologie (notamment la finitude) qu'avec des hypothèses de résolution des singularités, les F-isocristaux considérés jusqu'à présent sont des analogues p-adiques des systèmes locaux (ou faisceaux lisses) ℓ-adiques, et ne peuvent donc fournir qu'une sous-catégorie de celle qui est nécessaire pour développer le formalisme des six opérations de Grothendieck. On peut du reste penser que c'est précisément faute d'une bonne catégorie de coefficients que la cohomologie p-adique n'a pas jusqu'ici connu un développement comparable à celui de la cohomologie ℓ-adique.

Le développement, grâce notamment aux travaux de Kashiwara et de Mebkhout, de la théorie des $\mathcal{D}_X$-modules holonomes sur une variété complexe X, et la possibilité, qui en résulte par la correspondance de Riemann-Hilbert, de pouvoir interpréter algébriquement la catégorie $D^b_c(X)$ des complexes bornés de faisceaux de $\mathbb{C}$-vectoriels à cohomologie algébriquement constructible, ainsi que les six opérations sur cette catégorie (voir en particulier [6], [12], [13], [14], [15], [16], [17]), ont permis de remplir sur $\mathbb{C}$ le programme conjecturé par Grothendieck concernant les "coefficients de de Rham", programme qui apparaissait déjà dans [9] (voir aussi [10] p. 312, et l'introduction de [1], p. 22). Compte tenu des relations qui lient cohomologie rigide et cohomologie de de Rham, il semble donc naturel aujourd'hui de chercher à aborder le problème des coefficients de de Rham pour les variétés de caractéristique p au moyen de la théorie des $\mathcal{D}_X$-modules.

C'est la démarche que nous entreprenons dans le présent article. Soient $\mathcal{V}$ un anneau de valuation discrète complet d'inégales caractéristiques, de corps résiduel k. Le point de départ consiste à observer que, sur un $\mathcal{V}$-schéma formel lisse $\mathcal{X}$, de réduction X sur k, il existe un faisceau d'opérateurs différentiels $\mathcal{D}^\dagger_{\mathcal{X}}$ qui opère naturellement sur différents types de $(\mathcal{O}_{\mathcal{X}} \otimes \mathbb{Q})$-modules définis par la cohomologie rigide (par l'intermédiaire du morphisme de spécialisation sp : $\mathcal{X}_K \to \mathcal{X}$ de la fibre générique de $\mathcal{X}$ dans $\mathcal{X}$). De plus, la catégorie des $\mathcal{D}^\dagger_{\mathcal{X}}$-modules localement de présentation finie ne dépend pas vraiment de $\mathcal{X}$, mais seulement de sa réduction X sur k.

La principale difficulté de l'étude de $\mathcal{D}^\dagger_{\mathfrak{X}}$ vient alors de ce qu'il s'agit d'un faisceau d'opérateurs d'ordre infini, qui peut être vu comme un "complété faible" du faisceau usuel des opérateurs différentiels $\mathcal{D}_{\mathfrak{X}}$. En particulier, l'anneau des sections sur un ouvert affine du faisceau $\mathcal{D}^\dagger_{\mathfrak{X}\mathbf{Q}} := \mathcal{D}^\dagger_{\mathfrak{X}} \otimes \mathbf{Q}$ n'est pas en général un anneau nœthérien. Pour contourner cette difficulté, nous exposons dans la première partie une construction intrinsèque du faisceau $\mathcal{D}^\dagger_{\mathfrak{X}}$ comme limite inductive, pour $m \geq 0$, des complétés p-adiques $\widehat{\mathcal{D}}^{(m)}_{\mathfrak{X}}$ de sous-faisceaux $\mathcal{D}^{(m)}_{\mathfrak{X}}$ de $\mathcal{D}_{\mathfrak{X}}$, appelés faisceaux des opérateurs de niveau $\leq m$, et qui possèdent l'avantage d'être des faisceaux d'anneaux nœthériens. Ce point de vue permet en particulier de montrer que le faisceau $\mathcal{D}^\dagger_{\mathfrak{X}\mathbf{Q}}$ est un faisceau d'anneaux cohérent. La seconde partie est alors consacrée aux propriétés des modules cohérents sur des anneaux tels que $\widehat{\mathcal{D}}^{(m)}_{\mathfrak{X}}$, $\widehat{\mathcal{D}}^{(m)}_{\mathfrak{X}\mathbf{Q}}$, $\mathcal{D}^\dagger_{\mathfrak{X}\mathbf{Q}}$. Pour ne pas alourdir le présent article par trop de détails techniques, nous ne donnerons dans ces deux premières sections que des indications sommaires sur les démonstrations, en renvoyant à un article ultérieur [5] pour un exposé plus systématique et détaillé.

Le reste de cet article est consacré à divers exemples de $\mathcal{D}^\dagger_{\mathfrak{X}\mathbf{Q}}$-modules cohérents, permettant de faire le lien entre la théorie des $\mathcal{D}^\dagger_{\mathfrak{X}\mathbf{Q}}$-modules et la cohomologie rigide. Le premier exemple est celui des isocristaux convergents, vus comme $\mathcal{O}_{\mathfrak{X}\mathbf{Q}}$-modules munis d'une connexion intégrable. Nous montrons que la condition de convergence permet de les munir d'une structure canonique de $\mathcal{D}^\dagger_{\mathfrak{X}\mathbf{Q}}$-module, et qu'ils sont alors cohérents sur $\mathcal{D}^\dagger_{\mathfrak{X}\mathbf{Q}}$.

Nous considérons ensuite quelques exemples d'isocristaux surconvergents le long d'un fermé de $\mathfrak{X}$: en prenant l'image directe dérivée d'un tel isocristal par le morphisme de spécialisation, on obtient un complexe de $\mathcal{O}_{\mathfrak{X}\mathbf{Q}}$-modules, et nous montrons que ce complexe possède une structure naturelle de complexe de $\mathcal{D}^\dagger_{\mathfrak{X}\mathbf{Q}}$-modules. Nous examinerons en particulier le cas de l'isocristal "constant" $\mathcal{O}_{Y/K}$ sur un ouvert Y de X, qui donne naissance à des $\mathcal{D}^\dagger_{\mathfrak{X}\mathbf{Q}}$-modules notés $R^i v^\dagger_* \mathcal{O}_{\mathfrak{X}_K}$, jouant le rôle des images directes supérieures pour l'immersion $v : Y \hookrightarrow X$, ainsi que celui des faisceaux $\mathcal{H}^{\dagger i}_Z(\mathcal{O}_{\mathfrak{X}_K})$ définissant la cohomologie rigide à support dans le fermé complémentaire Z. Nous montrerons alors que, lorsque Z est lisse, ou est un diviseur à croisements normaux, ces $\mathcal{D}^\dagger_{\mathfrak{X}\mathbf{Q}}$-modules sont cohérents. Cette propriété reste conjecturale lorsque Z est une hypersurface quelconque, et devrait être un point clé pour la compréhension des propriétés de finitude de la cohomologie rigide.

Nous examinerons enfin le cas des isocristaux surconvergents qui interviennent dans la théorie p-adique des sommes exponentielles [3]. Le premier cas considéré, qui couvre en particulier le cas des caractères multiplicatifs de $\mathbb{F}_q$, $q = p^s$, est celui de l'isocristal surconvergent $\mathcal{K}_\alpha$ sur $\mathbb{G}_{mk} = \mathbb{P}^1_k - \{0, \infty\}$ défini, pour $\alpha \in \mathbb{Z}_p$, par le faisceau $\mathcal{O}_{\mathbb{G}_{mK}}$ sur $\mathbb{G}_{mK}$, muni de la connexion ∇ telle que $\nabla(1) = \alpha dt/t$: lorsque α n'est pas un nombre de Liouville, un calcul dû à Laumon montre que le $\mathcal{D}^\dagger_{\mathfrak{X}\mathbf{Q}}$-module $\mathrm{sp}_* \mathcal{K}_\alpha$ obtenu sur $\mathfrak{X} = \mathbb{P}^1_V$ est cohérent. Nous terminerons avec l'étude des isocristaux surconvergents sur $\mathbb{A}^1_k$ définis par la connexion $\nabla(1) = -\pi dt$ sur $\mathcal{O}_{\mathbb{A}^1_K}$, avec $\mathrm{ord}_p \pi = 1/(p-1)$, qui donnent en particulier les F-isocristaux de Dwork $\mathcal{L}_\psi$ associés aux caractères additifs ψ de $\mathbb{F}_q$: un tel isocristal, qui possède une singularité irrégulière à l'infini, définit encore sur $\mathcal{X} = \mathbb{P}^1_V$ un $\mathcal{D}^\dagger_{\mathcal{X}\mathbb{Q}}$-module cohérent.

Lors du colloque de Luminy "Systèmes différentiels et singularités" en Juillet 1983, j'avais eu l'occasion d'avoir plusieurs discussions avec Ph. Robba concernant l'anneau $\mathcal{D}^\dagger_{\mathcal{X}\mathbb{Q}}$; c'est avec émotion que je lui dédie le présent article. Les discussions et l'échange de correspondance que j'ai eu avec G. Laumon en 1985, dans lequel il a notamment mis en évidence la façon dont une partie de la théorie des $\mathcal{D}_X$-modules cohérents s'étend aux $\widehat{\mathcal{D}}^{(0)}_{\mathcal{X}\mathbb{Q}}$-modules (où $\widehat{\mathcal{D}}^{(0)}_{\mathcal{X}\mathbb{Q}}$ est le complété p-adique de l'anneau d'opérateurs différentiels engendré par les dérivations), ont été une étape importante pour m'aider à dégager le rôle que joue ici la notion de niveau d'un opérateur différentiel. Je remercie enfin Z. Mebkhout pour les discussions que nous avons eues à plusieurs reprises dans les années passées, qui ont été précieuses pour m'aider à mieux comprendre la théorie des $\mathcal{D}_X$-modules.

On désigne par p un nombre premier fixé dans tout l'article, et par $\mathbb{Z}_{(p)}$ le localisé de $\mathbb{Z}$ par rapport à l'idéal premier engendré par p. Tous les schémas formels considérés dans cet article seront des schémas formels pour la topologie p-adique.

1. Opérateurs différentiels de niveau fini

(1.0.1) Soit $f : X \to S$ un morphisme lisse de schémas (resp. schémas formels). Rappelons [EGA IV, § 16] rapidement quelques définitions concernant le calcul différentiel sur X ; pour simplifier les notations, il sera sous-entendu dans ce qui suit que toutes les notions considérées sont relatives à S, que nous ne ferons pas figurer dans les notations. Pour tout entier n, on note $\mathcal{P}^n_X$ l'algèbre du n-ième voisinage infinitésimal de X dans $X \times_S X$: si $\mathcal{I}$ est l'idéal de l'immersion diagonale, on a donc $\mathcal{P}^n_X := \mathcal{O}_{X\times X}/\mathcal{I}^{n+1}$. Il existe sur $\mathcal{P}^n_X$ deux structures de $\mathcal{O}_X$-algèbre, correspondant aux deux projections de $X \times_S X$ sur X ; sauf mention explicite du contraire, nous considèrerons $\mathcal{P}^n_X$ comme $\mathcal{O}_X$-algèbre par la structure gauche, correspondant à l'homomorphisme $a \longmapsto a \otimes 1$. Si $t_1, \ldots, t_d$ sont des coordonnées locales sur un ouvert U de X, relativement à S, et si on pose $\tau_i := 1 \otimes t_i - t_i \otimes 1 \in \mathcal{O}_{X\times X}$, $\mathcal{P}^n_X$ est alors libre sur U, de base les produits $\underline{\tau}^{\underline{k}}$ pour $|\underline{k}| \leq n$, avec $\underline{\tau}^{\underline{k}} := \tau_1^{k_1} \ldots \tau_d^{k_d}$ et $|\underline{k}| := k_1 + \ldots + k_d$.

Le faisceau $\mathcal{D}_{X n}$ des opérateurs différentiels d'ordre $\leq n$ sur X (relativement à S) est par définition le dual $\mathcal{H}om_{\mathcal{O}_X}(\mathcal{P}^n_X, \mathcal{O}_X)$, et on définit le faisceau $\mathcal{D}_X$ des opérateurs différentiels par $\mathcal{D}_X := \cup_n \mathcal{D}_{X n}$. Si on note $(\underline{\partial}^{[\underline{k}]})_{\underline{k}}$ la base de $\mathcal{D}_{X n}$ duale de la base $(\underline{t}^{\underline{k}})_{\underline{k}}$ de $\mathcal{P}^n_X$, les opérateurs $\underline{\partial}^{[\underline{k}]}$ vérifient les relations suivantes :

$$\forall\, \underline{k}', \underline{k}'' \in \mathbb{N}^d, \quad \underline{\partial}^{[\underline{k}']}\underline{\partial}^{[\underline{k}'']} = \binom{\underline{k}'+\underline{k}''}{\underline{k}'} \underline{\partial}^{[\underline{k}'+\underline{k}'']}, \tag{1.0.1.1}$$

$$\forall\, \underline{k} \in \mathbb{N}^d, \ \forall f \in \mathcal{O}_X, \quad \underline{\partial}^{[\underline{k}]} f = \sum_{\underline{k}'+\underline{k}''=\underline{k}} \underline{\partial}^{[\underline{k}']}(f)\underline{\partial}^{[\underline{k}'']}. \tag{1.0.1.2}$$

En particulier, si on note ∂_i les dérivations duales de la base dt_i de Ω^1_X, de sorte que, pour $\underline{1}_i = (0, \ldots, 1, \ldots, 0)$ avec 1 à la i-ième place, on a $\underline{\partial}^{[\underline{1}_i]} = \partial_i$, on voit que les $\underline{\partial}^{[\underline{k}]}$ jouent le rôle de puissances divisées des dérivations ∂_i.

Supposons que S soit un $\mathbb{Z}_{(p)}$-schéma. Il résulte alors des propriétés générales des puissances divisées que $\mathcal{D}_X$ est engendré comme algèbre sur $\mathcal{O}_X$ par les opérateurs $\partial_i^{[p^n]}$, pour $n \geq 0$. Par suite, même si X est nœthérien, les sections de $\mathcal{D}_X$ sur un ouvert affine de X ne forment pas un anneau nœthérien en général. Pour tout entier m, nous allons construire un autre faisceau d'opérateurs différentiels $\mathcal{D}_X^{(m)}$ sur X, qui sera engendré par les $\partial_i^{[p^n]}$, pour $n \leq m$, et dont les sections sur les ouverts affines seront des anneaux nœthériens. Cette construction repose sur la notion de *structure partielle d'idéal à puissances divisées.*

1.1. *Structures partielles d'idéal à puissances divisées*

Dans cet article, toutes les PD-structures considérées sur les $\mathbb{Z}_{(p)}$-algèbres seront supposées compatibles avec celles de l'idéal $p\mathbb{Z}_{(p)}$ [1, I 2.2.1].

(1.1.1) **Définitions.** Soient A une $\mathbb{Z}_{(p)}$-algèbre, $I \subset A$ un idéal, m un entier positif. On appelle *PD-structure partielle de niveau m* sur I, ou encore *m-PD-structure*, la donnée d'un PD-idéal $(J, \gamma) \subset I$ vérifiant les conditions suivantes :

(i) $\forall x \in I$, $x^{p^m} \in J$;

(ii) $pI \subset J$.

Un idéal muni d'une m-PD-structure sera appelé *m-PD-idéal.* Si (A, I, J, γ) et (A', I', J', γ') sont deux anneaux, munis de m-PD-idéaux, un *m-PD-morphisme* $f : (A, I, J, \gamma) \to (A', I', J', \gamma')$ est la donnée d'un PD-morphisme $f : (A, J, \gamma) \to (A', J', \gamma')$ tel que $f(I) \subset I'$.

Pour $m = 0$, cette notion coïncide avec la notion usuelle d'idéal à puissances divisées. Un exemple type d'idéal muni d'une m-PD-structure est fourni par l'idéal maximal $\mathfrak{m}$ d'un anneau de valuation discrète d'inégales caractéristiques, d'indice de ramification absolu e : si k est tel que $e/(p-1) \leq k \leq e + 1$, l'idéal $\mathfrak{m}^k$ est stable par les opérations $x \longmapsto x^n/n!$, et contient $p\mathfrak{m}$; il définit alors une m-PD-structure sur $\mathfrak{m}$ pour $m \geq \log_p k$.

Soient A une $\mathbb{Z}_{(p)}$-algèbre, $I \subset A$ un idéal. Le foncteur qui associe à toute $\mathbb{Z}_{(p)}$-algèbre A' munie d'un m-PD-idéal (I', J', γ') l'ensemble des homomorphismes $f : A \to A'$ tels que $f(I) \subset I'$ est représentable par une A-algèbre $\Delta_{(m)}(I)$, munie d'un m-PD-idéal $(I_1, J_1, {}^{[-]})$; on dira que $(\Delta_{(m)}(I), I_1, J_1, {}^{[-]})$ est la *m-PD-enveloppe de l'idéal I*. L'algèbre $\Delta_{(m)}(I)$ est l'enveloppe à puissances divisées usuelle de l'idéal $J = I^{(p^m)} + pA$ de A engendré par p et par les x^{p^m} pour $x \in I$; si on note par un $\tilde{}$ le PD-idéal engendré par un idéal, on a

$$J_1 = ((I^{(p^m)} + pI)\cdot\Delta_{(m)}(I))^{\sim},$$

et

$$I_1 = I\cdot\Delta_{(m)}(I) + J_1. \tag{1.1.1.1}$$

(1.1.2) Certains coefficients binômiaux modifiés jouent un rôle important dans ce qui suit. Nous utiliserons les notations suivantes : soient k', k'' deux entiers, et posons, pour un entier m fixé, $k' = p^m q' + r'$, $k'' = p^m q'' + r''$, et $k' + k'' = p^m q + r$, avec $0 \leq r, r', r'' < p^m$. On introduit alors les coefficients :

$$\left\{ {k'+k'' \atop k'} \right\}_{(m)} := \frac{q!}{q'!q''!}, \tag{1.1.2.1}$$

$$\left\langle {k'+k'' \atop k'} \right\rangle_{(m)} := \binom{k'+k''}{k'} \left\{ {k'+k'' \atop k'} \right\}_{(m)}^{-1}. \tag{1.1.2.2}$$

Le coefficient $\left\{ {k'+k'' \atop k'} \right\}_{(m)}$ est un entier, et le coefficient $\left\langle {k'+k'' \atop k'} \right\rangle_{(m)}$ un entier p-adique. S'il n'y a pas de confusion possible, nous omettrons l'indice (m).

Soit maintenant (A, I, J, γ) une $\mathbb{Z}_{(p)}$-algèbre munie d'un m-PD-idéal. Pour tout entier $k \in \mathbb{N}$, posons $k = p^m q + r$, avec $0 \leq r < p^m$. Pour tout élément $x \in I$, et tout $k \geq 0$, on pose

$$x^{\{k\}} := x^r \gamma_q(x^{p^m}), \tag{1.1.2.3}$$

ce qui est bien défini puisque $x^{p^m} \in J$. Si nécessaire, on précisera le niveau m par la notation $x^{\{k\}_{(m)}}$. Les opérations $x \longmapsto x^{\{k\}}$ seront appelées *puissances partiellement divisées de niveau m*. Elles vérifient des relations analogues à celles des puissances divisées usuelles :

(i) $\forall\, x \in I$, $x^{\{0\}} = 1$, et $x^{\{k\}} \in I$ pour tout $k \geq 1$;

(ii) $$\forall\, x \in I, \forall\, a \in A, \forall\, k \in \mathbb{N}, \quad (ax)^{\{k\}} = a^k x^{\{k\}} ; \tag{1.1.2.4}$$

(iii) $$\forall\, x, y \in I, \forall\, k \in \mathbb{N}, \quad (x+y)^{\{k\}} = \sum_{k'+k''=k} \left\langle {k \atop k'} \right\rangle_{(m)} x^{\{k'\}} y^{\{k''\}} ; \tag{1.1.2.5}$$

(iv) $$\forall\, x \in I, \forall\, k', k'' \in \mathbb{N}, \quad x^{\{k'\}} x^{\{k''\}} = \left\{ {k'+k'' \atop k'} \right\}_{(m)} x^{\{k'+k''\}} ; \tag{1.1.2.6}$$

(v) $$\forall\, x \in I, \forall\, k', k'' \in \mathbb{N}, \quad (x^{\{k'\}})^{\{k''\}} = \frac{q!}{(q'!)^{k''} q''!} x^{\{k'k''\}}, \tag{1.1.2.7}$$

avec ici $k'k'' = p^m q + r$, $0 \leq r < p^m$. Les opérations $x \longmapsto x^{\{k\}}$ jouent le rôle des opérations $x \longmapsto x^k/q!$.

Supposons que A soit une algèbre de polynômes, soit $A := R[t_1, \ldots, t_d]$, et $I := (t_1, \ldots, t_d)$. On montre alors que $\Delta_{(m)}(I)$ est un R-module libre ayant pour base les éléments $\underline{t}^{\{\underline{k}\}}$, $\underline{k} \in \mathbb{N}^d$.

(1.1.3) Les opérations précédentes permettent de définir sur tout idéal I d'une $\mathbb{Z}_{(p)}$-algèbre A muni d'une m-PD-structure (J, γ) une filtration canonique, analogue à la filtration I-adique, et à la filtration PD-adique sur un PD-idéal. Pour tout entier n, soit I_n l'idéal engendré par les éléments de la forme $x_1^{\{n_1\}} \ldots x_r^{\{n_r\}}$, avec $x_1, \ldots, x_r \in I$, et $\sum_i n_i \geq n$. On pose alors

$$I^{\{n\}} := I_n + (I_n \cap J)^{\sim}.$$

Cette filtration sera appelée *filtration m-PD-adique* de I. On dira que I est *m-PD-nilpotent* s'il existe un entier n tel que $I^{\{n\}} = 0$, et *topologiquement m-PD-nilpotent* si la filtration m-PD-adique est plus fine que la filtration $\mathfrak{m}$-adique. Pour que la m-PD-structure canonique d'un anneau de valuation discrète d'inégales caractéristiques soit topologiquement m-PD-nilpotente, il faut et suffit que $m > \log_p(e/(p-1))$.

Soient A une $\mathbb{Z}_{(p)}$-algèbre, I un idéal de A, $\Delta_{(m)}(I)$ la m-PD-enveloppe de I, I_1 le m-PD-idéal de $\Delta_{(m)}(I)$ défini en (1.1.1.1). On pose $\Delta_{(m)}^n(I) := \Delta_{(m)}(I)/I_1^{\{n+1\}}$. L'image de I_1 dans $\Delta_{(m)}^n(I)$ possède par passage au quotient une m-PD-structure canonique, et $\Delta_{(m)}^n(I)$ représente le foncteur sur la catégorie des $\mathbb{Z}_{(p)}$-algèbres A' munies d'un m-PD-idéal (I', J', γ') tel que $I'^{\{n+1\}} = 0$, qui associe à (A', I', J', γ') l'ensemble des homomorphismes $f : A \to A'$ tels que $f(I) \subset I'$.

Lorsque l'on part d'une algèbre de polynômes $R[t_1, \ldots, t_d]$, et de l'idéal $I := (t_1, \ldots, t_d)$, on vérifie que la filtration m-PD-adique sur $\Delta_{(m)}(I)$ est telle que $I_1^{\{n+1\}}$ soit l'idéal engendré par les $\underline{t}^{\{\underline{k}\}}$ pour $|\underline{k}| \geq n+1$. Il en résulte que $\Delta_{(m)}^n(I)$ est alors une R-algèbre finie libre de base les $\underline{t}^{\{\underline{k}\}}$ pour $|\underline{k}| \leq n$.

(1.1.4) Pour $m' \geq m$, toute m-PD-structure sur un idéal peut être considérée de manière évidente comme une m'-PD-structure. Si $k = p^m q + r = p^{m'} q' + r'$, avec $0 \leq r < p^m$, $0 \leq r' \leq p^{m'}$, on vérifie facilement que

$$x^{\{k\}_{(m')}} = (q!/q'!)x^{\{k\}_{(m)}}. \tag{1.1.4.1}$$

Il en résulte que, pour toute $\mathbb{Z}_{(p)}$-algèbre A et tout idéal $I \subset A$, les m-PD-enveloppes $\Delta_{(m)}(I)$ forment un système projectif pour m variable. De plus, les morphismes de transition sont compatibles aux filtrations m-PD-adiques, de sorte que, pour n fixé et m variable, les $\Delta_{(m)}^n(I)$ forment un système projectif, qui est en fait constant et égal à A/I^{n+1} pour $m > \log_p(n) - 1$.

engendré par les dérivations. Plus généralement, on déduit de ces relations que $\mathcal{D}_X^{(m)}$ est engendré comme $\mathcal{O}_X$-algèbre par les opérateurs $\partial_i^{\langle p^j \rangle}$ pour $j \leq m$.

(1.2.2) Pour m variable, les homomorphismes canoniques $\Delta^n_{(m+1)}(I) \to \Delta^n_{(m)}(I)$ définis en (1.1.4) fournissent par dualité des homomorphismes $\mathcal{D}^{(m)}_{X\,n} \to \mathcal{D}^{(m+1)}_{X\,n}$, qui sont des isomorphismes pour $m > \log_p(n) - 1$, le faisceau $\mathcal{D}^{(m)}_{X\,n}$ étant alors égal à $\mathcal{D}_{Xn}$. Soient $m' \geq m$, et $\underline{k} = p^m \underline{q} + \underline{r} = p^{m'} \underline{q}' + \underline{r}'$, avec $0 \leq r_i < p^m$, $0 \leq r_i' \leq p^{m'}$ pour tout i. On vérifie immédiatement que l'homomorphisme $\mathcal{D}^{(m)}_{X\,n} \to \mathcal{D}^{(m')}_{X\,n}$ envoie $\underline{\partial}^{\langle \underline{k} \rangle_{(m)}}$ sur $(\underline{q}!/\underline{q}'!)\underline{\partial}^{\langle \underline{k} \rangle_{(m')}}$. De plus, ces homomorphismes sont compatibles aux accouplements ; pour m variable, les faisceaux $\mathcal{D}_X^{(m)}$ forment donc un système inductif, et l'homomorphisme canonique

$$\varinjlim_m \mathcal{D}_X^{(m)} \longrightarrow \mathcal{D}_X,$$

qui envoie $\underline{\partial}^{\langle \underline{k} \rangle_{(m)}}$ sur $q!\,\partial^{[\underline{k}]}$, est un isomorphisme.

(1.2.3) **Proposition.** *Supposons que S soit nœthérien, et soit m un entier.*

(i) *Si S et X sont affines, l'anneau $D_X^{(m)} := \Gamma(X, \mathcal{D}_X^{(m)})$ est nœthérien.*

(ii) *Le faisceau $\mathcal{D}_X^{(m)}$ est un faisceau d'anneaux cohérents.*

On dispose en effet sur $D_X^{(m)}$ de la filtration par l'ordre, et le gradué associé $\mathrm{gr}\, D_X^{(m)}$ est commutatif d'après (1.2.1.2). Il est nœthérien, car il est engendré, comme $\Gamma(X, \mathcal{O}_X)$-algèbre, par les $\mathrm{gr}_i D_X^{(m)}$ pour $i \leq p^m$, qui sont des $\Gamma(X, \mathcal{O}_X)$-modules de type fini. L'assertion (i) en résulte par un argument classique.

Pour prouver (ii), il suffit alors de montrer que, lorsque X est affine assez petit, le foncteur qui associe à un $D_X^{(m)}$-module M le $\mathcal{D}_X^{(m)}$-module $\mathcal{D}_X^{(m)} \otimes_{D_X^{(m)}} M$ est exact. Comme $\mathcal{D}_X^{(m)}$ est un $\mathcal{O}_X$-module libre pour X assez petit, et que, X étant nœthérien, $\Gamma(X, -)$ commute aux sommes directes, l'homomorphisme $\mathcal{O}_X \otimes_{\Gamma(X,\, \mathcal{O}_X)} D_X^{(m)} \to \mathcal{D}_X^{(m)}$ est un isomorphisme, et l'assertion résulte de l'assertion analogue pour les $\mathcal{O}_X$-modules, valable également dans le cas formel.

1.3. *Opérateurs d'ordre infini*

(1.3.1) Soient maintenant $\mathcal{V}$ un anneau de valuation discrète complet d'inégales caractéristiques, d'idéal maximal $\mathfrak{m}$, et $\mathcal{X}$ un $\mathcal{V}$-schéma formel lisse. Appliquant les constructions précédentes à $S := \mathrm{Spf}\,\mathcal{V}$ et à $\mathcal{X}$, on obtient pour tout entier $m \geq 0$ un faisceau d'opérateurs différentiels $\mathcal{D}_{\mathcal{X}}^{(m)}$ sur $\mathcal{X}$. On notera $\widehat{\mathcal{D}}_{\mathcal{X}}^{(m)}$ (resp. $\widehat{\mathcal{D}}_{\mathcal{X}}$) le séparé complété de $\mathcal{D}_{\mathcal{X}}^{(m)}$ (resp. $\mathcal{D}_{\mathcal{X}}$) pour la topologie $\mathfrak{m}$-adique ; on a donc par définition

$$\widehat{\mathcal{D}}_{\mathcal{X}}^{(m)} := \varprojlim_n \mathcal{D}_{\mathcal{X}}^{(m)}/\mathfrak{m}^n \mathcal{D}_{\mathcal{X}}^{(m)} \simeq \varprojlim_n \mathcal{D}_{X_n}^{(m)},$$

1.2. *Opérateurs d'ordre fini*

(1.2.1) Soit maintenant $f : X \to S$ un morphisme lisse de $\mathbb{Z}_{(p)}$-schémas (resp. schémas formels). Appliquant les constructions qui précèdent à l'idéal $\mathcal{I}$ de la diagonale dans $X \times_S X$, on obtient des faisceaux d'anneaux $\Delta^n_{(m)}(\mathcal{I})$, qui seront notés $\Delta^n_{X/S(m)}$, ou encore $\Delta^n_{X(m)}$. Ces faisceaux sont munis des deux structures de $\mathcal{O}_X$-algèbre correspondant aux deux projections de $X \times_S X$ sur X, et nous les considérerons généralement comme $\mathcal{O}_X$-algèbres par la structure gauche. Si $t_1, \ldots, t_d$ sont des coordonnées locales sur un ouvert U de X, et si $\tau_i := 1 \otimes t_i - t_i \otimes 1$, $\Delta^n_{X(m)}$ est une $\mathcal{O}_X$-algèbre finie libre sur U, de base les $\underline{\tau}^{\{\underline{k}\}}$ pour $|\underline{k}| \leq n$: cela résulte de ce que $\Delta^n_{X(m)}$ ne dépend que de $\mathcal{O}_{X\times X}/\mathcal{I}^{n+1}$, qui est quotient sur U d'une algèbre de polynômes à coefficients dans $\mathcal{O}_X$.

On définit alors le *faisceau des opérateurs différentiels de niveau m et d'ordre n* comme étant le dual de $\Delta^n_{X(m)}$:

$$\mathcal{D}^{(m)}_{X\,n} := \mathcal{H}om_{\mathcal{O}_X}(\Delta^n_{X(m)}, \mathcal{O}_X).$$

Les homomorphismes surjectifs $\Delta^{n+1}_{X(m)} \to \Delta^n_{X(m)}$ définissent des injections

$$\mathcal{D}^{(m)}_{X\,n} \hookrightarrow \mathcal{D}^{(m)}_{X\,n+1},$$

et on définit le *faisceau* $\mathcal{D}^{(m)}_X$ *des opérateurs différentiels de niveau* m par

$$\mathcal{D}^{(m)}_X := \cup_n \mathcal{D}^{(m)}_{X\,n}.$$

Pour tous n, n', on dispose d'accouplements $\mathcal{D}^{(m)}_{X\,n} \times \mathcal{D}^{(m)}_{X\,n'} \to \mathcal{D}^{(m)}_{X\,n+n'}$ définis comme leurs analogues sur les $\mathcal{D}_{X n}$ dans [EGA IV, § 16]. On en déduit une structure d'algèbre sur $\mathcal{D}^{(m)}_X$. D'autre part, $\mathcal{D}^{(m)}_X$ opère sur $\mathcal{O}_X$ de la manière suivante : si $f \in \mathcal{O}_X$ et $P \in \mathcal{D}^{(m)}_{X\,n}$, la section $P(f) \in \mathcal{O}_X$ est l'image de f par l'application composée

$$\mathcal{O}_X \xrightarrow{d_2} \Delta^n_{X(m)} \xrightarrow{P} \mathcal{O}_X,$$

avec $d_2(f) := 1 \otimes f$.

Soient $t_1, \ldots, t_d$ des coordonnées locales sur un ouvert $U \subset X$, et $\tau_i := 1 \otimes t_i - t_i \otimes 1$. La base de $\mathcal{D}^{(m)}_{X\,n}$ duale de la base $(\underline{\tau}^{\{\underline{k}\}})_{\underline{k}}$ de $\Delta^n_{X(m)}$ sera notée $(\underline{\partial}^{\langle \underline{k}\rangle})_{\underline{k}}$. Si nécessaire, on précisera le niveau m par la notation $\underline{\partial}^{\langle \underline{k}\rangle_{(m)}}$. On déduit des relations de (1.1.2) les relations suivantes :

$$\forall\, \underline{k}', \underline{k}'' \in \mathbb{N}^d,\quad \underline{\partial}^{\langle \underline{k}'\rangle}\underline{\partial}^{\langle \underline{k}''\rangle} = \left\langle {\underline{k}'+\underline{k}'' \atop \underline{k}'} \right\rangle \underline{\partial}^{\langle \underline{k}'+\underline{k}''\rangle}, \tag{1.2.1.1}$$

$$\forall\, \underline{k} \in \mathbb{N}^d,\ \forall\, f \in \mathcal{O}_X,\quad \underline{\partial}^{\langle \underline{k}\rangle} f = \sum_{\underline{k}'+\underline{k}''=\underline{k}} \left\{ {\underline{k} \atop \underline{k}'} \right\} \underline{\partial}^{\langle \underline{k}'\rangle}(f)\,\underline{\partial}^{\langle \underline{k}''\rangle}. \tag{1.2.1.2}$$

Si $\underline{k} = p^m\underline{q} + \underline{r}$, avec $0 \leq r_i < p^m$ pour tout i, l'opérateur $\underline{\partial}^{\langle \underline{k}\rangle}$ joue donc le rôle de $\underline{q}!\,\underline{\partial}^{\underline{k}}/\underline{k}!$. Par exemple, lorsque $m = 0$, on a simplement $\underline{\partial}^{\langle \underline{k}\rangle} = \underline{\partial}^{\underline{k}}$, et le faisceau $\mathcal{D}^{(0)}_X$ est en posant $X_n := \mathscr{X} \times_S \operatorname{Spec}\mathcal{V}/\mathfrak{m}^n$. Les faisceaux $\widehat{\mathscr{D}}^{(m)}_{\mathscr{X}}$ sont des faisceaux d'opérateurs différentiels d'ordre infini : si $t_1, \ldots, t_d$ sont des coordonnées locales sur un ouvert $U \subset X$, et $\partial_1, \ldots, \partial_d$ les dérivations correspondantes, on a

$$\Gamma(U, \widehat{\mathscr{D}}^{(m)}_{\mathscr{X}}) = \{ \sum_{\underline{k}} a_{\underline{k}}\, \underline{\partial}^{\langle \underline{k} \rangle} \mid a_{\underline{k}} \in \Gamma(U, \mathcal{O}_{\mathscr{X}})\,;\ a_{\underline{k}} \to 0 \text{ pour } k \to \infty \}.$$

Comme $\widehat{\mathscr{D}}^{(m)}_{\mathscr{X}}$ s'obtient en complétant $\mathscr{D}^{(m)}_{\mathscr{X}}$ pour la topologie définie par un élément central, on déduit de (1.2.3) :

(1.3.2) **Proposition.** *Soit $U \subset \mathscr{X}$ un ouvert affine. Alors l'anneau $\widehat{D}^{(m)}_U := \Gamma(U, \widehat{\mathscr{D}}^{(m)}_{\mathscr{X}})$ est nœthérien.*

(1.3.3) Puisque $\mathcal{O}_{\mathscr{X}}$ est sans torsion, les homomorphismes $\mathscr{D}^{(m)}_{\mathscr{X}} \to \mathscr{D}^{(m')}_{\mathscr{X}} \to \mathscr{D}_{\mathscr{X}}$ sont injectifs, et il en est encore de même des homomorphismes $\widehat{\mathscr{D}}^{(m)}_{\mathscr{X}} \to \widehat{\mathscr{D}}^{(m')}_{\mathscr{X}} \to \widehat{\mathscr{D}}_{\mathscr{X}}$. On définit un sous-faisceau d'anneaux $\mathscr{D}^{\dagger}_{\mathscr{X}}$ de $\widehat{\mathscr{D}}_{\mathscr{X}}$ en posant

$$\mathscr{D}^{\dagger}_{\mathscr{X}} := \cup_m \widehat{\mathscr{D}}^{(m)}_{\mathscr{X}} \subset \widehat{\mathscr{D}}_{\mathscr{X}}.$$

Le faisceau $\mathscr{D}^{\dagger}_{\mathscr{X}}$ ainsi obtenu est étroitement lié à la notion de complétion faible de Monsky et Washnitzer, comme le montre la caractérisation locale des opérateurs de $\mathscr{D}^{\dagger}_{\mathscr{X}}$. Soient en effet U un ouvert affine de $\mathscr{X}$ sur lequel il existe des coordonnées locales, $\partial_1, \ldots, \partial_d$ les dérivations correspondantes, $A = \Gamma(U, \mathcal{O}_{\mathscr{X}})$, et $\| - \|$ la norme spectrale sur $A \otimes K$. Il est alors facile de voir que l'anneau $D^{\dagger}_U := \Gamma(U, \mathscr{D}^{\dagger}_{\mathscr{X}})$ possède la description suivante :

(1.3.4) **Proposition.** *Avec les notations précédentes, on a*

$$D^{\dagger}_U = \{ \sum_{\underline{k}} a_{\underline{k}}\, \underline{\partial}^{[\underline{k}]} \mid a_{\underline{k}} \in \Gamma(U, \mathcal{O}_{\mathscr{X}})\,;\ \exists\, c > 0,\ \eta < 1 \text{ tels que } \forall\, \underline{k},\ \|a_{\underline{k}}\| < c\eta^{|\underline{k}|} \}.$$

Il résulte de la construction de $\mathscr{D}^{\dagger}_{\mathscr{X}}$ que l'on a $\mathscr{D}^{\dagger}_{\mathscr{X}}/\mathfrak{m}^n\mathscr{D}^{\dagger}_{\mathscr{X}} \simeq \mathscr{D}_{X_n}$ pour tout n. Si U est un ouvert affine de $\mathscr{X}$, l'anneau $D^{\dagger}_U$ n'est donc pas nœthérien en général, et on peut montrer que l'anneau $D^{\dagger}_{U\mathbf{Q}} := D^{\dagger}_U \otimes \mathbf{Q}$ ne l'est pas non plus. On peut par contre montrer la cohérence des faisceaux $\widehat{\mathscr{D}}^{(m)}_{\mathscr{X}\mathbf{Q}} := \widehat{\mathscr{D}}^{(m)}_{\mathscr{X}} \otimes \mathbf{Q}$ et $\mathscr{D}^{\dagger}_{\mathscr{X}\mathbf{Q}} := \mathscr{D}^{\dagger}_{\mathscr{X}} \otimes \mathbf{Q}$:

(1.3.5) **Théorème.** *Les faisceaux $\widehat{\mathscr{D}}^{(m)}_{\mathscr{X}}$, $\widehat{\mathscr{D}}^{(m)}_{\mathscr{X}\mathbf{Q}}$, $\mathscr{D}^{\dagger}_{\mathscr{X}\mathbf{Q}}$ sont des faisceaux d'anneaux cohérents.*

Le cas des faisceaux $\widehat{\mathscr{D}}^{(m)}_{\mathscr{X}}$ et $\widehat{\mathscr{D}}^{(m)}_{\mathscr{X}\mathbf{Q}}$ se traite par les techniques considérées dans la section suivante. Pour ce qui est de $\mathscr{D}^{\dagger}_{\mathscr{X}\mathbf{Q}}$, nous nous limiterons ici à l'énoncé du point clé dont résulte la cohérence, en renvoyant à [5] pour la démonstration :

(1.3.6) **Proposition.** *Pour tout m, les homomorphismes canoniques* $\widehat{\mathscr{D}}^{(m)}_{\mathfrak{X}\,\mathbf{Q}} \to \widehat{\mathscr{D}}^{(m+1)}_{\mathfrak{X}\,\mathbf{Q}}$ *sont plats (à gauche et à droite).*

2. La catégorie des $\mathscr{D}^{\dagger}_{\mathfrak{X}\mathbf{Q}}$-modules cohérents

Dans toute la suite de cet article, on se place sous les hypothèses de (1.3.1).

2.1. *Les théorèmes A et B*

Nous explicitons ici les analogues des théorèmes A et B pour les faisceaux cohérents sur des anneaux tels que $\widehat{\mathscr{D}}^{(m)}_{\mathfrak{X}}$, ou $\mathscr{D}^{\dagger}_{\mathfrak{X}\mathbf{Q}}$. Les résultats qui suivent sont valables pour les $\mathscr{D}$-modules à gauche et à droite ; nous les énoncerons pour les $\mathscr{D}$-modules à gauche.

(2.1.1) On suppose maintenant que le $\mathscr{V}$-schéma formel $\mathfrak{X}$ est affine. On posera :

$$\widehat{D}^{(m)}_{\mathfrak{X}} := \Gamma(\mathfrak{X}, \widehat{\mathscr{D}}^{(m)}_{\mathfrak{X}}),\quad \widehat{D}^{(m)}_{\mathfrak{X}\,\mathbf{Q}} := \widehat{D}^{(m)}_{\mathfrak{X}} \otimes \mathbf{Q} \simeq \Gamma(\mathfrak{X}, \widehat{\mathscr{D}}^{(m)}_{\mathfrak{X}} \otimes \mathbf{Q}),$$
$$D^{\dagger}_{\mathfrak{X}} := \Gamma(\mathfrak{X}, \mathscr{D}^{\dagger}_{\mathfrak{X}}),\quad D^{\dagger}_{\mathfrak{X}\mathbf{Q}} := D^{\dagger}_{\mathfrak{X}} \otimes \mathbf{Q} \simeq \Gamma(\mathfrak{X}, \mathscr{D}^{\dagger}_{\mathfrak{X}} \otimes \mathbf{Q}),$$

les isomorphismes venant de ce que $\mathfrak{X}$ est un espace topologique nœthérien.

Désignons par $\mathscr{D}$ (resp. D) l'un des faisceaux $\widehat{\mathscr{D}}^{(m)}_{\mathfrak{X}}$, $\widehat{\mathscr{D}}^{(m)}_{\mathfrak{X}\,\mathbf{Q}}$, $\mathscr{D}^{\dagger}_{\mathfrak{X}}$, $\mathscr{D}^{\dagger}_{\mathfrak{X}\mathbf{Q}}$ (resp. l'un des anneaux $\widehat{D}^{(m)}_{\mathfrak{X}}$, $\widehat{D}^{(m)}_{\mathfrak{X}\,\mathbf{Q}}$, $D^{\dagger}_{\mathfrak{X}}$, $D^{\dagger}_{\mathfrak{X}\mathbf{Q}}$). Si M est un D-module à gauche de présentation finie, on lui associe un $\mathscr{D}$-module à gauche M^{Δ} de la façon suivante :

(i) *Cas où* $D = \widehat{D}^{(m)}_{\mathfrak{X}}$. Le D-module M est alors séparé et complet pour la topologie $\mathfrak{m}$-adique. Pour tout $n \geq 1$, soit $\widetilde{M}_n$ le $\mathscr{O}_{X_n}$-module quasi-cohérent défini par $M/\mathfrak{m}^n M$, qui est un $\mathscr{D}_{X_n}$-module de présentation finie. On pose

$$M^{\Delta} := \varprojlim_n \widetilde{M}_n.$$

(ii) *Cas où* $D = \widehat{D}^{(m)}_{\mathfrak{X}\,\mathbf{Q}}$. On peut trouver un $\widehat{D}^{(m)}_{\mathfrak{X}}$-module de présentation finie $\mathring{M}$ tel que $M \simeq \mathring{M} \otimes \mathbf{Q}$. On pose alors

$$M^{\Delta} := \mathring{M}^{\Delta} \otimes \mathbf{Q},$$

ce qui ne dépend pas du choix du réseau $\mathring{M}$.

(iii) *Cas où* $D = D^{\dagger}_{\mathfrak{X}}$. Comme $D^{\dagger}_{\mathfrak{X}} = \varinjlim_m \widehat{D}^{(m)}_{\mathfrak{X}}$, et que M est de présentation finie, il existe un entier m, et un $\widehat{D}^{(m)}_{\mathfrak{X}}$-module de présentation finie $M^{(m)}$ tels que

$$M \simeq D^{\dagger}_{\mathfrak{X}} \otimes_{\widehat{D}^{(m)}_{\mathfrak{X}}} M^{(m)}.$$

On définit alors M^{Δ} par

$$M^{\Delta} := \mathcal{D}^{\dagger}_{\mathfrak{X}} \otimes_{\hat{\mathcal{D}}^{(m)}_{\mathfrak{X}}} (M^{(m)})^{\Delta},$$

et M^{Δ} ne dépend pas de la façon dont on a descendu M en $M^{(m)}$.

(iv) *Cas où* $D = D^{\dagger}_{\mathfrak{X}\mathbf{Q}}$. Comme on a encore $D^{\dagger}_{\mathfrak{X}\mathbf{Q}} = \varinjlim_m \hat{D}^{(m)}_{\mathfrak{X}\mathbf{Q}}$, on procède comme dans le cas (iii) pour se ramener au cas (ii).

(2.1.2) **Théorème** (Théorème A). *Soient $\mathfrak{X}$ un $\mathcal{V}$-schéma formel affine et lisse, et D, $\mathcal{D}$ définis comme en* (2.1.1). *Soit $\mathcal{M}$ un $\mathcal{D}$-module. Les propriétés suivantes sont équivalentes :*

(i) *$\mathcal{M}$ est de présentation finie sur $\mathcal{D}$;*

(ii) *Il existe un D-module de présentation finie M tel que $\mathcal{M} \simeq M^{\Delta}$.*

(iii) *Le D-module $M := \Gamma(\mathfrak{X}, \mathcal{M})$ est de présentation finie, et l'homomorphisme canonique $\mathcal{D} \otimes_D M \to \mathcal{M}$ est un isomorphisme.*

De plus, si $\mathcal{D}$ est l'un des faiceaux $\hat{\mathcal{D}}^{(m)}_{\mathfrak{X}}$, $\hat{\mathcal{D}}^{(m)}_{\mathfrak{X}\mathbf{Q}}$ ou $\mathcal{D}^{\dagger}_{\mathfrak{X}\mathbf{Q}}$, le foncteur $M \longmapsto M^{\Delta}$ est exact.

Lorsque $\mathcal{D} = \hat{\mathcal{D}}^{(m)}_{\mathfrak{X}}$, $\hat{\mathcal{D}}^{(m)}_{\mathfrak{X}\mathbf{Q}}$ ou $\mathcal{D}^{\dagger}_{\mathfrak{X}\mathbf{Q}}$, il revient au même de dire que $\mathcal{M}$ est cohérent.

Dans le cas où $\mathcal{D} = \hat{\mathcal{D}}^{(m)}_{\mathfrak{X}}$, la démonstration est analogue à celle de [EGA I, § 10]. Le cas où $\mathcal{D} = \hat{\mathcal{D}}^{(m)}_{\mathfrak{X}\mathbf{Q}}$ se ramène au précédent grâce à l'existence, pour tout $\hat{\mathcal{D}}^{(m)}_{\mathfrak{X}\mathbf{Q}}$-module de présentation finie $\mathcal{M}$, d'un $\hat{\mathcal{D}}^{(m)}_{\mathfrak{X}}$-module de présentation finie $\mathring{\mathcal{M}}$ tel que $\mathcal{M} \simeq \mathring{\mathcal{M}} \otimes \mathbf{Q}$. Enfin, les cas où $\mathcal{D} = \mathcal{D}^{\dagger}_{\mathfrak{X}}$ ou $\mathcal{D} = \mathcal{D}^{\dagger}_{\mathfrak{X}\mathbf{Q}}$ se ramènent aux précédents par des arguments standards de passage à la limite inductive.

Lorsque $\mathcal{D} = \hat{\mathcal{D}}^{(m)}_{\mathfrak{X}}$, l'exactitude de $M \longmapsto M^{\Delta}$ provient de ce que $\hat{D}^{(m)}_{\mathfrak{X}}$ est nœthérien. Le cas de $\hat{\mathcal{D}}^{(m)}_{\mathfrak{X}\mathbf{Q}}$ en résulte en prenant des modèles entiers, et celui de $\mathcal{D}^{\dagger}_{\mathfrak{X}\mathbf{Q}}$ vient de la platitude de $\mathcal{D}^{\dagger}_{\mathfrak{X}\mathbf{Q}}$ sur les $\hat{\mathcal{D}}^{(m)}_{\mathfrak{X}\mathbf{Q}}$, d'après (1.3.6).

Le même type d'argument fournit le théorème B :

(2.1.3) **Théorème** (Théorème B). *Soient $\mathfrak{X}$ un $\mathcal{V}$-schéma formel affine et lisse, $\mathcal{D}$ l'un des faisceaux définis en* (2.1.1), *$\mathcal{M}$ un $\mathcal{D}$-module de présentation finie. Alors, pour tout $i \geq 1$, on a $H^i(\mathfrak{X}, \mathcal{M}) = 0$.*

2.2. *Indépendance par rapport au relèvement*

Soient respectivement $\boldsymbol{C}^{(m)}_{\mathfrak{X}}$, $\boldsymbol{C}^{(m)}_{\mathfrak{X}\mathbf{Q}}$, $\boldsymbol{C}^{\dagger}_{\mathfrak{X}}$, $\boldsymbol{C}^{\dagger}_{\mathfrak{X}\mathbf{Q}}$ les catégories des $\hat{\mathcal{D}}^{(m)}_{\mathfrak{X}}$, $\hat{\mathcal{D}}^{(m)}_{\mathfrak{X}\mathbf{Q}}$, $\mathcal{D}^{\dagger}_{\mathfrak{X}}$, $\mathcal{D}^{\dagger}_{\mathfrak{X}\mathbf{Q}}$-modules à gauche cohérents. On se propose de vérifier que les catégories $\boldsymbol{C}^{(m)}_{\mathfrak{X}}$, $\boldsymbol{C}^{(m)}_{\mathfrak{X}\mathbf{Q}}$, pour m assez grand, et $\boldsymbol{C}^{\dagger}_{\mathfrak{X}}$, $\boldsymbol{C}^{\dagger}_{\mathfrak{X}\mathbf{Q}}$, ne dépendent que de la réduction X de $\mathfrak{X}$ modulo $\mathfrak{m}$. Les énoncés sont encore les mêmes pour les $\mathcal{D}$-modules à droite.

(2.2.1) **Théorème.** *Soient $\mathscr{X}, \mathscr{X}'$ deux $\mathcal{V}$-schémas formels lisses ayant même réduction X modulo $\mathfrak{m}$. Si e est l'indice de ramification absolu de $\mathcal{V}$, et si $m > \log_p(e/(p-1))$, les catégories $\mathbf{C}^{(m)}_{\mathscr{X}}$ et $\mathbf{C}^{(m)}_{\mathscr{X}'}$ (resp. $\mathbf{C}^{(m)}_{\mathscr{X}\mathbf{Q}}$ et $\mathbf{C}^{(m)}_{\mathscr{X}'\mathbf{Q}}$) sont naturellement équivalentes.*

Lorsque $\mathscr{X}$ et $\mathscr{X}'$ sont affines, il existe un isomorphisme $\varphi : \mathscr{X}' \xrightarrow{\sim} \mathscr{X}$ induisant l'identité sur X. On en déduit pour tout m un isomorphisme φ^*-semi-linéaire ψ : $\widehat{\mathscr{D}}^{(m)}_{\mathscr{X}} \xrightarrow{\sim} \widehat{\mathscr{D}}^{(m)}_{\mathscr{X}'}$, et l'extension des scalaires correspondante définit une équivalence de catégories entre $\mathbf{C}^{(m)}_{\mathscr{X}}$ et $\mathbf{C}^{(m)}_{\mathscr{X}'}$.

Si $m > \log_p(e/(p-1))$, cette équivalence est en fait indépendante du choix de l'isomorphisme φ. En effet, soient $\varphi', \varphi'' : \mathscr{X}' \xrightarrow{\sim} \mathscr{X}$ deux isomorphismes induisant l'identité sur X, et $\psi', \psi'' : \widehat{\mathscr{D}}^{(m)}_{\mathscr{X}} \xrightarrow{\sim} \widehat{\mathscr{D}}^{(m)}_{\mathscr{X}'}$ les isomorphismes correspondants. Soient $t_1, \ldots, t_d$ des coordonnées locales sur $\mathscr{X}$, $t_i' := \varphi'^*(t_i)$, $t_i'' := \varphi''^*(t_i)$, ∂_i les dérivations correspondantes sur $\mathscr{X}$, $\partial_i' := \psi'(\partial_i)$, $\partial_i'' := \psi''(\partial_i)$. Comme $t_i'' - t_i' \in \mathfrak{m}\mathcal{O}_{\mathscr{X}'}$, et que cet idéal possède une m-PD-structure topologiquement m-PD-nilpotente grâce à la condition vérifiée par m, la série de Taylor

$$T := \sum_{\underline{k}} (\underline{t}'' - \underline{t}')^{\{\underline{k}\}} \underline{\partial}'^{\langle \underline{k} \rangle}$$

est définie et converge dans $\widehat{\mathscr{D}}^{(m)}_{\mathscr{X}'}$. On vérifie facilement qu'elle est indépendante du choix des coordonnées, et que, pour tout opérateur $P \in \widehat{\mathscr{D}}^{(m)}_{\mathscr{X}}$, on a $\psi''(P)T = T\psi'(P)$ dans $\widehat{\mathscr{D}}^{(m)}_{\mathscr{X}'}$. On définit alors un isomorphisme entre les deux foncteurs d'extension des scalaires définis par ψ' et ψ'' :

$$\varepsilon_{\varphi''\varphi'} : \widehat{\mathscr{D}}^{(m)}_{\mathscr{X}'\varphi''} \otimes_{\widehat{\mathscr{D}}^{(m)}_{\mathscr{X}}} \mathscr{M} \xrightarrow{\ \sim\ } \widehat{\mathscr{D}}^{(m)}_{\mathscr{X}'\varphi'} \otimes_{\widehat{\mathscr{D}}^{(m)}_{\mathscr{X}}} \mathscr{M}$$

en envoyant une section x de $\mathscr{M}$ sur $T \otimes x$. De plus, les isomorphismes $\varepsilon_{\varphi''\varphi'}$ vérifient la condition de cocycle, ce qui permet d'étendre la construction de l'équivalence de catégories $\mathbf{C}^{(m)}_{\mathscr{X}} \xrightarrow{\sim} \mathbf{C}^{(m)}_{\mathscr{X}'}$ au cas général. Le cas des catégories $\mathbf{C}^{(m)}_{\mathscr{X}\mathbf{Q}}$ et $\mathbf{C}^{(m)}_{\mathscr{X}'\mathbf{Q}}$ en résulte grâce à l'existence de modèles entiers pour les $\widehat{\mathscr{D}}^{(m)}_{\mathscr{X}\mathbf{Q}}$-modules cohérents.

Par passage à la limite inductive, on en déduit :

(2.2.2) **Corollaire.** *Soient $\mathscr{X}, \mathscr{X}'$ deux $\mathcal{V}$-schémas formels lisses ayant même réduction X modulo $\mathfrak{m}$. Les catégories $\mathbf{C}^{\dagger}_{\mathscr{X}}$ et $\mathbf{C}^{\dagger}_{\mathscr{X}'}$ (resp. $\mathbf{C}^{\dagger}_{\mathscr{X}\mathbf{Q}}$ et $\mathbf{C}^{\dagger}_{\mathscr{X}'\mathbf{Q}}$) sont naturellement équivalentes.*

(2.2.3) Nous renvoyons le lecteur à un article ultérieur [5] pour la description des opérations de fonctorialité en $\mathscr{X}$, et en X, des diverses catégories de $\mathscr{D}$-modules considérées ici.

3. Isocristaux convergents

(3.0.1) Rappelons brièvement la définition d'un isocristal convergent sur X (cf. [4, (2.3.2)], ou [2, (4.1)]).

On note K le corps des fractions de $\mathcal{V}$, et $\mathcal{X}_K$ la fibre générique du $\mathcal{V}$-schéma formel $\mathcal{X}$: c'est un espace analytique rigide sur K, muni d'une application continue $\mathrm{sp} : \mathcal{X}_K \to \mathcal{X}$, le morphisme de spécialisation [4, (0.2.2)]. Lorsque $\mathcal{X}$ est affine, soit $\mathcal{X} = \mathrm{Spf} A$, on a $\mathcal{X}_K = \mathrm{Spm}(A \otimes K)$; l'application sp associe alors à un idéal maximal de $A \otimes K$, définissant une extension finie $K(x)$ de K, l'idéal maximal de A formé des éléments dont l'image dans $K(x)$ est de valuation < 1 pour l'unique valuation sur $K(x)$ prolongeant celle de K. Pour tout ouvert $U \subset \mathcal{X}$, on a $\mathrm{sp}^{-1}(U) = U_K$.

Un isocristal convergent E sur X est défini par la donnée d'un $\mathcal{O}_{\mathcal{X}_K}$-module localement libre de rang fini $E_{\mathcal{X}}$ sur $\mathcal{X}_K$, muni d'une connexion intégrable

$$\nabla \; : \; E_{\mathcal{X}} \longrightarrow E_{\mathcal{X}} \otimes \Omega^1_{\mathcal{X}_K},$$

vérifiant la condition suivante [4, (2.2.14)] : pour tout ouvert affine $U = \mathrm{Spf} A$ de $\mathcal{X}$ sur lequel il existe un système de coordonnées locales $t_1, \ldots, t_d$, définissant des dérivations $\partial_1, \ldots, \partial_d$, toute section $e \in \Gamma(U_K, E_{\mathcal{X}})$, et tout $\eta < 1$, on a

$$\| \underline{\partial}^{[\underline{k}]} e \| \, \eta^{|\underline{k}|} \to 0 \text{ pour } |\underline{k}| \to \infty, \tag{3.0.1.1}$$

où $\underline{\partial}^{[\underline{k}]} e$ est défini par l'action des dérivations ∂_i fournie par la connexion ∇, et $\| \text{-} \|$ désigne une norme de Banach sur $\Gamma(U_K, E_{\mathcal{X}})$ [7, ch. 3, 3.7.3 prop. 3].

3.1. *Cohérence sur $\mathcal{D}^\dagger_{\mathcal{X}\mathbb{Q}}$ des isocristaux convergents*

(3.1.1) Soient E un $\mathcal{O}_{\mathcal{X}_K}$-module cohérent, et $\mathcal{E} = \mathrm{sp}_* E$. Utilisant le fait qu'un faisceau cohérent sur un espace affinoïde est déterminé par ses sections globales, et que, pour tout ouvert affine $U = \mathrm{Spf}\, A \subset \mathcal{X}$, $\mathrm{sp}^{-1}(U) = \mathrm{Spm}\, A \otimes K$ est affinoïde, on voit que $\mathrm{sp}^* \mathcal{E} \xrightarrow{\sim} E$, de sorte que la donnée de E équivaut à celle de $\mathcal{E}$. De même, comme $\mathrm{sp}^* \Omega^1_{\mathcal{X}} \xrightarrow{\sim} \Omega^1_{\mathcal{X}_K}$, on a un isomorphisme canonique $\mathcal{E} \otimes \Omega^1_{\mathcal{X}} \xrightarrow{\sim} \mathrm{sp}_*(E \otimes \Omega^1_{\mathcal{X}_K})$, et la donnée d'une connexion intégrable sur le $\mathcal{O}_{\mathcal{X}_K}$-module E équivaut à celle d'une connexion intégrable sur le $(\mathcal{O}_{\mathcal{X}} \otimes K)$-module $\mathcal{E}$. On notera $\mathcal{O}_{\mathcal{X}\mathbb{Q}} := \mathcal{O}_{\mathcal{X}} \otimes K$. La donnée d'un isocristal convergent sur X équivaut donc à celle d'un $\mathcal{O}_{\mathcal{X}\mathbb{Q}}$-module $\mathcal{E}$ localement projectif de type fini, muni d'une connexion intégrable ∇ telle que la condition de convergence (3.0.1.1) soit satisfaite ; on dira encore qu'une telle connnexion est *convergente*.

Le $\mathcal{O}_{\mathcal{X}\mathbb{Q}}$-module $\mathcal{E}$ est alors muni d'une structure canonique de $\mathcal{D}^\dagger_{\mathcal{X}\mathbb{Q}}$-module, prolongeant par continuité la structure de $\mathcal{D}_{\mathcal{X}\mathbb{Q}}$-module qui correspond à la connexion ∇. En effet, il suffit pour la définir de construire pour tout ouvert affine assez petit U de $\mathcal{X}$ un accouplement continu

$$\Gamma(U, \mathcal{D}^{\dagger}_{\mathcal{X}\mathbf{Q}}) \times \Gamma(U, \mathcal{E}) \longrightarrow \Gamma(U, \mathcal{E})$$

prolongeant celui que définit ∇. On peut supposer qu'il existe sur U des coordonnées locales, définissant une base de dérivations $\partial_1, \ldots, \partial_d$, et tout $P \in \Gamma(U, \mathcal{D}^{\dagger}_{\mathcal{X}\mathbf{Q}})$ s'écrit

$$P = \sum_{\underline{k}} a_{\underline{k}} \underline{\partial}^{[\underline{k}]},$$

où les $a_{\underline{k}} \in \Gamma(U, \mathcal{O}_{\mathcal{X}\mathbf{Q}})$ sont tels qu'il existe $c > 0$, $\eta < 1$ tels que $\|a_{\underline{k}}\| < c\eta^{|\underline{k}|}$ pour tout $\underline{k}$. Pour tout $e \in \Gamma(U, \mathcal{E})$, la relation (3.0.1.1) entraîne alors que $a_{\underline{k}}\underline{\partial}^{[\underline{k}]}e \to 0$ pour $|\underline{k}| \to \infty$, de sorte qu'on peut définir Pe comme somme de la série

$$Pe = \sum_{\underline{k}} a_{\underline{k}} \underline{\partial}^{[\underline{k}]} e.$$

Nous allons montrer que, muni de la structure de $\mathcal{D}^{\dagger}_{\mathcal{X}\mathbf{Q}}$-module ainsi définie, $\mathcal{E}$ est cohérent sur $\mathcal{D}^{\dagger}_{\mathcal{X}\mathbf{Q}}$. La démonstration s'appuie sur le résultat suivant d'Ogus [18, prop. 7.3], énoncé dans [18] pour $\widehat{\mathcal{D}}^{(0)}_{\mathcal{X}}$, mais qui reste valable sur $\widehat{\mathcal{D}}^{(m)}_{\mathcal{X}}$:

(3.1.2) **Proposition.** *Soient $\mathcal{X}$ un schéma formel lisse de type fini, de réduction X, E un isocristal convergent sur X, défini par un $\mathcal{O}_{\mathcal{X}\mathbf{Q}}$-module $\mathcal{E}$ localement projectif de type fini, muni d'une connexion intégrable et convergente, et $m \in \mathbb{N}$ un entier. Il existe alors un $\widehat{\mathcal{D}}^{(m)}_{\mathcal{X}}$-module $\mathring{\mathcal{E}}$, cohérent sur $\mathcal{O}_{\mathcal{X}}$, et un isomorphisme $\widehat{\mathcal{D}}^{(m)}_{\mathcal{X}\mathbf{Q}}$-linéaire $\mathring{\mathcal{E}} \otimes \mathbf{Q} \xrightarrow{\sim} \mathcal{E}$.*

Supposons d'abord $\mathcal{X}$ affine, avec des coordonnées locales $t_1, \ldots, t_d$; posons $\tau_i = 1 \otimes t_i - t_i \otimes 1$. Soient $A = \Gamma(\mathcal{X}, \mathcal{O}_{\mathcal{X}})$, $M = \Gamma(\mathcal{X}, \mathcal{E})$; notons $\Delta_{A_n(m)}$ la m-PD-enveloppe de l'idéal diagonal de $(A \otimes_{\mathcal{V}} A) \otimes \mathcal{V}/\mathfrak{m}^n$, et

$$\widehat{\Delta}_{A(m)} = \varprojlim_n \Delta_{A_n(m)} ;$$

$\widehat{\Delta}_{A(m)}$ est donc le complété p-adique du A-module libre de base les $\underline{\tau}^{\{\underline{k}\}}$. Posons $\eta = |p|^{p^{-m-1}}$, et

$$B = (A \widehat{\otimes} A\{T_1, \ldots, T_d\} / (pT_1 - \tau_1^{p^{m+1}}, \ldots, pT_d - \tau_d^{p^{m+1}})) \otimes \mathbf{Q}.$$

Observons tout d'abord qu'il existe un homomorphisme canonique $B \to \widehat{\Delta}_{A(m)} \otimes \mathbf{Q}$. En effet, un tel homomorphisme envoie nécessairement T_i sur $\tau_i^{p^{m+1}}/p$; comme on a $\tau_i^{p^{m+1}}/p = (p-1)!\, \tau_i^{\{p^{m+1}\}} \in \widehat{\Delta}_{A(m)}$, et que $\widehat{\Delta}_{A(m)}$ est séparé complet pour la topologie p-adique, l'assertion en résulte.

Comme M est de type fini sur $A \otimes \mathbf{Q}$, le produit tensoriel $M \otimes_{A\otimes\mathbf{Q}} B$ (pris ici pour la structure gauche, définie par $a \longmapsto a \otimes 1$) est séparé et complet. Pour tout $e \in M$, la série de Taylor

$$\theta(e) = \sum_{\underline{k}} \underline{\partial}^{[\underline{k}]} e \otimes \underline{\tau}^{\underline{k}} = \sum_{\underline{k}} \underline{\partial}^{\langle \underline{k} \rangle} e \otimes \underline{\tau}^{\{\underline{k}\}} \qquad (3.1.2.1)$$

converge dans $M \otimes_{A\otimes\mathbb{Q}} B$, car, si l'on pose $\underline{k} = p^{m+1}\underline{q} + \underline{r}$, avec $0 \le r_i < p^{m+1}$ pour tout i, on a $\underline{\tau}^{\underline{k}} = p^{|\underline{q}|}\underline{T}^{\underline{q}}\underline{\tau}^{\underline{r}}$, dans $M \otimes_A B$, et $\|p^{|\underline{q}|}\,\underline{\partial}^{[\underline{k}]}e\| \to 0$ pour $|\underline{k}| \to \infty$ d'après la condition de convergence (3.0.1.1). Il est facile de voir que l'application $\theta : M \to M \otimes_{A\otimes\mathbb{Q}} B$ ainsi obtenue est A-linéaire pour la structure droite de $M \otimes_{A\otimes\mathbb{Q}} B$, i. e. que pour tout $a \in A$, on a $\theta(ae) = (1\otimes a)\theta(e)$. Par composition avec l'extension des scalaires par l'homomorphisme $B \to \hat{\Delta}_{A(m)} \otimes \mathbb{Q}$ défini plus haut, on trouve donc une application A-linéaire pour la structure droite, encore notée θ,

$$\theta : M \longrightarrow M \otimes_{A\otimes\mathbb{Q}} (\hat{\Delta}_{A(m)} \otimes \mathbb{Q}).$$

Soit alors M' un sous-A-module de type fini de M tel que $M = M' \otimes \mathbb{Q}$, et posons

$$\mathring{M} = \theta^{-1}(M' \otimes_A \hat{\Delta}_{A(m)})\,;$$

il est clair que $\mathring{M} \otimes \mathbb{Q} \simeq M$. Comme le composé de θ avec l'extension des scalaires par l'homomorphisme d'augmentation $\hat{\Delta}_{A(m)} \to A$ est l'identité de M, on a $\mathring{M} \subset M'$ et $\mathring{M}$ est de type fini sur A. Comme A est nœthérien, M' de type fini sur A, et $\hat{\Delta}_{A(m)}$ le complété du A-module libre de base $\underline{\tau}^{\{\underline{k}\}}$, l'homomorphisme canonique

$$M' \otimes_A \hat{\Delta}_{A(m)} \to (\oplus_{\underline{k}} M')^{\wedge}$$

est un isomorphisme, et tout élément $x \in M' \otimes_A \hat{\Delta}_{A(m)}$ s'écrit de manière unique comme somme d'une série convergente

$$x = \sum_{\underline{k}} e_{\underline{k}} \otimes \underline{\tau}^{\{\underline{k}\}},$$

avec $e_{\underline{k}} \in M'$, et $e_{\underline{k}} \to 0$ si $|\underline{k}| \to \infty$. Par suite, d'après (3.1.2.1), on a

$$\mathring{M} = \{\, e \in M' \;\mid\; \forall\, \underline{k},\ \underline{\partial}^{\langle \underline{k}\rangle} e \in M' \,\}.$$

D'après (1.2.1.1), il en résulte que $\mathring{M}$ est un sous-$D_{\mathscr{X}}^{(m)}$-module de M. Le $\mathcal{O}_{\mathscr{X}}$-module cohérent $\mathring{\mathscr{E}}$ défini par $\mathring{M}$ est alors muni d'une structure de $\mathscr{D}_{\mathscr{X}}^{(m)}$-module ; comme le module de ses sections sur tout ouvert affine de $\mathscr{X}$ est séparé et complet, cette structure s'étend en une structure de $\hat{\mathscr{D}}_{\mathscr{X}}^{(m)}$-module, et l'isomorphisme $\mathring{\mathscr{E}} \otimes \mathbb{Q} \xrightarrow{\sim} \mathscr{E}$ est un isomorphisme de $\hat{\mathscr{D}}_{\mathscr{X}\,\mathbb{Q}}^{(m)}$-modules.

Dans le cas général (que nous n'utiliserons en fait pas ici), on choisit un recouvrement fini $(\mathscr{X}_i)$ de $\mathscr{X}$ du type précédent, et on recolle de proche en proche en reprenant l'argument d'Ogus [18, 7.5].

(3.1.3) **Proposition.** *Soit $\mathring{\mathscr{E}}$ un $\mathscr{D}_{\mathscr{X}}^{(m)}$-module, cohérent en tant que $\mathcal{O}_{\mathscr{X}}$-module. Alors :*

(i) *$\mathring{\mathscr{E}}$ est cohérent sur $\mathscr{D}_{\mathscr{X}}^{(m)}$;*

(ii) *l'homomorphisme canonique*

$$\mathring{\mathscr{E}} \longrightarrow \hat{\mathscr{D}}_{\mathscr{X}}^{(m)} \otimes_{\mathscr{D}_{\mathscr{X}}^{(m)}} \mathring{\mathscr{E}} \qquad (3.1.3.1)$$

est un isomorphisme.

On peut supposer que $\mathscr{X}$ est affine. Soient $A = \Gamma(\mathscr{X}, \mathcal{O}_{\mathscr{X}})$, $D_{\mathscr{X}}^{(m)} = \Gamma(\mathscr{X}, \mathscr{D}_{\mathscr{X}}^{(m)})$. Comme $\mathring{\mathscr{E}}$ est $\mathcal{O}_{\mathscr{X}}$-cohérent, il existe un homomorphisme surjectif $\mathcal{O}_{\mathscr{X}}^n \to \mathring{\mathscr{E}}$, donc un homomorphisme surjectif $(\mathscr{D}_{\mathscr{X}}^{(m)})^n \to \mathring{\mathscr{E}}$; soit $\mathscr{N}$ son noyau. La filtration par l'ordre sur $(\mathscr{D}_{\mathscr{X}}^{(m)})^n$ induit une filtration de $\mathscr{N}$ par des sous-$\mathcal{O}_{\mathscr{X}}$-modules $\mathscr{N}_i$, qui sont $\mathcal{O}_{\mathscr{X}}$-cohérents puisque noyaux d'homomorphismes entre $\mathcal{O}_{\mathscr{X}}$-modules cohérents. Comme le foncteur $\Gamma(\mathscr{X}, -)$ commute aux limites inductives, il en résulte que l'homomorphisme canonique $\mathcal{O}_{\mathscr{X}} \otimes_A \Gamma(\mathscr{X}, \mathscr{N}) \to \mathscr{N}$ est un isomorphisme ; il en est de même de $\mathcal{O}_{\mathscr{X}} \otimes_A D_{\mathscr{X}}^{(m)} \to \mathscr{D}_{\mathscr{X}}^{(m)}$. Comme, d'après (1.2.3), $D_{\mathscr{X}}^{(m)}$ est nœthérien, il existe un homomorphisme surjectif $(D_{\mathscr{X}}^{(m)})^r \to \Gamma(\mathscr{X}, \mathscr{N})$, qui induit donc un morphisme surjectif $(\mathscr{D}_{\mathscr{X}}^{(m)})^r \to \mathscr{N}$. La cohérence de $\mathring{\mathscr{E}}$ sur $\mathscr{D}_{\mathscr{X}}^{(m)}$ en résulte d'après (1.2.3).

Pour prouver l'assertion (ii), observons tout d'abord qu'on dispose d'une action naturelle de $\widehat{\mathscr{D}}_{\mathscr{X}}^{(m)}$ prolongeant par continuité celle de $\mathscr{D}_{\mathscr{X}}^{(m)}$, puisque, pour tout ouvert affine U de $\mathscr{X}$, $\Gamma(U, \mathring{\mathscr{E}})$ est séparé et complet pour la topologie p-adique. On obtient ainsi un homomorphisme $\widehat{\mathscr{D}}_{\mathscr{X}}^{(m)} \otimes_{\mathscr{D}_{\mathscr{X}}^{(m)}} \mathring{\mathscr{E}} \to \mathring{\mathscr{E}}$ dont le composé avec (3.1.3.1) est l'identité de $\mathring{\mathscr{E}}$. Il suffit donc de prouver que (3.1.3.1) est surjectif.

Pour cela, on peut supposer de plus que $\mathscr{X}$ possède un système de coordonnées locales $t_1, \ldots, t_d$. Pour $i = 1, \ldots, d$, notons ∂_i' l'opérateur $\partial_i^{\langle p^m \rangle}$. Pour tout entier $r < p^m$, et tout q, on déduit de (1.2.1.1) que $\partial_i^{\langle r \rangle} \partial_i'^{q} = u_{q,r} \partial_i^{\langle p^m q + r \rangle}$, où $u_{q,r}$ est une unité p-adique. Il en résulte que, pour tout ouvert affine $U \subset \mathscr{X}$, tout opérateur $P \in \Gamma(U, \widehat{\mathscr{D}}_{\mathscr{X}}^{(m)})$ s'écrit sous la forme

$$P = \sum_{\underline{k}} B_{\underline{k}} \underline{\partial}'^{\underline{k}}, \tag{3.1.3.2}$$

où les $B_{\underline{k}} \in \Gamma(U, \mathscr{D}_{\mathscr{X}}^{(m)})$ sont des opérateurs différentiels d'ordre $< p^m$ par rapport à chacun des ∂_i, et $B_{\underline{k}} \to 0$ pour $|\underline{k}| \to \infty$.

Soit $e \in \Gamma(U, \mathring{\mathscr{E}})$. Comme $\mathring{\mathscr{E}}$ est cohérent sur $\mathcal{O}_{\mathscr{X}}$, le sous-$\mathcal{O}_{\mathscr{X}}$-module engendré par les $\partial_i'^k e$, pour un i fixé et k variable, est engendré par un nombre fini d'entre eux. On peut donc trouver des sections $a_{ij} \in \Gamma(U, \mathcal{O}_{\mathscr{X}})$ donnant une relation de la forme

$$(\partial_i'^{n_i} - \sum_{j < n_i} a_{ij} \partial_i'^{j}) e = 0. \tag{3.1.3.3}$$

Par récurrence sur k, tout opérateur $\partial_i^{\langle r \rangle} \partial_i'^{k}$, où $0 \leq r < p^m$, peut s'écrire sous la forme

$$\partial_i^{\langle r \rangle} \partial_i'^{k} = Q_{kr} (\partial_i'^{n_i} - \sum_{j < n_i} a_{ij} \partial_i'^{j}) + R_{kr}, \tag{3.1.3.4}$$

où les R_{kr} sont d'ordre $< p^m n_i$ par rapport à ∂_i. C'est en effet clair si $k < n_i$, et, pour $k \geq n_i$, on peut écrire

$$\partial_i^{\langle r \rangle} \partial_i'^{k} = \partial_i^{\langle r \rangle} \partial_i'^{k - n_i} (\partial_i'^{n_i} - \sum_{j < n_i} a_{ij} \partial_i'^{j}) + \partial_i^{\langle r \rangle} \partial_i'^{k - n_i} (\sum_{j < n_i} a_{ij} \partial_i'^{j}),$$

le deuxième terme étant un opérateur d'ordre $\leq p^m (k-1) + r$. On en déduit que, pour tout multi-indice $\underline{k}$, et tout opérateur $B_{\underline{k}} \in \Gamma(U, \mathscr{D}_{\mathscr{X}}^{(m)})$, d'ordre $< p^m$ par rapport à chacun des ∂_i, il existe des opérateurs $R_{\underline{k}} \in \Gamma(U, \mathscr{D}_{\mathfrak{X}}^{(m)})$, d'ordre $< p^m(n_1 + \ldots + n_d)$, et $Q_{i\underline{k}} \in \Gamma(U, \mathscr{D}_{\mathfrak{X}}^{(m)})$ tels que

$$B_{\underline{k}}\underline{\partial}'^{\underline{k}} = \sum_i Q_{i\underline{k}}(\partial_i'^{n_i} - \sum_{j<n_i} a_{ij}\partial_i'^{j}) + R_{\underline{k}}.$$

De plus, les calculs s'effectuant dans $\Gamma(U, \mathscr{D}_{\mathfrak{X}}^{(m)})$, on a $Q_{i\underline{k}} \to 0$, $R_{\underline{k}} \to 0$ si $B_{\underline{k}} \to 0$.

Appliquant ce qui précède à la suite $B_{\underline{k}}\underline{\partial}'^{\underline{k}}$ de somme P, on peut alors réécrire P sous la forme

$$P = \sum_i (\sum_{\underline{k}} Q_{i\underline{k}})(\partial_i'^{n_i} - \sum_{j<n_i} a_{ij}\partial_i'^{j}) + \sum_{\underline{k}} R_{\underline{k}}.$$

Comme $\sum_{\underline{k}} R_{\underline{k}}$ est d'ordre fini, cela permet d'écrire dans $\widehat{\mathscr{D}}_{\mathfrak{X}}^{(m)} \otimes_{\mathscr{D}_{\mathfrak{X}}^{(m)}} \mathring{\mathscr{E}}$

$$P \otimes e = (\sum_{\underline{k}} R_{\underline{k}}) \otimes e = 1 \otimes (\sum_{\underline{k}} R_{\underline{k}})e,$$

ce qui montre la surjectivité de (3.1.3.1).

(3.1.4) **Proposition.** *Soit E un isocristal convergent sur X, défini par un $\mathcal{O}_{\mathfrak{X}\mathbf{Q}}$-module $\mathscr{E}$ localement projectif de type fini, muni d'une connexion intégrable et convergente. Alors :*

(i) *Pour tout entier m, l'homomorphisme canonique*

$$\mathscr{E} \longrightarrow \widehat{\mathscr{D}}_{\mathfrak{X}}^{(m)}{}_{\mathbf{Q}} \otimes_{\mathscr{D}_{\mathfrak{X}\mathbf{Q}}} \mathscr{E} \qquad (3.1.4.1)$$

est un isomorphisme, et il en est de même de l'homomorphisme canonique

$$\mathscr{E} \longrightarrow \mathscr{D}_{\mathfrak{X}\mathbf{Q}}^{\dagger} \otimes_{\mathscr{D}_{\mathfrak{X}\mathbf{Q}}} \mathscr{E}. \qquad (3.1.4.2)$$

(ii) *Pour la structure de $\mathscr{D}_{\mathfrak{X}\mathbf{Q}}^{\dagger}$-module (resp. $\widehat{\mathscr{D}}_{\mathfrak{X}}^{(m)}{}_{\mathbf{Q}}$-module) définie en* (3.1.1), *$\mathscr{E}$ est cohérent sur $\mathscr{D}_{\mathfrak{X}\mathbf{Q}}^{\dagger}$ (resp. $\widehat{\mathscr{D}}_{\mathfrak{X}}^{(m)}{}_{\mathbf{Q}}$).*

Soit $m \geq 0$, et choisissons, grâce à (3.1.2), un $\widehat{\mathscr{D}}_{\mathfrak{X}}^{(m)}$-module $\mathring{\mathscr{E}}$, cohérent sur $\mathcal{O}_{\mathfrak{X}}$, et un isomorphisme $\mathring{\mathscr{E}} \otimes \mathbf{Q} \xrightarrow{\sim} \mathscr{E}$. L'isomorphisme (3.1.4.1) résulte alors de (3.1.3.1) pour $\mathring{\mathscr{E}}$ en tensorisant par $\mathbf{Q}$. Passant à la limite inductive sur m, on en déduit l'isomorphisme (3.1.4.2).

D'autre part, $\mathring{\mathscr{E}}$ est $\mathscr{D}_{\mathfrak{X}}^{(m)}$-cohérent d'après (3.1.3). Si on considère une présentation de $\mathring{\mathscr{E}}$ sur $\mathscr{D}_{\mathfrak{X}}^{(m)}$:

$$(\mathscr{D}_{\mathfrak{X}}^{(m)})^s \longrightarrow (\mathscr{D}_{\mathfrak{X}}^{(m)})^r \longrightarrow \mathring{\mathscr{E}} \longrightarrow 0,$$

on obtient par tensorisation avec $\mathbf{Q}$, puis avec $\mathscr{D}_{\mathfrak{X}\mathbf{Q}}^{\dagger}$ (resp. $\widehat{\mathscr{D}}_{\mathfrak{X}}^{(m)}{}_{\mathbf{Q}}$) sur $\mathscr{D}_{\mathfrak{X}\mathbf{Q}}$, une présentation analogue de $\mathscr{D}_{\mathfrak{X}\mathbf{Q}}^{\dagger} \otimes_{\mathscr{D}_{\mathfrak{X}\mathbf{Q}}} \mathscr{E}$ (resp. $\widehat{\mathscr{D}}_{\mathfrak{X}}^{(m)} \otimes_{\mathscr{D}_{\mathfrak{X}\mathbf{Q}}} \mathscr{E}$), donc de $\mathscr{E}$ d'après (i). La cohérence de $\mathscr{E}$ résulte alors de (1.3.5).

3.2. *Le complexe de de Rham d'un $\mathcal{D}^\dagger_{\mathfrak{X}\mathbf{Q}}$-module*

L'exemple le plus important d'isocristal convergent est l'isocristal "constant" $\mathcal{O}_{X/K}$, défini par $\mathcal{O}_{\mathfrak{X}\mathbf{Q}}$ muni de la connexion trivale. Dans ce cas, on dispose comme dans le cas classique de la résolution localement libre sur $\mathcal{D}^\dagger_{\mathfrak{X}\mathbf{Q}}$ fournie par le complexe de Spencer associé au faisceau tangent [17, I (2.1.17)].

Observons d'abord que la division par ∂ est possible dans $\widehat{\mathcal{D}}^{(m)}_{\mathfrak{X}\mathbf{Q}}$ et $\mathcal{D}^\dagger_{\mathfrak{X}\mathbf{Q}}$.

(3.2.1) **Proposition.** *Soient U un ouvert de $\mathfrak{X}$ sur lequel il existe un système de coordonnées locales $t_1, \ldots, t_d$, définissant des dérivations $\partial_1, \ldots, \partial_d$.*

(i) *Soit $P = \sum_{\underline{k}} a_{\underline{k}} \underline{\partial}^{\langle \underline{k} \rangle} \in \Gamma(U, \widehat{\mathcal{D}}^{(m)}_{\mathfrak{X}})$; si $a_{\underline{k}} = 0$ pour tout $\underline{k}$ tel que $k_1 = 0$, il existe un unique opérateur $Q \in \Gamma(U, \widehat{\mathcal{D}}^{(m)}_{\mathfrak{X}})$ tel que $p^m P = Q\,\partial_1$.*

(ii) *Soit $P = \sum_{\underline{k}} a_{\underline{k}} \underline{\partial}^{[\underline{k}]} \in \Gamma(U, \mathcal{D}^\dagger_{\mathfrak{X}\mathbf{Q}})$; si $a_{\underline{k}} = 0$ pour tout $\underline{k}$ tel que $k_1 = 0$, il existe un unique opérateur $Q \in \Gamma(U, \mathcal{D}^\dagger_{\mathfrak{X}\mathbf{Q}})$ tel que $P = Q\,\partial_1$.*

L'unicité de Q résulte de ce que $\widehat{\mathcal{D}}^{(m)}_{\mathfrak{X}}$ est sans torsion. Soit $\underline{k}$ tel que $k_1 \neq 0$. Dans $\widehat{\mathcal{D}}^{(m)}_{\mathfrak{X}}$, on dispose de la relation (1.2.1.1)

$$\left\langle {k_1 \atop 1} \right\rangle \partial_1^{\langle k_1 \rangle} = \partial_1^{\langle k_1 - 1 \rangle} \partial_1,$$

avec :

$$\left\langle {k_1 \atop 1} \right\rangle = \begin{cases} k \text{ si } p^m \nmid k, \\ p^m \text{ si } p^m \mid k. \end{cases} \tag{3.2.1.1}$$

Quel que soit k_1, $\left\langle {k_1 \atop 1} \right\rangle$ divise donc p^m dans $\mathbb{Z}_p$, ce qui permet de définir Q en posant

$$Q = \sum_{k_1 \neq 0} \left(p^m \left\langle {k_1 \atop 1} \right\rangle^{-1}\right) a_{\underline{k}} \underline{\partial}^{\langle \underline{k} - \underline{1}_1 \rangle},$$

en notant $\underline{1}_1 = (1, 0, \ldots, 0)$.

L'assertion (ii) résulte de (i) par passage à la limite, mais peut aussi être prouvée directement : on a $a_{\underline{k}} \partial_1^{[k_1]} = (a_{\underline{k}}/k_1)\partial_1^{[k_1 - 1]}\partial_1$, et il existe $c > 0$, et $\eta < 1$ tels que $\|a_{\underline{k}}\| < c\,\eta^{|\underline{k}|}$; comme $|1/k_1| = O(k_1) = O(|\underline{k}|)$, l'existence de Q en découle.

(3.2.2) **Proposition.** *Soit $\mathcal{T}_{\mathfrak{X}} = \Omega^{1\vee}_{\mathfrak{X}}$ le faisceau tangent sur $\mathfrak{X}$. L'homomorphisme $\mathcal{D}^\dagger_{\mathfrak{X}\mathbf{Q}} \to \mathcal{O}_{\mathfrak{X}\mathbf{Q}}$ envoyant $P \in \mathcal{D}^\dagger_{\mathfrak{X}\mathbf{Q}}$ sur $P \cdot 1 \in \mathcal{O}_{\mathfrak{X}\mathbf{Q}}$ fait du complexe de Spencer*

$$0 \longrightarrow \mathcal{D}^\dagger_{\mathfrak{X}\mathbf{Q}} \otimes_{\mathcal{O}_{\mathfrak{X}}} \overset{d}{\wedge} \mathcal{T}_{\mathfrak{X}} \longrightarrow \ldots \longrightarrow \mathcal{D}^\dagger_{\mathfrak{X}\mathbf{Q}} \otimes_{\mathcal{O}_{\mathfrak{X}}} \overset{2}{\wedge} \mathcal{T}_{\mathfrak{X}} \longrightarrow \mathcal{D}^\dagger_{\mathfrak{X}\mathbf{Q}} \otimes_{\mathcal{O}_{\mathfrak{X}}} \mathcal{T}_{\mathfrak{X}} \longrightarrow \mathcal{D}^\dagger_{\mathfrak{X}\mathbf{Q}}$$

une résolution $\mathcal{D}^\dagger_{\mathfrak{X}\mathbf{Q}}$-linéaire de $\mathcal{O}_{\mathfrak{X}\mathbf{Q}}$. En particulier, si $\partial_1, \ldots, \partial_d$ sont les dérivations correspondant à un système de coordonnées locales, on a un isomorphisme

$$\mathscr{D}^\dagger_{\mathscr{X}\mathbb{Q}} / \sum_{i=1}^{d} \mathscr{D}^\dagger_{\mathscr{X}\mathbb{Q}} \cdot \partial_i \xrightarrow{\sim} \mathcal{O}_{\mathscr{X}\mathbb{Q}}.$$

L'énoncé analogue sur $\widehat{\mathscr{D}}^{(m)}_{\mathscr{X}\,\mathbb{Q}}$ est vrai pour tout $m \in \mathbb{N}$.

Rappelons que la différentielle $d : \mathscr{D}^\dagger_{\mathscr{X}\mathbb{Q}} \otimes_{\mathcal{O}_\mathscr{X}} \overset{k}{\wedge} \mathscr{T}_\mathscr{X} \longrightarrow \mathscr{D}^\dagger_{\mathscr{X}\mathbb{Q}} \otimes_{\mathcal{O}_\mathscr{X}} \overset{k-1}{\wedge} \mathscr{T}_\mathscr{X}$ est définie par

$$d(P \otimes \partial_1 \wedge \ldots \wedge \partial_k) = \sum_{i=1}^{k} (-1)^{i-1} P\partial_i \otimes \partial_1 \wedge \ldots \wedge \widehat{\partial}_i \wedge \ldots \wedge \partial_k$$
$$+ \sum_{i<j} (-1)^{i+j} P \otimes [\partial_i, \partial_j] \wedge \partial_1 \wedge \ldots \wedge \widehat{\partial}_i \wedge \ldots \wedge \widehat{\partial}_j \wedge \ldots \wedge \partial_k,$$

pour $P \in \mathscr{D}^\dagger_{\mathscr{X}\mathbb{Q}}$, $\partial_1, \ldots, \partial_k \in \mathscr{T}_\mathscr{X}$.

On peut supposer qu'il existe sur $\mathscr{X}$ un système de coordonnées locales ; si ∂_1, ..., ∂_d sont les dérivations correspondantes, les sections $1 \otimes \partial_{i_1} \wedge \ldots \wedge \partial_{i_k}$ forment une base de $\mathscr{D}^\dagger_{\mathscr{X}\mathbb{Q}} \otimes_{\mathcal{O}_\mathscr{X}} \wedge \mathscr{T}_\mathscr{X}$. Comme la multiplication à droite par ∂_i est injective sur $\mathscr{D}^\dagger_{\mathscr{X}\mathbb{Q}}$, et que $[\partial_i, \partial_j] = 0$, on voit facilement que le complexe

$$0 \to \mathscr{D}^\dagger_{\mathscr{X}\mathbb{Q}} \otimes_{\mathcal{O}_\mathscr{X}} \overset{d}{\wedge} \mathscr{T}_\mathscr{X} \to \ldots \to \mathscr{D}^\dagger_{\mathscr{X}\mathbb{Q}} \otimes_{\mathcal{O}_\mathscr{X}} \mathscr{T}_\mathscr{X} \to \mathscr{D}^\dagger_{\mathscr{X}\mathbb{Q}} \to \mathscr{D}^\dagger_{\mathscr{X}\mathbb{Q}} / \sum_{i=1}^{d} \mathscr{D}^\dagger_{\mathscr{X}\mathbb{Q}} \cdot \partial_i \to 0$$

est acyclique. Il est clair d'autre part que le noyau de $\mathscr{D}^\dagger_{\mathscr{X}\mathbb{Q}} \to \mathcal{O}_{\mathscr{X}\mathbb{Q}}$ est l'ensemble des opérateurs de la forme $\sum_{\underline{k} \neq \underline{0}} a_{\underline{k}} \underline{\partial}^{[\underline{k}]} \in \mathscr{D}^\dagger_{\mathscr{X}\mathbb{Q}}$, et un tel opérateur appartient à $\sum_i \mathscr{D}^\dagger_{\mathscr{X}\mathbb{Q}} \cdot \partial_i$ d'après (3.2.1).

Remarque. On aurait aussi pu partir de la résolution de Spencer de $\mathcal{O}_\mathscr{X}$ sur $\mathscr{D}^{(0)}_\mathscr{X}$, puis tensoriser sur $\mathscr{D}^{(0)}_\mathscr{X}$ avec $\widehat{\mathscr{D}}^{(m)}_{\mathscr{X}\mathbb{Q}}$ et $\mathscr{D}^\dagger_{\mathscr{X}\mathbb{Q}}$, qui sont plats sur $\mathscr{D}^{(0)}_\mathscr{X}$, et appliquer les isomorphismes (3.1.4.1) et (3.1.4.2). Par contre, pour $m > 0$, l'énoncé est faux en général pour $\mathcal{O}_\mathscr{X}$ sur $\mathscr{D}^{(m)}_\mathscr{X}$ et $\widehat{\mathscr{D}}^{(m)}_\mathscr{X}$: par exemple, si $\dim(X) = 1$, on a

$$\mathcal{O}_\mathscr{X} \simeq \mathscr{D}^{(m)}_\mathscr{X} / \sum_{j=0}^{m} \mathscr{D}^{(m)}_\mathscr{X} \cdot \partial^{\langle p^j \rangle} \simeq \widehat{\mathscr{D}}^{(m)}_\mathscr{X} / \sum_{j=0}^{m} \widehat{\mathscr{D}}^{(m)}_\mathscr{X} \cdot \partial^{\langle p^j \rangle},$$

et, pour $j > 0$, les $\partial^{\langle p^j \rangle}$ ne sont pas divisibles par ∂.

(3.2.3) **Corollaire.** *Soient $\mathscr{M}$ un $\mathscr{D}^\dagger_{\mathscr{X}\mathbb{Q}}$-module à gauche, et $\mathscr{M} \otimes \Omega^\cdot_\mathscr{X}$ le complexe de de Rham de $\mathscr{M}$, défini par la connexion sous-jacente. Il existe un isomorphisme canonique dans la catégorie dérivée des faisceaux de K-vectoriels sur $\mathscr{X}$*

$$\mathbb{R}\mathscr{H}om_{\mathscr{D}^\dagger_{\mathscr{X}\mathbb{Q}}}(\mathcal{O}_{\mathscr{X}\mathbb{Q}}, \mathscr{M}) \simeq \mathscr{M} \otimes \Omega^\cdot_\mathscr{X}.$$

Il suffit en effet d'utiliser la résolution de (3.2.2) pour calculer le complexe $\mathbb{R}\mathscr{H}om_{\mathscr{D}^\dagger_{\mathscr{X}\mathbb{Q}}}(\mathcal{O}_{\mathscr{X}\mathbb{Q}}, \mathscr{M})$.

(3.2.4) **Corollaire.** *Soient E un isocristal convergent sur X, défini par un $\mathcal{O}_{\mathfrak{X}\mathbb{Q}}$-module $\mathcal{E}$ localement projectif de type fini, muni d'une connexion intégrable et convergente. Il existe des isomorphismes canoniques*

$$H^i_{\mathrm{conv}}(X, E) \simeq \mathrm{Ext}^i_{\mathcal{D}^\dagger_{\mathfrak{X}\mathbb{Q}}}(\mathcal{O}_{\mathfrak{X}\mathbb{Q}}, \mathcal{E}).$$

En particulier, si X est propre sur k, il existe des isomorphismes canoniques

$$H^i_{\mathrm{rig}}(X, E) \simeq \mathrm{Ext}^i_{\mathcal{D}^\dagger_{\mathfrak{X}\mathbb{Q}}}(\mathcal{O}_{\mathfrak{X}\mathbb{Q}}, \mathcal{E}).$$

Soit $E_{\mathfrak{X}}$ le $\mathcal{O}_{\mathfrak{X}_K}$-module à connexion intégrable définissant E. Par définition de la cohomologie convergente, on a

$$H^i_{\mathrm{conv}}(X, E) = H^i(\mathfrak{X}_K, E_{\mathfrak{X}} \otimes \Omega^{\cdot}_{\mathfrak{X}_K}).$$

Pour tout ouvert affine $U \subset \mathfrak{X}$, $\mathrm{sp}^{-1}(U)$ est un ouvert affinoïde de $\mathfrak{X}_K$. Comme les faisceaux $E_{\mathfrak{X}} \otimes \Omega^i_{\mathfrak{X}_K}$ sont cohérents sur $\mathcal{O}_{\mathfrak{X}_K}$, on a

$$\mathbb{R}\mathrm{sp}_*(E_{\mathfrak{X}} \otimes \Omega^{\cdot}_{\mathfrak{X}_K}) \simeq \mathrm{sp}_*(E_{\mathfrak{X}} \otimes \Omega^{\cdot}_{\mathfrak{X}_K}) \simeq \mathcal{E} \otimes \Omega^{\cdot}_{\mathfrak{X}},$$

et l'énoncé en résulte. Lorsque X est de plus propre sur k, la cohomologie rigide coïncide par construction avec la cohomologie convergente.

4. Isocristaux surconvergents : le cas constant

Soient Y un ouvert de X, Z le fermé complémentaire. Alors que les isocristaux convergents sur X considérés dans la section précédente fournissent des $\mathcal{D}^\dagger_{\mathfrak{X}\mathbb{Q}}$-modules "sans singularités sur X", analogues des fibrés à connexion intégrable en caractéristique 0, les isocristaux convergents sur Y qui sont surconvergents le long de Z vont donner des exemples de $\mathcal{D}^\dagger_{\mathfrak{X}\mathbb{Q}}$-modules avec des singularités le long de Z. Nous allons d'abord montrer que l'image directe dérivée d'un tel isocristal par le morphisme de spécialisation possède une structure canonique de complexe de la catégorie dérivée $D^b(\mathcal{D}^\dagger_{\mathfrak{X}\mathbb{Q}})$. Nous montrerons ensuite, lorsque E est l'isocristal $\mathcal{O}_{Y/K}$, et Z une sous-variété lisse, ou un diviseur à croisements normaux, que les faisceaux de cohomologie de ce complexe sont des $\mathcal{D}^\dagger_{\mathfrak{X}\mathbb{Q}}$-modules cohérents, et qu'il en est de même pour les faisceaux de cohomologie locale de $\mathcal{O}_{\mathfrak{X}_K}$ à support dans Z ; de plus, tous ces modules ont sur $\mathcal{D}^\dagger_{\mathfrak{X}\mathbb{Q}}$ la même présentation que les $\mathcal{D}_X$-modules analogues en caractéristique 0.

(4.0.1) Nous rappelerons d'abord rapidement ce qu'est un isocristal sur Y, surconvergent le long de Z [4, (2.3.2)]. Pour tout fermé $Z \subset X$, le tube de Z dans $\mathfrak{X}$ est par définition le sous-ensemble $]Z[_{\mathfrak{X}} := \mathrm{sp}^{-1}(Z) \subset \mathfrak{X}_K$ [4, (1.1.2)]. C'est un ouvert pour la topologie rigide de $\mathfrak{X}_K$: s'il existe des sections $f_1, \ldots, f_r \in \Gamma(\mathfrak{X}, \mathcal{O}_{\mathfrak{X}})$, de réductions

$\bar{f}_1, \ldots, \bar{f}_r, \in \Gamma(X, \mathcal{O}_X)$ telles que $Z = V(\bar{f}_1, \ldots, \bar{f}_r)$, alors $]Z[_{\mathfrak{X}}$ est l'ouvert de $\mathfrak{X}_K$ défini par les conditions

$$]Z[_{\mathfrak{X}} = \{ x \in \mathfrak{X}_K \mid \forall i, \ |f_i(x)| < 1 \}.$$

Plus généralement, si Z est l'ensemble des zéros modulo $\mathfrak{m}$ d'une famille de sections $f_1, \ldots, f_r$ de $\mathcal{O}_{\mathfrak{X}}$, on pose, pour tout $\lambda \leq 1$,

$$]Z[_\lambda := \{ x \in \mathfrak{X}_K \mid \forall i, \ |f_i(x)| < \lambda \},$$
$$V_\lambda := \mathfrak{X}_K -]Z[_\lambda.$$

Pour λ assez près de 1, $]Z[_\lambda$ ne dépend pas du choix des f_i, ce qui permet de le définir par recollement lorsqu'il n'existe pas de sections globales définissant Z. De plus, le système inductif des $]Z[_\lambda$ ne dépend pas de la structure de sous-schéma fermé de X mise sur Z. On note $v : Y \hookrightarrow \mathfrak{X}$, $v_\lambda : V_\lambda \hookrightarrow \mathfrak{X}_K$ les inclusions, et on pose, pour tout faisceau abélien F sur $\mathfrak{X}_K$,

$$v^\dagger F := \varinjlim_{\lambda \to 1^-} v_{\lambda *} v_\lambda^* F, \tag{4.0.1.1}$$
$$\underline{\Gamma}^\dagger_{]Z[}(F) := \mathrm{Ker}\,[F \to v^\dagger F]. \tag{4.0.1.2}$$

Les foncteur $v^\dagger$ et $\underline{\Gamma}^\dagger_{]Z[}$ sont exacts, et, si F est un $v^\dagger\mathcal{O}_{\mathfrak{X}_K}$-module, l'homomorphisme canonique $F \to v^\dagger F$ est un isomorphisme [4, (2.1.3)]. Si $v' : Y' \hookrightarrow \mathfrak{X}'$ est l'inclusion d'un ouvert dans un second schéma formel, et $f : \mathfrak{X}' \to \mathfrak{X}$ un morphisme tel que $f(Y') \subset Y$, définissant $f_K : \mathfrak{X}'_K \to \mathfrak{X}_K$, il existe, pour tout faisceau F sur $\mathfrak{X}_K$, un morphisme canonique

$$v^\dagger F \to f_{K*} v'^\dagger f_K^* F. \tag{4.0.1.3}$$

Un isocristal sur Y, surconvergent le long de Z, est défini par la donnée d'un $v^\dagger\mathcal{O}_{\mathfrak{X}_K}$-module $E_{\mathfrak{X}}$ localement libre de rang fini, muni d'une connexion intégrable ∇, telle que, si $v'^\dagger$ est le foncteur analogue à $v^\dagger$ sur $]X[_{\mathfrak{X}^2}$, et

$$p_1, p_2 : (]X[_{\mathfrak{X}^2}, v'^\dagger\mathcal{O}_{\mathfrak{X}^2_K}) \rightrightarrows (\mathfrak{X}_K, v^\dagger\mathcal{O}_{\mathfrak{X}_K})$$

les deux projections du tube de la diagonale dans $\mathfrak{X}_K \times \mathfrak{X}_K$ sur $\mathfrak{X}_K$, il existe un isomorphisme de $v'^\dagger\mathcal{O}_{\mathfrak{X}^2_K}$-modules $\varepsilon : p_2^* E_{\mathfrak{X}} \xrightarrow{\sim} p_1^* E_{\mathfrak{X}}$ induisant sur le complété formel le long de la diagonale l'isomorphisme $\hat{\varepsilon}$ que définit la série de Taylor de ∇, donné en coordonnées locales par :

$$\hat{\varepsilon}(1 \otimes e) = \sum_{\underline{k}} \underline{\partial}^{[\underline{k}]} e \otimes \underline{\tau}^{\underline{k}}.$$

On dit alors que ∇ est *surconvergente*. La surconvergence est une propriété locale sur $\mathfrak{X}$.

Soit E un $v^\dagger\mathcal{O}_{\mathfrak{X}_K}$-module localement libre de rang fini, muni d'une connexion intégrable ∇. D'après [4, (2.2.3)], il existe, pour λ assez près de 1, un $\mathcal{O}_{V_\lambda}$-module localement libre de rang fini E_0, muni d'une connexion intégrable ∇_0, et un isomorphisme horizontal $E \xrightarrow{\sim} v^\dagger E_0$. Supposons que Z soit un diviseur de X. Si U est un ouvert affine de $\mathscr{X}$ tel que $Z = V(f)$, avec $f \in \Gamma(U, \mathcal{O}_{\mathscr{X}})$, alors, pour tout λ, V_λ est défini par la condition $|f(x)| \geq \lambda$, et l'ouvert $\mathrm{sp}^{-1}(U) \cap V_\lambda = U_K \cap V_\lambda$ est un ouvert affinoïde de $\mathscr{X}_K$. Pour que ∇ soit surconvergente, il faut et suffit, d'après [4, (2.2.13)], que, pour tout ouvert affine $U \subset \mathscr{X}$ vérifiant la condition précédente, et sur lequel il existe un système de coordonnées locales, et pour tout $\eta < 1$, il existe $\lambda_\eta < 1$ tel que l'on ait, pour tout $e \in \Gamma(U_K \cap V_\lambda, E_0)$, avec $\lambda_\eta \leq \lambda < 1$,

$$\|\underline{\partial}^{[\underline{k}]} e\| \, \eta^{|\underline{k}|} \to 0 \ \text{ pour } |\underline{k}| \to \infty, \tag{4.0.1.4}$$

en notant $\|\text{-}\|$ une norme de Banach sur $\Gamma(U_K \cap V_\lambda, E_0)$.

4.1. *Image directe par spécialisation d'un isocristal surconvergent*

Il est facile de voir qu'il n'existe pas en général d'action du faisceau $\mathrm{sp}^{-1}(\mathscr{D}^\dagger_{\mathscr{X}\mathbf{Q}})$ sur $\mathcal{O}_{\mathscr{X}_K}$ prolongeant par continuité celle de $\mathrm{sp}^{-1}(\mathscr{D}_{\mathscr{X}\mathbf{Q}}) = \mathscr{D}_{\mathscr{X}_K}$. Pour définir l'action de $\mathscr{D}^\dagger_{\mathscr{X}\mathbf{Q}}$ sur le complexe $\mathbb{R}\mathrm{sp}_* E_{\mathscr{X}}$, où E est un isocristal sur Y surconvergent le long de Z, nous utiliserons un représentant de ce complexe associé à des recouvrements de $\mathscr{X}$ et Y. Commençons par un énoncé général fournissant une résolution à la Čech d'un faisceau de la forme $v^\dagger E$.

(4.1.1) Soit $\mathfrak{X} = (\mathscr{X}_i)_{i \in I}$ un recouvrement ouvert fini de $\mathscr{X}$, et, pour tout i, soit $\mathfrak{Y}_i = (Y_{ij})_{j \in J_i}$ un recouvrement ouvert fini de $Y_i := Y \cap \mathscr{X}_i$. Pour tout $\underline{i} = (i_0, \ldots, i_h)$, on note $\mathscr{X}_{\underline{i}} = \mathscr{X}_{i_0} \cap \ldots \cap \mathscr{X}_{i_h}$, et $u_{\underline{i}}$ l'inclusion de $\mathscr{X}_{\underline{i}K}$ dans $\mathscr{X}_K$; on recouvre $Y_{\underline{i}} := Y \cap \mathscr{X}_{\underline{i}}$ par le recouvrement $\mathfrak{Y}_{\underline{i}} := (Y_{\underline{i}\underline{j}})_{\underline{j}}$, où $\underline{j} = (j_0, \ldots, j_h) \in J_{i_0} \times \ldots \times J_{i_h}$, et

$$Y_{\underline{i}\underline{j}} := Y_{i_0 j_0} \cap \ldots \cap Y_{i_h j_h}.$$

L'intersection de $k+1$ ouverts du recouvrement $\mathfrak{Y}_{\underline{i}}$ sera notée $Y_{\underline{i}\underline{\underline{j}}}$, avec

$$\underline{\underline{j}} = \begin{pmatrix} \underline{j}_0 \\ \ldots \\ \underline{j}_k \end{pmatrix} = \begin{pmatrix} j_{00} & \cdots & j_{0h} \\ \ldots & \ldots & \ldots \\ j_{k0} & \cdots & j_{kh} \end{pmatrix},$$

$$Y_{\underline{i}\underline{\underline{j}}} = \bigcap_{\alpha=0}^{k} \bigcap_{\beta=0}^{h} Y_{i_\beta j_{\alpha\beta}}.$$

On notera $v_{\underline{i}}$ (resp. $v_{\underline{i}\underline{\underline{j}}}$) l'inclusion de $Y_{\underline{i}}$ (resp. $Y_{\underline{i}\underline{\underline{j}}}$) dans $\mathscr{X}_{\underline{i}}$, et $v^\dagger_{\underline{i}}$, $v^\dagger_{\underline{i}\underline{\underline{j}}}$ les analogues du foncteur (4.0.1.1) pour les ouverts $Y_{\underline{i}}$, $Y_{\underline{i}\underline{\underline{j}}}$ de $\mathscr{X}_{\underline{i}}$.

Sur $\mathscr{X}_{\underline{i}K}$, on dispose du complexe de Čech pour les foncteurs $v^\dagger_{\underline{i}\underline{\underline{j}}}$ défini en [4, (2.1.8)]. Appliqué à $E_{\underline{i}} := u^*_{\underline{i}}(E)$, il fournit un complexe

$$\check{C}^{\dagger\cdot}(\mathfrak{Y}_{\underline{i}}, E_{\underline{i}}) \ : \ \prod_{\underline{j}} v^\dagger_{\underline{i}\underline{j}} E_{\underline{i}} \longrightarrow \prod_{\underline{j}_0, \underline{j}_1} v^\dagger_{\underline{i}\underline{j}_0\underline{j}_1} E_{\underline{i}} \longrightarrow \ \ldots \ \longrightarrow \prod_{\underline{\underline{j}}} v^\dagger_{\underline{i}\underline{\underline{j}}} E_{\underline{i}} \longrightarrow \ \ldots,$$

qui est une résolution de $v_{\underline{i}}^{\dagger}E_{\underline{i}}$.

Si $\underline{i} = (i_0, \ldots, i_k)$, et $\underline{i}_{\hat{\alpha}} := (i_0, \ldots, \hat{\imath}_{\alpha}, \ldots, i_k)$, il existe un morphisme de complexes

$$\rho^{\cdot}_{\underline{i}\,\underline{i}_{\hat{\alpha}}} : u_{\underline{i}_{\hat{\alpha}}*}\check{C}^{\dagger\cdot}(\mathfrak{Y}_{\underline{i}_{\hat{\alpha}}}, E_{\underline{i}_{\hat{\alpha}}}) \longrightarrow u_{\underline{i}*}\check{C}^{\dagger\cdot}(\mathfrak{Y}_{\underline{i}}, E_{\underline{i}})$$

prolongeant le morphisme $u_{\underline{i}_{\hat{\alpha}}*}E_{\underline{i}_{\hat{\alpha}}} \to u_{\underline{i}*}E_{\underline{i}}$, défini comme suit. Pour tout $\underline{j} = (j_0, \ldots, j_k)$, notons $\underline{j}_{\hat{\alpha}} := (j_{0\hat{\alpha}}, \ldots, j_{k\hat{\alpha}})$. Pour toute section s de $u_{\underline{i}_{\hat{\alpha}}*}\check{C}^{\dagger\cdot}(\mathfrak{Y}_{\underline{i}_{\hat{\alpha}}}, E_{\underline{i}_{\hat{\alpha}}})$, on pose

$$\rho^{\cdot}_{\underline{i}\,\underline{i}_{\hat{\alpha}}}(s)_{\underline{i}\,\underline{j}} := \rho_{\underline{j}\,\underline{j}_{\hat{\alpha}}}(s_{\underline{i}_{\hat{\alpha}}\,\underline{j}_{\hat{\alpha}}}),$$

où $\rho_{\underline{j}\,\underline{j}_{\hat{\alpha}}}$ est l'image par $u_{\underline{i}_{\hat{\alpha}}*}$ du morphisme de fonctorialité (4.0.1.3) défini par l'immersion $\mathcal{X}_{\underline{i}} \hookrightarrow \mathcal{X}_{\underline{i}_{\hat{\alpha}}}$, qui envoie $Y_{\underline{i}\,\underline{j}}$ dans $Y_{\underline{i}_{\hat{\alpha}}\,\underline{j}_{\hat{\alpha}}}$. Pour $\underline{i}$ variable, ces morphismes donnent naissance à un bicomplexe qui sera noté $\check{C}^{\dagger\cdot\cdot}(\mathfrak{X}, (\mathfrak{Y}_i)_i, E)$; en bidegré (h, k), il est donc défini par

$$\check{C}^{\dagger hk}(\mathfrak{X}, (\mathfrak{Y}_i)_i, E) = \prod_{\underline{i}=(i_0,\ldots,i_h)} u_{\underline{i}*}\Big(\prod_{\underline{j}=(j_0,\ldots,j_k)} v^{\dagger}_{\underline{i}\,\underline{j}}E_{\underline{i}}\Big).$$

(4.1.2) **Lemme.** *Soient $\mathcal{X}'$ un ouvert de $\mathcal{X}$, Y' un ouvert de $\mathcal{X}'$ tel que $Y' \subset Y$, soient $u : \mathcal{X}'_K \hookrightarrow \mathcal{X}_K$, $v' : Y' \hookrightarrow \mathcal{X}'$ les immersions correspondantes, et notons par l'exposant $\dagger$ les foncteurs* (4.0.1.1) *relatifs aux immersions ouvertes v et v'. Alors :*

(i) *Pour tout $\mathcal{O}_{\mathcal{X}_K}$-module F, l'homomorphisme canonique*

$$u_*v'^{\dagger}F \longrightarrow v^{\dagger}u_*v'^{\dagger}F$$

est un isomorphisme ;

(ii) *Si $Y' = Y \cap \mathcal{X}'$, l'homomorphisme canonique*

$$v^{\dagger}u_*F \longrightarrow v^{\dagger}u_*v'^{\dagger}F$$

est également un isomorphisme.

L'assertion (i) résulte de ce que $u_*v'^{\dagger}F$ est un $v^{\dagger}\mathcal{O}_{\mathcal{X}_K}$-module, par l'intermédiaire de l'homomorphisme canonique $v^{\dagger}\mathcal{O}_{\mathcal{X}_K} \to u_*v'^{\dagger}\mathcal{O}_{\mathcal{X}'_K}$. Si $Y' = Y \cap \mathcal{X}'$, on a, pour tout ouvert affinoïde W de $\mathcal{X}_K$,

$$\begin{aligned}\Gamma(W, v^{\dagger}\, u_*\, v'^{\dagger}\, F) &= \varinjlim_{\lambda,\lambda'} \Gamma(W \cap V_{\lambda} \cap \mathcal{X}'_K \cap V'_{\lambda'}, F)\\ &= \varinjlim_{\lambda} \Gamma(W \cap V_{\lambda} \cap \mathcal{X}'_K, F) = \Gamma(W, v^{\dagger}\, u_*\, F),\end{aligned}$$

car les ouverts V'_{λ} définissant le foncteur $v'^{\dagger}$ sur $\mathcal{X}'_K$ sont égaux aux $V_{\lambda} \cap \mathcal{X}'_K$.

(4.1.3) **Proposition.** *Soit E un faisceau abélien sur $\mathcal{X}_K$. Avec les hypothèses et les notations de* (4.1.1), *le complexe simple associé au bicomplexe $\check{C}^{\dagger\cdot\cdot}(\mathfrak{X}, (\mathfrak{Y}_i)_i, E)$ est une résolution de $v^{\dagger}E$.*

Pour tout k, il existe un morphisme canonique $v^\dagger E \to \check{C}^{\dagger 0k}(\mathfrak{X}, (\mathfrak{Y}_i)_i, E)$, défini par la famille des morphismes $v^\dagger E \to u_{\underline{i}*}(v^\dagger_{\underline{i}\underline{j}} E_{\underline{i}})$. Pour k variable, ces morphismes forment un morphisme de complexes du complexe $v^\dagger E \to v^\dagger E \to v^\dagger E \to \ldots$ dont le terme général est $v^\dagger E$, et la différentielle 0 en degré pair, et $\mathrm{Id}_{v^\dagger E}$ en degré impair, dans le complexe $\check{C}^{\dagger 0\cdot}(\mathfrak{X}, (\mathfrak{Y}_i)_i, E)$. Il suffit donc de prouver que, pour tout k, le complexe

$$\check{C}^{\dagger 0k}(\mathfrak{X}, (\mathfrak{Y}_i)_i, E) \longrightarrow \check{C}^{\dagger 1k}(\mathfrak{X}, (\mathfrak{Y}_i)_i, E) \longrightarrow \ldots \qquad (4.1.3.1)$$

est une résolution de $v^\dagger E$.

Comme les $\mathcal{X}_{iK}$ forment un recouvrement admissible de $\mathcal{X}_K$, il suffit de prouver cette assertion après localisation au-dessus des $\mathcal{X}_{iK}$; on a alors $u_i^*(v^\dagger E) \simeq v_i^\dagger E_i$. Sur $\mathcal{X}_{iK}$, le foncteur $\check{C}^{\dagger\cdot}(\mathfrak{Y}_i, -)$ appliqué au complexe $u_i^*(\check{C}^{\dagger\cdot k}(\mathfrak{X}, (\mathfrak{Y}_i)_i, E))$ donne un bicomplexe $\check{C}^{\dagger\cdot}(\mathfrak{Y}_i, u_i^*(\check{C}^{\dagger\cdot k}(\mathfrak{X}, (\mathfrak{Y}_i)_i, E)))$. Pour tout faisceau F sur $\mathcal{X}_{iK}$, le complexe $\check{C}^{\dagger\cdot}(\mathfrak{Y}_i, F)$ est une résolution de $v_i^\dagger F$ d'après [4, (2.1.8)] ; le complexe simple associé au bicomplexe $\check{C}^{\dagger\cdot}(\mathfrak{Y}_i, u_i^*(\check{C}^{\dagger\cdot k}(\mathfrak{X}, (\mathfrak{Y}_i)_i, E)))$ est donc quasi-isomorphe à $v_i^\dagger u_i^*(\check{C}^{\dagger\cdot k}(\mathfrak{X}, (\mathfrak{Y}_i)_i, E))$. De même, le complexe $\check{C}^{\dagger\cdot}(\mathfrak{Y}_i, v_i^\dagger E_i)$ est une résolution de $v_i^\dagger v_i^\dagger E_i$. Comme, pour tout $v_i^\dagger \mathcal{O}_{\mathcal{X}_{iK}}$-module F, on a $F \xrightarrow{\sim} v_i^\dagger F$ [4, (2.1.3)], on voit qu'il suffit de prouver les deux assertions suivantes :

a) pour tout h, on a

$$u_i^* \check{C}^{\dagger hk}(\mathfrak{X}, (\mathfrak{Y}_i)_i, E) \xrightarrow{\sim} v_i^\dagger u_i^* \check{C}^{\dagger hk}(\mathfrak{X}, (\mathfrak{Y}_i)_i, E) ;$$

b) pour tout $\underline{j} = (j_0, \ldots, j_k)$, le complexe

$$v^\dagger_{i\underline{j}}\, u_i^* \check{C}^{\dagger 0k}(\mathfrak{X}, (\mathfrak{Y}_i)_i, E) \longrightarrow v^\dagger_{i\underline{j}}\, u_i^* \check{C}^{\dagger 1k}(\mathfrak{X}, (\mathfrak{Y}_i)_i, E) \longrightarrow \ldots$$

est une résolution de $v^\dagger_{i\underline{j}}\, v_i^\dagger E_i$.

Pour tout $\underline{i} = (i_0, \ldots, i_h)$, soit $\underline{i}' = (i, i_0, \ldots, i_h)$. Notons $u'_{\underline{i}}$ l'inclusion de $\mathcal{X}_{\underline{i}'K} = \mathcal{X}_{iK} \cap \mathcal{X}_{\underline{i}K}$ dans $\mathcal{X}_{iK}$, u'_i celle de $\mathcal{X}_{\underline{i}'K}$ dans $\mathcal{X}_{\underline{i}K}$, $Y'_{\underline{i}\underline{j}} = \mathcal{X}_{iK} \cap Y_{\underline{i}\underline{j}}$, et $v'_{\underline{i}\underline{j}} : Y'_{\underline{i}\underline{j}} \hookrightarrow \mathcal{X}_{\underline{i}'K}$. L'assertion a) résulte des isomorphismes

$$\begin{aligned} v_i^\dagger u_i^* \check{C}^{\dagger hk}(\mathfrak{X}, (\mathfrak{Y}_i)_i, E) &= v_i^\dagger u_i^*\Big(\prod_{\underline{i}=(i_0,\ldots,i_h)} u_{\underline{i}*}\Big(\prod_{\underline{j}=(j_0,\ldots,j_k)} v^\dagger_{\underline{i}\underline{j}} E_{\underline{i}}\Big)\Big) \\ &\simeq \prod_{\underline{i}=(i_0,\ldots,i_h)} \prod_{\underline{j}=(j_0,\ldots,j_k)} v_i^\dagger u'_{\underline{i}*} u_i'^{*} v^\dagger_{\underline{i}\underline{j}} E_{\underline{i}} \\ &\simeq \prod_{\underline{i}=(i_0,\ldots,i_h)} \prod_{\underline{j}=(j_0,\ldots,j_k)} v_i^\dagger u'_{\underline{i}*} v'^\dagger_{\underline{i}\underline{j}} E_{\underline{i}'} \\ &\simeq \prod_{\underline{i}=(i_0,\ldots,i_h)} \prod_{\underline{j}=(j_0,\ldots,j_k)} u'_{\underline{i}*} v'^\dagger_{\underline{i}\underline{j}} E_{\underline{i}'} \\ &\simeq u_i^* \check{C}^{\dagger hk}(\mathfrak{X}, (\mathfrak{Y}_i)_i, E), \end{aligned}$$

provenant des relations évidentes $u_i^* u_{\underline{i}*} \simeq u'_{\underline{i}*} u_i'^*$ et $u_i'^* v_{i\underline{i}}^\dagger \simeq v_{i\underline{i}}'^\dagger u_i'^*$, et de (4.1.2) (i) appliqué aux ouverts $\mathscr{X}_{i'K} \subset \mathscr{X}_{iK}$ et $Y'_{i\underline{i}} \subset Y_i$.

Pour prouver l'assertion b), on construit une homotopie sur le complexe

$$0 \longrightarrow v_{i\underline{i}}^\dagger v_i^\dagger E_i \longrightarrow v_{i\underline{i}}^\dagger u_i^* \check{C}^{\dagger 0 k}(\mathfrak{X}, (\mathfrak{Y}_i)_i, E) \longrightarrow v_{i\underline{i}}^\dagger u_i^* \check{C}^{\dagger 1 k}(\mathfrak{X}, (\mathfrak{Y}_i)_i, E) \longrightarrow \ldots.$$

On définit en effet un homomorphisme

$$\kappa : v_{i\underline{i}}^\dagger u_i^* \Big(\prod_{\underline{i}=(i_0,\ldots,i_{h+1})} u_{\underline{i}*}\Big(\prod_{\underline{\underline{j}}=(\underline{j}_0,\ldots,\underline{j}_k)} v_{\underline{i}\underline{\underline{j}}}^\dagger E_{\underline{i}}\Big)\Big) \longrightarrow v_{i\underline{i}}^\dagger u_i^* \Big(\prod_{\underline{i}=(i_0,\ldots,i_h)} u_{\underline{i}*}\Big(\prod_{\underline{\underline{j}}=(\underline{j}_0,\ldots,\underline{j}_k)} v_{\underline{i}\underline{\underline{j}}}^\dagger E_{\underline{i}}\Big)\Big)$$

de la manière suivante. Pour tout $\underline{\underline{j}} = (\underline{j}_0, \ldots, \underline{j}_k)$, avec $\underline{j}_\alpha = (j_{\alpha 0}, \ldots, j_{\alpha h})$, on pose

$$\underline{\underline{j}}' = \begin{pmatrix} j_0 & j_{00} & \cdots & j_{0h} \\ \cdots & \cdots & \cdots & \cdots \\ j_k & j_{k0} & \cdots & j_{kh} \end{pmatrix}.$$

On observe alors que, quels que soient $\underline{i}$, $\underline{\underline{j}}$, l'homomorphisme

$$v_{i\underline{i}}^\dagger u_i^* u_{\underline{i}*} v_{\underline{i}\underline{\underline{j}}}^\dagger E_{\underline{i}} \longrightarrow v_{i\underline{i}}^\dagger u_i^* u_{\underline{i}'*} v_{\underline{i}'\underline{\underline{j}}'}^\dagger E_{\underline{i}'}$$

est un isomorphisme : en effet, en gardant les notations précédentes, on a comme plus haut

$$v_{i\underline{i}}^\dagger u_i^* u_{\underline{i}*} v_{\underline{i}\underline{\underline{j}}}^\dagger E_{\underline{i}} \simeq v_{i\underline{i}}^\dagger u'_{\underline{i}*} u_i'^* v_{\underline{i}\underline{\underline{j}}}^\dagger E_{\underline{i}} \simeq v_{i\underline{i}}^\dagger u'_{\underline{i}*} v_{\underline{i}\underline{\underline{j}}}'^\dagger E_{\underline{i}'},$$

et d'autre part

$$v_{i\underline{i}}^\dagger u_i^* u_{\underline{i}'*} v_{\underline{i}'\underline{\underline{j}}'}^\dagger E_{\underline{i}'} \simeq v_{i\underline{i}}^\dagger u'_{\underline{i}*} v_{\underline{i}'\underline{\underline{j}}'}^\dagger E_{\underline{i}'}.$$

Comme $Y_{\underline{i}'\underline{\underline{j}}'} = Y'_{i\underline{i}} \cap Y'_{\underline{i}\underline{\underline{j}}}$, on a $v_{\underline{i}'\underline{\underline{j}}'}^\dagger \simeq v_{i\underline{i}}'^\dagger v_{\underline{i}\underline{\underline{j}}}'^\dagger$ d'après [4, (2.1.7)]. Le lemme (4.1.2) (ii) fournit donc les isomorphismes

$$v_{i\underline{i}}^\dagger u'_{\underline{i}*} v_{\underline{i}'\underline{\underline{j}}'}^\dagger E_{\underline{i}'} \simeq v_{i\underline{i}}^\dagger u'_{\underline{i}*} v_{i\underline{i}}'^\dagger v_{\underline{i}\underline{\underline{j}}}'^\dagger E_{\underline{i}'} \simeq v_{i\underline{i}}^\dagger u'_{\underline{i}*} v_{\underline{i}\underline{\underline{j}}}'^\dagger E_{\underline{i}'},$$

d'où l'assertion. Cet isomorphisme permet alors de définir κ en posant, pour $s \in v_{i\underline{i}}^\dagger u_i^* \check{C}^{\dagger h+1\, k}(\mathfrak{X}, (\mathfrak{Y}_i)_i, E)$,

$$\kappa(s)_{\underline{i}\underline{\underline{j}}} := s_{\underline{i}'\underline{\underline{j}}'}.$$

On vérifie immédiatement que κ est bien une homotopie du complexe considéré.

(4.1.4) **Lemme.** *Soient $\mathscr{X}$ un $\mathcal{V}$-schéma formel séparé, de réduction X sur k, $\mathscr{X}'$ un ouvert affine de $\mathscr{X}$, $u : \mathscr{X}'_K \hookrightarrow \mathscr{X}_K$ l'immersion ouverte correspondante, $f \in \Gamma(\mathscr{X}', \mathcal{O}_{\mathscr{X}})$, $Y = D(f) \subset \mathscr{X}'$, $v^\dagger$ le foncteur (4.0.1.1) sur $\mathscr{X}'_K$ associé à l'ouvert Y de $\mathscr{X}'$.*

(i) *Si E est un $v^\dagger\mathcal{O}_{\mathscr{X}'_K}$-module de présentation finie, alors u_*E est acyclique pour le foncteur sp_*, où $\mathrm{sp} : \mathscr{X}_K \to \mathscr{X}$ est le morphisme de spécialisation.*

(ii) *Si de plus E est muni d'une connexion intégrable, surconvergente le long de $X - Y$, alors $\mathrm{sp}_* u_* E$ est muni d'une structure canonique de $\mathscr{D}_{\mathscr{X}\mathbb{Q}}^\dagger$-module, foncto-*

rielle en E. Si $\mathscr{X}'_1$ est un ouvert affine de $\mathscr{X}'$, avec $u_1 : \mathscr{X}'_{1K} \hookrightarrow \mathscr{X}_K$, et si $Y_1 \subset \mathscr{X}'_1 \cap Y$ est un ouvert de $\mathscr{X}'_1$, de la forme $Y_1 = D(f_1)$, avec $f_1 \in \Gamma(\mathscr{X}'_1, \mathcal{O}_{\mathscr{X}})$, définissant un foncteur $v_1^\dagger$ sur $\mathscr{X}'_{1K}$, l'homomorphisme canonique

$$\mathrm{sp}_*(u_*E) \longrightarrow \mathrm{sp}_*(u_{1*}v_1^\dagger u_1^* E)$$

est $\mathscr{D}^\dagger_{\mathscr{X}\mathbf{Q}}$-linéaire.

Comme E est de présentation finie, il existe $\lambda < 1$, et un $\mathcal{O}_{\mathscr{X}_K}$-module cohérent E_0 sur V_λ tels que $E \simeq v^\dagger E_0$ [4, (2.1.10)]. Pour tout ouvert affine U de $\mathscr{X}$, et tout i, on a alors

$$H^i(\mathrm{sp}^{-1}(U), u_*E) = H^i((U \cap \mathscr{X}')_K, v^\dagger E_0).$$

Puisque $\mathscr{X}$ est séparé, $U \cap \mathscr{X}'$ est affine, et $(U \cap \mathscr{X}')_K$ un ouvert affinoïde de $\mathscr{X}'_K$. Comme Y est un ouvert de $\mathscr{X}'$ de la forme $D(f)$, il résulte de [4, (3.1.7)] que les $H^i(\mathrm{sp}^{-1}(U), u_*E)$ sont nuls pour $i \geq 1$, d'où l'assertion (i).

Supposons maintenant que E soit muni d'une connexion intégrable et surconvergente ∇. Il existe alors $\lambda_0 < 1$, et un $\mathcal{O}_{\mathscr{X}_K}$-module cohérent E_0 sur V_{λ_0}, muni d'une connexion intégrable ∇_0, tels que $(E, \nabla) \simeq v^\dagger(E_0, \nabla_0)$. Soient U un ouvert affine de $\mathscr{X}$ sur lequel il existe des coordonnées locales, et $P = \sum_{\underline{k}} a_{\underline{k}} \partial^{[\underline{k}]} \in \Gamma(U, \mathscr{D}^\dagger_{\mathscr{X}\mathbf{Q}})$. Soient $c > 0$, $\eta < 1$ tels que $\| a_{\underline{k}} \| < c\,\eta^{|\underline{k}|}$. D'après (4.0.1.4), la surconvergence de ∇ entraîne l'existence de λ_1, avec $\lambda_0 \leq \lambda_1 < 1$, tel que, pour tout $e \in \Gamma((U \cap \mathscr{X}')_K \cap V_\lambda, E_0)$, avec $\lambda_1 \leq \lambda < 1$, on ait $\| \underline{\partial}^{[\underline{k}]} e \| \, \eta^{|\underline{k}|} \to 0$. On peut donc définir $Pe \in \Gamma((U \cap \mathscr{X}')_K \cap V_\lambda, E_0)$ comme somme de la série convergente $\sum_{\underline{k}} a_{\underline{k}} (\underline{\partial}^{[\underline{k}]} e)$. Puisque $(U \cap \mathscr{X}')_K$ est affinoïde, on a

$$\Gamma((U \cap \mathscr{X}')_K, E) \simeq \varinjlim_\lambda \Gamma((U \cap \mathscr{X}')_K \cap V_\lambda, E_0),$$

et l'action de P sur $\Gamma((U \cap \mathscr{X}')_K, E) = \Gamma(U, \mathrm{sp}_* u_* E)$ en résulte par passage à la limite inductive. Il est facile de voir qu'elle ne dépend pas des choix effectués, et, pour P variable, définit une structure de $\Gamma(U, \mathscr{D}^\dagger_{\mathscr{X}\mathbf{Q}})$-module sur $\Gamma(U, \mathrm{sp}_* u_* E)$. On obtient ainsi une structure de $\mathscr{D}^\dagger_{\mathscr{X}\mathbf{Q}}$-module sur $\mathrm{sp}_* u_* E$.

De plus, si (E', ∇') est un second $v^\dagger \mathcal{O}_{\mathscr{X}_K}$-module localement libre de rang fini, et $\varphi : E \to E'$ un homomorphisme horizontal, il existe, pour λ_0 assez près de 1, un $\mathcal{O}_{\mathscr{X}_K}$-module cohérent E'_0 sur V_{λ_0}, muni d'une connexion intégrable ∇'_0, tels que $(E', \nabla') \simeq v^\dagger(E'_0, \nabla'_0)$, et un homomorphisme horizontal $\varphi_0 : E_0 \to E'_0$ tel que $\varphi = v^\dagger(\varphi_0)$. Pour tout $\lambda \geq \lambda_0$, l'application $\Gamma((U \cap \mathscr{X}')_K \cap V_\lambda, E_0) \to \Gamma((U \cap \mathscr{X}')_K \cap V_\lambda, E'_0)$ est continue, donc commute à l'action de $P \in \Gamma(U, \mathscr{D}^\dagger_{\mathscr{X}\mathbf{Q}})$ qu'on vient de définir pour λ assez près de 1. La première partie de l'assertion (ii) en résulte ; la deuxième partie vient de ce que, pour tout λ, l'homomorphisme de restriction

$$\Gamma((U \cap \mathscr{X}')_K \cap V_\lambda, E_0) \longrightarrow \Gamma((U \cap \mathscr{X}'_1)_K \cap V_{1\lambda}, E_0),$$

où $V_{1\lambda} = \mathscr{X}'_{1K} -]\mathscr{X}'_1 - Y_1[_\lambda$, est continu, donc compatible à l'action de $P \in \Gamma(U, \mathscr{D}^\dagger_{\mathscr{X}\mathbf{Q}})$.

Notant toujours $v : Y \hookrightarrow X$ l'inclusion d'un ouvert dans X, nous poserons, pour tout faisceau abélien E sur $\mathscr{X}_K$,

$$\mathbb{R}v_*^\dagger E := \mathbb{R}\mathrm{sp}_*(v^\dagger E), \qquad R^i v_*^\dagger E := R^i\mathrm{sp}_*(v^\dagger E).$$

Si E est un $v^\dagger\mathcal{O}_{\mathscr{X}_K}$-module, on a donc simplement $\mathbb{R}v_*^\dagger E = \mathbb{R}\mathrm{sp}_* E$. Soit d'autre part T un fermé de X ; nous noterons

$$\mathbb{R}\underline{\Gamma}_T^\dagger E := \mathbb{R}\mathrm{sp}_*(\underline{\Gamma}_{]T[}^\dagger E), \qquad \mathscr{H}_T^{\dagger i}(E) := R^i\mathrm{sp}_*(\underline{\Gamma}_{]T[}^\dagger E).$$

Comme le suggèrent ces notations, les foncteurs $\mathbb{R}v_*^\dagger$ et $\mathbb{R}\underline{\Gamma}_T^\dagger$ jouent le rôle d'une image directe de Y dans X, et d'une cohomologie locale à support dans T, pour les "fonctions à singularités surconvergentes".

(4.1.5) **Proposition.** *Soient $\mathscr{X}$ un $\mathcal{V}$-schéma formel séparé, de type fini, de réduction X sur k, Y un ouvert de X, de complémentaire Z, T un fermé de X, E un isocristal sur Y, surconvergent le long de Z. Il existe sur le complexe $\mathbb{R}v_*^\dagger E_{\mathscr{X}}$ (resp. $\mathbb{R}\underline{\Gamma}_T^\dagger E_{\mathscr{X}}$) une structure naturelle de complexe de $\mathscr{D}_{\mathscr{X}\mathbb{Q}}^\dagger$-modules, qui en fait un complexe de $D^b(\mathscr{D}_{\mathscr{X}\mathbb{Q}}^\dagger)$ fonctoriel en E. Si on note $Y' = Y - (T \cap Y)$, et $v' : Y' \hookrightarrow X$ l'immersion ouverte correspondante, la suite exacte de faisceaux sur $\mathscr{X}_K$*

$$0 \longrightarrow \underline{\Gamma}_{]T[}^\dagger E \longrightarrow E \longrightarrow v'^\dagger E \longrightarrow 0$$

définit un triangle distingué de $D^b(\mathscr{D}_{\mathscr{X}\mathbb{Q}}^\dagger)$, fonctoriel en E,

$$\begin{array}{ccc} & \mathbb{R}v_*'^\dagger E & \\ \swarrow & & \nwarrow \\ \mathbb{R}\underline{\Gamma}_T^\dagger E_{\mathscr{X}} & \longrightarrow & \mathbb{R}v_*^\dagger E_{\mathscr{X}}. \end{array}$$

Soit $\mathfrak{X} = (\mathscr{X}_i)_{i\in I}$ un recouvrement affine fini de $\mathscr{X}$, et, pour tout i, soit $\mathfrak{Y}_i = (Y_{ij})_{j\in J_i}$ un recouvrement fini de $Y_i := Y \cap \mathscr{X}_i$ par des ouverts de la forme $D(f_{ij})$, où $f_{ij} \in \Gamma(\mathscr{X}_i, \mathcal{O}_{\mathscr{X}})$. Notons $v : Y \hookrightarrow \mathscr{X}$; d'après (4.1.3), le complexe simple $\check{C}^{\dagger\cdot}(\mathfrak{X}, (\mathfrak{Y}_i)_i, E_{\mathscr{X}})$ associé au bicomplexe $\check{C}^{\dagger\cdot\cdot}(\mathfrak{X}, (\mathfrak{Y}_i)_i, E_{\mathscr{X}})$ est une résolution de $v^\dagger E_{\mathscr{X}}$, qui est égal à $E_{\mathscr{X}}$ puisque $E_{\mathscr{X}}$ est un $v^\dagger\mathcal{O}_{\mathscr{X}_K}$-module. D'autre part, avec les notations de (4.1.1), les ouverts $\mathscr{X}_{\underline{i}}$ sont affines, et les $Y_{\underline{i}\underline{j}}$ de la forme $D(f_{\underline{i}\underline{j}})$, avec $f_{\underline{i}\underline{j}} = \prod_{\alpha,\beta} f_{i_\beta j_{\alpha\beta}}$. Le lemme (4.1.4) entraîne alors que le complexe $\mathrm{sp}_*\check{C}^{\dagger\cdot}(\mathfrak{X}, (\mathfrak{Y}_i)_i, E_{\mathscr{X}})$ est un représentant de $\mathbb{R}\mathrm{sp}_* E_{\mathscr{X}}$. De plus, chacun des faisceaux $\mathrm{sp}_*(u_{\underline{i}*} v_{\underline{i}\underline{j}}^\dagger u_{\underline{i}}^* E_{\mathscr{X}})$ est muni d'une structure canonique de $\mathscr{D}_{\mathscr{X}\mathbb{Q}}^\dagger$-module, fonctorielle en E, et le complexe $\mathrm{sp}_*\check{C}^{\dagger\cdot}(\mathfrak{X}, (\mathfrak{Y}_i)_i, E_{\mathscr{X}})$ est alors un complexe de $\mathscr{D}_{\mathscr{X}\mathbb{Q}}^\dagger$-modules. Enfin, le passage à des recouvrements $\mathfrak{X}'$, $\mathfrak{Y}_{i'}'$ plus fins induit un morphisme de complexes

$$\mathrm{sp}_*\check{C}^{\dagger\cdot}(\mathfrak{X}, (\mathfrak{Y}_i)_i, E_{\mathscr{X}}) \to \mathrm{sp}_*\check{C}^{\dagger\cdot}(\mathfrak{X}', (\mathfrak{Y}_{i'}')_{i'}, E_{\mathscr{X}})$$

qui est compatible à l'action de $\mathscr{D}_{\mathscr{X}\mathbb{Q}}^\dagger$ d'après (4.1.4) (ii). En associant à E la classe dans $D(\mathscr{D}^\dagger_{\mathfrak{X}\mathbb{Q}})$ du complexe $\mathrm{sp}_*\check{C}^{\dagger\cdot}(\mathfrak{X}, (\mathfrak{Y}_i)_i, E_{\mathfrak{X}})$, on obtient un foncteur qui redonne le foncteur $\mathbb{R}v^\dagger_*$ par restriction des scalaires de $\mathscr{D}^\dagger_{\mathfrak{X}\mathbb{Q}}$ à $\mathcal{O}_{\mathfrak{X}\mathbb{Q}}$. Qu'il soit à valeurs dans $D^b(\mathscr{D}^\dagger_{\mathfrak{X}\mathbb{Q}})$ résulte de la finitude des recouvrements employés, et de ce que l'on peut remplacer les complexes de cochaînes par des complexes de cochaînes alternées.

Pour traiter le cas du complexe $\mathbb{R}\underline{\Gamma}^\dagger_T E_{\mathfrak{X}}$, on peut supposer que, pour tout i, le recouvrement $\mathfrak{Y}_i$ est choisi de telle sorte qu'il existe une sous-famille $\mathfrak{Y}'_i$ de $\mathfrak{Y}_i$ formant un recouvrement de $Y' \cap X_i$. Le faisceau $\underline{\Gamma}^\dagger_{]T[} E_{\mathfrak{X}}$ admet pour résolution le complexe $E_{\mathfrak{X}} \to v'^\dagger E_{\mathfrak{X}}$, qui admet lui-même pour résolution le complexe simple associé au bicomplexe

$$\check{C}^{\dagger\cdot}(\mathfrak{X}, (\mathfrak{Y}_i)_i, E_{\mathfrak{X}}) \longrightarrow \check{C}^{\dagger\cdot}(\mathfrak{X}, (\mathfrak{Y}'_i)_i, E_{\mathfrak{X}}).$$

Par suite, le complexe $\mathbb{R}\underline{\Gamma}^\dagger_T E_{\mathfrak{X}}$ admet pour représentant le complexe

$$\mathrm{sp}_*\check{C}^{\dagger\cdot}(\mathfrak{X}, (\mathfrak{Y}_i)_i, E_{\mathfrak{X}}) \to \mathrm{sp}_*\check{C}^{\dagger\cdot}(\mathfrak{X}, (\mathfrak{Y}'_i)_i, E_{\mathfrak{X}}).$$

Ce dernier est muni d'une structure naturelle de complexe de $\mathscr{D}^\dagger_{\mathfrak{X}\mathbb{Q}}$-modules d'après ce qui précède, et on obtient encore ainsi un complexe de $D^b(\mathscr{D}^\dagger_{\mathfrak{X}\mathbb{Q}})$ indépendant du choix des recouvrements et fonctoriel en E. Comme c'est le cône du morphisme de complexes de $\mathscr{D}^\dagger_{\mathfrak{X}\mathbb{Q}}$-modules représentant $\mathbb{R}v^\dagger_* E_{\mathfrak{X}} \to \mathbb{R}v'^\dagger_* E_{\mathfrak{X}}$, l'énoncé en découle.

(4.1.6) **Proposition.** *Sous les hypothèses de* (4.1.5), *il existe un isomorphisme canonique*

$$\mathbb{R}v^\dagger_*(E_{\mathfrak{X}} \otimes \Omega^\cdot_{\mathfrak{X}_K}) \simeq \mathbb{R}\mathscr{H}om_{\mathscr{D}^\dagger_{\mathfrak{X}\mathbb{Q}}}(\mathcal{O}_{\mathfrak{X}\mathbb{Q}}, \mathbb{R}v^\dagger_* E_{\mathfrak{X}})$$

(resp.
$$\mathbb{R}\underline{\Gamma}^\dagger_T(E_{\mathfrak{X}} \otimes \Omega^\cdot_{\mathfrak{X}_K}) \simeq \mathbb{R}\mathscr{H}om_{\mathscr{D}^\dagger_{\mathfrak{X}\mathbb{Q}}}(\mathcal{O}_{\mathfrak{X}\mathbb{Q}}, \mathbb{R}\underline{\Gamma}^\dagger_T E_{\mathfrak{X}})).$$

Choisissons des recouvrements $\mathfrak{X} = (\mathscr{X}_i)_{i \in I}$ de $\mathscr{X}$ (resp. (resp. $\mathfrak{Y}_i = (Y_{ij})_{j \in J_i}$ de $Y_i := Y \cap \mathscr{X}_i$), comme dans la démonstration de (4.1.5). Soit E un isocristal sur Y, surconvergent le long de Z. D'après le lemme (4.1.4) (i), le complexe simple $\check{C}^{\dagger\cdot}(\mathfrak{X}, (\mathfrak{Y}_i)_i, E_{\mathfrak{X}} \otimes \Omega^\cdot_{\mathfrak{X}_K})$ associé au tricomplexe $\check{C}^{\dagger\cdot\cdot}(\mathfrak{X}, (\mathfrak{Y}_i)_i, E_{\mathfrak{X}} \otimes \Omega^\cdot_{\mathfrak{X}_K})$ est une résolution du complexe $E_{\mathfrak{X}} \otimes \Omega^\cdot_{\mathfrak{X}_K}$, à termes acycliques pour sp_*, et fournissant donc un isomorphisme

$$\mathbb{R}v^\dagger_*(E_{\mathfrak{X}} \otimes \Omega^\cdot_{\mathfrak{X}_K}) = \mathbb{R}\mathrm{sp}_*(E_{\mathfrak{X}} \otimes \Omega^\cdot_{\mathfrak{X}_K}) \simeq \mathrm{sp}_*\check{C}^{\dagger\cdot}(\mathfrak{X}, (\mathfrak{Y}_i)_i, E_{\mathfrak{X}} \otimes \Omega^\cdot_{\mathfrak{X}_K}).$$

D'autre part, le complexe $\mathbb{R}\mathrm{sp}_*E$, en tant que complexe de $\mathscr{D}^\dagger_{\mathfrak{X}\mathbb{Q}}$-modules, est représenté par construction par le complexe $\mathrm{sp}_*\check{C}^{\dagger\cdot}(\mathfrak{X}, (\mathfrak{Y}_i)_i, E_{\mathfrak{X}})$. En utilisant la résolution de $\mathcal{O}_{\mathfrak{X}\mathbb{Q}}$ construite en (3.2.2), on obtient l'isomorphisme

$$\mathbb{R}\mathscr{H}om_{\mathscr{D}^\dagger_{\mathfrak{X}\mathbb{Q}}}(\mathcal{O}_{\mathfrak{X}\mathbb{Q}}, \mathbb{R}\mathrm{sp}_* E_{\mathfrak{X}}) \simeq \mathrm{sp}_*\check{C}^{\dagger\cdot}(\mathfrak{X}, (\mathfrak{Y}_i)_i, E_{\mathfrak{X}}) \otimes_{\mathcal{O}_{\mathfrak{X}}} \Omega^\cdot_{\mathfrak{X}}.$$

Pour achever la démonstration, il suffit donc de construire un isomorphisme

$$\mathrm{sp}_*\check{C}^{\dagger\cdot}(\mathfrak{X}, (\mathfrak{Y}_i)_i, E_{\mathfrak{X}}) \otimes_{\mathcal{O}_{\mathfrak{X}}} \Omega^\cdot_{\mathfrak{X}} \xrightarrow{\ \sim\ } \mathrm{sp}_*\check{C}^{\dagger\cdot}(\mathfrak{X}, (\mathfrak{Y}_i)_i, E_{\mathfrak{X}} \otimes \Omega^\cdot_{\mathfrak{X}_K}),$$

c'est à dire, avec les notations de (4.1.1), de construire pour tous $\underline{i}$, $\underline{j}$ un isomorphisme commutant aux restrictions

$$(\mathrm{sp}_* u_{\underline{i}*} v^\dagger_{\underline{i}\underline{j}} u^*_{\underline{i}} E_{\mathfrak{X}}) \otimes_{\mathcal{O}_{\mathfrak{X}}} \Omega^{\cdot}_{\mathfrak{X}} \xrightarrow{\sim} \mathrm{sp}_*(u_{\underline{i}*} v^\dagger_{\underline{i}\underline{j}} u^*_{\underline{i}}(E_{\mathfrak{X}} \otimes \Omega^{\cdot}_{\mathfrak{X}_K})),$$

ce qu'on déduit de ce que $\mathrm{sp}^* \Omega^{\cdot}_{\mathfrak{X}} \simeq \Omega^{\cdot}_{\mathfrak{X}_K}$, et de ce que $\Omega^{\cdot}_{\mathfrak{X}}$ est localement libre de rang fini.

De même, si l'on suppose, comme dans la démonstration de (4.1.5), que les recouvrements $\mathfrak{Y}_i$ sont choisis de telle sorte qu'il existe des sous-recouvrements $\mathfrak{Y}'_i$ formant des recouvrements des ouverts $Y' \cap \mathcal{X}_i$, où $Y' = Y - (T \cap Y)$, on obtient des isomorphismes

$$\mathbb{R}\underline{\Gamma}^\dagger_T(E_{\mathfrak{X}} \otimes \Omega^{\cdot}_{\mathfrak{X}_K})) \simeq \mathrm{sp}_*[\check{C}^{\dagger\cdot}(\mathfrak{X}, (\mathfrak{Y}_i)_i, E_{\mathfrak{X}} \otimes \Omega^{\cdot}_{\mathfrak{X}_K}) \longrightarrow \check{C}^{\dagger\cdot}(\mathfrak{X}, (\mathfrak{Y}'_i)_i, E_{\mathfrak{X}} \otimes \Omega^{\cdot}_{\mathfrak{X}_K})],$$

$$\mathbb{R}\mathcal{H}om_{\mathcal{D}^\dagger_{\mathfrak{X}\mathbb{Q}}}(\mathcal{O}_{\mathfrak{X}\mathbb{Q}}, \mathbb{R}\underline{\Gamma}^\dagger_T E_{\mathfrak{X}}) \simeq \mathrm{sp}_*[\check{C}^{\dagger\cdot}(\mathfrak{X}, (\mathfrak{Y}_i)_i, E_{\mathfrak{X}}) \longrightarrow \check{C}^{\dagger\cdot}(\mathfrak{X}, (\mathfrak{Y}'_i)_i, E_{\mathfrak{X}})] \otimes_{\mathcal{O}_{\mathfrak{X}}} \Omega^{\cdot}_{\mathfrak{X}},$$

et on conclut comme précédemment.

(4.1.7) **Corollaire.** *Supposons X propre et lisse sur k, et soient* $v : Y \hookrightarrow X$ *une immersion ouverte, T un fermé de X, et E un isocristal surconvergent sur Y. Si l'on munit les complexes* $\mathbb{R}v^\dagger_* E_{\mathfrak{X}}$ *et* $\mathbb{R}\underline{\Gamma}^\dagger_T E_{\mathfrak{X}}$ *des structures de complexes de* $D^b(\mathcal{D}^\dagger_{\mathfrak{X}\mathbb{Q}})$ *définies en* (4.1.5), *il existe des isomorphismes canoniques*

$$H^i_{\mathrm{rig}}(Y/K, E) \simeq \mathrm{Ext}^i_{\mathcal{D}^\dagger_{\mathfrak{X}\mathbb{Q}}}(\mathcal{O}_{\mathfrak{X}\mathbb{Q}}, \mathbb{R}v^\dagger_* E_{\mathfrak{X}}),$$

$$H^i_T(Y/K, E) \simeq \mathrm{Ext}^i_{\mathcal{D}^\dagger_{\mathfrak{X}\mathbb{Q}}}(\mathcal{O}_{\mathfrak{X}\mathbb{Q}}, \mathbb{R}\underline{\Gamma}^\dagger_T E_{\mathfrak{X}}).$$

Comme on a par définition

$$H^i_{\mathrm{rig}}(Y/K, E) := H^i(\mathcal{X}_K, E_{\mathfrak{X}} \otimes \Omega^{\cdot}_{\mathfrak{X}_K}) \simeq H^i(\mathcal{X}, \mathbb{R}v^\dagger_*(E_{\mathfrak{X}} \otimes \Omega^{\cdot}_{\mathfrak{X}_K})),$$

$$H^i_T(Y/K, E) := H^i(\mathcal{X}_K, \underline{\Gamma}^\dagger_{]T[}(E_{\mathfrak{X}} \otimes \Omega^{\cdot}_{\mathfrak{X}_K})) \simeq H^i(\mathcal{X}, \mathbb{R}\underline{\Gamma}^\dagger_T(E_{\mathfrak{X}} \otimes \Omega^{\cdot}_{\mathfrak{X}_K})),$$

le corollaire résulte immédiatement de (4.1.6).

4.2. *Cohomologie locale à support dans un diviseur lisse*

On suppose ici que Z est un diviseur lisse dans X, et on garde les notations de (4.0.1). Nous nous limiterons au cas d'un isocristal constant sur Y, renvoyant à un article ultérieur pour des résultats plus généraux, en liaison avec l'analogue pour les $\mathcal{D}^\dagger_{\mathfrak{X}\mathbb{Q}}$-modules du théorème de Kashiwara sur les $\mathcal{D}$-modules à support dans une sous-variété lisse [6, VI 7.11]. On observera que, Z étant un diviseur, les faisceaux $R^i v^\dagger_* \mathcal{O}_{\mathfrak{X}_K}$ sont nuls pour $i > 0$, et les faisceaux $\mathcal{H}^{\dagger i}_Z(\mathcal{O}_{\mathfrak{X}_K})$ sont nuls pour $i \neq 1$ [4]. Si $U = \mathrm{Spf} A$ est un ouvert affine de X, et $f \in \Gamma(U, \mathcal{O}_{\mathfrak{X}})$ une équation de $Z \cap U$ modulo $\mathfrak{m}$, on a

$$\Gamma(U, \mathrm{sp}_* \mathcal{O}_{\mathfrak{X}_K}) = A[1/f]^\dagger \otimes K, \tag{4.2.0.1}$$

$$\Gamma(U, \mathcal{H}^{\dagger 1}_Z(\mathcal{O}_{\mathfrak{X}_K})) = (A[1/f]^\dagger/A) \otimes K. \tag{4.2.0.2}$$

Nous aurons à utiliser le lemme suivant :

(4.2.1) **Lemme.** *Soient* $\mathfrak{X} = \mathrm{Spf} A$ *un* $\mathcal{V}$*-schéma formel affine réduit,* $\|\cdot\|$ *une norme de Banach sur* $A \otimes K$, $f \in A$, $B = A_{\{f\}} \otimes K$, *où* $A_{\{f\}}$ *est le séparé complété de* A_f, *et* $(a_k)_{k \geq 0}$ *une suite d'éléments de* $A \otimes K$ *vérifiant les conditions*

(i) *Il existe des constantes* $c \in \mathbb{R}$, $\eta < 1$ *telles que, pour tout* k, $\|a_k\| \leq c\,\eta^k$;

(ii) *Dans* B, *on a*

$$\sum_{k \geq 0} a_k f^{-k} = 0.$$

Si, pour tout $j \geq 0$, *on pose*

$$b_j := \sum_{k=0}^{j} a_k f^{j-k},$$

il existe c', *et* $\eta' < 1$, *pour lesquels on a* $\|b_j\| \leq c'\eta'^j$.

Si l'énoncé est vrai pour une norme sur $A \otimes K$, il l'est pour toute norme équivalente. Comme $A \otimes K$ est réduit, la semi-norme spectrale sur $\mathrm{Spm}(A \otimes K)$ est une norme de Banach, équivalente à toute norme de Banach [7, 6.2.4, th.1], ce qui permet de supposer que $\|\cdot\|$ est la norme spectrale. Soit $\eta' \in \mathbb{R}$ un élément dont une puissance appartient au groupe de la valuation de K, et tel que $\eta < \eta' < 1$. On considère le recouvrement de $\mathfrak{X}_K$ par les deux ouverts affinoïdes

$$V_1 := \{x \in \mathfrak{X}_K \mid |f(x)| \leq \eta'\}, \quad V_2 := \{x \in \mathfrak{X}_K \mid |f(x)| \geq \eta'\},$$

et il suffit de majorer les normes spectrales de b_j sur V_1 et V_2. Pour $x \in V_1$, on a

$$|b_j(x)| = \Big|\sum_{k=0}^{j} a_k(x) f(x)^{j-k}\Big| \leq \sup_{k \leq j} \|a_k\|\, \eta'^{j-k} \leq c\,\eta'^j.$$

Dans $\Gamma(V_2, \mathcal{O}_{\mathfrak{X}_K})$, on a $|a_k(x) f^{-k}(x)| \leq c\,(\eta/\eta')^k$. Comme A est réduit, il en est de même de $\Gamma(V_2, \mathcal{O}_{\mathfrak{X}_K})$ [7, 7.3.2, cor. 10]. La norme spectrale est donc une norme de Banach sur $\Gamma(V_2, \mathcal{O}_{\mathfrak{X}_K})$, si bien que la série $\sum_k a_k f^{-k}$ converge vers $b \in \Gamma(V_2, \mathcal{O}_{\mathfrak{X}_K})$. Comme b est d'image nulle dans B, on a $|f(x)| < 1$ en tout point du support de b, et le principe du maximum entraîne que, quitte à augmenter η', on peut supposer que $\sum_k a_k f^{-k} = 0$ dans $\Gamma(V_2, \mathcal{O}_{\mathfrak{X}_K})$. On obtient alors sur V_2 la majoration

$$|b_j(x)| = \Big|-\sum_{k > j} a_k(x) f(x)^{j-k}\Big| \leq \sup_{k > j} \|a_k\|\, \eta'^{j-k} \leq c\,\eta'^j.$$

(4.2.2) **Proposition.** *Supposons que Z soit un diviseur lisse dans X, et soit v l'immersion de $Y = X - Z$ dans X. Alors :*

(i) *Le $\mathscr{D}^\dagger_{\mathfrak{X}\mathbf{Q}}$-module $v^\dagger_* \mathcal{O}_{\mathfrak{X}_K}$ est cohérent ;*

(ii) *Si U est un ouvert de $\mathfrak{X}$ sur lequel il existe des coordonnées locales $t_1, \ldots, t_d$ telles que $Z = V(t_1)$ modulo $\mathfrak{m}$, et si $\partial_1, \ldots, \partial_d$ sont les dérivations correspondantes, la suite*

$$(\mathscr{D}^\dagger_{\mathfrak{X}\mathbf{Q}})^d \xrightarrow{\ \psi\ } \mathscr{D}^\dagger_{\mathfrak{X}\mathbf{Q}} \xrightarrow{\ \varphi\ } v^\dagger_* \mathcal{O}_{\mathfrak{X}_K} \longrightarrow 0, \tag{4.2.2.1}$$

où $\varphi(P) = P\cdot(1/t_1)$, et ψ est défini par

$$\psi(P_1, \ldots, P_d) = P_1 \partial_1 t_1 + \sum_{i=2}^{d} P_i \partial_i, \tag{4.2.2.2}$$

est exacte.

Comme $\mathscr{D}^\dagger_{\mathfrak{X}\mathbf{Q}}$ est cohérent, la première assertion résulte de la seconde, qu'il suffit de vérifier lorsque U est affine. Posons $A = \Gamma(U, \mathcal{O}_{\mathfrak{X}})$, et $D^\dagger_{U\mathbf{Q}} = \Gamma(U, \mathscr{D}^\dagger_{\mathfrak{X}\mathbf{Q}})$. Observons d'abord que l'homomorphisme $\varphi : D^\dagger_{U\mathbf{Q}} \to \Gamma(U, v^\dagger_* \mathcal{O}_{\mathfrak{X}_K}) = A[1/t_1]^\dagger$ est surjectif : tout élément f de $A[1/t_1]^\dagger$ s'écrit sous la forme $\sum_{k \geq 0} a_k t_1^{-k-1}$, pour une suite d'éléments $a_k \in A$ telle que $\|a_k\| < c\,\eta^k$ pour un certain $\eta < 1$; l'opérateur $P = \sum_{k \geq 0} (-1)^k a_k \partial^{[k]}$ appartient alors à $D^\dagger_{U\mathbf{Q}}$ et vérifie $P\cdot(1/t_1) = f$.

Soit $P \in D^\dagger_{U\mathbf{Q}}$ tel que $P\cdot(1/t_1) = 0$. Si $P = \sum_{\underline{k}} a_{\underline{k}} \underline{\partial}^{[\underline{k}]}$, on peut écrire $P = \sum'_{\underline{k}} a_{\underline{k}} \underline{\partial}^{[\underline{k}]} + \sum''_{\underline{k}} a_{\underline{k}} \underline{\partial}^{[\underline{k}]}$, en regroupant dans $\sum'$ (resp. $\sum''$) l'ensemble des termes relatifs aux indices $\underline{k}$ tels que $k_i = 0$ pour $i > 1$ (resp. tels qu'il existe $i > 1$ pour lequel $k_i \neq 0$). D'après (3.2.1), l'opérateur $\sum''_{\underline{k}} a_{\underline{k}} \underline{\partial}^{[\underline{k}]}$ appartient à l'idéal à gauche engendré par $\partial_2, \ldots, \partial_d$, donc à l'image de ψ. On peut ainsi supposer que P est de la forme $P = \sum_{k \geq 0} a_k \partial_1^{[k]}$.

Puisque $P \in \mathrm{Ker}(\varphi)$, on a dans $A[1/t_1]^\dagger$ la relation

$$\sum_{k \geq 0} (-1)^k a_k t_1^{-k-1} = 0. \tag{4.2.2.3}$$

Pour tout $j \geq 0$, soit $b_j \in A$ l'élément défini par

$$b_j := (-1)^{j+1} \sum_{k=0}^{j-1} (-1)^k a_k t_1^{j-k-1}.$$

Comme $P \in D^\dagger_{U\mathbf{Q}}$, il existe c, et $\eta < 1$, tels que $\|a_k\| \leq c\,\eta^k$. D'après (4.2.1), il existe donc c', et $\eta' < 1$, tels que $\|b_j\| \leq c' \eta'^j$, et on peut définir un opérateur $Q \in D^\dagger_{U\mathbf{Q}}$ en posant

$$Q := \sum_{j \geq 0} b_j \partial_1^{[j]}.$$

On a alors $Qt_1 = P$: il suffit en effet de le vérifier dans $\Gamma(U \cap Y, \mathscr{D}^\dagger_{\mathfrak{X}\mathbf{Q}}) \supset D^\dagger_{U\mathbf{Q}}$, et, d'après (4.2.2.3), on a alors pour tout $j \geq 0$ l'égalité

$$b_j = (-1)^j \sum_{k \geq j} (-1)^k a_k t_1^{j-k-1},$$

ce qui permet d'écrire :

$$\begin{aligned} Qt_1 &= \sum_{j\geq 0}\big(\sum_{k\geq j}(-1)^{k-j}a_k t_1^{j-k-1}\big)\partial_1^{[j]}t_1 \\ &= \sum_{k\geq 0}a_k\big(\sum_{j=0}^{k}(-1)^{k-j}t_1^{j-k-1}\partial_1^{[j]}\big)t_1 \\ &= \sum_{k\geq 0}a_k(\partial_1^{[k]}t_1^{-1})t_1 \\ &= P. \end{aligned}$$

Comme $b_0 = 0$, il existe d'après (3.2.1) $Q \in D^\dagger_{U\mathbf{Q}}$ tel que $Q = Q_1\partial_1$, d'où l'énoncé.

(4.2.3) **Corollaire.** *Sous les hypothèses de* (4.2.2) *:*

(i) *Le* $\mathscr{D}^\dagger_{\mathfrak{X}\mathbf{Q}}$*-module* $\mathscr{H}^{\dagger 1}_Z(\mathcal{O}_{\mathfrak{X}_K})$ *est cohérent ;*

(ii) *Sur* U*, la suite*

$$(\mathscr{D}^\dagger_{\mathfrak{X}\mathbf{Q}})^d \xrightarrow{\ \psi\ } \mathscr{D}^\dagger_{\mathfrak{X}\mathbf{Q}} \xrightarrow{\ \varphi\ } \mathscr{H}^{\dagger 1}_Z(\mathcal{O}_{\mathfrak{X}_K}) \longrightarrow 0, \qquad (4.2.3.1)$$

où $\varphi(P) = P\cdot(1/t_1)$*, et* ψ *est défini par*

$$\psi(P_1,\dots,P_d) = P_1t_1 + \sum_{i=2}^{d} P_i\partial_i, \qquad (4.2.3.2)$$

est exacte.

Par image directe par le morphisme sp : $\mathfrak{X}_K \to \mathfrak{X}$, la suite exacte

$$0 \longrightarrow \underline{\Gamma}^\dagger_{]Z[}(\mathcal{O}_{\mathfrak{X}_K}) \longrightarrow \mathcal{O}_{\mathfrak{X}_K} \longrightarrow v^\dagger\mathcal{O}_{\mathfrak{X}_K} \longrightarrow 0$$

fournit une suite exacte de $\mathscr{D}^\dagger_{\mathfrak{X}\mathbf{Q}}$-modules

$$0 \longrightarrow \mathcal{O}_{\mathfrak{X}\mathbf{Q}} \longrightarrow v^\dagger_*\mathcal{O}_{\mathfrak{X}_K} \longrightarrow \mathscr{H}^{\dagger 1}_Z(\mathcal{O}_{\mathfrak{X}_K}) \longrightarrow 0.$$

La cohérence de $\mathscr{H}^{\dagger 1}_Z(\mathcal{O}_{\mathfrak{X}_K})$ résulte donc de celle de $\mathcal{O}_{\mathfrak{X}\mathbf{Q}}$ et $v^\dagger_*\mathcal{O}_{\mathfrak{X}_K}$. Si $U = \mathrm{Spf}A$ est un ouvert affine de $\mathfrak{X}$ muni de coordonnées $t_1, \dots, t_d$ telles que $Z = V(t_1)$ modulo $\mathfrak{m}$, alors $\Gamma(U, \mathscr{H}^{\dagger 1}_Z(\mathcal{O}_{\mathfrak{X}_K}))$ s'identifie à $(A[1/t_1]^\dagger/A) \otimes K$. Comme, en tant que sous-$D^\dagger_{\mathfrak{X}\mathbf{Q}}$-module de $A[1/t_1]^\dagger \otimes K$, on a $A = (D^\dagger_{\mathfrak{X}\mathbf{Q}}t_1)\cdot(1/t_1)$, l'assertion (ii) résulte de (4.2.2) (ii).

4.3. *Cohomologie locale à support dans un diviseur à croisements normaux*

(4.3.0) Soient toujours $Z \subset X$ un diviseur, Y l'ouvert complémentaire, $v : Y \hookrightarrow X$ l'immersion correspondante. Il paraît raisonnable de conjecturer que, quel que soit Z, les faisceaux $v^\dagger_*\mathcal{O}_{\mathfrak{X}_K}$ et $\mathscr{H}^{\dagger 1}_Z(\mathcal{O}_{\mathfrak{X}_K})$ sont cohérents sur $\mathscr{D}^\dagger_{\mathfrak{X}\mathbf{Q}}$. Nous allons montrer que c'est le cas lorsque Z est un diviseur à croisements normaux, réunion de sous-variétés lisses se coupant transversalement. Cela permet aussi de traiter le cas des faisceaux $R^i v_*^\dagger \mathcal{O}_{\mathfrak{X}_K}$ et $\mathcal{H}_Z^{\dagger i}(\mathcal{O}_{\mathfrak{X}_K})$ quand Z est une sous-variété lisse de codimension r.

(4.3.1) **Lemme.** *Soient A une $\mathcal{V}$-algèbre formellement lisse, possédant un système de coordonnées locales $t_1, \ldots, t_d$, définissant des dérivations $\partial_1, \ldots, \partial_d$, et r, j, k trois entiers positifs, tels que $r \le d$, $j < k$. On pose $f = t_1 \ldots t_r$, et, pour tout $i > 0$, on note Q_i l'opérateur défini par*

$$Q_i := \prod_{\alpha=1}^{r} (\partial_\alpha^{[i-1]} - \partial_\alpha^{[i]} t_\alpha) - \prod_{\alpha=1}^{r} \partial_\alpha^{[i-1]}. \tag{4.3.1.1}$$

On a alors

$$\prod_{\alpha=1}^{r} \partial_\alpha^{[j]} = (-1)^{r(k-j)} f^{k-j} \prod_{\alpha=1}^{r} \partial_\alpha^{[k]} - \sum_{i=j+1}^{k} (-1)^{r(i-j-1)} f^{i-j-1} Q_i. \tag{4.3.1.2}$$

Pour tout $j \ge 0$, et tout α, on a

$$t_\alpha \partial_\alpha^{[j+1]} = \partial_\alpha^{[j+1]} t_\alpha - \partial_\alpha^{[j]}.$$

On en déduit la relation

$$(-1)^r f \prod_{\alpha=1}^{r} \partial_\alpha^{[j+1]} = \prod_{\alpha=1}^{r} (\partial_\alpha^{[j]} - \partial_\alpha^{[j+1]} t_\alpha) = \prod_{\alpha=1}^{r} \partial_\alpha^{[j]} + Q_{j+1},$$

et la relation (4.3.1.2) en résulte par récurrence.

(4.3.2) **Proposition.** *Supposons que Z vérifie les hypothèses de (4.3.0). Alors :*

(i) *Le $\mathcal{D}_{\mathfrak{X}\mathbf{Q}}^\dagger$-module $v_*^\dagger \mathcal{O}_{\mathfrak{X}_K}$ est cohérent ;*

(ii) *Si U est un ouvert de $\mathfrak{X}$ sur lequel il existe des coordonnées locales $t_1, \ldots, t_d$ telles que $Z = V(t_1 \ldots t_r)$ modulo $\mathfrak{m}$, et si $\partial_1, \ldots, \partial_d$ sont les dérivations correspondantes, la suite*

$$(\mathcal{D}_{\mathfrak{X}\mathbf{Q}}^\dagger)^d \xrightarrow{\ \psi\ } \mathcal{D}_{\mathfrak{X}\mathbf{Q}}^\dagger \xrightarrow{\ \varphi\ } v_*^\dagger \mathcal{O}_{\mathfrak{X}_K} \longrightarrow 0, \tag{4.3.2.1}$$

où $\varphi(P) = P \cdot (1/t_1 \ldots t_r)$, et ψ est défini par

$$\psi(P_1, \ldots, P_d) = \sum_{i=1}^{r} P_i \partial_i t_i + \sum_{i=r+1}^{d} P_i \partial_i, \tag{4.3.2.2}$$

est exacte.

L'assertion (i) résulte encore de (ii). Soit $U = \operatorname{Spf} A$ un ouvert affine de $\mathfrak{X}$ vérifiant les conditions de (ii) ; posons $f = t_1 \ldots t_r$. Un élément de $\Gamma(U, v_*^\dagger \mathcal{O}_{\mathfrak{X}_K})$ s'écrit sous la forme

$$g = \sum_{k \ge 0} a_k f^{-k-1},$$

la suite des éléments $a_k \in A_K$ étant telle qu'il existe $c > 0$, $\eta < 1$ tels que $\|a_k\| \leq c\,\eta^k$. On peut alors définir un opérateur $P \in \Gamma(U, \mathscr{D}^\dagger_{\mathscr{X}\mathbf{Q}})$ en posant $P = \sum_k (-1)^{rk} a_k \partial_1^{[k]} \ldots \partial_r^{[k]}$; on obtient $\varphi(P) = P\cdot(1/t_1 \ldots t_r) = g$, d'où la surjectivité de φ.

Soit $I \subset \mathscr{D}^\dagger_{\mathscr{X}\mathbf{Q}}$ l'idéal à gauche engendré par $(\partial_1 t_1, \ldots, \partial_r t_r, \partial_{r+1}, \ldots, \partial_n)$. Il est clair que $I \subset \mathrm{Ker}(\varphi)$. Soit $P = \sum_{\underline{k}} a_{\underline{k}} \underline{\partial}^{[\underline{k}]} \in \Gamma(U, \mathrm{Ker}(\varphi))$. En procédant comme dans la démonstration de (4.2.2), on d'abord supposer que P est de la forme $P = \sum_{\underline{k}} a_{\underline{k}} \underline{\partial}^{[\underline{k}]}$, avec $\underline{\partial} = (\partial_1, \ldots, \partial_r)$, $\underline{k} = (k_1, \ldots, k_r)$.

On peut ensuite se ramener au cas où P est de la forme $\sum_k a_k \partial_1^{[k]} \ldots \partial_r^{[k]}$, la suite (a_k) étant telle qu'il existe c, $\eta < 1$, tels que $\|a_k\| \leq c\,\eta^k$. En effet, si $\underline{k}$ est un multi-indice, et si $k = \sup_\alpha k_\alpha$, la relation (4.3.1.2) permet d'écrire, pour tout α,

$$\partial_\alpha^{[k_\alpha]} = (-1)^{k-k_\alpha} t_\alpha^{k-k_\alpha} \partial_\alpha^{[k]} + \sum_{i=k_\alpha+1}^{k} (-1)^{(i-k_\alpha-1)} t_\alpha^{i-k_\alpha-1} \partial_\alpha^{[i]} t_\alpha .$$

Pour tout $k \in \mathbb{N}$, soit I_k l'ensemble des $\underline{k}$ tels que $k_\alpha \leq k$ pour tout α, et qu'il existe α pour lequel $k_\alpha = k$. On réécrit P sous la forme

$$P = \sum_{k \geq 0} \sum_{\underline{k} \in I_k} a_{\underline{k}} \prod_{\alpha=1}^{r} ((-1)^{k-k_\alpha} t_\alpha^{k-k_\alpha} \partial_\alpha^{[k]} + \sum_{i_\alpha = k_\alpha+1}^{k} (-1)^{(i_\alpha-k_\alpha-1)} t_\alpha^{i_\alpha-k_\alpha-1} \partial_\alpha^{[i_\alpha]} t_\alpha).$$

Posons alors

$$P_0 = \sum_{k \geq 0} \sum_{\underline{k} \in I_k} a_{\underline{k}} \prod_{\alpha=1}^{r} (-1)^{k-k_\alpha} t_\alpha^{k-k_\alpha} \partial_\alpha^{[k]} .$$

L'opérateur P_0 ainsi défini est bien dans $D^\dagger_{U\mathbf{Q}}$, car, pour tout $\underline{k} \in I_k$, on a $k \leq |\underline{k}|$, de sorte que $|a_{\underline{k}}| \leq c\,\eta^{|\underline{k}|} \leq c\,\eta^k = c(\eta^{1/r})^{rk}$. D'autre part, en désignant par J un sous-ensemble variable de $\{1, \ldots, r\}$, on peut écrire $P - P_0$ comme une combinaison linéaire à coefficients dans A d'opérateurs de la forme

$$a_{\underline{k}} \prod_{\alpha \notin J} \partial_\alpha^{[k]} \prod_{\alpha \in J} \partial_\alpha^{[i_\alpha]} t_\alpha = (a_{\underline{k}} \prod_{\alpha \notin J} \partial_\alpha^{[k]} \prod_{\alpha \in J} \partial_\alpha^{[i_\alpha - 1]} / i_\alpha) \prod_{\alpha \in J} \partial_\alpha t_\alpha$$

avec $k_\alpha < i_\alpha \leq k$, $J \neq \{1, \ldots, r\}$ (car l'un des k_α est égal à k), et $J \neq \emptyset$. Cette dernière condition montre que ces opérateurs sont dans l'idéal engendré par $\partial_1 t_1, \ldots, \partial_r t_r$, donc dans I. L'ordre de chacun des opérateurs

$$a_{\underline{k}} \prod_{\alpha \notin J} \partial_\alpha^{[k]} \prod_{\alpha \in J} \partial_\alpha^{[i_\alpha - 1]}$$

est compris entre k et rk, tandis que le coefficient $a_{\underline{k}}$ vérifie comme plus haut la majoration $\|a_{\underline{k}}\| \leq c(\eta^{1/r})^{rk}$. Comme, pour k fixé, ces opérateurs sont en nombre fini, on en déduit (compte tenu de (3.2.1)) que $P - P_0$ est somme d'une famille sommable qui converge dans I.

On peut donc supposer que P est de la forme $P = \sum_{k \geq 0} a_k \partial_1^{[k]} \ldots \partial_r^{[k]}$, avec

$$\sum_{k \geq 0} (-1)^{rk} a_k f^{-k-1} = P\cdot(1/f) = 0. \tag{4.3.2.3}$$

Pour tout $i \geq 1$, soit Q_i l'opérateur différentiel défini en (4.3.1.1). Pour tout k, la relation (4.3.1.2) permet d'écrire, en notant $\underline{\partial}^{[j]} = \prod_{\alpha=1}^{r} \partial_\alpha^{[j]}$,

$$\sum_{j=0}^{k} a_j \underline{\partial}^{[j]} = (\sum_{j=0}^{k}(-1)^{r(k-j)} a_j f^{k-j})\underline{\partial}^{[k]} + \sum_{i=1}^{k}(\sum_{j=0}^{i-1}(-1)^{r(i-j-1)} a_j f^{i-j-1})Q_i. \quad (4.3.2.4)$$

D'après (4.2.1), la relation (4.3.2.3) entraîne qu'il existe c' et $\eta' < 1$ tels que, quel que soit k, on ait

$$\|\sum_{j=0}^{k}(-1)^{r(k-j)} a_j f^{k-j}\| \leq c'\eta'^{k},$$

de sorte que, pour $k \to \infty$, on a

$$(\sum_{j=0}^{k}(-1)^{r(k-j)} a_j f^{k-j})\underline{\partial}^{[k]} \to 0$$

dans $D^\dagger_{U\mathbf{Q}}$, et même dans $\widehat{D}^{(m)}_{U\mathbf{Q}}$ pour m assez grand. De même, on a pour tout i

$$\|\sum_{j=0}^{i-1}(-1)^{r(i-j-1)} a_j f^{i-j-1}\| \leq c'\eta'^{i-1}.$$

Or l'opérateur Q_i est une combinaison à coefficients entiers, indépendants de i, d'opérateurs du type

$$\prod_{\alpha \notin J} \partial_\alpha^{[i-1]} \prod_{\alpha \in J} \partial_\alpha^{[i]} t_\alpha,$$

où $J \subset \{1, \ldots, r\}$ est un sous-ensemble non vide. Pour tout J, la série de terme général

$$(\sum_{j=0}^{i-1}(-1)^{r(i-j-1)} a_j f^{i-j-1}) \prod_{\alpha \notin J} \partial_\alpha^{[i-1]} \prod_{\alpha \in J} \partial_\alpha^{[i]} t_\alpha$$

converge donc dans $D^\dagger_{U\mathbf{Q}}$ vers un opérateur qui, d'après (3.2.1), peut s'écrire sous la forme $Q_J \prod_{\alpha \in J} \partial_\alpha t_\alpha$, donc appartient à I. Par passage à la limite pour $k \to \infty$, la relation (4.3.2.4) montre donc que $P \in I$.

(4.3.3) **Corollaire.** *Sous les hypothèses de* (4.3.2) *:*

(i) *Le* $\mathscr{D}^\dagger_{\mathfrak{X}\mathbf{Q}}$*-module* $\mathscr{H}^{\dagger 1}_Z(\mathcal{O}_{\mathfrak{X}_K})$ *est cohérent ;*

(ii) *Sur* U, *la suite*

$$(\mathscr{D}^\dagger_{\mathfrak{X}\mathbf{Q}})^{d+1} \xrightarrow{\ \psi\ } \mathscr{D}^\dagger_{\mathfrak{X}\mathbf{Q}} \xrightarrow{\ \varphi\ } \mathscr{H}^{\dagger 1}_Z(\mathcal{O}_{\mathfrak{X}_K}) \longrightarrow 0, \quad (4.3.3.1)$$

où $\varphi(P) = P\cdot(1/t_1 \ldots t_r)$, *et* ψ *est défini par*

$$\psi(P_0, \ldots, P_d) = P_0 t_1 \ldots t_r + \sum_{i=1}^{r} P_i \partial_i t_i + \sum_{i=r+1}^{d} P_i \partial_i, \quad (4.3.3.2)$$

est exacte.

La démonstration est identique à celle de (4.2.3), en remplaçant t_1 par $t_1 \ldots t_r$.

(4.3.4) **Corollaire.** *Soit Z une sous-variété lisse de codimension r dans X, et soit v l'immersion de Y = X − Z dans X. Alors :*

(i) *Pour tout i, les $\mathcal{D}^{\dagger}_{\mathfrak{X}\mathbf{Q}}$-modules $R^i v^{\dagger}_* \mathcal{O}_{\mathfrak{X}_K}$ et $\mathcal{H}^{\dagger i}_Z(\mathcal{O}_{\mathfrak{X}_K})$ sont cohérents ;*

(ii) *Si U est un ouvert de $\mathfrak{X}$ sur lequel il existe des coordonnées locales $t_1, \ldots, t_d$ telles que $Z = V(t_1, \ldots, t_r)$ modulo $\mathfrak{m}$, et si $\partial_1, \ldots, \partial_d$ sont les dérivations correspondantes, la suite*

$$(\mathcal{D}^{\dagger}_{\mathfrak{X}\mathbf{Q}})^d \xrightarrow{\ \psi\ } \mathcal{D}^{\dagger}_{\mathfrak{X}\mathbf{Q}} \xrightarrow{\ \varphi\ } \mathcal{H}^{\dagger r}_Z(\mathcal{O}_{\mathfrak{X}_K}) \longrightarrow 0, \tag{4.3.4.1}$$

où $\varphi(P) = P\cdot(1/t_1 \ldots t_r)$, et ψ est défini par

$$\psi(P_1, \ldots, P_d) = \sum_{i=1}^{r} P_i t_i + \sum_{i=r+1}^{d} P_i \partial_i, \tag{4.3.4.2}$$

est exacte.

Si $r = 1$, l'énoncé a été montré en (4.2.2) et (4.2.3). Pour $r \geq 2$, la suite exacte longue des $R^i\mathrm{sp}_*$ et la nullité des $\mathcal{H}^{\dagger i}_Z(\mathcal{O}_{\mathfrak{X}_K})$ pour $i \neq r$ montrent que $\mathcal{O}_{\mathfrak{X}_K} \simeq v^{\dagger}_*\mathcal{O}_{\mathfrak{X}_K}$, $R^i v^{\dagger}_* \mathcal{O}_{\mathfrak{X}_K} = 0$ pour $i \neq 0, r-1$, et $R^{r-1} v^{\dagger}_* \mathcal{O}_{\mathfrak{X}_K} \simeq \mathcal{H}^{\dagger r}_Z(\mathcal{O}_{\mathfrak{X}_K})$.

Rappelons d'abord le calcul de $\mathcal{H}^{\dagger r}_Z(\mathcal{O}_{\mathfrak{X}_K})$ sur un ouvert affine U vérifiant les conditions de (ii) [4]. Posons $Y_i = D(t_i) \subset X$, et soient $v_{i_0 \ldots i_k} : Y_{i_0} \cap \ldots \cap Y_{i_k} \hookrightarrow X$ les immersions ouvertes correspondantes. D'après (4.1.3) et (4.1.4), on peut calculer les complexes $\mathbb{R}v^{\dagger}_*\mathcal{O}_{\mathfrak{X}_K}$ et $\mathbb{R}\underline{\Gamma}^{\dagger}_Z\mathcal{O}_{\mathfrak{X}_K}$ en utilisant le recouvrement de Y par les Y_i, avec $i = 1, \ldots, r$: $\mathbb{R}v^{\dagger}_*\mathcal{O}_{\mathfrak{X}_K}$ est représenté par le complexe commençant en degré 0 :

$$\prod_{i=1}^{d} \mathrm{sp}_*(v^{\dagger}_i \mathcal{O}_{\mathfrak{X}_K}) \longrightarrow \prod_{i_0<i_1} \mathrm{sp}_*(v^{\dagger}_{i_0 i_1}\mathcal{O}_{\mathfrak{X}_K}) \longrightarrow \ldots \longrightarrow \mathrm{sp}_*(v^{\dagger}_{1\ldots r}\mathcal{O}_{\mathfrak{X}_K}) \longrightarrow 0,$$

et $\mathbb{R}\underline{\Gamma}^{\dagger}_Z\mathcal{O}_{\mathfrak{X}_K}$ par le complexe commençant en degré 0 :

$$\mathcal{O}_{\mathfrak{X}\mathbf{Q}} \longrightarrow \prod_{i=1}^{r} \mathrm{sp}_*(v^{\dagger}_i \mathcal{O}_{\mathfrak{X}_K}) \longrightarrow \prod_{i_0<i_1} \mathrm{sp}_*(v^{\dagger}_{i_0 i_1}\mathcal{O}_{\mathfrak{X}_K}) \longrightarrow \ldots \longrightarrow \mathrm{sp}_*(v^{\dagger}_{1\ldots r}\mathcal{O}_{\mathfrak{X}_K}) \longrightarrow 0.$$

Pour toute suite $i_0, \ldots, i_k$, et tout ouvert affine $U = \mathrm{Spf} A \subset \mathfrak{X}$, on a

$$\Gamma(U, \mathrm{sp}_*(j^{\dagger}_{i_0\ldots i_k}\mathcal{O}_{\mathfrak{X}_K})) = A_K[1/t_{i_0}\ldots t_{i_k}]^{\dagger},$$

avec $A_K := A \otimes K$, de sorte qu'on obtient un isomorphisme

$$A_K[1/t_1\ldots t_r]^{\dagger} / \sum_{i=1}^{r} A_K[1/t_1 .. \hat{t}_i .. t_r]^{\dagger} \xrightarrow{\ \sim\ } \mathcal{H}^{\dagger r}_Z(\mathcal{O}_{\mathfrak{X}_K}).$$

D'après (4.3.2), l'application $D^{\dagger}_{U\mathbf{Q}} \to A_K[1/t_1\ldots t_r]^{\dagger}$ envoyant 1 sur $1/t_1\ldots t_r$ est surjective, de sorte qu'il en est de même de φ. De même, l'image de $A_K[1/t_1 .. \hat{t}_i .. t_r]^{\dagger}$ dans $A_K[1/t_1\ldots t_r]^{\dagger}$ est le sous-$D^{\dagger}_{U\mathbf{Q}}$-module engendré par $1/t_1 .. \hat{t}_i .. t_r$, soit encore le sous-$D^{\dagger}_{U\mathbf{Q}}$-module $(D^{\dagger}_{U\mathbf{Q}} t_i)\cdot 1/t_1\ldots t_r$. La suite exacte (4.3.2.1) implique donc l'assertion (ii).

5. Isocristaux surconvergents associés aux caractères

Nous donnerons maintenant deux exemples de $\mathcal{D}^{\dagger}_{\mathfrak{X}\mathbf{Q}}$-modules cohérents définis par des isocristaux surconvergents non constants. Dans les cas les plus importants, ces isocristaux sont respectivement facteurs de l'image directe de l'isocristal constant par un revêtement de Kummer et d'Artin-Schreier [3], et sont à la base de la théorie p-adique des sommes exponentielles développée par Dwork, Adolphson-Sperber, Robba, etc, dans laquelle ils correspondent aux caractères multiplicatifs et additifs d'un corps fini.

5.1. *Isocristaux associés aux caractères multiplicatifs*

On prend ici $\mathcal{V} = \mathbb{Z}_p$, et le schéma formel $\mathfrak{X}$ est la droite projective formelle $\widehat{\mathbb{P}}^1_{\mathcal{V}}$. Nous désignerons par t la coordonnée canonique sur la droite affine formelle $\widehat{\mathbb{A}}^1_{\mathcal{V}} = \widehat{\mathbb{P}}^1_{\mathcal{V}} - \{\infty\} \subset \widehat{\mathbb{P}}^1_{\mathcal{V}}$, ∂ la dérivation par rapport à t ; sur l'ouvert $\widehat{\mathbb{P}}^1_{\mathcal{V}} - \{0\}$, nous utiliserons la coordonnée $t' = 1/t$, et nous noterons ∂' la dérivation par rapport à t'.

(5.1.1) On pose $Y = \mathbb{P}^1_k - \{0, \infty\}$, et $v : Y \hookrightarrow X$. Pour tout $\alpha \in \mathbb{Z}_p$, on note $\mathcal{K}_\alpha$ l'isocristal surconvergent sur Y dont la réalisation sur $\mathfrak{X}_K = \mathbb{P}^{1an}_K$ est donnée par $v^\dagger \mathcal{O}_{\mathfrak{X}_K}$, muni de la connexion ∇ définie sur $\mathbb{A}^{1an}_K - \{0\}$ par

$$\nabla(1) = \alpha t^{-1} \otimes dt$$

(cf. [3, (2.3)]). Rappelons que, si $\mathbb{F}_q$ est un corps fini de caractéristique p, n un diviseur de $q-1$, et χ la puissance i-ième du caractère de Teichmüller sur le groupe $\mu_n(\mathbb{F}_q)$ des racines n-ièmes de l'unité, χ correspond dans la théorie p-adique des sommes exponentielles à l'isocristal $\mathcal{K}_\chi := \mathcal{K}_{\alpha_\chi}$, avec $\alpha_\chi = i/n$.

(5.1.2) **Proposition** (Laumon). *Supposons que $\alpha \in \mathbb{Z}_p$ soit non Liouville. Alors :*

(i) *Le $\mathcal{D}^{\dagger}_{\mathfrak{X}\mathbf{Q}}$-module $v^\dagger_* \mathcal{K}_\alpha$ est cohérent.*

(ii) *Supposons que $\alpha \notin \mathbb{N}$. Sur $\widehat{\mathbb{A}}^1_{\mathcal{V}}$, la suite de $\mathcal{D}^{\dagger}_{\mathfrak{X}\mathbf{Q}}$-modules*

$$0 \longrightarrow \mathcal{D}^{\dagger}_{\mathfrak{X}\mathbf{Q}} \xrightarrow{\ \psi\ } \mathcal{D}^{\dagger}_{\mathfrak{X}\mathbf{Q}} \xrightarrow{\ \varphi\ } v^\dagger_* \mathcal{K}_\alpha \longrightarrow 0 \qquad (5.1.2.1)$$

définie par $\varphi(P) = P\cdot 1$, et $\psi(Q) = Q(t\partial - \alpha)$, est exacte.

Lorsque $\alpha \in \mathbb{Z}$, $v^\dagger_* \mathcal{K}_\alpha$ est isomorphe en tant que $\mathcal{D}^{\dagger}_{\mathfrak{X}\mathbf{Q}}$-module à $v^\dagger_* \mathcal{O}_{\mathfrak{X}_K}$ muni de la connexion triviale, par l'isomorphisme envoyant 1 sur $t^{-\alpha}$. On peut donc supposer que $\alpha \notin \mathbb{N}$. Notons aussi qu'il suffit de prouver l'assertion (ii), car, avec la coordonnée t', ∇ est définie par $\nabla(1) = -\alpha t'^{-1} \otimes dt'$, et on dispose sur $\widehat{\mathbb{P}}^1_{\mathcal{V}} - \{0\}$ d'une suite analogue à (5.1.2.1) obtenue en changeant α en $-\alpha$, t en t', ∂ en ∂'.

Soit $U = \mathrm{Spf} A \subset \widehat{\mathbb{A}}^1_{\mathcal{V}}$ un ouvert affine ; on a alors $\Gamma(U, v_*^\dagger \mathcal{K}_\alpha) = A_K[1/t]^\dagger$, et, pour tout $n \in \mathbb{Z}$,

$$\partial \cdot t^n = n t^{n-1} + t^n \partial \cdot 1 = (n+\alpha) t^{n-1}.$$

Si $P = \sum_k a_k \partial^{[k]}$, on a donc

$$\varphi(P) = \sum_{k \geq 0} \frac{\alpha(\alpha-1)\ldots(\alpha-k+1)}{k!} a_k t^{-k} = \sum_{k\geq 0} \binom{\alpha}{k} a_k t^{-k},$$

où $\binom{\alpha}{k}$ est un entier p-adique, puisque $\alpha \in \mathbb{Z}_p$. Un élément $f \in A_K[1/t]^\dagger$ peut s'écrire sous la forme $f = \sum_k b_k t^{-k}$, avec $b_k \in A_K$ et $\|b_k\| \leq c\,\eta^k$ pour un $\eta < 1$. Comme $\alpha \notin \mathbb{N}$, on peut poser

$$a_k := \binom{\alpha}{k}^{-1} b_k.$$

Soit η' tel que $\eta < \eta' < 1$. Comme α est non Liouville, il existe $c' > 0$ tel que $|\binom{\alpha}{k}| \geq c'\eta'^k$ pour tout k [8, lemme 3.1], de sorte que

$$\|a_k\| \leq c\,\eta^k \left|\binom{\alpha}{k}\right|^{-1} \leq (c/c')(\eta/\eta')^k.$$

L'opérateur $P = \sum_k a_k \partial^{[k]}$ appartient donc à $D^\dagger_{U\mathbb{Q}}$, d'où la surjectivité de φ.

Soit $Q \in D^\dagger_{U\mathbb{Q}}$ un opérateur tel que $Q(t\partial - \alpha) = 0$. Si $Q = \sum_{k\geq 0} b_k \partial^{[k]}$, on a

$$Q(t\partial - \alpha) = \sum_{k\geq 0} ((k-\alpha) b_k + k\,t\,b_{k-1})\partial^{[k]}, \tag{5.1.2.2}$$

d'où la nullité des b_k par récurrence sur k, et l'injectivité de ψ.

Il est clair que $\mathrm{Im}(\psi) \subset \mathrm{Ker}(\varphi)$. Soit $P \in \mathrm{Ker}(\varphi)$, avec $P = \sum_k a_k \partial^{[k]}$. On a donc

$$\sum_{k\geq 0} \binom{\alpha}{k} a_k t^{-k} = 0.$$

Pour tout $j \geq 0$, posons

$$b'_j = \sum_{k=0}^{j} \binom{\alpha}{k} a_k t^{j-k}, \quad b_j = \left[\binom{\alpha}{j}(j-\alpha)\right]^{-1} b'_j.$$

D'après (4.2.1), il existe c', $\eta' < 1$, tels que $\|b'_j\| \leq c'\eta'^j$. Soit η'' tel que $\eta' < \eta''^2 < 1$. Comme α est non Liouville, il existe c_1 tel que, pour tout j, on ait $|j-\alpha| > c_1\eta''^j$. D'autre part, il existe c_2 tel que $|\binom{\alpha}{k}| \geq c_2\eta''^k$. On obtient donc une majoration de la forme $\|b_j\| \leq c''(\eta'/\eta''^2)^j$, qui permet de définir un opérateur $Q \in D^\dagger_{U\mathbb{Q}}$ en posant $Q = \sum_j b_j \partial^{[j]}$. On a alors

$$Q(t\partial - \alpha) = \sum_{j\geq 0} ((j-\alpha) b_j + j\,t\,b_{j-1})\partial^{[j]}$$

avec

$$(j-\alpha) b_j + j\,t\,b_{j-1} = \binom{\alpha}{j}^{-1} \sum_{k=0}^{j} \binom{\alpha}{k} a_k t^{j-k} + \binom{\alpha}{j-1}^{-1} j(j-\alpha-1)^{-1} t \sum_{k=0}^{j-1} \binom{\alpha}{k} a_k t^{j-k-1},$$

d'où $(j-\alpha) b_j + j\,t\,b_{j-1} = a_j$, et $Q(t\partial - \alpha) = P$, si bien que $\mathrm{Im}(\psi) = \mathrm{Ker}(\varphi)$.

Remarque. On observera que la condition introduite sur α est précisément celle qui assure la finitude du groupe de cohomologie rigide $H^1_{\mathrm{rig}}(U/K, \mathcal{K}_\alpha)$.

5.2. *Isocristaux associés aux caractères additifs*

Soient $q = p^s$, $\mathcal{V}_0 = W(\mathbb{F}_q)$ l'anneau des vecteurs de Witt à coefficients dans $\mathbb{F}_q$, $K_0 = \mathrm{Frac}(\mathcal{V}_0)$, $K = K_0(\mu_p)$, où $\mu_p \subset \overline{\mathbb{Q}}_p$ est le groupe des racines p-ièmes de l'unité, $\mathcal{V}$ l'anneau des entiers de K. On notera encore $\mathcal{X}$ la droite projective formelle $\widehat{\mathbb{P}}^1_{\mathcal{V}}$, t la coordonnée canonique sur la droite affine formelle $\widehat{\mathbb{A}}^1_{\mathcal{V}} = \widehat{\mathbb{P}}^1_{\mathcal{V}} - \{\infty\} \subset \widehat{\mathbb{P}}^1_{\mathcal{V}}$, $t' = 1/t$, ∂ la dérivation par rapport à t, ∂' la dérivation par rapport à t'.

(5.2.1) On pose ici $Y = \mathbb{A}^1_k$, et on note encore v l'immersion de Y dans X. Rappelons [3] que, suivant Dwork, on associe à un caractère additif ψ de $\mathbb{F}_q$ un élément $\pi \in \mathcal{V}$, vérifiant

$$\mathrm{ord}_p\, \pi \;=\; 1/(p-1). \tag{5.2.1.1}$$

Au caractère ψ correspond alors dans la théorie p-adique des sommes exponentielles un F-isocristal surconvergent sur Y, noté $\mathcal{L}_\psi$, ou encore $\mathcal{L}_\pi$, dont la réalisation sur $\mathcal{X}_K = \mathbb{P}^{1an}_K$ est donnée par $v^\dagger \mathcal{O}_{\mathcal{X}_K}$, muni de la connexion ∇ définie sur $\mathbb{A}^{1an}_K$ par

$$\nabla(1) \;=\; -\,\pi \otimes dt.$$

Pour établir la cohérence sur $\mathcal{D}^\dagger_{\mathcal{X}\mathbb{Q}}$ de $v^\dagger_* \mathcal{L}_\pi$, la seule propriété de π que nous aurons à utiliser sera la relation (5.2.1.1). Rappelons tout d'abord le lemme suivant :

(5.2.2) **Lemme.** *Soient A une K-algèbre de Tate, $\|\text{-}\|$ une norme de Banach sur A, (a_n) une suite d'élément de A, et $\pi \in K$ un élément tel que* $\mathrm{ord}_p \pi = 1/(p-1)$. *Alors les deux conditions suivantes sont équivalentes :*

(i) *Il existe $c \in \mathbb{R}$, et $\eta < 1$ tels que, pour tout n, on ait $\|a_n\| \le c\,\eta^n$;*

(ii) *Il existe $c' \in \mathbb{R}$, et $\eta' < 1$ tels que, pour tout $k \le n$, on ait $\|a_n k!/\pi^k\| \le c'\,\eta'^n$.*

Si $k = \sum_i a_i p^i$ est le développement de k en base p, et $\sigma(k) = \sum_i a_i$, on a

$$\mathrm{ord}_p\, k! \;=\; (k - \sigma(k))/(p-1).$$

Par suite, $\mathrm{ord}_p\, (\pi^k/k!) = \sigma(k)/(p-1) \ge 0$, d'où (ii) $\Rightarrow$ (i). Inversement, on a

$$\sigma(k) \;\le\; (p-1)(1+\log_p k) \;\le\; (p-1)(1+\log_p n),$$

d'où $|k!/\pi^k| \le p\,n$, et (i) $\Rightarrow$ (ii).

(5.2.3) **Proposition.** *Avec les notations de* (5.2.1), *soient* $\pi \in \mathcal{V}$ *un élément tel que* $\mathrm{ord}_p(\pi) = 1/(p-1)$, *et* $\mathcal{L}_\pi = v^\dagger \mathcal{O}_{\mathfrak{X}_K}$, *muni de la connexion* ∇ *telle que* $\nabla(1) = -\pi \otimes dt$. *Alors :*

(i) *Le* $\mathcal{D}^\dagger_{\mathfrak{X}\mathbf{Q}}$*-module* $v^\dagger_* \mathcal{L}_\pi$ *est cohérent.*

(ii) *Sur l'ouvert* $\widehat{\mathbb{P}}^1_{\mathcal{V}} - \{0\}$, *la suite de* $\mathcal{D}^\dagger_{\mathfrak{X}\mathbf{Q}}$*-modules*

$$0 \longrightarrow \mathcal{D}^\dagger_{\mathfrak{X}\mathbf{Q}} \xrightarrow{\ \psi\ } \mathcal{D}^\dagger_{\mathfrak{X}\mathbf{Q}} \xrightarrow{\ \varphi\ } v^\dagger_* \mathcal{L}_\pi \longrightarrow 0 \tag{5.2.3.1}$$

définie par $\varphi(P) = P\cdot 1$, *et* $\psi(Q) = Q(t'^2\partial' - \pi)$, *est exacte.*

Sur $\widehat{\mathbb{A}}^1_{\mathcal{V}} = \widehat{\mathbb{P}}^1_{\mathcal{V}} - \{\infty\}$, le $\mathcal{D}^\dagger_{\mathfrak{X}\mathbf{Q}}$-module $v^\dagger_* \mathcal{L}_\pi$ est cohérent d'après (3.1.4), et il est en fait facile de voir qu'on a un isomorphisme

$$\mathcal{D}^\dagger_{\mathfrak{X}\mathbf{Q}} / \mathcal{D}^\dagger_{\mathfrak{X}\mathbf{Q}}(\partial + \pi) \xrightarrow{\ \sim\ } v^\dagger_* \mathcal{L}_\pi.$$

Il suffit donc de prouver l'assertion (ii).

Soient $U \subset \widehat{\mathbb{P}}^1_{\mathcal{V}} - \{0\}$ un ouvert affine, avec $U = \mathrm{Spf} A$, et $U' = U - \{\infty\}$; on a donc $\Gamma(U', \mathcal{O}_{\mathfrak{X}}) = A\{1/t'\} = A\{t\}$; on notera $D^\dagger_{U\mathbf{Q}} = \Gamma(U, \mathcal{D}^\dagger_{\mathfrak{X}\mathbf{Q}})$, $D^\dagger_{U'\mathbf{Q}} = \Gamma(U', \mathcal{D}^\dagger_{\mathfrak{X}\mathbf{Q}})$, et on a une inclusion $D^\dagger_{U\mathbf{Q}} \subset D^\dagger_{U'\mathbf{Q}}$. Comme $\mathcal{L}_\pi = v^\dagger \mathcal{O}_{\mathfrak{X}_K}$ en tant que $\mathcal{O}_{\mathfrak{X}_K}$-module, on a aussi

$$\Gamma(U, v^\dagger_* \mathcal{L}_\pi) = A[1/t']^\dagger \subset A\{1/t'\} = \Gamma(U', v^\dagger_* \mathcal{L}_\pi).$$

Observons d'abord que, dans $D^\dagger_{U'\mathbf{Q}}$, les dérivations ∂ et ∂' sont liées par la relation $t'\partial' = -t\partial$. En écrivant, pour $k \geq 1$,

$$(k+1)t'^{k+1}\partial'^{[k+1]} = (k+1)t'(t'^k\partial'^{[k]})\partial' = -(k+1)t'(t'^k\partial'^{[k]})t^2\partial,$$

on en déduit, par récurrence sur k, les relations

$$t'^k\partial'^{[k]} = (-1)^k \sum_{i=1}^{k} \binom{k-1}{i-1} t^i \partial^{[i]}, \tag{5.2.3.2}$$

$$t^k\partial^{[k]} = (-1)^k \sum_{i=1}^{k} \binom{k-1}{i-1} t'^i \partial'^{[i]}. \tag{5.2.3.3}$$

Comme la connexion de $\mathcal{L}_\pi$ est telle que $\partial\cdot 1 = -\pi$, donc $\partial^{[i]}\cdot 1 = (-\pi)^i/i!$, on en déduit la formule

$$\partial'^{[k]}\cdot 1 = (-1)^k \sum_{i=1}^{k} \binom{k-1}{i-1} t'^{-k-i}(-\pi)^i/i!,$$

de sorte que, pour $P = \sum_{k\geq 0} a_k \partial'^{[k]} \in D^\dagger_{U\mathbf{Q}}$, on a

$$\varphi(P) = \sum_{k\geq 0} (-1)^k a_k \Big(\sum_{i=1}^{k} \binom{k-1}{i-1} t'^{-k-i}(-\pi)^i/i!\Big). \tag{5.2.3.4}$$

Soit alors $f \in \Gamma(U, v^\dagger_* \mathcal{L}_\pi) = A[1/t']^\dagger$. Grâce au lemme (5.2.2), on peut écrire f sous la forme

$$f = \sum_{k\geq 0} \pi^k a_k / k! t'^k,$$

la suite d'éléments $a_k \in A$ étant telle que $\|a_k\| \leq c\eta^k$, avec $\eta < 1$. Considérons alors l'opérateur

$$P = \sum_{k\geq 0} (-1)^k a_k t^k \partial^{[k]} = a_0 + \sum_{k\geq 1} a_k (\sum_{i=1}^{k} \binom{k-1}{i-1} t'^i \partial'^{[i]})$$

$$= a_0 + \sum_{i\geq 1} (\sum_{k\geq i} \binom{k-1}{i-1} a_k) t'^i \partial'^{[i]}.$$

Cette dernière expression montre que $P \in D^\dagger_{U\mathbf{Q}}$, et la première montre que $P\cdot 1 = f$. Par suite, φ est surjectif.

Pour prouver l'injectivité de ψ, il suffit de montrer que la multiplication à droite par $t'^2\partial' - \pi = -(\partial + \pi)$ est injective dans $D^\dagger_{U'\mathbf{Q}}$. Or tout élément Q de $D^\dagger_{U'\mathbf{Q}}$ peut s'écrire comme somme d'une série $\sum_{k\geq 0} b_k \partial^{[k]}$, où les b_k forment une suite d'éléments de $A\{1/t'\}$ telle qu'il existe $c \in \mathbb{R}$, $\eta < 1$, avec $\|b_k\| \leq c\eta^k$. On a alors

$$Q(\partial + \pi) = (\sum_{k\geq 0} b_k \partial^{[k]})(\partial + \pi) = \sum_{k\geq 0} (k\, b_{k-1} + \pi b_k)\partial^{[k]},$$

de sorte que la relation $Q(\partial + \pi) = 0$ entraîne que, pour tout k, $b_0 = (-\pi)^k b_k / k!$. On déduit alors de (5.2.2) la nullité de $\|b_0\|$, donc de tous les b_k.

Soit $I = \mathrm{Im}(\psi)$ l'idéal à gauche engendré dans $D^\dagger_{U\mathbf{Q}}$ par $t'^2\partial' - \pi$; il est clair que $I \subset \mathrm{Ker}(\varphi)$. Pour prouver l'inclusion réciproque, nous allons d'abord observer que tout opérateur $P = \sum_{k\geq 0} a_k \partial'^{[k]} \in D^\dagger_{U\mathbf{Q}}$ est congru modulo I à un opérateur de la forme $P' = \sum_{k\geq 0} c_k t'^k \partial'^{[k]}$, où les $c_k \in A$ sont tels qu'il existe $c \in \mathbb{R}$, $\eta < 1$, avec $\|c_k\| \leq c\eta^k$. Pour cela, on montre que, dans $D^\dagger_{U\mathbf{Q}}$, on peut écrire $P = P' + P''$, avec P' du type voulu et $P'' \in I$, en posant :

$$P' = \sum_{k\geq 0} k!\, a_k \partial'^{[k]} \partial^{[k]} / (-\pi)^k, \quad P'' = \sum_{k\geq 1} a_k \partial'^{[k]} (1 - k!\, \partial^{[k]} / (-\pi)^k).$$

En effet, on a pour tout $k \geq 1$,

$$\partial'^{[k]} \partial^{[k]} = \partial'^{[k]} t'^k t^k \partial^{[k]} = (-1)^k \sum_{1\leq i\leq k} \binom{k-1}{i-1} \partial'^{[k]} t'^{k+i} \partial'^{[i]},$$

d'où l'on déduit que $\partial'^{[k]}\partial^{[k]}$ est un opérateur de la forme

$$\partial'^{[k]} \partial^{[k]} = \sum_{1\leq j\leq 2k} \alpha_{jk} t'^j \partial'^{[j]},$$

avec $\alpha_{jk} \in \mathbb{Z}$. On peut alors écrire, en posant $a'_k = k!\, a_k / (-\pi)^k$,

$$P' = a'_0 + \sum_{j\geq 1} (\sum_{2k\geq j} \alpha_{jk} a'_k) t'^j \partial'^{[j]},$$

ce qui montre, compte tenu de (5.2.2), que P' est un opérateur de $D^\dagger_{U\mathbf{Q}}$ de la forme voulue. D'autre part, on a pour tout $k \geq 1$

$$1 - k!\,\partial^{[k]}/(-\pi)^k \;=\; 1-(-\partial/\pi)^k \;=\; (\sum_{i=0}^{k-1}(-\partial/\pi)^i)(1+\partial/\pi),$$

de sorte que, pour montrer que $P'' \in I$, il suffit de prouver que la série

$$Q'' = \sum_{k\geq 1} a_k \partial'^{[k]}(\sum_{i=0}^{k-1}(-\partial/\pi)^i) \;=\; \sum_{0\leq i<k} a_k (i!/(-\pi)^i)\partial'^{[k]}\partial^{[i]}$$

converge dans $D^{\dagger}_{U\mathbf{Q}}$. Or on voit comme précédemment que, pour $i \leq k$, l'opérateur $\partial'^{[k]}\partial^{[i]}$ est de la forme

$$\partial'^{[k]}\partial^{[i]} \;=\; \sum_{1\leq j\leq k+i} \alpha_{jki}\, t'^{\,j-k+i}\partial'^{[j]},$$

où $\alpha_{jki} \in \mathbb{Z}$, et $\alpha_{jki} = 0$ pour $j < k-i$. On obtient donc

$$Q'' = \sum_{j\geq 1}(\sum_{\substack{i<k\\ j\leq k+i}} \alpha_{jki}(i!/(-\pi)^i)a_k t'^{\,j-k+i})\partial'^{[j]}.$$

D'après (5.2.2), il existe $c' \in \mathbb{R}$, et $\eta' < 1$ tels que, pour tous $i \leq k$, on ait

$$\|(i!/\pi^i)a_k\| \;\leq\; c'\eta'^k.$$

Comme, dans la sommation définissant le coefficient de $\partial'^{[j]}$, on a $j \leq k+i < 2k$, il en résulte que $Q'' \in D^{\dagger}_{U\mathbf{Q}}$.

Il reste alors à démontrer que si un opérateur de la forme $P = \sum_{k\geq 0} a_k t'^k \partial'^{[k]}$, avec $\|a_k\| \leq c\,\eta^k$, $\eta < 1$, est dans $\mathrm{Ker}(\varphi)$, il peut s'écrire sous la forme $Q(t'^2\partial' - \pi)$, avec $Q \in D^{\dagger}_{U\mathbf{Q}}$. D'après (5.2.3.4), on a dans $A\{1/t'\}$ la relation

$$\sum_{k\geq 0}(-1)^k a_k (\sum_{i=1}^{k}\binom{k-1}{i-1} t'^{-i}(-\pi)^i/i!) \;=\; 0. \qquad (5.2.3.5)$$

Posons alors $c_0 = a_0$, et, pour $i \geq 1$,

$$c_i \;=\; ((-\pi)^i/i!)\sum_{k\geq i}(-1)^k a_k\binom{k-1}{i-1},$$

de sorte que la relation (5.2.3.5) s'écrit

$$\sum_{i\geq 0} c_i t'^{-i} \;=\; 0. \qquad (5.2.3.6)$$

On observera que, pour tout i, $|c_i| \leq c\,\eta^i$. Si on pose, pour tout $j \geq 0$,

$$b_j \;=\; \sum_{i=0}^{j} c_i t'^{\,j-i},$$

il résulte de (4.2.1) qu'il existe $c' \in \mathbb{R}$, et $\eta' < 1$, tels que l'on ait $|b_j| \leq c'\eta'^j$ pour tout j. D'autre part, on peut grâce à (5.2.3.2) écrire dans $D^{\dagger}_{U'\mathbf{Q}}$ l'opérateur P sous la forme

$$P \;=\; \sum_{k\geq 0} a_k t'^k\partial'^{[k]} \;=\; a_0 + \sum_{k\geq 1}(-1)^k a_k(\sum_{i=1}^{k}\binom{k-1}{i-1} t^i\partial^{[i]})$$

$$= \sum_{k\geq 1} (-1)^k a_k (\sum_{i=1}^{k} \binom{k-1}{i-1} t^i (\partial^{[i]} - (-\pi)^i/i!)),$$

la deuxième égalité résultant de (5.2.3.5). Comme on a

$$\partial^{[i]} - (-\pi)^i/i! = -((-\pi)^i/i!)(\sum_{j=0}^{i-1}(-\partial/\pi)^j)(1+\partial/\pi),$$

il suffit de montrer que l'opérateur

$$Q = \sum_{k\geq 1} (-1)^k a_k (\sum_{i=1}^{k} \binom{k-1}{i-1} t^i ((-\pi)^i/i!)(\sum_{j=0}^{i-1}(-\partial/\pi)^j))$$

appartient à $D^{\dagger}_{U\mathbf{Q}}$. Or on peut encore écrire Q sous la forme

$$Q = \sum_{j\geq 0}(\sum_{i>j}((-\pi)^i/i!)(\sum_{k\geq i}(-1)^k a_k \binom{k-1}{i-1}) t'^{-i})(-\partial/\pi)^j,$$

c'est à dire

$$Q = \sum_{j\geq 0}(\sum_{i>j} c_i t'^{-i})(-\partial/\pi)^j.$$

Compte tenu de (5.2.3.6), on a donc aussi

$$Q = -\sum_{j\geq 0}(\sum_{i\leq j} c_i t'^{-i})(-\partial/\pi)^j.$$

Pour $j \geq 1$, on a d'après (5.2.3.3)

$$(-\partial/\pi)^j = (j!/\pi^j) t'^j \sum_{h=1}^{j} \binom{j-1}{h-1} t'^h \partial'^{[h]}.$$

On en déduit donc

$$Q = -c_0 - \sum_{j\geq 1}(\sum_{i\leq j} c_i t'^{j-i})(j!/\pi^j)(\sum_{h=1}^{j}\binom{j-1}{h-1} t'^h \partial'^{[h]})$$

$$= -c_0 - \sum_{h\geq 1}(\sum_{j\geq h}\binom{j-1}{h-1}(j!/\pi^j) b_j) t'^h \partial'^{[h]}.$$

D'après l'estimation vue plus haut pour $\|b_j\|$ et le lemme (5.2.2), il en résulte que $Q \in D^{\dagger}_{U\mathbf{Q}}$, ce qui achève la démonstration.

Remarque. Si on prend un élément π tel que $\mathrm{ord}_p \pi < 1/(p-1)$, la connexion correspondante sur $\mathcal{L}_\pi$ ne vérifie plus la condition de convergence (3.0.1.1), de sorte que $\mathrm{sp}_* \mathcal{L}_\pi$ n'a plus de structure de $\mathcal{D}^{\dagger}_{\mathfrak{X}\mathbf{Q}}$-module. D'autre part, si $\mathrm{ord}_p \pi > 1/(p-1)$, le rayon de convergence de $\exp \pi t$ est > 1, et $\exp \pi t$ définit un isomorphisme $\mathcal{L}_\pi \simeq \mathcal{O}_{X/K}$.

Bibliographie

[EGA] A. Grothendieck (avec la collaboration de J. Dieudonné), *Eléments de Géométrie Algébrique*, Publ. Math. I.H.E.S. **4, 8, 11, 17, 20, 24, 28, 32**.

[SGA4] M. Artin, A. Grothendieck, J.-L. Verdier, *Théorie des topos et cohomologie étale des schémas*, Lecture Notes in Math. **269, 270, 305**, Springer-Verlag (1972).

[1] P. Berthelot, *Cohomologie cristalline des schémas de caractéristique* $p > 0$, Lecture Notes in Math. **407**, Springer Verlag (1974).

[2] P. Berthelot, *Géométrie rigide et cohomologie des variétés algébriques de caractéristique p*, Journées d'analyse p-adique (1982), in *Introduction aux cohomologies p-adiques*, Bull. Soc. Math. France, Mémoire **23**, p. 7–32 (1986).

[3] P. Berthelot, *Cohomologie rigide et théorie de Dwork : le cas des sommes exponentielles*, Astérisque **119-120**, p. 17–49 (1984).

[4] P. Berthelot, *Cohomologie rigide et cohomologie rigide à support propre*, à paraître dans Astérisque.

[5] P. Berthelot, *$\mathscr{D}^\dagger$-modules cohérents*, en préparation.

[6] A. Borel et al., *Algebraic D-modules*, Perspectives in Math. **2**, Academic Press (1987).

[7] S. Bosch, U. Güntzer, R. Remmert, *Non-archimedean analysis*, Grundlehren des math. Wissenschaften **261**, Springer-Verlag (1984).

[8] G. Christol, *Un théorème de transfert pour les disques singuliers réguliers*, Astérisque **119-120**, p. 151–168 (1984).

[9] A. Grothendieck, *On the de Rham cohomology of algebraic varieties*, Publ. Math. I.H.E.S. **29**, p. 351–359 (1966).

[10] A. Grothendieck, *Crystals and the de Rham cohomology of schemes*, in *Dix exposés sur la cohomologie des schémas*, p. 306–358, North-Holland (1968).

[11] R. Hartshorne, *Residues and Duality*, Lecture Notes in Math. **20**, Springer Verlag (1966).

[12] M. Kashiwara, *Faisceaux constructibles et systèmes holonomes d'équations aux dérivées partielles à points singuliers réguliers*, Sém. Goulaouic-Schwarz, 1979-80, exp. 19, Ecole Polytechnique (1981).

[13] M. Kashiwara, *The Riemann-Hilbert problem for holonomic systems*, Publ. R.I.M.S. **437**, Kyoto University (1983).

[14] Z. Mebkhout, *Cohomologie locale des espaces analytiques complexes*, Thèse Université Paris VII (1979).

[15] Z. Mebkhout, *Une équivalence de catégories*, Comp. Math. **51**, p. 51–62 (1984).

[16] Z. Mebkhout, *Une autre équivalence de catégories*, Comp. Math. **51**, p. 63–88 (1984).

[17] Z. Mebkhout, *Le formalisme des six opérations de Grothendieck pour les $\mathscr{D}_X$-modules cohérents*, Travaux en cours **35**, Hermann (1989).

[18] A. Ogus, *The convergent topos in characteristic p*, Grothendieck Festschrift, Progress in Math., Birkhäuser (1990).

EXTENSIONS DE D-MODULES ET GROUPES DE GALOIS DIFFERENTIELS

par D. BERTRAND

Univ. Paris VI, Maths, T. 46,
4, Place Jussieu, F-75252 Paris Cedex O5 .

La théorie des équations différentielles linéaires à second membre peut être vue comme un analogue différentiel de la théorie de Kummer. Pour tout entier n donné, celle-ci décrit les extensions d'un corps K engendrées par les racines n-ièmes des éléments de K ; elle repose sur une bonne connaissance de l'extension cyclotomique d'ordre n de K. Celle-là étudie les équations de la forme $Ly = g$, où L est un opérateur différentiel donné; on sait (méthode de variations des constantes) le rôle qu'y joue l'équation homogène correspondante.

Inspiré par le travail de Bashmakov et de Ribet [23] en théorie de Kummer sur les groupes algébriques, je montre ici que cette analogie formelle se transporte sans peine à l'étude des groupes de Galois *non réductifs* associés à ces deux situations.

La première partie traite de la notion d'extension de D-modules, qui englobe celle d'équation à second membre, et dont on décrit les propriétés élémentaires. On détermine dans la deuxième partie l'image des représentations galoisiennes définies par de telles extensions. La troisième partie passe en revue quelques applications, en particulier aux équations hypergéométriques et à des questions de transcendance.

Je remercie R. Coleman et N. Katz, dont les travaux [11] et [17] m'ont fourni le cadre de la présentation qui suit, F. Beukers, qui m'en a indiqué le lien avec la théorie de Shidlovsky, ainsi que Z. Mebkhout et J-P. Ramis pour d'utiles discussions.

§1. Extensions de D-modules.

Soient X un ouvert propre de la droite projective $\mathbf{P}_1(\mathbf{C})$, et O_X

son algèbre affine. La O_X-algèbre $D = D_X$ des opérateurs différentiels sur X est engendrée par une dérivation ∂ de O_X. Par D-module, on entend ici un *fibré à connexion* V sur X. Vu la forme de X, il revient au même de considérer un O_X-module libre de type fini $V^\sim$ muni d'une action $\partial_{(V)}$ de ∂ vérifiant les propriétés usuelles des dérivations. Dans une base B de $V^\sim$ sur O_X, $\partial_{(V)}$ est représentée par un opérateur $\partial + A_{(B)}$, où $A_{(B)}$ est une matrice carrée à coefficients dans O_X. On notera $H^*(V)$ les groupes de cohomologie du complexe de de Rham algébrique $0 \to V \to V \to 0$ défini par $\partial_{(V)}$. En d'autres termes, $H^0(V)$ est formé des sections horizontales (globales) de V, tandis que $H^1(V)$ s'identifie au co-noyau de $\partial_{(V)}$ (les autres groupes sont nuls). Ce sont des espaces vectoriels sur **C**, de dimensions $h^*(V)$ finies.

Soient maintenant V, V' deux D-modules. On note V^* le D-module $\mathrm{Hom}(V,O_X)$ (muni de la connexion duale), de sorte qu'un morphisme de V vers V' est une section horizontale (globale) du D-module $V'\otimes V^*$ (voir par exemple [13], [2]). Une extension de V par V' est une suite exacte $0 \to V' \to E \to V \to 0$ dans la catégorie des D-modules; par abus de language, on la notera encore E. Le procédé usuel de Baer (ou des techniques plus fonctorielles, voir [22]), permet de munir l'ensemble $\mathrm{Ext}_D(V,V')$ des classes d'isomorphismes d'extensions de V par V' d'une structure de groupe, qui vérifie :

Lemme 1 (Coleman [11]): *les groupes* $\mathrm{Ext}_D(V,V')$ *et* $H^1(V'\otimes V^*)$ *sont canoniquement isomorphes.*

<u>Démonstration</u>: soit E un (représentant d'un) élément de $\mathrm{Ext}_D(V,V')$. Puisque X est affine, la suite exacte de fibrés associée à l'extension $E^\sim$ est scindable. Choisissons-en une section $s : V^\sim \to E^\sim$. Alors, l'application $\partial_{(E)}s - s\partial_{(V)}$ est un O_X-homomorphisme de $V^\sim$ vers $V'^\sim$, dont la classe ξ_E dans $H^1(V'\otimes V^*)$ ne dépend pas du choix de s, ni du représentant de E. L'application

$$\xi : \mathrm{Ext}_D(V,V') \to H^1(V'\otimes V^*) : E \to \xi(E) = \xi_E$$

définit l'isomorphisme escompté, comme on le vérifie aisément (dans des bases convenables -et si l'on convient d'écrire les éléments de $(O_X)^n$ en colonne -, la transposée de la matrice d'un représentant de ξ_E apparaît dans celle de $\partial_{(E)}$ sous la forme d'un rectangle en position inférieure gauche).

On peut préciser cet énoncé lorsque $V' = O_X$ (muni de sa connexion canonique), en convenant d'appeler extension vectorielle universelle du D-module V la classe d'isomorphisme des extensions E de V

par une somme directe finie U de copies du D-module O_X , telles toute extension $0 \to U' \to E' \to V \to 0$ de ce type se déduise de E par image directe sous un unique morphisme de U dans U' (voir [12] pour une généralisation de cette définition, et [15] pour un analogue dans la catégorie des modules de Drin'feld). Une telle extension existe : c'est l'extension de V par U = $Hom(H^1(V^*),O_X)$ qui correspond, dans l'isomorphisme fourni par le lemme 1, à l'élément identité de $H^1(U \otimes V^*) = Hom(H^1(V^*),H^1(V^*))$. Le rang $h^1(V^*)$ de U est donné par le lemme suivant, où S désigne l'ensemble des points à l'infini de X , et Irr(V) la somme des irrégularités de V aux différents points de S .

Lemme 2 : *soit* V *un* D*-module de rang* n . *Alors,*

$$h^1(V^*) = (card(S) - 2)\, n \; + \; Irr(V) + h^0(V^*) \;.$$

En particulier, si V *est un module irréductible non isomorphe à* O_X *(resp. si* V = O_X *) , il existe exactement* (card(S) - 2) n + Irr(V) *(resp.* card(S) - 1*) extensions de* V *par* O_X *linéairement indépendantes sur* **C** .

Démonstration : le premier énoncé résulte immédiatement du théorème de comparaison de Deligne ([13], p. 111; voir aussi [17]) :

$$h^0(V) - h^1(V) = n\, \chi(X) - Irr(V) \;,$$

appliquée au D-module V* , dont l'irrégularité vaut celle de V . Pour le second, noter que si V est irréductible, V* l'est aussi, et n'admet donc pas de section horizontale globale, à moins d'être isomorphe à O_X (de façon générale, tout morphisme non nul entre deux D-modules irréductibles est un isomorphisme, les fonctions rationnelles dont la dérivée logarithmique appartient à O_X étant des unité de O_X).

Remarque 1 : Une extension E de V par V' fournit par dualité une extension E* de V'* par V* . Le lemme 2 permet donc également de décrire les extensions de O_X par V : elles sont paramétrées par $H^1(V)$.

Pour les applications, de type rationnel, que nous avons en vue, il s'avère commode de travailler dans la catégorie des **D**-modules, où $\mathbf{D} = D_K$ = $K[\partial]$ désigne l'anneau des opérateurs différentiels à coefficients dans le corps K = $\mathbf{C}(\mathbf{P}_1)$ des fonctions rationnelles sur X , et où les notations précédentes (et suivantes) s'étendent de façon naturelle . Tout D-module V définit par extensions des scalaires un **D**-module, qu'on notera V_K , ou encore V . La remarque concluant la démonstration précédente, jointe à un calcul local aux différentes places de X , montre en effet que si deux D-modules V, V' sont isomorphes en tant que **D**-modules, il le sont déjà en tant que D-modules. De plus :

Lemme 3 : *l' application naturelle* : $Ext_D(V,V') \to Ext_{\mathbf{D}}(V,V')$ *est injective (mais non surjective)* .

<u>Démonstration</u> : posant $W = V'\otimes V^*$, on se ramène à montrer que l'application naturelle de $H^1(W)$ dans $H^1(W_K) = W_K/\partial_{(W)}W_K$ est injective, c'est-à-dire qu'un élément f de W_K dont l'image par $\partial_{(W)}$ appartient à W est nécessairement dans W . Cela se vérifie par un calcul local (noter qu'en toute place de O_X , de paramètre local t , le générateur ∂ de D est égal à d/dt à une unité locale près).

Explicitons ces énoncés dans le cadre des equations différentielles. Si M est un élément unitaire de $D = O_X[\partial]$, le O_X-module D/DM est naturellement muni d'une action (à gauche) de ∂ , qui en fait un D-module V(M) , et d'une base canonique B(M), formée par les classes des premières puissances ≥ 0 de ∂ . Le dual de V(M) est isomorphe à D/DM* , où M* désigne l'opérateur différentiel adjoint de M (voir [16], 1.5, ainsi que [17], 2.9 et [18] pour une présentation plus intrinsèque), et les solutions de M s'identifient aux vecteurs horizontaux de ce dual . De même, l'application $^t(0, ..., g) \to g$ établit un **C**-isomorphisme de $H^1(V(M))$ sur le conoyau O_X/M^*O_X de M* dans O_X.

Soit alors M = L'L un opérateur décomposé, où L , L' sont des éléments unitaires de D d'ordres > 0 . Il définit une extension :

$$E(L,L') : 0 \to V(L') \to V(M) \to V(L) \to 0 ,$$

où la seconde flèche désigne la multiplication à droite par L , et la troisième la surjection canonique. L'élément de $Ext_D(V(L),V(L'))$ associé correspond à la classe dans $H^1(Hom(V(L),V(L')))$ du O_X-homomorphisme $\xi_{L,L'}$ qui envoie le dernier vecteur de B(L) sur le premier de B(L') et annule les autres vecteurs de B(L) (pour s'en convaincre, choisir {B(L')L, B(L)} pour base de M). Lorsque $L' = \partial$, le composé Ξ des isomorphismes $Ext_D(V(L),O_X) \to H^1(V(L^*)) \to O_X/LO_X$ fournit le lien annoncé dans l'introduction avec les équations à second membre. Plus précisément, on a :

Lemme 4 : i) *si* g *est un élément non nul de* O_X *, le* **D**-*module* $E(L,\partial-g^{-1}\partial g)$ *est isomorphe à un élément* $E_g(L)$ *de* $Ext_D(V(L),O_X)$ *dont l'image par* Ξ *coïncide avec la classe de* g *dans* O_X/LO_X .

ii) *supposons que* V(L) *soit irréductible et non isomorphe à* O_X . *Alors, l'extension* $E(L,\partial)$ *est triviale dans* $Ext_D(V(L),O_X)$ *si et seulement si* ∂L *est divisible à droite dans* **D** *par un opérateur de la forme* $\partial-f^{-1}\partial f$, *où* f *appartient à* O_X .

<u>Démonstration</u> : i) la multiplication à droite par g définit un isomorphisme de $\mathbf{D}/\mathbf{D}(\partial-g^{-1}\partial g)$ sur $\mathbf{D}/\mathbf{D}\partial = (O_X)_K$, et l'image directe $E'_g(L)$ de la **D**-extension $E(L, \partial-g^{-1}\partial g)$ de V(L) par O_X sous cet isomorphisme est donnée dans $H^1(Hom_K(V_K(L),K)$ par le composé par la multiplication à gauche

par g du K-homomorphisme $\xi_{L,\ \partial-\partial g/g}$ de $V_K(L)$ dans K. Comme ce composé provient d'un morphisme de $V(L)^{\sim}$ dans O_X, $E'_g(L)$ provient en fait d'une D-extension $E_g(L)$. (En terme concret, on effectuera la transformation de cisaillement de matrice $\mathrm{diag}(g, \mathrm{Id}_n)$ sur la transposée de la matrice représentative de $\partial_{E(L,\partial-\partial g/g)}$ dans la base $\{L, B(L)\}$).

ii) D'après ce qui précède, $E(L,\partial)$ est triviale si et seulement si l'équation $Ly = 1$ admet une solution f dans O_X. Alors, ∂L appartient à l'idéal à gauche engendré dans **D** par son annulateur minimal $\partial - f^{-1}\partial f$. Inversément, si cette condition est vérifiée, f est une solution de ∂L et Lf est une constante. De l'hypothèse faite sur $V(L)$, on déduit comme au lemme 2 que cette constante n'est pas nulle, et l'équation $Lf = 1$ est bien résoluble dans O_X.

Remarque 2 : On prendra garde au fait qu'un élément L de D peut être irréductible dans D *sans* l'être dans **D** (il n'y a pas ici de lemme de Gauss !). C'est par exemple le cas de l'opérateur hypergéométrique sur $X = \mathbf{P}_1 - \{0,1,\infty\}$ associé, pour des paramètres a et c génériques, et un entier $n > 1$, à la fonction $z^{1-c}{}_2F_1(1+a-c, 1-n-c, 2-c; z)$. En revanche, $V(L)$ est irréductible si et seulement si $V_K(L)$ l'est (l'intersection avec V d'un **D**-sous-module **W** de V_K est un réseau W stable sous $\partial_{(V)}$).

Convenons enfin d'appeler *équivalents* deux opérateurs unitaires L, N de **D** tels les **D**-modules $V_K(L)$, $V_K(N)$ soient isomorphes. Le critère de trivialité ci-dessus se généralise de la façon suivante :

Lemme 5 : *soient* L, L' *deux éléments unitaires de* D *tels qu'aucun sous-module non nul de* $V(L)$ *ne soit isomorphe à un sous-module de* $V(L')$. *Alors, l'extension* $E(L,L')$ *est triviale si et seulement s'il existe deux éléments* N, N' *de* **D**, *respectivement équivalents à* L, L', *tels que* $L'L = NN'$.

<u>Démonstration</u> : si $E = E(L,L')$ est triviale, E admet un sous-D-module V' isomorphe à $V(L')$, et tel que E/V' soit un D-module isomorphe à $V(L)$. On conclut en étendant les scalaires à K, et en notant comme dans [24], 2.3, que l'image de 1 dans E/V' en est un vecteur cyclique. Inversément, les suites exactes de **D**-modules fournies par les décompositions $L'L$, NN' permettent de construire, comme dans [19] et [24], 2.2, un **D**-homomorphisme du sous-module $V_K(N)$ de E_K dans $V_K(L)$, qui ne peut éviter d'être injectif (donc bijectif, et de fournir alors une section de E_K) que s'il contient un élément non nul de $V_K(L')$. Il existe alors un **D**-morphisme, donc aussi un D-morphisme, non trivial d'un sous-module de $V(L)$ dans $V(L')$, ce que notre hypothèse interdit. Par conséquent, $E(L,L')$ est triviale dans $\mathrm{Ext}_{\mathbf{D}}(V(L),V(L'))$, et donc dans $\mathrm{Ext}_D(V(L),V(L'))$.

Remarque 3 : la première partie du raisonnement ci-dessus entraîne la version relative suivante du lemme du vecteur cyclique : si L (resp. L') est donné, les éléments de $E_{\mathbf{D}}(V_K(L), V_K(L'))$ sont *tous* représentés par des extensions de la forme E(N,N') , où N (resp. N') parcourt l'ensemble des éléments de **D** équivalents à L (resp. L') . On peut donc se limiter dans ces questions à l'étude des opérateurs décomposés dans **D** .

§2. Représentations galoisiennes.

On reprend les notations du paragraphe précédent. On fixe de plus une extension différentielle minimale K' de K , de corps de constantes **C** , dans laquelle tous les éléments de D admettent un système fondamental de solutions, et on note $G = Gal_\partial(K'/K)$ le groupe de Galois différentiel de cette extension. C'est un groupe pro-algébrique, dont on n'aura en fait à considérer que des quotients de dimension finie (et en nombre fini) du type suivant.

Soit V est un D-module de rang n . On note V^∂ le **C**-espace vectoriel de dimension n de V(K') formé des vecteurs horizontaux de $V\otimes K'$. L'extension minimale K_V de K où sont définis les éléments de V^∂ est une extension de Picard-Vessiot, dont le groupe de Galois différentiel $G_V = Gal_\partial(K_V/K)$ s'identifie à l'image de la représentation naturelle de G sur V^∂ ; on désignera par $H_V = Gal_\partial(K'/K_V)$ le noyau de cette représentation. Les groupes de cohomologie que nous considérerons seront relatifs à ces représentations, et à des cochaines continues. Le dual V* de V fournit la même extension de Picard-Vessiot K_V que V , et une représentation de G_V sur $(V^*)^\partial = (V^\partial)^*$ duale de celle de V . De même, pour un second D-module V' , la représentation $(V'\otimes V^*)^\partial = V'^\partial\otimes V^{\partial *}$ de G annule $H_V \cap H_{V'} = H_{V\oplus V'}$, et $K_{V'\otimes V^*}$ est contenu dans le compositum $K_V.K_{V'} = K_{V\oplus V'}$ des corps K_V et $K_{V'}$.

Soient alors E une extension de V par V' . Comme la représentation E^∂ de G est une extension de V^∂ par V'^∂ , le corps K_E contient $K_{V\oplus V'}$. Le groupe de Galois différentiel

$$G^u(E) = Gal_\partial(K_E/K_{V\oplus V'})$$

de l'extension $K_E/K_V.K_{V'}$ est nul si E est triviale. De façon générale, il s'identifie à l'image de $H_V \cap H_{V'}$ dans $Aut(E^\partial)$, et donc à un sous-espace vectoriel de $Hom(V^\partial, V^{*\partial})$. C'est cette image qu'on cherche ici à décrire. Pour alléger, on désigne par W le D-module $V'\otimes V^*$, par π les projections canoniques $E \to V$ et $E^\partial \to V^\partial$, et par $G^{ss}(E) = G_{V\oplus V'} = Gal_\partial(K_{V\oplus V'}/K)$ le quotient de G_E par $G^u(E)$.

Soit g un représentant dans W de l'image de E dans $H^1(W)$. Trouver le corps de définition K_E d'une base de E^∂ (i.e. d'un "système fondamental" de vecteurs horizontaux de E) revient à rechercher la plus petite extension différentielle de $K_{V\oplus V'}$ où l'extension

$$0 \to V' \otimes K_{V\oplus V'} \to E \otimes K_{V\oplus V'} \to V \otimes K_{V\oplus V'} \to 0$$

se trivialise; en effet, $V \otimes K_{V\oplus V'}$ et $V' \otimes K_{V\oplus V'}$ sont déjà, par définition de $K_{V\oplus V'}$, isomorphes à des sommes directes de copies du $K_{V\oplus V'}\otimes D$ - module trivial $K_{V\oplus V'}$. C'est donc le corps $K_E = K_{V\oplus V'}(f)$ engendré sur $K_{V\oplus V'}$ par une solution f de l'équation

$$\partial_W(f) = g .$$

Comme en théorie de Kummer (voir [23], dont ce qui suit est fortement inspiré), ce corps est bien indépendant de la solution f choisie, puisque deux solutions diffèrent d'un élément de W^∂ , et que l'extension de Picard-Vessiot K_W de W est contenue dans $K_{V\oplus V'}$. Dans ces conditions, l'homomorphisme

$$\psi_E : H_{V\oplus V'} \to W^\partial : \sigma \mapsto \psi_E(\sigma) = \sigma f - f$$

ne dépend que de E . Il fournit l'injection canonique, mentionnée plus haut, de son image $G^u(E)$ dans $\mathrm{Hom}(V^\partial, V'^\partial)$. Quant à l'application **C**- linéaire

$$\psi : \mathrm{Ext}_D(V,V') \to \mathrm{Hom}(H_{V\oplus V'}, W^\partial) : E \mapsto \psi(E) = \psi_E ,$$

on l'obtient par construction en composant l'isomorphisme ξ avec l'application : $H^1(W) \to H^1(G,W^\partial)$ qui, à la classe de g , associe la classe de cohomologie de l'homomorphisme croisé : $\sigma \mapsto \sigma g - g$ de G dans W^∂, suivi de la restriction $H^1(G,W^\partial) \to H^1(H_{V\oplus V'},W^\partial)$ de G à $H_{V\oplus V'}$. Quitte à remplacer K' par une clôture différentielle de K , on peut encore décrire ψ de la façon suivante. La suite exacte de G-modules: $0 \to W^\partial \to W(K') \to W(K') \to 0$, où la troisième flèche est $\partial_{(V)}$, donne naissance à un opérateur cobord :

$$W(K')^G/\partial_{(W)}(W(K')^G) = H^1(W_K) \to H^1(G,W^\partial) ,$$

et ψ se déduit de cette flèche par les homomorphismes décrits aux lemmes 1 et 3 d'une part, et par la suite d'inflation-restriction :

$$0 \to H^1(G^{ss}(E), W^\partial) \to H^1(G,W^\partial) \to \mathrm{Hom}_{G^{ss}(E)}(H_{V\oplus V'},W^\partial)$$

d'autre part. En particulier, l'image $G^u(E)$ de ψ_E est un sous-espace de W^∂ stable sous l'action de G_W , et l'application ψ est injective dès que $H^1(G_{V\oplus V'}, W^\partial)$ est nul. Le lemme suivant permet, dans de nombreux cas, de vérifier cette dernière hypothèse, et joue ainsi le rôle du théorème 90 de Hilbert (voir [26] pour un analogue p-adique plus souple).

Lemme 6 : *soit* G *un groupe algébrique réductif sur* **C** , *et* V *une représentation rationnelle de* G . *Alors,* $H^1(G,V) = 0$.

<u>Démonstration</u> : la cohomologie de G s'injectant dans celle de sa composantre neutre, on peut supposer G connexe. Par récurrence sur la dimension de V, on se ramène au cas où V est une représentation triviale ou irréductible. Dans le premier cas , on note que G n'admet pas de quotient unipotent. Dans le second, le centre de G est formé d'homothéties (Schur). S'il n'est pas réduit à l'identité, le lemme de Sah montre que $H^1(G,V)$ est nul. Sinon, G, et son algèbre de Lie $\mathbf{g}$, sont semi-simples. En particulier (voir [8] §6, exercice 1.b), $H^1(\mathbf{g},V) = 0$. Mais la cohomologie (par cochaines continues) de G s'injecte dans dans la cohomologie de $\mathbf{g}$, et $H^1(G,V)$ est encore nul. (Lorsque G est semi-simple, on peut également établir ce lemme en considérant le produit semi-direct de G par V associé à la représentation V, et en notant que d'après le théorème de Levi- Malcev - voir [8], 6.8, corollaire 3; [28], théorème 3.1.18 -, il n'admet qu'une classe de conjugaisons de sections de sa projection sur G .)

Les remarques précédentes ramènent l'étude de $G^u(L)$ à celle des sous-D-modules de W . Plutôt qu'un énoncé général, nous donnons maintenant quelques exemples simples d'applications de ce principe.

Théorème 1 :*soient* V , V' *deux* D-*modules tels que* $V \otimes V'^*$ *soit un* D-*module irréductible, et soit* E *une extension non triviale de* V *par* V'. *Alors,* $Gal_\partial(K_E/K_{V\oplus V'})$ *est isomorphe à* $Hom(V^\partial, V'^\partial)$.

<u>Démonstration</u> : puisque $W = V \otimes V'^*$ est irréductible, V , V' et les représentations W^∂ , V^∂ , V'^∂ le sont également. Par conséquent, le groupe $G_{V\oplus V'}$, dont $V \oplus V'$ est une représentation fidèle complètement réductible , est réductif, et les remarques précédentes, jointes à l'hypothèse faite sur l'extension E , entraînent que l'homomorphisme ψ_E n'est pas nul. Son image $G^u(E)$ est ainsi un sous-G_W-module de W^∂ non nul, et remplit donc tout W^∂ .

Il est facile d'étendre cette proposition à des situations semi-simples, en considérant les images directes ou inverses de E associées aux différents facteurs de V et de V' . Voici une illustration typique de cette démarche.

Théorème 2 : *soient* V *un* D-*module irréductible , et* E *une extension de* V *par une somme directe* $V' = (O_X)^r$ *de copies de* O_X . *On suppose que les images directes de* E *sous les projections de* V' *sur ses différents facteurs* O_X *sont des éléments de* $Ext_D(V,O_X)$ *linéairement indépendants sur* **C** . *Alors,* $Gal_\partial(K_E/K_V)$ *est isomorphe à* $(V^{*\partial})^r$.

<u>Démonstration</u> : soient π_i , $i = 1$, ..., r , les projections en question (pour les D-modules, et pour leurs espaces de sections), et E_i l'extension $(\pi_i)_*E$. La construction de l'homomorphisme ψ étant fonctorielle, $\psi(E_i)$ coïncide

avec $\pi_i \circ \psi_E$, et l'injectivité de ψ entraîne que ces r homomorphismes de H_V dans $V^{*\partial}$ sont linéairement indépendants sur **C** . Considérons alors l'image $G^u(E)$ de H_V dans $(V^{*\partial})^r$ par leur produit. C'en est un G_V-module. Mais puisque $V^{*\partial}$ est par hypothèse G_V-irréductible, $G^u(E)$ ne peut différer de $(V^{*\partial})^r$ que s'il existe des G_V- endomorphismes (c'est à dire des scalaires, d'après Schur) $a_1, ..., a_r$ non tous nuls tels que $a_1\psi(E_r) + ...+ a_r\,\psi(E_r)$ annule H_V . Ceci contredit l'indépendance linéaire des $\psi(E_i)$.

En termes concrets, on a démontré, avec le notations de la première partie :

Corollaire : *soit* L *un opérateur différentiel à coefficients dans* K , *de rang* n , *et irréductible. Notons* K_L *l'extension de Picard-Vessiot de* L *sur* K , *et soient* $f_1, ..., f_r$ *des éléments de* K' *tels que* $Lf_1, ..., Lf_r$ *soient des éléments de* K *linéairement indépendants sur* **C** *modulo* LK . *Alors, le degré de transcendance de l'extension* $K_L(\partial^i f_j\ ;\ j=1,...,r,\ i=0,...,n-1)$ *de* K_L *est égal à* nr .

Terminons par une version tordue du théorème 2 :

Théorème 3 : *soient* V *un* D*-module irréductible* , L(1), ... , L(r) *des* D*-modules de rang* 1 *deux à deux non isomorphes, et* E *une extension de* V *par* V' = L(1)⊕... ⊕ L(r) . *On suppose que l'image directe* E(i) *de* E *sous la projection de* V' *sur chacun des facteurs* L(i) *est un élément non nul de* $Ext_D(V,L(i))$. *Alors, le groupe de Galois différentiel du corps* K_E *sur* $K_{V\oplus L(1)\oplus ... \oplus L(r)}$ *est isomorphe à la somme directe des groupes* $V^{*\partial}\otimes L(i)^\partial$.

Démonstration : les représentations de G associées aux D-modules Hom(V, L(i)) sont les produits de sa représentation $V^{*\partial}$ par des caractères χ_i distincts. Elles sont donc irréductibles et deux à deux non isomorphes. Si la conclusion du théorème était incorrecte, la projection du groupe de Galois étudié sur l'un des facteurs $V^{*\partial}\otimes L(i)^\partial$ serait donc nulle, et l'extension E(i) correspondante serait triviale.

Le lecteur averti rapprochera l'énoncé du théorème 2 ou de son corollaire, et plus encore sa démonstration, du théorème 1.2 de [23] . Les hypothèses de semi-simplicité jouent dans les deux cas un rôle fondamental. Mais Ribet étend dans [23] sa méthode à des 1-motifs de type "doublement"mixtes. Cela correspond, dans la situation différentielle, à supposer que V est lui-même une extension non triviale de D-modules semi-simples, et l'on peut encore obtenir des analogues des résultats de [23] dans ce cas (voir la remarque 5 de la dernière partie). En fait, le cadre d'étude naturel des équations différentielles semble être celui des motifs à n poids (voir par exemple [25]). Nous reviendrons sur ces remarques ultérieurement.

§3. Applications.

Les énoncés précédents permettent d'établir de façon unifiée divers résultats (pour l'essentiel bien connus) d'indépendance algébrique de fonctions classiques, en choisissant convenablement le D-module V. Nous en donnons ici quatre exemples. Les deux premiers concernent des équations fuchsiennes, et ne reflètent donc que des propriétés standard de la monodromie. Des singularités irrégulières apparaissent dans les deux derniers.

a) <u>Le cas</u> $V = O_X$.

D'après le lemme 2, le **C**-espace vectoriel $Ext_D(O_X, O_X)$ est de dimension card(S) - 1. Par ailleurs, le groupe $O_X^*/\mathbf{C}^*$ est un **Z**-module libre de rang card(S) - 1 (considérer l'application

$$Rés_S : O_X^* \to Hom(S,\mathbf{Z}) : u \mapsto \{s \mapsto \text{résidu en s de } du/u\},$$

et appliquer le théorème des résidus). Un lien direct entre ce groupe et cet espace vectoriel est fourni par l'application qui, à une unité u de O_X, associe l'extension $E_{du(\partial)/u}(\partial)$ de O_X par O_X (notations du lemme 4); celle-ci définit en effet un homomorphisme de groupes

$$\lambda : O_X^*/\mathbf{C}^* \to Ext_D(O_X, O_X),$$

dont on vérifie aisément le caractère injectif. Mais mieux :

Lemme 7 : *l'homomorphisme* λ *s'étend en un homomorphisme injectif (donc un isomorphisme)* $\lambda\otimes 1$ *de* $(O_X^*/\mathbf{C}^*)\otimes\mathbf{C}$ *dans* $Ext_D(O_X, O_X)$.

[L'injectivité de λ exprime que les classes modulo $2i\pi\mathbf{Z}$ des logarithmes d'éléments de O_X^* multiplicativement indépendants sont linéairement indépendantes sur **Z**, et le lemme 7 qu'elles le sont même sur **C**.]

<u>Démonstration</u> : l'énoncé revient à affirmer que l'image de $(O_X^*/\mathbf{C}^*)$ par $Rés_S$ est un réseau du **C**-espace vectoriel qu'elle engendre dans $Hom(S,\mathbf{C})$, ce qui est clair. Dans l'esprit du point b) ci-dessous, on peut encore interpréter cet énoncé de la façon suivante. Soit X^h l'espace analytique associé à X. La suite de cohomologie déduite de la suite exacte de l'exponentielle sur X^h fournit un isomorphisme du groupe $(O_{X^h})^*/exp(O_{X^h})$ dans $H^1(X^h, 2i\pi\mathbf{Z})$. En second lieu, l'intersection de O_X^* avec $exp(O_{X^h})$ est formée d'éléments indéfiniment divisibles dans O_X^*, d'après (par exemple !) l'argument de monodromie de Shafarevich-Manin (cf. [20], p.215), et est donc réduite aux constantes. Enfin, comme $V = O_X$ n'a que des singularités régulières, $H^1(X^h,\mathbf{C}) = H^1(X^h,(V^h)^\partial) = H^1(X^h,V^h)$ est isomorphe à $H^1(V)$ (cf. [22], §0),

donc à $Ext_D(O_X,O_X)$. Il reste à joindre à ces remarques l'isomorphisme de $H^1(X^h,2i\pi\mathbf{Z})\otimes\mathbf{C}$ avec $H^1(X^h,\mathbf{C})$ pour obtenir l'application λ , et le lemme 7 .

En combinant le lemme 7 au théorème 2 (ou plutôt, la traduction de l'un au corollaire de l'autre), et en les étendant pour plus de généralité au cas des revêtements finis de X , on a finalement établi :

Théorème 4 : *soient* $\ell_1, ..., \ell_r$ *des fonctions holomorphes sur un ouvert de* $\mathbf{C}$, *telles que* $\exp(\ell_1), ..., \exp(\ell_r)$ *soient des fonctions algébriques multiplicativement indépendantes modulo* $\mathbf{C}^*$. *Alors,* $\ell_1, ..., \ell_r$ *sont algébriquement indépendantes sur* $\mathbf{C}(x)$.

Bien sûr, il ne s'agit là que d'un cas particulier du théorème d'Ax [2] sur l'analogue fonctionnel de la conjecture de Schanuel. Mais il est intéressant de noter que l'essentiel de sa preuve s'est concentrée au lemme 7, qu'on peut voir comme un analogue fonctionnel de la conjecture de Leopoldt.

b) Le cas des équations de Picard-Fuchs .

Pour alléger, nous nous limitons ici à décrire la théorie de Manin [20], telle que l'a récemment interprétée Coleman [11], dans le cas d'un schéma elliptique A sur X , qu'on suppose ne devenir constant sur aucun revêtement fini de X (autrement dit, en notant encore A la courbe elliptique obtenue par extension des scalaires à K , l'invariant j(A/K) *n'est pas constant*) . La connexion de Gauss-Manin munit alors le fibré $H^1{}_{dR}(A/X)$, de rang 2 sur X , d'une structure de D-module irréductible. En particulier, son groupe de Galois différentiel est isomorphe à $SL_2(\mathbf{C})$ (Picard-Lefschetz, ou Gauss), et toute section non nulle ω de $\Omega^1{}_{A/X}$ en est un vecteur cyclique.

Soient $L = L_{A/X,\omega}$ l'équation de Picard-Fuchs correspondant à ω , et $x \mapsto \{\omega_1(x), \omega_2(x)\}$ une base de solutions de L formée d'un couple de périodes fondamentales de ω_x sur A_x . Pour tout point rationnel P sur A(K) (c'est-à-dire pour pour toute section P de A/X), l'intégration sur un chemin de A_x allant de 0 à P_x permet de définir une section de classe de Nilsson (au sens de [13], p.122) du fibré tangent relatif $Lie_{A/X}$. Dans la base duale de ω , elle s'identifie à une fonction u_P , bien définie à l'addition d'un élément du groupe $\Omega = \mathbf{Z}\,\omega_1 + \mathbf{Z}\,\omega_2$ près . Dans ces conditions, Lu_P est une fonction uniforme sur X , à croissance modérée à l'infini, donc un élément m_P de O_X , et l'application

$$\mu : A(K)/A_{tor}(K) \to Ext_D(V(L),O_X) ,$$

qui associe à la classe de P l'extension de V(L) par O_X correspondant à la classe de m_P dans O_X/LO_X (lemme 4), est un homomorphisme de groupe. Le théorème du noyau de Manin ([20], §6; [11], Théorème 1.4.3) en affirme le caractère injectif. (Combiné avec la généralisation du lemme 2 à une base X quelconque, ce théorème fournit d'ailleurs une majoration bien connue du rang des groupes de Mordell-Weil des courbes elliptiques sur les corps de fonctions.) Mais mieux :

Lemme 8 : *l'homomorphisme* $\mu\otimes 1$ *de* $A(K)\otimes\mathbf{C}$ *dans* $\mathrm{Ext}_D(V(L),\mathcal{O}_X)$ *est injectif* .

[Les classes modulo Ω des logarithmes elliptiques d'éléments de A(K) linéairement indépendants sur $\mathbf{Z}$ le sont aussi. L'injectivité de μ exprime que leurs classes modulo $\Omega\otimes\mathbf{C}+\mathcal{O}_X$ le sont encore, et le lemme 7 qu'elles le sont même sur $\mathbf{C}$. Rappelons à ce propos qu'un schéma elliptique non isoconstant n'a pas de multiplication complexe.]

Démonstration : pour l'injectivité de μ lui-même, on note que si $\mu(P)$ est nul dans $H^1(V(L), \mathcal{O}_X)$ sans que P le soit dans $A(K)\otimes\mathbf{Q}$, il existe un élément f de $\mathcal{O}_X$ tel que u_P - f soit une combinaison $\mathbf{C}$-linéaire de ω_1, ω_2 , et le théorème de Picard- Lefschetz montre que c'en est alors une combinaison $\mathbf{Q}$-linéaire (voir [19], p.192) ; par l'argument rappelé au lemme 7 , un multiple non nul de P serait alors indéfiniment divisible dans A(K) , ce que la théorie des hauteurs interdit. (Voir [11] pour une présentation plus intrinsèque de ces arguments, et pour une démonstration de nature algébrique.) Quant à $\mu\otimes 1$, on peut le traiter de la façon élémentaire suivante. Soient $a_1, ..., a_r$ des nombres complexes linéairements indépendants sur $\mathbf{Q}$, et P(1), ..., P(r) des points d'ordre infini de A(K) tels que $a_1u_{P(1)} + ... + a_ru_{P(r)}$ appartienne à $\Omega\otimes\mathbf{C} + \mathcal{O}_X$. Le théorème de Picard-Lefschetz montre de nouveau que quitte à translater les $u_{P(i)}$ par des éléments de $\Omega\otimes\mathbf{Q}$, cette expression appartient à $\mathcal{O}_X$. Mais d'après le théorème 1 et l'injectivité de μ , chacun des groupes de Galois $G^u(\mu(P(i)))$ est isomorphe à $\Omega\otimes\mathbf{C}$. Les D-modules $\mu(P(i))$ étant fuchsiens, l'action d'une monodromie convenalement choisie dans $\mathrm{Gal}_\partial(K'/K)$ fournit alors une relation de dépendance $\mathbf{C}$-linéaire absurde entre ω_1 et ω_2.

En combinant le lemme 8 au théorème 2 (étendus une fois encore aux revêtements finis de X), on a finalement établi :

Théorème 5 : *considérons deux fonctions algébriques* $g_2(x)$ *et* $g_3(x)$ *telles que* $(g_2)^3/(g_3)^2$ *ne soit pas constante . Soient* $\wp_x$ *la fonction de Weierstrass d'invariants* $\{g_2(x),g_3(x)\}$, $\{\omega_1(x), \omega_2(x)\}$ *une base localement holomorphe de son réseau de périodes, et* $u_1, ..., u_r$ *des fonctions holomorphes sur un ouvert de* $\mathbf{C}$ *, telles que les fonctions* $u_1, ..., u_r, \omega_1, \omega_2$ *soient linéairement indépendantes sur* $\mathbf{Z}$, *et que* $\wp_x(u_1(x)),..., \wp_x(u_r(x))$ *soient des fonctions* *algébriques. Alors, les* 2r *fonctions* $u_1(x)$, $(d/dx)u_1(x)$, ..., $u_r(x)$, $(d/dx)u_r(x)$ *sont algébriquement indépendantes sur le corps* $\mathbf{C}(x,\omega_1(x),\omega_2(x),(d/dx)\omega_1(x))$.

Noter que cet énoncé n'est pas de même nature que les résultats de [10] . Ceux-ci concernent en effet une fonction de Weierstrass $\wp$ fixée, c'est-à-dire un schéma elliptique A/X constant.

Remarque 4 : une extension d'un D-module du type $V(L_{A/X,\omega})$ par O_X apparaît également dans l'interprétation donnée par Beukers ([4], prop.1) de la méthode d'Apéry pour l'irrationalité de $\zeta(2)$. Idem pour l'étude de Log2 , avec une extension par O_X d'un D-module de rang 1 , trivial sur un revêtement fini de X . Existe-t-il un lien direct entre ces extensions et ces nombres irrationnels ?

c) Le cas hypergéométrique .

Les groupes de Galois des opérateurs hypergéométriques L irréductibles sont entièrement connus, grâce à l'extension faite par Gabber et Katz [17] des résultats de [6] . Ces articles reposent sur des arguments purement algébriques (ou formels, à l'infini). Pour L confluent, une méthode analytique, due par Ramis et développée dans [21], consiste à étudier le groupe de Galois de $(K_\infty)^h \otimes V(L)$ sur le corps $(K_\infty)^h$ des fonctions *méromorphes* à l'infini, en interprétant les matrices de Stokes (définies au moyen d'un procédé de resommation canonique) comme des automorphismes différentiels. Elle permet en principe de traiter également les opérateurs hypergéométriques réductibles (voir [14] pour les opérateurs de rang petit). Mais il est plus commode de faire appel dans ce cas à la méthode (purement algébrique) du §2 . Voici comment.

Pour alléger, nous nous bornerons au cas confluent. On peut alors choisir $X = \mathbf{G}_m$ pour base, sans modifier la convention du §1 sur les D-modules (voir [17] pour une présentation plus adéquate). On pose :

$$\partial = x(d/dx) \ , \ O_X(\alpha) = V(\partial - \alpha) \ ,$$

où α est un nombre complexe quelconque, et on rappelle que l'application $\alpha \mapsto O_X(\alpha)$ établit un isomorphisme de $\mathbf{C}/\mathbf{Z}$ sur le groupe des classes d'isomorphisme de D-modules fuchsiens de rang 1 sur X .

Soient $n > m$ deux entiers ≥ 1 , et $\{\alpha_1, ..., \alpha_m\}$, $\{\beta_1, ..., \beta_n\}$ deux familles de nombres complexes. On note

$$L = L_{m,n}(\{\alpha_i\}_i, \{\beta_j\}_j) = \Pi_{j=1,...,n}(\partial + \beta_j - 1) - z\,\Pi_{i=1,...,m}(\partial + \alpha_i)$$

l'opérateur hypergéométrique généralisé d'ordre n correspondant. Il admet une singularité régulière en 0 , où les exposants valent $\{1- \beta_j\}_j$, et une singularité irrégulière en ∞ , où l'irrégularité vaut 1 , la partie de pente nulle a une longueur m , et les exposants correspondants valent $\{\alpha_i\}_i$. Par ailleurs, le D-module V(L) est irréductible si et seulement si aucun des α_i n'est congru à l'un des β_j modulo **Z** (voir [6], lemme 4.2; [17], corollaire 3.2.1) , et $h^1(V(L))$ est alors égal à 1 (lemme 2). En particulier, il n'existe essentiellement *qu'une* extension non triviale d'un D-module hypergéométrique irréductible $V(L_{m,n}(\{\alpha_i\}_i,\{\beta_j\}_j))$ par O_X . Elle se trouve être encore donnée par un opérateur hypergéométrique $M = L_{m+1,n+1}(\{\alpha_i\}_{i=1,\dots,m+1}, \{\beta_j\}_{j=1,\dots n+1})$, où α_{m+1} et β_{n+1} désigne deux entiers rationnels convenablement choisis. Plus précisément (voir [9], proposition 1) :

(a) si $\alpha_{m+1} \geq \beta_{n+1}$, V(M) définit une extension de V(L) par O_X (et une extension de O_X par V(L) dans le cas contraire) ;

(b) pour L irréductible, cette extension est donc triviale (en vertu du lemme 4 (ii)) si et seulement s'il existe un paramètre entier β_j de L majorant strictement α_{m+1} , ou un paramètre entier α_i de L minorant strictement β_{n+1} .

Le théorème 1, joint aux résultats de [17] (et à une torsion éventuelle par un module du type $O_X(-\alpha)$) fournit donc le groupe de Galois de tous les opérateurs hypergéométriques M dont le semi-simplifié possède deux facteurs (voir [9], théorème 2).

Plus généralement, soit M un opérateur hypergéométrique d'ordre q , réductible dans l'anneau **D** . Puisque card(S) vaut ici 2 , on déduit du lemme 2 et de l'additivité de $h^0 - h^1$ sur les suites exactes de D-modules que le semi-simplifié de V(M) est une somme directe d'un module hypergéométrique irréductible V(L) de rang $n < q$, et de $r = q - n$ modules de type $O_X(\alpha)$ (voir [17], cor. 3.7.5.2). Supposons de plus que V(L) soit un quotient de V(M) , et que ces r nombres α ne soient pas congrus modulo **Z** . Comme $\mathrm{Ext}(O_X(\alpha),O_X(\alpha')) = \mathrm{Ext}(O_X(\alpha-\alpha'), O_X) = 0$ si $\alpha - \alpha'$ n'est pas entier , V(M) est alors une extension de V(L) par une somme directe de D-modules de rang 1 , pour laquelle on peut énoncer :

Théorème 6 (K. Boussel [9]): *soient* $q > p > r$ *des entiers* ≥ 1, *et* $\alpha_1, \dots, \alpha_p$, $\beta_1, \dots, \beta_q$ *des nombres complexes vérifiant les conditions suivantes* :

i) $\alpha_i - \beta_i$ *est un entier* ≥ 0 *pour* $i = 1, \dots, r$;

ii) *aucune des différences* $\alpha_i - \alpha_{i'}$ $(1 \leq i < i' \leq p)$, $\beta_j - \beta_{j'}$ $(1 \leq j < j' \leq q)$, $\alpha_i - \beta_j$ $(i,j > r)$ *n'est entière* .

Alors, l'extension de Picard-Vessiot de $M = L_{p,q}(\{\alpha_i\}_{1 \leq i \leq p}, \{\beta_j\}_{1 \leq j \leq q})$ *contient le compositum du corps* $\mathbf{C}(x, x^{\alpha_i}, 1 \leq i \leq r)$ *et de l'extension de Picard-Vessiot de* $L = L_{p-r,q-r}(\{\alpha_i\}_{i>r}, \{\beta_j\}_{j>r})$, *et le groupe de Galois différentiel relatif correspondant est isomorphe à* $\mathbf{C}^{(q-r)r}$.

<u>Démonstration</u> : les hypothèses faites sur les r premiers paramètres de M entraînent, d'après le point (a) ci-dessus, que $V(M)$ est une extension de $V(L)$ par $O_X(\alpha_1)\oplus \ldots\oplus O_X(\alpha_r)$. On tire de ii) et du point (b) que $V(L)$ est irréductible, et que l'extensions de $V(L)$ par $O_X(\alpha_i)$ déduites de $V(M)$ par la projection canonique n'est triviale pour aucun indice $i \leq r$. Le théorème 3 permet de conclure.

Remarque 5 : On trouvera dans [9] une analyse exhaustive des types de décomposition possibles des opérateurs hypergéométriques réductibles, et de leurs groupes de Galois. Par exemple, si l'hypothèse de positivité de i) est mise en défaut par les $r' < r$ premiers couples de paramètres, $V(M)$ est une extension par $V' = O_X(\alpha_{r'+1}) \oplus \ldots\oplus O_X(\alpha_r)$ d'une extension E' de $V'' = O_X(\alpha_1) \oplus \ldots\oplus O_X(\alpha_{r'})$ par $V(L)$: c'est la situation mentionnée à la fin du § 2 . Sous l'hypothèse ii), la dimension du groupe de Galois relatif y croît de $(r-r')r'$ unités (le point clef étant, comme dans [23], 3.3, que le seul sous-G-module de E'^{∂} se projetant surjectivement sur $(V'')^{\partial}$ est E'^{∂} lui-même). En revanche, des dégénerescences apparaissent lorsque ii) n'est plus satisfaite.

d) <u>Lien avec la méthode de Shidlovsky</u> .

Soient $\mathbf{E}$ un D-module , et $\mathbf{f}$ une section horizontale de l'extension de $\mathbf{E}$ à l'espace analytique X^h . Supposons que S soit défini sur la clôture algébrique $\mathbf{Q}^a$ de $\mathbf{Q}$ dans $\mathbf{C}$, et que $\mathbf{E}$ et $\mathbf{f}$ vérifient les hypothèses d'arithméticité de la théorie des E-fonctions de Siegel, en un point base algébrique α_0 de $X \cup S$. En particulier, $\mathbf{E}$ et ses fibres $\mathbf{E}_\alpha$ aux points α de $X(\mathbf{Q}^a)$ sont munies d'une $\mathbf{Q}^a$-structure. Dans sa généralisation bien connue des résultats de Siegel, Shidlovsky a démontré que si la puissance symétrique $S^t\mathbf{f}$ de $\mathbf{f}$ engendre le D-module $S^t\mathbf{E}$ pour *tout* entier $t \geq 1$, il en est de même, en tout point algébrique $\alpha \neq \alpha_0$ de X , des puissances symétriques de $\mathbf{f}(\alpha)$ vis à vis des puissances symétriques du $\mathbf{Q}^a$-espace vectoriel $\mathbf{E}_\alpha$ (voir [27] pour des raffinements de cette assertion). En d'autres termes, après choix d'une base de $\mathbf{E}$ sur $\mathbf{Q}^a[X]$, les composantes de $\mathbf{f}(\alpha)$ sont 'homogènement' algébriquement indépendantes sur $\mathbf{Q}^a$ dès que celles de $\mathbf{f}$ le sont sur $\mathbf{Q}^a(X)$ - ou sur $\mathbf{C}(X)$, cela revient au même en vertu de la définition des E-fonctions . Cette dernière hypothèse équivaut encore à dire que l'orbite de $S^t\mathbf{f}$ sous l'action naturelle du groupe $G_{\mathbf{E}} = \mathrm{Gal}_\partial(K_{\mathbf{E}}/K)$ sur $(S^t\mathbf{E})^\partial = S^t(\mathbf{E}^\partial)$ engendre $S^t(\mathbf{E}^\partial)$ sur $\mathbf{C}$, pour tout $t \geq 1$.

Dans [6], Beukers, Brownawell et Heckman ont montré qu'à condition de supposer $G_{\mathbf{E}}$ *réductif* , le résultat de Shidlovsky peut se déduire de la méthode originale de Siegel, et retrouve de ce fait un caractère effectif. Cette hypothèse limite néanmoins beaucoup le champ d'application de leur méthode. La démarche du §2 permet dans de nombreux cas de remédier à cette difficulté, grâce à l'énoncé suivant, où on reprend les notations du §2 relativement à une extension $0 \to V' \to \mathbf{E} \to V \to 0$ de D-modules .

Théorème 7 (Beukers [5]): *soient* t *un entier* ≥ 1 , *et* **f** *un élément de* $\mathbf{E}^{\partial}$, *de projection* $v = \pi(\mathbf{f})$ *sur* V^{∂} , *tel que* :

i) *l'orbite de* $S^t v$ *sous l'action de* G_V *engendre* $S^t(V^{\partial})$;

ii) *l'orbite de* v *sous l'action de* $G^u(\mathbf{E})$ *coïncide avec* $(V')^{\partial}$.

Alors, l'orbite de $S^t\mathbf{f}$ *sous l'action de* $G_{\mathbf{E}}$ *engendre* $S^t(\mathbf{E}^{\partial})$.

La démonstration du théorème 7 repose sur la filtration naturelle de $S^t\mathbf{E}$ par les D-modules $S^i\mathbf{E}\otimes S^{t-i}V'$, et sur une récurrence sur i . Son application au théorème de Shidlovsky est claire sous l'hypothèse du théorème 1 , où sa condition ii) est satisfaite dès que v n'est pas nulle. Dans des cas plus généraux, on considèrera une filtration $\{\mathbf{E}^j\}$ de $\mathbf{E}$ de gradués irréductibles, et on procédera par récurrence sur j , en combinant le théorème 7 à l'argument mentionné à la remarque 5 ci-dessus.

Signalons, pour conclure sur un ton plus arithmétique, que les résultats de Bézivin et Robba sur les opérateurs de Polya L (et leur application à l'indépendance linéaire sur $\mathbf{Q}$ de valeurs de fonctions classiques) s'interprètent également en termes d'extensions de D -modules. Le théorème principal de [7] (voir aussi [1]) peut ainsi se lire comme un critère local-global de trivialité pour certains éléments de $H^1(V(L)^*) = \mathrm{Ext}_D(V(L),O_X)$, l'expression 'local' se référant ici aux équations différentielle p-adiques associées à L . Il serait très intéressant de développer l'étude 'kummérienne' du §2 dans ce cadre, en recherchant un analogue différentiel des groupes de Tate-Shafarevich.

Références

[1] Y. ANDRE : Critères de rationalité et d'algébricité; exposé à Luminy, Mai 88.

[2] J. AX : On Schanuel's conjecture; Ann. Maths., 93, 1971, 252-268.

[3] D. BABBITT - V. VARADARAJAN : *Local moduli for meromorphic differential equations* ; Astérisque; 169-170, 1989.

[4] F. BEUKERS : Irrationality of π^2, periods of an elliptic curve, and $\Gamma_1(5)$; Birkhäuser Prog. Maths, 31, 1983, 47-66.

[5] F. BEUKERS : communication orale, Juin 1989.

[6] F. BEUKERS-D. BROWNAWELL-G. HECKMAN : Siegel normality ; Ann. Maths, 127, 1988, 279-308.

[7] J-P. BEZIVIN - P. ROBBA : Rational solutions of linear differential equations; J. Austr. Math. Soc., 46, 1989, 184-196.

[8] N. BOURBAKI : *Groupes et algèbres de Lie* , Chap. 1 , Paris , 1971.

[9] K. BOUSSEL : Groupes de Galois des équations hypergéométriques ...; CRAS Paris, 309,1989, 587-589 (et article en préparation).

[10] D. BROWNAWELL - K. KUBOTA : Algebraic independence of Weierstrass functions; Acta Arith., 33, 1977, 111-149.

[11] R. COLEMAN : Manin's proof of the Mordell-Weil conjecture over function fields; manuscrit, Berkeley, Mars 1989.

[12] R. COLEMAN : lettre à B. Mazur, Mars 1987.

[13] P. DELIGNE : *Equations différentielles à points singuliers réguliers* ; Springer LN. 163, 1970 .

[14] A. DUVAL - C. MITSCHI : Matrices de Stokes et groupes de Galois des équations hypergéométriques ...; Pacific J. Math., 135, 1988 .

[15] E. GEKELER : De Rham cohomology and the Gauss-Manin connection for Drinfeld modules; ce volume .

[16] N. KATZ : On the calculation of some differential Galois groups; Inv. math., 87, 1987, 13-61.

[17] N. KATZ : *Exponential sums and differential equations* ; Princeton UP, à paraître.

[18] LE D. T. - Z. MEBKHOUT : Introduction to linear differential systems; Proc. Symp. Pure Maths, 40, 1983, 2, 31-63.

[19] B. MALGRANGE : Sur la réduction formelle des équations différentielles à singularités irrégulières; manuscrit, Grenoble, 1979.

[20] Y. MANIN : Rational points of algebraic curves over function fields;Izv. 27, 1963, 1395-1440 = AMS Transl., 37, 1966, 189-234.

[21] J. MARTINET - J-P. RAMIS : Elementary acceleration and multisummability; manuscrit, Luminy, Septembre 1989 .

[22] Z. MEBKHOUT : Le théorème de comparaison ...; Publ. math. 69, 1989, 47-89.

[23] K. RIBET : Kummer theory on extensions of abelian arieties by tori; Duke Math. J., 46, 1979, 745-761.

[24] P. ROBBA : Lemmes de Hensel pour les opérateurs différentiels ... ; Ens. math., 26, 1980, 279-311.

[25] T. SCHOLL : Remarks on special values of L-functions; manuscrit.

[26] J-P. SERRE : Sur les groupes de congruence des variétés abéliennes; Izv. AN SSSR, 28, 1964, 3-18.

[27] A. SHIDLOVSKI : *Nombres transcendants* [en russe]; Nauka, 1987.

[28] V. VARADARAJAN : *Lie groups, Lie algebras, and their representations* ; Springer GTM, 102, 1984.

Duality in Rigid Analysis

Bruno Chiarellotto *
Dipartimento di Matematica Pura e Applicata
Universita' di Padova
Via Belzoni 7 , 35131 Padova Italy

Throughout the article K denotes an algebraically closed field, complete for an ultrametric valuation (We will call such a field an ultrametric field).

**

In this paper we want to give a partial answer to the question:

is it possible to have a Serre-type duality in the rigid analytic setting ?

i.e. Given X a regular rigid analytic variety of dimension n and $\mathcal{M}$ a coherent sheaf of $\mathcal{O}_X$-modules is it possible to define topological pairings:

$$H_c^\bullet(X,\mathcal{M}) \times Ext^\bullet(X,\mathcal{M},\Omega_X^n) \longrightarrow K$$

$$H^\bullet(X,\mathcal{M}) \times Ext_c^\bullet(X,\mathcal{M},\Omega_X^n) \longrightarrow K$$

in such a way that each space is the strong dual of the other (via the involved pairing) ([SED], [RR], [SUO])?

We need some definitions and tools to solve this problem: a notion of cohomology with "compact" support, a "canonical" topology on the cohomology groups and a notion of "residue".

In this paper we will construct the duality for a rigid Stein space X [LU] and a coherent sheaf $\mathcal{M}$ on it. Here is an outline of the contents. In §1 we introduce the notion of a family of compact (or, better, co-compact) supports. In §2 we will prove the duality for $X = \mathbf{A}_K^n$ and the structural sheaf by introducing a residue map (2.8) (Note 2.12 deals with the problem of the unicity of this map). In §3 we will provides the Serre duality for a coherent sheaf on $\mathbf{A}_K^n$. In §4, theorem (4.21) gives the duality for a coherent sheaf on a Stein space. The key point in the proof is the existence of a closed immersion for a Stein space in an affine rigid space [LU]. Finally in §5 we will prove that the notion of residue for a Stein space arising from §4 is independent of the closed immersion.

Of course there are still some gaps in the theory. One is the fact that we deal, here, only with non-singular spaces. On the other hand little is known about a rigid analytic variety viewed as a topos. Furthemore our notion of compact support can not work for affinoid spaces (see (1.6)). It does, however, seen to work for " rigid spaces without boundary" (see [LUT]) and we plan further investigation along these lines.

Some interesting results on cohomology with supports in the rigid analytic setting can be found in Y. Morita's article in this volume (see also [MOS]).

---

* Supported by C.N.R. Italy.

The most important part of this work was prepared during our visit to the "Equipe L.A.212" of CNRS (Université Paris VII): we thank that institution. We must also thank Professor Z. Mebkhout because he gave us the basic ideas of this paper and much assistance during its preparation. We should also mention Professor J. Fresnel, Professor M. van der Put and Professor P. Berthelot for much help and many suggestions we had from them.

§1

We refer to [BGR], [FRE], [F-vdP] for the basic notion about rigid analytic varieties. Throughout the article we will employ the "strong" (saturated) topology on them.

It's natural to associate a situs to a rigid analytic variety X (based on the strong Grothendieck topology) and finally to consider the associated topos, which we will indicate by $Rig(X)$ (as in the case of the étale topos, one can find a topologically generating family ([SGAIV]VII1.7, 3.1, III4.1: the admissible affinoids)). All the morphisms of the situs are injections: it follows that the elements of the topos associated with an admissible open set are open objects of $Rig(X)$ ([SGAIV]IV2.1).

We want to introduce a family of supports which will play the role of the family of compact sets in the classical case. In $Rig(X)$ it is, perhaps, easier to define a "dual" notion i.e. a family of co-supports (this was in the mind of Grothendieck in the first edition of [SGAIV]):

Definition 1.1. *A family of co-supports is a subset $\widetilde{\phi}$ of the family of open sets of $Rig(X)$ (i.e. subobjects of the final object), subject to the rules:*

i. The intersection (in the sense of $Rig(X)$) of a finite number of elements in $\widetilde{\phi}$ is still in $\widetilde{\phi}$.

ii. Any open set which contains an element of $\widetilde{\phi}$ belongs to $\widetilde{\phi}$.

Our point is that a finite union of admissible affinoid subdomains of X should be considered a "compact" (note that a finite union of admissible affinoids is an admissible open set([BGR]9.1.4)). We assert:

Proposition 1.2. *(Fresnel) If X is a separated rigid analytic variety and Y is a finite union of admissible affinoids of X then $X \setminus Y$ (as set) is an admissible open subset of X (thus associated to an open object of $Rig(X)$).*

Proof. *i.* Suppose $X = SpmA$ is an affinoid space and Y is a rational subset:

$$Y = \{x \in X\,;\quad |\, f_i(x)\,| \leq |\, f_0(x)\,|,\quad i = 1\ldots,n\}$$

where $f_i \in A$, $\sum_{i=0}^{n} f_i A = A$. It is possible to find a $\pi \in K^*$ such that $|\, f_0(x)\,| > |\,\pi\,|$ for every $x \in Y$. If

$$X_+ = \{x \in X\,;\quad |\, f_0(x)\,| \geq |\,\pi\,|\,\}$$
$$X_- = \{x \in X\,;\quad |\, f_0(x)\,| \leq |\,\pi\,|\,\}$$

then $\{X_+\,,\ X_-\}$ is an admissible covering of X and $f_0(x)$ is invertible in X_+.

We put

$$X_+^i = \{x \in X_+\,;\quad \left|\frac{f_i(x)}{f_0(x)}\right| > 1\}.$$

Then $\bigcup_{i=1}^{n} X_+^i$ is an admissible open of X_+ ([BGR]9.1.4) , and of X . On the other hand

$$(X \setminus Y) \cap X_- = X_-$$

$$(X \setminus Y) \cap X_+ = \bigcup_{i=1}^{n} X_+^i$$

and it follows that $X \setminus Y$ is an admissible of X ([BGR]9.1.2) .

ii. Suppose $X = SpmA$ and $Y \subseteq X$ is a finite union of admissible affinoids . Then

$$Y = \bigcup_{i=1}^{s} Y_i$$

is a finite union of Y_i rational subdomains in X ([BGR]7.3.5), so

$$X \setminus Y = \bigcap_{i=1}^{s} (X \setminus Y_i)$$

By *i.* ,we have that each $X \setminus Y_i$ is admissible, and thus, so too is the finite intersection.

iii. In the generic case take $\{X_i\}_{i\in I}$ an admissible affinoid covering of X. For each i,

$$(X \setminus Y) \cap X_i = X_i \setminus (Y \cap X_i).$$

But $Y \cap X_i$ is a finite union of admissible affinoids of X_i ([BGR] p.393 , since X is separated) : $X_i \setminus (Y \cap X_i)$ is an admissible of X_i (by *ii.*) and, of X ([BGR]p.354). Thus being true for each X_i of the admissible covering $\{X_i\}$ it follows, from the definition of strong topology, that $X \setminus Y$ is admissible ([BGR]9.1.2) **Q.E.D.**

So far we can talk about the open object of *Rig*(X) associated with an admissible open which is the complement of a finite union of admissible affinoids.

Proposition(Definition)1.3. *In Rig(X), with X a separated rigid analytic variety, the family of open objects given by "the open objects of Rig(X) containing an open set which is the complement of a finite union of admissible affinoids" is a family of co-supports. We will indicate it by $[c$.*

Proof. We apply the previous proposition and we note that in a separated rigid analytic variety the intersection of two admissible affinoids is still an admissible affinoid ([BGR]9.6.1).**Q.E.D.**

Remark. We don't know if each open object in *Rig*(X) is associated with an admissible open of X. We don't know if the family $[c$ is a local family.

Naturally we are going to define the sections with compact support in the complement of a co-compact using "the exact sequence of cohomology with supports" ([SGAIV]V6.5). We recall that to each open object is associated a closed one ([SGAIV]IV). Finally for $\mathcal{L} \in obRig(X)$ we have

$$\underline{H}_c^0(\mathcal{L}) = \varinjlim_{U\in[c} \underline{H}_{X\setminus U}^0(\mathcal{L})$$

(if U_1 is an open object of *Rig*(X), $\mathcal{L}(U_1) = Hom(U_1, \mathcal{L})$) and the sections

$$H_c^0(\mathcal{L}) = \varinjlim_{U\in[c} H_{X\setminus U}^0(\mathcal{L})$$

([SGAIV]V.6).

It is, then, possible to define the derived functors (cfr. [AR],[BER],[SGAIV]V.6) . Using the fact that the limit is filtering, for $\mathcal{F} \in obRig(X)_{ab}$ (category of abelian sheaves) we get:

$$\underline{H}^p_c(\mathcal{F}) = \varinjlim_{U \in [c} \underline{H}^p_{X \setminus U}(\mathcal{F})$$

and

$$H^p_c(\mathcal{F}) = \varinjlim_{U \in [c} H^p_{X \setminus U}(\mathcal{F})$$

Remark 1.4. In the same way, we can define $Ext^{\bullet}_{\mathcal{O}_X,c}$ and $\mathcal{E}xt^{\bullet}_{\mathcal{O}_X,c}$: they too satisfy the direct limit construction.

Remark 1.5. The set given by the open objects which are the complement of a finite union of affinoids are a co-final family for the direct limit.

Remark 1.6. If X is a compact (proper over K) rigid analytic variety, the compact support cohomology is the usual one. On the other hand, for $X = SpmA$, with A an affinoid algebra, the compact support cohomology and the usual one coincide. As we will see, affinoid spaces won't appear in the duality theorems (2.11).

One of the tools we need is the Yoneda-Cartier pairing for the $Ext^{\bullet}$ in the rigid setting ([CA],[SUO], [HARD]). To this end, we state the following extension lemma ([SUO]2.1]):

Lemma 1.7. *If $\mathcal{I}$ is an injective $\mathcal{O}_X$-module and*

$$\mathcal{N} \longrightarrow \mathcal{M}$$

an injective map of $\mathcal{O}_X$-modules, if V is an open in $Rig(X)$ and

$$\phi \in Ext^0_{\mathcal{O}_X, X\setminus V}(X, \mathcal{N}, \mathcal{I})$$

then it is possible to find an extension

$$\overline{\phi} \in Ext^0_{\mathcal{O}_X, X\setminus V}(X, \mathcal{M}, \mathcal{I}).$$

Proof. We need only note that ([SGAIV]V6.8.6)

$$Ext^0_{\mathcal{O}_X, X\setminus V}(X, \mathcal{P}, \mathcal{F}) = Ext^0_{\mathcal{O}_X}(X, \mathcal{P}, \underline{H}^0_{X\setminus V}(\mathcal{F}))$$

for $\mathcal{P}, \mathcal{F}$ $\mathcal{O}_X$-modules and use the fact that if $\mathcal{I}$ is an injective then $\underline{H}^0_{X\setminus V}(\mathcal{I}))$ is injective ([SGAIV]V4.11). **Q.E.D.**

Finally we can define the Yoneda-Cartier pairing

$$Ext^p_{\mathcal{O}_X}(X, \mathcal{M}, \mathcal{F}) \times Ext^q_{\mathcal{O}_X,c}(X, \mathcal{F}, \mathcal{G}) \longrightarrow Ext^{p+q}_{\mathcal{O}_X,c}(X, \mathcal{M}, \mathcal{G})$$

Remark. The same argument works for $\mathcal{E}xt_{\mathcal{O}_X}$.

§2

This paragraph deals with the simplest case of $Serre-type$ duality : the structural sheaf of the affine rigid space $\mathbf{A}^n_K$.

To accomplish this goal, we need to define topologies and a residue formula in such a way that the pairing

$$H^0(\mathbf{A}_K^n, \mathcal{O}_{\mathbf{A}_K^n}) \times H_c^n(\mathbf{A}_K^n, \Omega_{\mathbf{A}_K^n}^n) \longrightarrow H_c^n(\mathbf{A}_K^n, \Omega_{\mathbf{A}_K^n}^n) \xrightarrow{res} K$$

gives a perfect duality (each space is the strong dual of the other). We recall that $\mathbf{A}_K^n$ is a quasi-Stein space [KI] (for the notion of Stein space in rigid analysis see [LU]).

Definition. A rigid separated variety is a quasi-Stein space if there exists an admissible affinoid covering $\{ U_i \}_{i \in \mathbf{N}}$ (countable) such that

$$U_i \subseteq U_{i+1}$$

and

$$\mathcal{O}_X(U_{i+1}) \longrightarrow \mathcal{O}_X(U_i)$$

is dense.

Remark. For $\mathbf{A}_K^n$ such an admissible covering is given, for example , by $\{B_{\epsilon^i}\}_{i \in \mathbf{N}}$

$$B_{\epsilon^i} = \{(x_j) \in \mathbf{A}_K^n \quad | \, x_j \,| \leq \epsilon^i\}$$

with $\epsilon \in | K^* |, \epsilon > 1$.

Cartan's theorems A and B hold [KI] , since $\mathcal{O}_{\mathbf{A}_K^n}$ is a coherent sheaf

$$H^i(\mathbf{A}_K^n, \mathcal{O}_{\mathbf{A}_K^n}) = 0 \qquad i > 0.$$

We recall the definition of coherent sheaf in the rigid setting:

Definition 2.0. [BGR] In a rigid analytic variety X a sheaf of $\mathcal{O}_X$-modules $\mathcal{M}$ is coherent if there exists an admissible affinoid covering $\{U_i\}_{i \in I}$ such that the restriction of $\mathcal{M}$ to each $U_i = SpmA_i$ is a sheaf associated to a finite A_i-module M_i

$$\mathcal{M}_{|U_i} = \tilde{M}_i$$

Remark. One can prove that for a coherent $\mathcal{O}_X$-module $\mathcal{M}$ and for every $U = SpmA$ admissible affinoid of X, the restriction $\mathcal{M}_{|U}$ is a sheaf associated to a finite A-module M_U :

$$\mathcal{M}_{|U} = \tilde{M}_U.$$

Each affinoid algebra A is a Banach space: every submodule of a finite A-module is closed and every A-linear map between finite A-modules is continuous ([BGR]3.7.3 , 6.1.1). It follows that

$$M_U = \frac{A^n}{\mathcal{A}} = \mathcal{M}(U)$$

has a canonical (by the open mapping theorem [BGR],[BOU]) topology as Banach space. On the other hand an affinoid algebra A is of countable type : it follows that M_U is of countable type, again by [BGR].

We wish to endow $H^0(\mathbf{A}_K^n, \mathcal{O}_{\mathbf{A}_K^n})$ with a canonical topology. First of all we notice that for $\mathbf{A}_K^n$ and, in general, for a quasi-Stein space, the admissible countable affinoid coverings form a co-final system for the set of all the coverings. (Each admissible countable affinoid covering of an affinoid space admits a refinement which is a finite admissible affinoid covering.

A quasi-Stein space has a countable affinoid covering) . Take an admissible countable affinoid covering , $\mathcal{U}$, and consider its Čech-complex in $\mathcal{O}_{\mathbf{A}^n_K}$:

$$C^\bullet(\mathcal{U}, \mathcal{O}_{\mathbf{A}^n_K})$$

From the acyclicity of the affinoid subdomains, its cohomology will be $H^0(\mathbf{A}^n_K, \mathcal{O}_{\mathbf{A}^n_K})$. On the other hand the sections of the structural sheaf on each affinoid form a Banach space: the Čech-complex $C^\bullet(\mathcal{U}, \mathcal{O}_{\mathbf{A}^n_K})$ is a complex of Fréchet spaces ([BOU]I3.2). We can put the induced topology (of Fréchet space) on $H^0(\mathbf{A}^n_K, \mathcal{O}_{\mathbf{A}^n_K})$. If $\mathcal{V}$ is an admissible countable affinoid covering which is a refinement of $\mathcal{U}$, there will be continuous maps between Fréchet spaces:

$$C^\bullet(\mathcal{U}, \mathcal{O}_{\mathbf{A}^n_K}) \xrightarrow{\varphi_\bullet} C^\bullet(\mathcal{V}, \mathcal{O}_{\mathbf{A}^n_K})$$

which induce isomorphisms in cohomology. But the open mapping theorem (still valid in the ultrametric setting [BOU]) allows us to conclude that this isomorphism in cohomology is, actually, a homeomorphism ([RR] lemma1) for the induced topologies :

$$\tilde{\varphi}_0 : H^0(\mathcal{U}, \mathcal{O}_{\mathbf{A}^n_K}) \longrightarrow H^0(\mathcal{V}, \mathcal{O}_{\mathbf{A}^n_K}).$$

This gives a canonical topology as Fréchet space of countable type to $H^0(\mathbf{A}^n_K, \mathcal{O}_{\mathbf{A}^n_K})$ (the sections of the stuctural sheaf on each affinoid subdomain constitute an affinoid algebra which is a Banach algebra of countable type. Every subspace of a space of countable type is of countable type) . In particular

$$H^0(\mathbf{A}^n_K, \mathcal{O}_{\mathbf{A}^n_K}) = \{ \sum_{\alpha \in \mathbf{N}^n} a_\alpha x^\alpha \quad a_\alpha \in K \quad \lim_{|\alpha| \to +\infty} | a_\alpha | \epsilon^{|\alpha|} = 0 \quad \epsilon \in | K^* | \ \} \tag{2.1}$$

We now treat the space $H^n_c(\mathbf{A}^n_K, \Omega^n_{\mathbf{A}^n_K})$. Once we have fixed a system of coordinates on $\mathbf{A}^n_K$, $x_1, \dots, x_n$, a co-final system of compacts is given by:

$$\mathcal{B} = \{ B_\epsilon = \{ x \in \mathbf{A}^n_K, \quad | x_i | \leq \epsilon \quad i = 1, \dots, n \} \qquad \epsilon \in | K^* | \ \} \tag{2.2}$$

(In fact on every affinoid the maximum modulus principle holds ([BGR]6.2.1)). We can write

$$H^p_c(\mathbf{A}^n_K, \Omega^n_{\mathbf{A}^n_K}) = \varinjlim_{B_\epsilon \in \mathcal{B}} H^p_{B_\epsilon}(\mathbf{A}^n_K, \Omega^n_{\mathbf{A}^n_K})$$

Proposition 2.3. *Under the previous hyphothesis*

$$H^p_c(\mathbf{A}^n_K, \Omega^n_{\mathbf{A}^n_K}) = 0 \qquad p \neq n$$

For $p = n$

$$H^n_c(\mathbf{A}^n_K, \Omega^n_{\mathbf{A}^n_K}) = \{ \sum_{\alpha \in (\mathbf{N}^\bullet)^n} \frac{a_\alpha}{x^\alpha} dx_1 \wedge \dots \wedge dx_n \quad a_\alpha \in K \ \exists \epsilon \in | K^* | \quad \lim_{|\alpha| \to +\infty} | a_\alpha | \epsilon^{|\alpha|} = 0 \}$$

Furthemore, we can endow this space with a canonical topology.

Proof. Once we have fixed a global system of coordinates we can use the results of Morita [MOS] for the vanishing of the cohomology. In fact the cohomology will be the direct limit of

$$H^n_{B_\epsilon}(\mathbf{A}^n_K, \Omega^n_{\mathbf{A}^n_K}) \simeq H^{n-1}(\mathbf{A}^n_K \setminus B_\epsilon, \Omega^n_{\mathbf{A}^n_K})$$

($\mathbf{A}^n_K$ is a quasi-Stein and $\Omega^n_{\mathbf{A}^n_K}$ is coherent). On $H^{n-1}(\mathbf{A}^n_K \backslash B_\epsilon, \Omega^n_{\mathbf{A}^n_K})$ there exists a canonical topology. Indeed $\Omega^n_{\mathbf{A}^n_K}$ is, in particular, a coherent sheaf: the countable admissible affinoid coverings of $\mathbf{A}^n_K \backslash B_\epsilon$ form a co-final system for the coverings (even if $\mathbf{A}^n_K \backslash B_\epsilon$ is not quasi-Stein, but it is finite union of quasi-Stein ...) : the Čech-complex of $\Omega^n_{\mathbf{A}^n_K}$ of each such a covering consists of Fréchet spaces.The same argument we used for $\mathcal{O}_{\mathbf{A}^n_K}$ shows that there exists a canonical topology (not a priori separated) for $H^{n-1}(\mathbf{A}^n_K \setminus B_\epsilon, \Omega^n_{\mathbf{A}^n_K})$ i.e. independent of the covering ([RR]lemma1). For $H^n_c(\mathbf{A}^n_K, \Omega^n_{\mathbf{A}^n_K})$ we will take the direct limit of these canonical topologies. The fact that $\mathcal{B}$ (2.2) is co-final is independent of the coordinates: the resultant topology is canonical **Q.E.D**.

Among all the coverings which one can use in order to calculate $H^{n-1}(\mathbf{A}^n_K \setminus B_\epsilon, \Omega^n_{\mathbf{A}^n_K})$ there is the following

$$\mathcal{U} = \{ U_{\epsilon,i} = \{x \in \mathbf{A}^n_K ; \mid x_i \mid > \epsilon \} \quad i = 1, \dots, n\} \tag{2.4}$$

Every $U_{\epsilon,i}$ is quasi-Stein and the sections of a coherent sheaf on a quasi-Stein form a Fréchet space (If $\mathcal{F}$ is a coherent sheaf, on each affinoid its sections form a Banach space. Taking an admissible countable affinoid covering $\mathcal{V}$, the global sections will form a closed subspace of $C^\bullet(\mathcal{V}, \mathcal{F})$ which is a Fréchet space. The topology is canonical (see the argument for $H^0(\mathbf{A}^n_K, \mathcal{O}_{\mathbf{A}^n_K})$ [RR]).). On the other hand the intersections of the elements of $\mathcal{U}$ are still quasi-Stein spaces: the Cěch-complex $C^\bullet(\mathcal{U}, \Omega^n_{\mathbf{A}^n_K})$ is a complex of Fréchet spaces, the induced topology in cohomology is the canonical topology in $H^{n-1}(\mathbf{A}^n_K \setminus B_\epsilon, \Omega^n_{\mathbf{A}^n_K})$. In particular this topology on $H^n_c(\mathbf{A}^n_K, \Omega^n_{\mathbf{A}^n_K})$ is given as the direct limit of the Banach spaces

$$A_\epsilon = \{ \sum_{\alpha \in (bfN^\bullet)^n} \frac{a_\alpha}{x^\alpha} dx_1 \wedge \dots \wedge dx_n \, ; a_\alpha \in K \, , \quad \lim_{|\alpha| \to +\infty} \mid a_\alpha \mid \epsilon^{|\alpha|} = 0 \quad \}$$

with $\epsilon \in \mid K^* \mid$.

From (2.3) it is possible to give a residue formula for $H^n_c(\mathbf{A}^n_K, \Omega^n_{\mathbf{A}^n_K})$:

$$res_x \; : \quad H^n_c(\mathbf{A}^n_K, \Omega^n_{\mathbf{A}^n_K}) \longrightarrow K \tag{2.6}$$

$$\sum_{\alpha \in (\mathbf{N}^\bullet)^n} \frac{a_\alpha}{x^\alpha} dx_1 \wedge \dots \wedge dx_n \longrightarrow a_{(1,\dots 1)}$$

"A priori" such a continuous map depends on the choice of coordinates. We will see later that it is unique up to a constant (2.8).

We can prove:

Theorem 2.7. *The spaces*

$$H^n_c(\mathbf{A}^n_K, \Omega^n_{\mathbf{A}^n_K}) \quad and \quad H^0(\mathbf{A}^n_K, \mathcal{O}^n_{\mathbf{A}^n_K})$$

endowed with the canonical topologies are in perfect duality (one is the strong dual of the other) using the pairing:

$$H^n_c(\mathbf{A}^n_K, \Omega^n_{\mathbf{A}^n_K}) \times H^0(\mathbf{A}^n_K, \mathcal{O}^n_{\mathbf{A}^n_K}) \longrightarrow H^n_c(\mathbf{A}^n_K, \Omega^n_{\mathbf{A}^n_K}) \xrightarrow{res_x} K$$

In particular $H^0(\mathbf{A}^n_K, \mathcal{O}^n_{\mathbf{A}^n_K})$ is a reflexive Fréchet space of countable type and $H^n_c(\mathbf{A}^n_K, \Omega^n_{\mathbf{A}^n_K})$ is a strong dual of a Fréchet space, complete, reflexive and Hausdorff.

Proof. This is an adaptation of some results about duality between direct and inverse limits of Banach and Fréchet spaces in the ultrametric setting ([MO-SCH], [MOH] Note the different techniques used in the case that the field is, or is not , spherically complete). The only remark consists in noticing that the canonical topology on $H_c^n(\mathbf{A}_K^n, \Omega_{\mathbf{A}_K^n}^n)$ can be (formally) defined as the direct limit of bounded series:

$$A_{\epsilon,b} = \{ \sum_{\alpha \in (bfN^*)^n} \frac{a_\alpha}{x^\alpha} dx_1 \wedge \ldots \wedge dx_n \; a_\alpha \in K \quad \sup_{|\alpha| \to +\infty} \mid a_\alpha \mid \epsilon^{|\alpha|} < +\infty \}$$

($\epsilon \in \mid K^* \mid$). **Q.E.D.**

Using the previous theorem we can prove that our notion of residue is essentially unique [AK].

Proposition 2.8. *On $H_c^n(\mathbf{A}_K^n, \Omega_{\mathbf{A}_K^n}^n)$ and $H^0(\mathbf{A}_K^n, \mathcal{O}_{\mathbf{A}_K^n}^n)$, endowed with the canonical topologies, suppose there exists*

$$\widetilde{res} \; : \; H_c^n(\mathbf{A}_K^n, \Omega_{\mathbf{A}_K^n}^n) \longrightarrow K$$

which gives a perfect duality between the two spaces. Then $\widetilde{res}$ is equal to res_x (2.6) up to a multiplication by a constant in K

Proof. In particular res_x will be an element of the dual of $H_c^n(\mathbf{A}_K^n, \Omega_{\mathbf{A}_K^n}^n)$ with the canonical topology. By the perfect duality for $\widetilde{res}$, there will be $a \in H^0(\mathbf{A}_K^n, \mathcal{O}_{\mathbf{A}_K^n}^n)$ such that:

$$\begin{array}{ccc} H_c^n(\mathbf{A}_K^n, \Omega_{\mathbf{A}_K^n}^n) & \xrightarrow{a} & H_c^n(\mathbf{A}_K^n, \Omega_{\mathbf{A}_K^n}^n) \\ & res_x \searrow \quad \swarrow \widetilde{res} & \\ & K & \end{array}$$

will be commutative. On the other hand the same argument works for $\widetilde{res}$ respect to res_x , using a $b \in H^0(\mathbf{A}_K^n, \mathcal{O}_{\mathbf{A}_K^n}^n)$. Finally , there is the commutative diagram:

$$\begin{array}{ccccc} H_c^n(\mathbf{A}_K^n, \Omega_{\mathbf{A}_K^n}^n) & \xrightarrow{a} & H_c^n(\mathbf{A}_K^n, \Omega_{\mathbf{A}_K^n}^n) & \xrightarrow{b} & H_c^n(\mathbf{A}_K^n, \Omega_{\mathbf{A}_K^n}^n) \\ & res_x \searrow & \widetilde{res} \downarrow & \swarrow res_x & \\ & & K & & \end{array}$$

Using again the fact that res_x gives a perfect duality, we conclude that

$$ab = 1$$

in $H^0(\mathbf{A}_K^n, \mathcal{O}_{\mathbf{A}_K^n}^n)$. But for entire analytic functions (2.1) this is true if and only if a, b are constant ([BGR]5.1.3,[AM]). **Q.E.D.**

We will call the generic residue (up to a constant) in $\mathbf{A}_K^n$: res_n.

Remark 2.9. The same duality works in the case of a free sheaf $\mathcal{O}_{\mathbf{A}_K^n}^p$, $p \in \mathbf{N}$, or if we substitute the whole affine space with:

$$\overset{\circ}{B}_\epsilon = \{ x \in \mathbf{A}_K^n \quad \mid \quad \mid x_i \mid < \epsilon \; i = 1, \ldots , n \; \} \tag{2.10}$$

Remark 2.11. If we take $X = SpmK < x >$; [BGR], $H^0(X, \mathcal{O}_X) = K < x >$ and this is not self-dual . This shows that our definition is not valid for rigid affinoid spaces.

Note 2.12. We want to check the compatibility of res_n in $\mathbf{A}^n_K$ and the algebraic residue in $H^n(\mathbf{P}^n_K, \Omega^n_{\mathbf{P}^n_K})$ (via G.A.G.A. in the rigid setting [KO]). This will show the invariance of res_n for automorphisms of $\mathbf{P}^n_K$. By lemma 3.7 one has:

$$H^i_{B_\epsilon}(\mathbf{A}^n_K, \Omega^n_{\mathbf{A}^n_K}) \cong H^i_{B_\epsilon}(\mathbf{P}^n_K, \Omega^n_{\mathbf{P}^n_K}).$$

Consider the maps:

$$\gamma : H^{n-1}(\mathbf{A}^n_K \setminus B_\epsilon, \Omega^n_{\mathbf{A}^n_K}) \xrightarrow{\sim} H^n_{B_\epsilon}(\mathbf{A}^n_K, \Omega^n_{\mathbf{A}^n_K})$$

and

$$\Gamma : H^n_{B_\epsilon}(\mathbf{A}^n_K, \Omega^n_{\mathbf{A}^n_K}) \,(\xrightarrow{\sim} H^n_{B_\epsilon}(\mathbf{P}^n_K, \Omega^n_{\mathbf{P}^n_K})) \longrightarrow H^n(\mathbf{P}^n_K, \Omega^n_{\mathbf{P}^n_K}).$$

To construct these maps one takes an injective resolution:

$$0 \longrightarrow \Omega^n_{\mathbf{P}^n_K} \xrightarrow{i} J_0 \xrightarrow{d} J_1 \xrightarrow{d} J_2 \to \ldots\ldots \tag{2.13}$$

its restriction to $\mathbf{A}^n_K$ will be an injective resolution of $\Omega^n_{\mathbf{A}^n_K}$. For the sheaf $\Omega^n_{\mathbf{A}^n_K}$ one obtains the following exact sequence:

$$0 \to H^\bullet_{B_\epsilon}(\mathbf{A}^n_K, J_\bullet) \longrightarrow H^\bullet(\mathbf{A}^n_K, J_\bullet) \longrightarrow H^\bullet(\mathbf{A}^n_K \setminus B_\epsilon, J_\bullet) \to 0.$$

For γ we proceed as follows : take $l \in H^{n-1}(\mathbf{A}^n_K \setminus B_\epsilon, \Omega^n_{\mathbf{A}^n_K})$, it can be considered as a section of J_{n-1} on $\mathbf{A}^n_K \setminus B_\epsilon$, then there exists $\bar{l} \in H^0(\mathbf{A}^n_K, J_{n-1})$ whose restriction to $\mathbf{A}^n_K \setminus B_\epsilon$ is l. Applying d of (2.13), $d\bar{l} \in H^0(\mathbf{A}^n_K, J_n)$ and, in particular, $d\bar{l} \in H^0_{B_\epsilon}(\mathbf{A}^n_K, J_n)$:

$$\gamma(\text{class of } l) = \text{class of } d\bar{l}.$$

The extension with zero $(d\bar{l})_e \in H^0(\mathbf{P}^n_K, J_n)$ gives :

$$\Gamma(\text{class of } d\bar{l}) = \text{class of } (d\bar{l})_e \in H^n(\mathbf{P}^n_K, \Omega^n_{\mathbf{P}^n_K}).$$

On the other hand we can associate to the following quasi-Stein admissible covering of $\mathbf{P}^n_K$ (we will indicate by $x_0, \ldots, x_n$ the coordinates of $\mathbf{P}^n_K$; $\mathbf{A}^n_K$ is viewed into $\mathbf{P}^n_K$ as the set $U_0 = \{(x_k) \mid x_0 \neq 0\}$):

$$\mathcal{U}_0 = \{U_0, U_{i,\epsilon} \;\; i = 1, \ldots, n\}$$

$(U_{i,\epsilon} = \{(x_k) \mid x_i \neq 0 \,,\, |\frac{x_0}{x_i}| < \frac{1}{\epsilon}\})$ a quasi-Stein admissible covering of $\mathbf{A}^n_K \setminus B_\epsilon$:

$$\mathcal{U} = \{V_i = U_0 \cap U_{i,\epsilon} \quad i = 1, \ldots, n\}$$

$(V_i = U_0 \cap U_{i,\epsilon} = \{(x_k) \mid x_0 \neq 0,\, |\frac{x_i}{x_0}| > \epsilon\}$. For the associate alternating Čech-complex

$$\mathcal{C}^{n-1}(\mathcal{U}, \Omega^n_{\mathbf{A}^n_K}) = \mathcal{C}^n(\mathcal{U}_0, \Omega^n_{\mathbf{P}^n_K})$$

and for each $k \in [0, n-1]$ there exists:

$$\mathcal{C}^k(\mathcal{U}, \Omega^n_{\mathbf{A}^n_K}) \xrightarrow{f_k} \mathcal{C}^{k+1}(\mathcal{U}_0, \Omega^n_{\mathbf{P}^n_K})$$

$$(g_{\sigma_0, \ldots, \sigma_k}) \longrightarrow \begin{cases} g_{\sigma_0 \ldots \sigma_k, 0} \\ 0 \text{ otherwize} \end{cases}$$

($\sigma_i \in [1,n]$), which associates the same section if the open set U_0 appears in the nerve of $\mathcal{U}_0$ (in this case we will always impose 0 in the last position), 0 in the other cases. With these definitions the diagrams ($k \in [0, n-1]$)

$$\begin{array}{ccc} \mathcal{C}^k(\mathcal{U}, \Omega^n_{\mathbf{A}^n_K}) & \xrightarrow{\delta} & \mathcal{C}^{k+1}(\mathcal{U}, \Omega^n_{\mathbf{A}^n_K}) \\ \downarrow f_k & & \downarrow f_{k+1} \\ \mathcal{C}^{k+1}(\mathcal{U}_\prime, \Omega^n_{\mathbf{P}^n_K}) & \xrightarrow{\delta_0} & \mathcal{C}^{k+2}(\mathcal{U}_\prime, \Omega^n_{\mathbf{P}^n_K}) \end{array} \tag{2.14}$$

are commutative i.e. :

$$f_{k+1} \circ \delta = \delta_0 \circ f_k.$$

It is then possible to define a map:

$$f : \check{H}^{n-1}(\mathcal{U}, \Omega^n_{\mathbf{A}^n_K}) \longrightarrow \check{H}^n(\mathcal{U}_0, \Omega^n_{\mathbf{P}^n_K}).$$

Finally the main point of this note consists in showing that the following diagramm:

$$\begin{array}{ccccccc} & & \check{H}^{n-1}(\mathcal{U}, \Omega^n_{\mathbf{A}^n_K}) & \longrightarrow & H^{n-1}(\mathbf{A}^n_K \setminus B_\epsilon, \Omega^n_{\mathbf{A}^n_K}) & & \\ & \swarrow^{res_n} & & & & \searrow^{\gamma} & \\ K & \Xi_1 & \downarrow f & & \Xi_2 & & H^n_{B_\epsilon}(\mathbf{A}^n_K, \Omega^n_{\mathbf{A}^n_K}) \\ & \nwarrow_{res} & & & & \swarrow_{\Gamma} & \\ & & \check{H}^n(\mathcal{U}_0, \Omega^n_{\mathbf{P}^n_K}) & \longrightarrow & H^n(\mathbf{P}^n_K, \Omega^n_{\mathbf{P}^n_K}) & & \end{array}$$

is commutative, too. The symbol res represents the algebraic residue in $\mathbf{P}^n_K$ ([HARA]): in the rigid setting the G.A.G.A type theorems hold [KO].

Take $\xi \in \check{H}^{n-1}(\mathcal{U}, \Omega^n_{\mathbf{A}^n_K})$ (we will denote in the same way its representative $\xi \in \mathcal{Z}^{n-1}(\mathcal{U}, \Omega^n_{\mathbf{A}^n_K}) = \mathcal{C}^{n-1}(\mathcal{U}, \Omega^n_{\mathbf{A}^n_K})$): its residue, $res_n\xi$, is given by the coefficient of $\frac{1}{x_1 \ldots x_k}$. The same value is assumed by res on $f(\xi)$ i.e. for ξ viewed in $\mathcal{Z}^{n-1}(\mathcal{U}_0, \Omega^n_{\mathbf{P}^n_K}) = \mathcal{C}^n(\mathcal{U}_0, \Omega^n_{\mathbf{P}^n_K})$. This shows that the Ξ_1 part of the diagramm is commutative. We now study the part Ξ_2. Consider $\xi \in \mathcal{C}^n(\mathcal{U}_0, \Omega^n_{\mathbf{P}^n_K}) = \mathcal{C}^{n-1}(\mathcal{U}, \Omega^n_{\mathbf{A}^n_K})$ (i.e. we are taking an element of $H^{n-1}(\mathbf{A}^n_K \setminus B_\epsilon, \Omega^n_{\mathbf{A}^n_K})$ and the element of $H^n(\mathcal{U}_0, \Omega^n_{\mathbf{P}^n_K})$ associated to it by f), we want to show that the related elements of $H^n(\mathbf{P}^n_K, \Omega^n_{\mathbf{P}^n_K})$ and $H^{n-1}(\mathbf{A}^n_K \setminus B_\epsilon, \Omega^n_{\mathbf{A}^n_K})$ are in the relationship expressed by Ξ_2. Take $i(\xi) \in \mathcal{Z}^{n-1}(\mathcal{U}, J_0) = \mathcal{Z}^n(\mathcal{U}_0, J_0)$ (2.13). The injective sheaf J_0 is acyclic [SGAIV] : there exists $\xi_1 \in \mathcal{C}^{n-2}(\mathcal{U}, \Omega^n_{\mathbf{A}^n_K})$ such that $\delta(\xi_1) = i(\xi)$ (2.14). Using the map f_{n-2} (which is , actually, an injection) by (2.14):

$$\delta_0(f_{n-2}(\xi_1)) = i(\xi) \in \mathcal{Z}^n(\mathcal{U}_0, J_0),$$

this means that $f_{n-2}(\xi_1)$ realizes $i(\xi)$ as a bord in $\mathcal{C}^\bullet(\mathcal{U}_0, J_0)$. By (2.13) :

$$\delta(d\xi_1) = d\delta\xi_1 = di\xi = 0$$

$$\delta_0(df_{n-2}\xi_1) = d\delta_0 f_{n-2}\xi_1 = di\xi = 0.$$

It follows that

$$d\xi_1 \in \mathcal{Z}^{n-2}(\mathcal{U}, J_1) \qquad df_{n-2}\xi_1 = f_{n-2}d\xi_1 \in \mathcal{Z}^{n-1}(\mathcal{U}_0, J_1).$$

We can repeat the same argument for J_1 : there will be $\xi_2 \in \mathcal{C}^{n-3}(\mathcal{U}, J_1)$ such that $\delta\xi_2 = d\xi_1$. For $\mathbf{P}^n_K$ we take $f_{n-3}(\xi_2) \in \mathcal{C}^{n-2}(\mathcal{U}_0, J_1)$:

$$\delta_0 f_{n-3}\xi_2 = f_{n-2}\delta\xi_2 = f_{n-2}d\xi_1 = df_{n-2}\xi_1,$$

as before $d\xi_2 \in \mathcal{Z}^{n-3}(\mathcal{U}, J_2)$ and $df_{n-3}\xi_2 = f_{n-3}d\xi_2 \in \mathcal{Z}^{n-2}(\mathcal{U}_0, J_2)$. At the end we obtain $d\xi_{n-1} \in \mathcal{Z}^0(\mathcal{U}, J_{n-1})$ and we put $d\xi_{n-1} = l \in H^0(\mathbf{A}^n_K \backslash B_\epsilon, J_{n-1})$; of course $dl = d^2\xi_{n-1} = 0$ so $l \in H^{n-1}(\mathbf{A}^n_K \setminus B_\epsilon, \Omega^n_{\mathbf{A}^n_K})$. For $\mathbf{P}^n_K$ we can represent $d\xi_{n-1} = (\lambda_i)^n_{i=1}$, $\lambda_i \in \Gamma(U_0 \cap U_{i,\epsilon}, J_{n-1})$, then

$$f_0 d\xi_{n-1} = df_0\xi_{n-1} = (\lambda_{1,0}, \dots, \lambda_{n,0}, 0_{1,1}, \dots 0_{n,n}) \in \mathcal{Z}^1(\mathcal{U}_0, J_{n-1})$$

$(\lambda_{i,0} = \lambda_i)$. By the acyclicity of J_{n-1} there will exist $\tilde{\xi}_n \in \mathcal{C}^0(\mathcal{U}_0, J_{n-1})$ such that

$$\delta_0(\tilde{\xi}_n) = df_0\xi_{n-1} = f_0 d\xi_{n-1}.$$

To construct $\tilde{\xi}_n$ one takes an extension $\bar{l} \in H^0(\mathbf{A}^n_K, J_{n-1})$ of $l \in H^0(\mathbf{A}^n_K \setminus B_\epsilon, J_{n-1})$ and by our choice of coordinates $\bar{l} \in \Gamma(U_0, J_{n-1})$; thus

$$\tilde{\xi}_n = (\bar{l}, 0_1, \dots, 0_n) \in \mathcal{C}^0(\mathcal{U}_0, J_{n-1}).$$

Using (2.13)

$$d\tilde{\xi}_n = (d\bar{l}, 0_1, \dots, 0_n) \in \mathcal{Z}^0(\mathcal{U}_0, J_n)$$

whence we obtain $(d\bar{l}, 0_1, \dots, 0_n) = (d\bar{l})_e \in H^n(\mathbf{P}^n_K, \Omega^n_{\mathbf{P}^n_K})$ as in the construction of Γ and γ. This concludes the proof of the commutativity of Ξ_2. As we said, this note gives a " unicity statement " to our notion of residue. In fact by the compatibility we can derive the invariance for automorphisms of $\mathbf{P}^n_K$.

§3

We want to extend the duality to a coherent $\mathcal{O}_{\mathbf{A}^n_K}$-module (cf. (2.0) and [BGR], [FRE]). One of the most outstanding features of coherent $\mathcal{O}_{\mathbf{A}^n_K}$-modules on a rigid analytic variety X is the fact that for this thick category the rigid variety has sufficiently many points i.e. if $\mathcal{M}$ is a coherent $\mathcal{O}_{\mathbf{A}^n_K}$-module

$$\mathcal{M}_x = 0 \quad \forall x \in X$$

then $\mathcal{M} = 0$ ([FRE],[BGR]9.4).

As we said $\mathbf{A}^n_K$ is a quasi-Stein space, it follows that for a coherent $\mathcal{O}_{\mathbf{A}^n_K}$-module $\mathcal{M}$:

$$H^p(\mathbf{A}^n_K, \mathcal{M}) = 0 \qquad if\ p \neq 0$$

[KI].

In the first part of this paragraph we will study the pairing for $\mathcal{M}$:

$$H^0(\mathbf{A}^n_K, \mathcal{M}) \times Ext^n_{\mathcal{O}_{\mathbf{A}^n_K},c}(\mathbf{A}^n_K, \mathcal{M}, \Omega^n_{\mathbf{A}^n_K}) \longrightarrow K \tag{3.1}$$

In $H^0(\mathbf{A}^n_K, \mathcal{M})$ (for a coherent $\mathcal{O}_{\mathbf{A}^n_K}$-module $\mathcal{M}$) it is possible to introduce a canonical topology. Once we have noted that the sections over an affinoid admissible open of a coherent sheaf form a Banach space of countable type (see after (2.0)) the same arguments we used for $\mathcal{O}_{\mathbf{A}^n_K}$ (§2) allow us to conclude that $H^0(\mathbf{A}^n_K, \mathcal{M})$ is a Fréchet space of countable type [RR].

We don't know if our family of supports is a local family: for this reason we cannot give a canonical topology on $Ext^\bullet_{\mathcal{O}_{\mathbf{A}^n_K},c}(\mathbf{A}^n_K, \mathcal{M}, \Omega^n_{\mathbf{A}^n_K})$.

Later, (3.14), we will consider the pairing ($\mathcal{M}$ ia a coherent $\mathcal{O}_{\mathbf{A}^n_K}$-module)

$$H^p_c(\mathbf{A}^n_K, \mathcal{M}) \times Ext^{n-p}_{\mathcal{O}_{\mathbf{A}^n_K}}(\mathbf{A}^n_K, \mathcal{M}, \Omega^n_{\mathbf{A}^n_K}) \longrightarrow K \quad p = 0, \dots, n \tag{3.2}$$

We want to study the topologies on these spaces. For $H^\bullet_c(\mathbf{A}^n_K, \mathcal{M})$ as usual (§2) we have :

$$H^\bullet_c(\mathbf{A}^n_K, \mathcal{M}) = \varinjlim_{\epsilon \in |K^*|} H^\bullet_{B_\epsilon}(\mathbf{A}^n_K, \mathcal{M}).$$

Since the space $\mathbf{A}_K^n$ is quasi Stein we get

$$H^p_{B_\epsilon}(\mathbf{A}_K^n, \mathcal{M}) \simeq H^{p-1}(\mathbf{A}_K^n \setminus B_\epsilon, \mathcal{M}) \qquad p \geq 2$$

$$H^1_{B_\epsilon}(\mathbf{A}_K^n, \mathcal{M}) \simeq \frac{H^0(\mathbf{A}_K^n \setminus B_\epsilon, \mathcal{M})}{H^0(\mathbf{A}_K^n, \mathcal{M})}$$

and $H^0_{B_\epsilon}(\mathbf{A}_K^n, \mathcal{M})$ is a closed subspace of $H^0(\mathbf{A}_K^n, \mathcal{M})$.

For each $H^\bullet(\mathbf{A}_K^n \setminus B_\epsilon, \mathcal{M})$ and $H^0(\mathbf{A}_K^n, \mathcal{M})$ there is a canonical topology (see proofs of (2.3), (3.1)), which will induce a topology on $H^\bullet_{B_\epsilon}(\mathbf{A}_K^n, \mathcal{M})$. The direct limit of these topological spaces $H^\bullet_{B_\epsilon}(\mathbf{A}_K^n, \mathcal{M})$ will give the canonical topology on $H^\bullet_c(\mathbf{A}_K^n, \mathcal{M})$.

We want now to introduce a canonical topology on $Ext^\bullet_{\mathcal{O}_{\mathbf{A}_K^n}}(\mathbf{A}_K^n, \mathcal{M}, \Omega^n_{\mathbf{A}_K^n})$. We need some results:

Proposition 3.3. *If $\mathcal{G}$, $\mathcal{F}$ are coherent $\mathcal{O}_X$-modules in a rigid analytic variety X, then*

$$\mathcal{E}xt^i_{\mathcal{O}_X}(\mathcal{F}, \mathcal{G})$$

are coherent $\mathcal{O}_X$-modules for each i and

$$\mathcal{E}xt^i_{\mathcal{O}_X}(\mathcal{F}, \mathcal{G})_y = Ext^i_{\mathcal{O}_{X,y}}(\mathcal{F}_y, \mathcal{G}_y).$$

Proof. For the first claim take $U = SpmA$ an admissible affinoid open of X. The fact that the restriction of an injective $\mathcal{O}_X$-module $\mathcal{I}$ to U, $\mathcal{I}_{|U}$, is still an injective $\mathcal{O}_U$-module ([KO]3.8), implies:

$$\mathcal{E}xt^i_{\mathcal{O}_X}(\mathcal{F}, \mathcal{G})_{|U} = \mathcal{E}xt^i_{\mathcal{O}_{X|U}}(\mathcal{F}_{|U}, \mathcal{G}_{|U}).$$

But $\mathcal{F}_{|U} = \widetilde{F}$ where F is a finite A-module. It is possible to write

$$0 \longrightarrow F_1 \longrightarrow A^m \longrightarrow F \longrightarrow 0$$

and F_1 is a finite A-module (A is noetherian). The relative sequence of associated $\mathcal{O}_U$-modules is still exact:

$$0 \longrightarrow \widetilde{F_1} \longrightarrow \widetilde{A^m} \longrightarrow \widetilde{F} = \mathcal{F}_{|U} \longrightarrow 0 \tag{3.4}$$

[BGR]. In particular $\widetilde{F_1}$ is a coherent $\mathcal{O}_U$-module. Finally , by induction on the cohomology, using the fact that

$$\mathcal{E}xt^0_{\mathcal{O}_{X|U}}(\mathcal{F}_{|U}, \mathcal{G}_{|U})$$

is coherent if $\mathcal{F}$,$\mathcal{G}$ are coherent [LIU] and the "thickness" of the category of the coherent $\mathcal{O}_U$-modules we çan reach the desired conclusion.

For the second statement the point is, as in the classical case, that the fiber $\mathcal{I}_y$, $y \in X$, of an injective $\mathcal{O}_X$-module $\mathcal{I}$, is an injective $\mathcal{O}_{X,y}$-module (the proof is as in ([GROT]IV) using the fact that an affinoid algebra is noetherian [BGR]). **Q.E.D.**

Remark 3.5. The second statement is true for every $\mathcal{O}_X$ -module $\mathcal{G}$.

Corollary 3.6. *If $X = \mathbf{A}_K^n$ and $\mathcal{F}$,$\mathcal{G}$ are coherent $\mathcal{O}_{\mathbf{A}_K^n}$-modules, then*

$$\mathcal{E}xt^i_{\mathcal{O}_{\mathbf{A}_K^n}}(\mathcal{F}, \mathcal{G}) = 0 \qquad if\ i > n.$$

Proof. From the proposition (3.4), we know that

$$\mathcal{E}xt^i_{\mathcal{O}_{\mathbf{A}_K^n}}(\mathcal{F}, \mathcal{G})$$

are coherent sheaves : it follows that the points are sufficient to separate these sheaves. On the other hand (3.4):

$$\mathcal{E}xt^i_{\mathcal{O}_{\mathbf{A}^n_K}}(\mathcal{F}\,,\mathcal{G})_y \;=\; \mathcal{E}xt^i_{\mathcal{O}_{\mathbf{A}^n_K},y}(\mathcal{F}_y\,,\mathcal{G}_y)$$

for every $y \in \mathbf{A}^n_K$. But $\mathcal{O}_{\mathbf{A}^n_K,y}$ is a regular, noetherian local ring of dimension n and $\mathcal{F}_y\,,\mathcal{G}_y$ are finite $\mathcal{O}_{\mathbf{A}^n_K,y}$-modules. This implies that $\forall\; y \in \mathbf{A}^n_K$:

$$\mathcal{E}xt^i_{\mathcal{O}_{\mathbf{A}^n_K},y}(\mathcal{F}_y\,,\mathcal{G}_y) = 0$$

if $i > n$ **Q.E.D.**

To conclude the study of (3.2) we note that, since $\mathbf{A}^n_K$ is quasi Stein and $\mathcal{E}xt^\bullet_{\mathcal{O}_{\mathbf{A}^n_K}}(\mathcal{M},\Omega^n_{\mathbf{A}^n_K})$ are coherent $\mathcal{O}_{\mathbf{A}^n_K}$-modules (3.4) :

$$Ext^p_{\mathcal{O}_{\mathbf{A}^n_K}}(\mathbf{A}^n_K,\mathcal{M},\Omega^n_{\mathbf{A}^n_K}) \;=\; H^0(\mathbf{A}^n_K,\mathcal{E}xt^p_{\mathcal{O}_{\mathbf{A}^n_K}}(\mathcal{M},\Omega^n_{\mathbf{A}^n_K}))$$

and thus it is possible to give a canonical topology to $Ext^p_{\mathcal{O}_{\mathbf{A}^n_K}}(\mathbf{A}^n_K,\mathcal{M},\Omega^n_{\mathbf{A}^n_K})$ (global sections of a coherent sheaf): Fréchet spaces of countable type (3.1) (for the spectral sequence see ([SGAIV]V4.10)).

We have defined canonical topologies on the objects appearing in the pairing (3.2) and (3.1). Before stating the duality theorems we need an "excision lemma" in the rigid setting.

Lemma 3.7. *Under the previous hypotheses for $\mathcal{M}$, $\mathcal{O}_{\mathbf{A}^n_K}$-module*

$$H^0_{B_\epsilon}(\mathring{B}_{\epsilon_i},\mathcal{M}) \;=\; H^0_{B_\epsilon}(\mathring{B}_{\epsilon_j},\mathcal{M}) \;=\; H^0_{B_\epsilon}(\mathbf{A}^n_K,\mathcal{M})$$

with $\epsilon < \epsilon_i < \epsilon_j$; $\epsilon, \epsilon_i, \epsilon_j \in \mid K^* \mid$.

Proof. For notations see (2.2), (2.10). For the proof see ([MOS]lemma1). **Q.E.D.**

Remark 3.8. One can prove that the extentions of the previous isomorphisms (3.7) to the cohomological groups are homeomorphisms for the canonical topologies when $\mathcal{M}$ is a coherent $\mathcal{O}_{\mathbf{A}^n_K}$-module. It is only necessary to observe that a countable admissible affinoid covering of $\mathbf{A}^n_K \setminus B_\epsilon$ will induce a countable admissible covering of $\mathring{B}_{\epsilon_i} \setminus B_\epsilon$ consisting of quasi-Stein spaces (an affinoid is quasi-Stein).

Remark 3.9.f3.2 We recall that by our definition:

$$Ext^p_{\mathcal{O}_{\mathbf{A}^n_K},c}(\mathbf{A}^n_K,\mathcal{M},\Omega^n_{\mathbf{A}^n_K}) \;=\; \varinjlim_{\epsilon\in|K^*|} Ext^p_{\mathcal{O}_{\mathbf{A}^n_K},B_\epsilon}(\mathbf{A}^n_K,\mathcal{M},\Omega^n_{\mathbf{A}^n_K})$$

for every $p \in \mathbf{N}$.

By (3.1) we can, now, state:

Theorem 3.10. *Under the previous hypotheses, the space $H^0(\mathbf{A}^n_K,\mathcal{M})$ endowed with the natural topology given after (3.1) is a reflexive Fréchet space of countable type. The pairing:*

$$H^0(\mathbf{A}^n_K,\mathcal{M}) \times Ext^n_{\mathcal{O}_{\mathbf{A}^n_K},c}(\mathbf{A}^n_K,\mathcal{M},\Omega^n_{\mathbf{A}^n_K}) \longrightarrow K \tag{3.1}$$

makes $Ext^n_{\mathcal{O}_{\mathbf{A}^n_K},c}(\mathbf{A}^n_K,\mathcal{M},\Omega^n_{\mathbf{A}^n_K})$ the dual of the previous space. Furthemore

$$Ext^p_{\mathcal{O}_{\mathbf{A}^n_K},c}(\mathbf{A}^n_K,\mathcal{M},\Omega^n_{\mathbf{A}^n_K}) \;=\; 0 \qquad if\; p \neq n.$$

Proof. We know that

$$H^0(\mathbf{A}^n_K, \mathcal{M})$$

is a Fréchet space of countable type. What we need to show is that it is reflexive, that there is a bijection (as sets) given by the pairing:

$$Ext^n_{\mathcal{O}_{\mathbf{A}^n_K},c}(\mathbf{A}^n_K, \mathcal{M}, \Omega^n_{\mathbf{A}^n_K}) \longrightarrow (H^0(\mathbf{A}^n_K, \mathcal{M}))', \tag{3.11}$$

and the vanishing of the other groups of cohomology for $Ext^\bullet_{\mathcal{O}_{\mathbf{A}^n_K},c}(\mathbf{A}^n_K, \mathcal{M}, \Omega^n_{\mathbf{A}^n_K})$.

To calculate $H^0(\mathbf{A}^n_K, \mathcal{M})$ one can take the admissible affinoid covering $\{B_{\epsilon^i}\}_{i\in\mathbf{N}}$, $\epsilon \in \mid K^* \mid$, $\epsilon > 1$ and

$$H^0(\mathbf{A}^n_K, \mathcal{M}) = \lim_{\leftarrow} H^0(B_{\epsilon^i}, \mathcal{M}).$$

Each B_{ϵ^i} is an affinoid admissible open and $\mathcal{M}_{|B_{\epsilon^i}}$ is an associated module. We note that

$$\mathcal{O}_{\mathbf{A}^n_K}(B_{\epsilon^i}) = \{\sum_{\alpha\in\mathbf{N}^n} a_\alpha x^\alpha \quad a_\alpha \in K \quad \lim_{|\alpha|\to+\infty} \mid a_\alpha \mid \epsilon^{i|\alpha|} = 0\}$$

is a Banach space (in particular it is an affinoid K-algebra) under the norm

$$\| \sum a_\alpha x^\alpha \|_i = sup \mid a_\alpha \mid \epsilon^{i|\alpha|}.$$

In the previous situation (2.0)

$$\mathcal{M}(B_{\epsilon^i}) = \frac{\mathcal{O}_{\mathbf{A}^n_K}(B_{\epsilon^i})^n}{\mathcal{A}}$$

where $\mathcal{A}$ is a sub $\mathcal{O}_{\mathbf{A}^n_K}(B_{\epsilon^i})$-module. We consider two cases. If the field K is spherically complete [MOH], then one can prove that the maps of Banach spaces

$$\mathcal{O}_{\mathbf{A}^n_K}(B_{\epsilon^i}) \longrightarrow \mathcal{O}_{\mathbf{A}^n_K}(B_{\epsilon^{i-1}}) \qquad i \in \mathbf{N}$$

are c-compact (i.e. the image of the unit ball is relatively c-compact : the proof is given using the analogous result for bounded functions [MOH],[MOS]). From the facts that

$$H^0(B_{\epsilon^i}, \mathcal{M}) = \frac{\mathcal{O}_{\mathbf{A}^n_K}(B_{\epsilon^i})^n}{\mathcal{A}}$$

and that $\mathcal{M}$ is coherent:

$$H^0(B_{\epsilon^{i-1}}, \mathcal{M}) = \frac{\mathcal{O}_{\mathbf{A}^n_K}(B_{\epsilon^i})^n}{\mathcal{A}} \otimes \mathcal{O}_{\mathbf{A}^n_K}(B_{\epsilon^{i-1}}) = \frac{\mathcal{O}_{\mathbf{A}^n_K}(B_{\epsilon^{i-1}})^n}{\widetilde{\mathcal{A}}}$$

As in the classical case (i.e. over $\mathbf{C}$) , the image of a c-compact set by a continuous map is still c-compact. We can conclude that all the maps of Banach spaces

$$H^0(B_{\epsilon^i}, \mathcal{M}) \longrightarrow H^0(B_{\epsilon^{i-1}}, \mathcal{M})$$

are c-compact. On the other hand by [KI] (the covering $\{B_{\epsilon^i}\}_{i\in\mathbf{N}}$ gives the structure of a quasi-Stein space to $\mathbf{A}^n_K$):

$$H^0(\mathbf{A}^n_K, \mathcal{M}) \longrightarrow H^0(B_{\epsilon^i}, \mathcal{M})$$

has dense image for every i. It follows from ([MOH]3.4) that $H^0(\mathbf{A}^n_K, \mathcal{M})$ is a reflexive Fréchet space of countable type and its dual is

$$H^0(\mathbf{A}^n_K, \mathcal{M})' = \varinjlim H^0(B_{\epsilon^i}, \mathcal{M})'$$

If the field K is not spherically complete we can reach the same conclusion. In fact, in this case each Fréchet space of countable type is reflexive ([SCH]9.9) : in particular this is true for $H^0(\mathbf{A}^n_K, \mathcal{M})$. Again , the fact that each

$$H^0(\mathbf{A}^n_K, \mathcal{M}) \longrightarrow H^0(B_{\epsilon^i}, \mathcal{M})$$

is dense, allows us to write

$$H^0(\mathbf{A}^n_K, \mathcal{M})' = \varinjlim H^0(B_{\epsilon^i}, \mathcal{M})'.$$

So, for general K ultrametric field, $H^0(\mathbf{A}^n_K, \mathcal{M})$ is reflexive and

$$H^0(\mathbf{A}^n_K, \mathcal{M})' = \varinjlim_{i\in\mathbf{N}} H^0(B_{\epsilon^i}, \mathcal{M})'.$$

One can have a homeomorphism of topological spaces (2.10)

$$H^0(\mathbf{A}^n_K, \mathcal{M}) = \varprojlim_{i\in\mathbf{N}} H^0(\overset{\circ}{B}_{\epsilon^i}, \mathcal{M})$$

and the fact that

$$H^0(\mathbf{A}^n_K, \mathcal{M})' = \varinjlim_{i\in\mathbf{N}} H^0(\overset{\circ}{B}_{\epsilon^i}, \mathcal{M})'.$$

(In fact each $H^0(\mathbf{A}^n_K, \mathcal{M}) \to H^0(\overset{\circ}{B}_{\epsilon^i}, \mathcal{M})$ is dense. The set $\overset{\circ}{B}_{\epsilon^i}$ (2.10) is quasi-Stein : to show this fact one can choose a covering $\{B_{\epsilon_n}\}_{n\in\mathbf{N}}$, $\epsilon_n \to \epsilon^i$, $\epsilon_n \in | K^* |$. In general every sequence $\{\eta_n\}_{n\in\mathbf{N}}$, $\eta_n \in | K^* |$, $\eta_n \to +\infty$ can give an admissible affinoid covering of $\mathbf{A}^n_K$, $\{B_{\eta_n}\}_{n\in\mathbf{N}}$, which makes $\mathbf{A}^n_K$ quasi-Stein). We prove now that the dual of $H^0(\overset{\circ}{B}_{\epsilon^i}, \mathcal{M})$ is given by $Ext^n_{\mathcal{O}_{\mathbf{A}^n_K},c}(\overset{\circ}{B}_{\epsilon^i}, \mathcal{M}, \Omega^n_{\mathbf{A}^n_K})$ via the pairing

$$H^0(\overset{\circ}{B}_{\epsilon^i}, \mathcal{M}) \times Ext^n_{\mathcal{O}_{\mathbf{A}^n_K},c}(\overset{\circ}{B}_{\epsilon^i}, \mathcal{M}, \Omega^n_{\mathbf{A}^n_K}) \longrightarrow K \qquad (3.11)_i$$

and that

$$Ext^p_{\mathcal{O}_{\mathbf{A}^n_K},c}(\overset{\circ}{B}_{\epsilon^i}, \mathcal{M}, \Omega^n_{\mathbf{A}^n_K}) = 0 \qquad if\, p \neq n.$$

With

$$Ext^\bullet_{\mathcal{O}_{\mathbf{A}^n_K},c}(\overset{\circ}{B}_{\epsilon^i}, \mathcal{M}, \Omega^n_{\mathbf{A}^n_K}) = \varinjlim_{\eta<\epsilon^i} Ext^\bullet_{\mathcal{O}_{\mathbf{A}^n_K},B_\eta}(\overset{\circ}{B}_{\epsilon^i}, \mathcal{M}, \Omega^n_{\mathbf{A}^n_K})$$

(In fact we don't know if the notion of compact support we gave is a local one. Morally we are working with the compacts inside $\overset{\circ}{B}_{\epsilon^i}$).

And from (3.7)

$$\varinjlim_{\eta<\epsilon^i} Ext^p_{\mathcal{O}_{\mathbf{A}^n_K},B_\eta}(\mathbf{A}^n_K, \mathcal{M}, \Omega^n_{\mathbf{A}^n_K}) = \varinjlim_{\eta<\epsilon^i} Ext^p_{\mathcal{O}_{\mathbf{A}^n_K},B_\eta}(\overset{\circ}{B}_{\epsilon^i}, \mathcal{M}, \Omega^n_{\mathbf{A}^n_K}). \qquad (3.7)_i$$

To prove $(3.11)_i$ one should first note that on $\overset{\circ}{B}_{\epsilon^i}$ (open polydisk) there exists a finite free resolution of $\mathcal{M}$ (on $\overset{\circ}{B}_{\epsilon^i}$, $\mathcal{M}$ is associated to a finite $\mathcal{O}_{\mathbf{A}^n_K}(\overset{\circ}{B}_{\epsilon^i})$-module) [GRU]:

$$0 \longrightarrow \mathcal{O}^{n_l}_{\overset{\circ}{B}_{\epsilon^i}} \longrightarrow \ldots\ldots \longrightarrow \mathcal{O}^{n_1}_{\overset{\circ}{B}_{\epsilon^i}} \longrightarrow \mathcal{M}_{|\overset{\circ}{B}_{\epsilon^i}} \longrightarrow 0. \tag{3.12}$$

We will prove the two assertions in $(3.11)_i$ by induction on the length $l(\mathcal{M})$ of the free resolution (3.12). If $l(\mathcal{M}) = 1$, $\mathcal{M}$ is free (of course in $\overset{\circ}{B}_{\epsilon^i}$) and one can see the duality as in (2.9). Suppose all assertions true for length $< n-1$ and take $\mathcal{M}$, $l(\mathcal{M}) = n$. Then

$$0 \longrightarrow \mathcal{A} \longrightarrow \mathcal{O}^q_{\overset{\circ}{B}_{\epsilon^i}} \longrightarrow \mathcal{M}_{|\overset{\circ}{B}_{\epsilon^i}} \longrightarrow 0.$$

where $\mathcal{A}$ is coherent in $\overset{\circ}{B}_{\epsilon^i}$ and $l(\mathcal{A}) = n-1$. But since $\overset{\circ}{B}_{\epsilon^i}$ is a quasi-Stein space, the short sequence of reflexive Fréchet spaces of countable type :

$$0 \longrightarrow H^0(\overset{\circ}{B}_{\epsilon^i}, \mathcal{A}) \longrightarrow H^0(\overset{\circ}{B}_{\epsilon^i}, \mathcal{O}^q_{\overset{\circ}{B}_{\epsilon^i}}) \longrightarrow H^0(\overset{\circ}{B}_{\epsilon^i}, \mathcal{M}) \longrightarrow 0$$

is exact. Take the dual sequence : it is still exact (if K is spherically complete the argument is as in [SED] using the Hahn-Banach theorem, if K is not spherically complete the "countable type" hypothesis allows us to have a Hahn-Banach type extention theorem [SCH]):

$$0 \longleftarrow H^0(\overset{\circ}{B}_{\epsilon^i}, \mathcal{A})' \longleftarrow H^0(\overset{\circ}{B}_{\epsilon^i}, \mathcal{O}^q_{\overset{\circ}{B}_{\epsilon^i}})' \longleftarrow H^0(\overset{\circ}{B}_{\epsilon^i}, \mathcal{M})' \longleftarrow 0.$$

By induction the sequence

$$0 \leftarrow Ext^n_{\mathcal{O}_{\mathbf{A}^n_K},c}(\overset{\circ}{B}_{\epsilon^i}, \mathcal{A}, \Omega^n_{\mathbf{A}^n_K})' \longleftarrow Ext^n_{\mathcal{O}_{\mathbf{A}^n_K},c}(\overset{\circ}{B}_{\epsilon^i}, \mathcal{O}^q_{\overset{\circ}{B}_{\epsilon^i}}, \Omega^n_{\mathbf{A}^n_K})' \longleftarrow Ext^n_{\mathcal{O}_{\mathbf{A}^n_K},c}(\overset{\circ}{B}_{\epsilon^i}, \mathcal{M}, \Omega^n_{\mathbf{A}^n_K})' \leftarrow 0$$

is isomorphic to the previous one and still exact. We have proved $(3.11)_i$. Using $(3.7)_i$ we find that:

$$\varinjlim_i (Ext^\bullet_{\mathcal{O}_{\mathbf{A}^n_K},c}(\overset{\circ}{B}_{\epsilon^i}, \mathcal{M}, \Omega^n_{\mathbf{A}^n_K})) = \varinjlim_i (\lim_{\eta < \epsilon^i} Ext^\bullet_{\mathcal{O}_{\mathbf{A}^n_K},B_\eta}(\mathbf{A}^n_K, \mathcal{M}, \Omega^n_{\mathbf{A}^n_K})) =$$

$$= Ext^\bullet_{\mathcal{O}_{\mathbf{A}^n_K},c}(\mathbf{A}^n_K, \mathcal{M}, \Omega^n_{\mathbf{A}^n_K}).$$

By the compatibility of the pairing $(3.11)_i$ at each i, the dual of

$$H^0(\mathbf{A}^n_K, \mathcal{M})$$

is

$$\varinjlim_i (Ext^n_{\mathcal{O}_{\mathbf{A}^n_K},c}(\overset{\circ}{B}_{\epsilon^i}, \mathcal{M}, \Omega^n_{\mathbf{A}^n_K})) = Ext^n_{\mathcal{O}_{\mathbf{A}^n_K},c}(\mathbf{A}^n_K, \mathcal{M}, \Omega^n_{\mathbf{A}^n_K}).$$

Q.E.D.

Remark 3.13. We didn't put a topology on $Ext^\bullet_{\mathcal{O}_{\mathbf{A}^n_K},c}$. Of course, imposing the strong dual topology deriving from (3.1) on it, the pairing will become a perfect pairing: one space is the strong dual of the other.

We now study (3.2).

Theorem 3.14. *Under the previous hypotheses the spaces* $Ext^{n-p}_{\mathcal{O}_{\mathbf{A}^n_K}}(\mathbf{A}^n_K,\mathcal{M},\Omega^n_{\mathbf{A}^n_K})$ *and* $H^p_c(\mathbf{A}^n_K,\mathcal{M})$ *endowed with the canonical topologies are, respectively, a reflexive Fréchet countable space and a complete reflexive, Hausdorff strong dual of a Fréchet space. The pairing :*

$$H^p_c(\mathbf{A}^n_K,\mathcal{M})\times Ext^{n-p}_{\mathcal{O}_{\mathbf{A}^n_K}}(\mathbf{A}^n_K,\mathcal{M},\Omega^n_{\mathbf{A}^n_K})\longrightarrow K\quad p=0,\dots,n \tag{3.2}$$

is a perfect pairing (each one is the strong dual of the other) .

Proof. The spaces

$$Ext^{\bullet}_{\mathcal{O}_{\mathbf{A}^n_K}}(\mathbf{A}^n_K,\mathcal{M},\Omega^n_{\mathbf{A}^n_K})$$

endowed with the canonical topology are Fréchet spaces of countable type : see after corollary (3.6). The first step, as usual, is to show the duality (3.2) in the case $\overset{\circ}{B}_{\epsilon^i}$ ($\epsilon\in|K^*|,\epsilon>1$)(this time with given topology on both sides).

By [GRU] there exists a finite free resolution of $\mathcal{M}_{|\overset{\circ}{B}_{\epsilon^i}}$. We make our induction on the length $l(\mathcal{M})$ of this resolution. For $l(\mathcal{M})=1$, $\mathcal{M}$ is free on $\overset{\circ}{B}_{\epsilon^i}$ and we use (2.9). Suppose

$$0\longrightarrow\mathcal{A}\longrightarrow\mathcal{O}^q_{\overset{\circ}{B}_{\epsilon^i}}\longrightarrow\mathcal{M}_{|\overset{\circ}{B}_{\epsilon^i}}\longrightarrow 0.$$

and $l(\mathcal{M})-1=l(\mathcal{A})$. By induction we may suppose the result known for $\mathcal{A}$ (i.e. theorem (3.14) at level $\overset{\circ}{B}_{\epsilon^i}$). We obtain the long exact sequence

$$\begin{aligned}0\longrightarrow Ext^0_{\mathcal{O}_{\mathbf{A}^n_K}}(\overset{\circ}{B}_{\epsilon^i},\mathcal{M},\Omega^n_{\mathbf{A}^n_K})&\xrightarrow{\psi}Ext^0_{\mathcal{O}_{\mathbf{A}^n_K}}(\overset{\circ}{B}_{\epsilon^i},\mathcal{O}^q_{\overset{\circ}{B}_{\epsilon^i}},\Omega^n_{\mathbf{A}^n_K})\xrightarrow{\varphi}\\ \to Ext^0_{\mathcal{O}_{\mathbf{A}^n_K}}(\overset{\circ}{B}_{\epsilon^i},\mathcal{A},\Omega^n_{\mathbf{A}^n_K})&\xrightarrow{\delta}Ext^1_{\mathcal{O}_{\mathbf{A}^n_K}}(\overset{\circ}{B}_{\epsilon^i},\mathcal{M},\Omega^n_{\mathbf{A}^n_K})\longrightarrow 0\end{aligned}\tag{3.15}$$

and

$$Ext^i_{\mathcal{O}_{\mathbf{A}^n_K}}(\overset{\circ}{B}_{\epsilon^i},\mathcal{M},\Omega^n_{\mathbf{A}^n_K})\simeq Ext^{i-1}_{\mathcal{O}_{\mathbf{A}^n_K}}(\overset{\circ}{B}_{\epsilon^i},\mathcal{A},\Omega^n_{\mathbf{A}^n_K})\qquad i\geq 2 \tag{3.15$_i$}$$

(We know that $Ext^p_{\mathcal{O}_{\mathbf{A}^n_K}}(\overset{\circ}{B}_{\epsilon^i},\mathcal{M},\mathcal{F})=0$ for $p\geq n+1$ and for every $\mathcal{M}$,$\mathcal{F}$ coherent) . By induction, we can write the strong dual sequence for the middle part in (3.15):

$$H^n_c(\overset{\circ}{B}_{\epsilon^i},\mathcal{O}^q_{\overset{\circ}{B}_{\epsilon^i}})\xleftarrow{\varphi^t}H^n_c(\overset{\circ}{B}_{\epsilon^i},\mathcal{A}) \tag{3.17}$$

with the canonical topologies (As in the proof of (3.10) for a $\mathcal{F}$, $\mathcal{O}_{\mathbf{A}^n_K}$-module : $H^{\bullet}_c(\overset{\circ}{B}_{\epsilon^i},\mathcal{F})=\varinjlim_{\eta\leq\epsilon^i}H^{\bullet}_{B_\eta}(\overset{\circ}{B}_{\epsilon^i},\mathcal{F})$.).

The map (3.17) is a part of a long exact sequence:

$$\begin{aligned}0\longrightarrow H^{n-1}_c(\overset{\circ}{B}_{\epsilon^i},\mathcal{M})&\xrightarrow{\delta^t}H^n_c(\overset{\circ}{B}_{\epsilon^i},\mathcal{A})\xrightarrow{\varphi^t}\\ \to H^n_c(\overset{\circ}{B}_{\epsilon^i},\mathcal{O}^q_{\overset{\circ}{B}_{\epsilon^i}})&\xrightarrow{\psi^t}H^n_c(\overset{\circ}{B}_{\epsilon^i},\mathcal{M})\longrightarrow 0\end{aligned}\tag{3.18}$$

and

$$H^i_c(\overset{\circ}{B}_{\epsilon^i},\mathcal{M})\simeq H^{i+1}_c(\overset{\circ}{B}_{\epsilon^i},\mathcal{A})\qquad for\quad i\leq n-2 \tag{3.18$_i$}$$

(by the inductive hypothesis $H_c^p(\mathring{B}_{\epsilon^i}, \mathcal{A}) = 0$ if $p \geq n+1$). We need to know that (3.18), and $(3.18)_i$ endowed with the canonical topologies are the strong dual sequences of (3.15) and $(3.15)_i$.

We divide the proof of this fact in two parts: α) we will prove that taking the transposed sequence of (3.15) and $(3.15)_i$ we get (3.18) and $(3.18)_i$ as sets but the topologies in $H_c^{n-1}(\mathring{B}_{\epsilon^i}, \mathcal{M})$, $H_c^n(\mathring{B}_{\epsilon^i}, \mathcal{M})$ and $H_c^i(\mathring{B}_{\epsilon^i}, \mathcal{M})$ $(i \leq n-2)$ will be the induced by (respectively) (3.17) and $H_c^{i+1}(\mathring{B}_{\epsilon^i}, \mathcal{A})$; β) we will prove that ψ^t, δ^t in (3.18) and the various isomorphisms in $(3.18)_i$ are homeomorphisms in the images for the canonical topologies.

Then, using the fact that, by induction, we know that the strong dual of

$$Ext^0_{\mathcal{O}_{\mathbf{A}^n_K}}(\mathring{B}_{\epsilon^i}, \mathcal{O}^q_{\mathring{B}_{\epsilon^i}}, \Omega^n_{\mathbf{A}^n_K}) \xrightarrow{\varphi} Ext^0_{\mathcal{O}_{\mathbf{A}^n_K}}(\mathring{B}_{\epsilon^i}, \mathcal{A}, \Omega^n_{\mathbf{A}^n_K})$$

and

$$Ext^p_{\mathcal{O}_{\mathbf{A}^n_K}}(\mathring{B}_{\epsilon^i}, \mathcal{A}, \Omega^n_{\mathbf{A}^n_K}) \qquad p \leq 1$$

are realized by (3.17) and $H_c^{n-p}(\mathring{B}_{\epsilon^i}, \mathcal{A})$ with canonical topologies, we conclude that the strong dual of $Ext^\bullet_{\mathcal{O}_{\mathbf{A}^n_K}}(\mathring{B}_{\epsilon^i}, \mathcal{M}, \Omega^n_{\mathbf{A}^n_K})$ is $H_c^{n-\bullet}(\mathring{B}_{\epsilon^i}, \mathcal{M})$ via the pairing.

For α), one can take a sequence $\{\eta_n\}_{n \in \mathbf{N}}$, $\eta_n \to \epsilon^i$, $\eta_n \in | K^* |$ and to obtain (3.10):

$$Ext^0_{\mathcal{O}_{\mathbf{A}^n_K}}(\mathring{B}_{\epsilon^i}, \mathcal{O}^q_{\mathring{B}_{\epsilon^i}}, \Omega^n_{\mathbf{A}^n_K}) = \varprojlim_{\eta_n} Ext^0_{\mathcal{O}_{\mathbf{A}^n_K}}(B_{\eta_n}, \mathcal{O}^q_{\mathring{B}_{\epsilon^i}}, \Omega^n_{\mathbf{A}^n_K}) = \varprojlim_{n} A_n$$

and each element of the projective limit is a Banach space of countable type. We will denote by $\{B_n\}$ the analogous Banach sequence relative to the space

$$Ext^0_{\mathcal{O}_{\mathbf{A}^n_K}}(\mathring{B}_{\epsilon^i}, \mathcal{A}, \Omega^n_{\mathbf{A}^n_K})$$

and by φ_n the maps of Banach spaces

$$A_n \xrightarrow{\varphi_n} B_n \tag{3.19}$$

(continuous because $\mathcal{O}_{\mathbf{A}^n_K}(B_{\eta_n})$-linear and A_n, B_n are finite $\mathcal{O}_{\mathbf{A}^n_K}(B_{\eta_n})$-modules). Since the φ_n of (3.19) are $\mathcal{O}_{\mathbf{A}^n_K}(B_{\eta_n})$-linear, they have closed images [BGR]. It follows that

$$0 \longrightarrow ker\varphi_n \longrightarrow A_n \xrightarrow{\varphi_n} B_n \longrightarrow Coker\varphi_n \longrightarrow 0 \tag{3.20$_n$}$$

is an exact sequence of Banach spaces of countable type which are finite $\mathcal{O}_{\mathbf{A}^n_K}(B_{\eta_n})$-modules. If the field K is spherically complete (as in the proof of (3.10)) the maps in the projective limit $\{\varphi_n\}$ are c-compact : it follows that all the relative maps

$$Ker\varphi_n \longrightarrow Ker\varphi_{n-1} \quad , \quad Coker\varphi_n \longrightarrow Coker\varphi_{n-1}$$

are c-compact.One can write the transpose of $(3.20)_n$ with the Banach dual topology:

$$0 \longleftarrow ker\varphi_n{}' \longleftarrow A'_n \xleftarrow{\varphi_n{}'} B'_n \longleftarrow Coker\varphi_n{}' \longleftarrow 0 \tag{3.20$'_n$}$$

By the Hahn-Banach theorem, which is valid because K is spherically complete [IN] , $(3.20)'_n$ is still exact and continuous. If K is not spherically complete, the fact that $(3.20)_n$ is made up of Banach spaces of countable type allows us to conclude that $(3.20)'_n$ is still exact (there is, in fact, an extension theorem without identity for the norm [SCH]) and $A'_n/\varphi^t B'_n \simeq Ker\varphi_n{}'$, $Ker\varphi^t \simeq Coker\varphi_n{}'$ with the induced topologies (in the spherically complete case this isomorphism will be isometric. Here, by the open mapping theorem [BGR], we can only say that it is a homeomorphism). On the other hand the countability hypothesis implies that each element of $(3.20)_n$ is a reflexive space [SCH],[vRO].

We note also (independently of whether or not the field is spherically complete) that each map which forms the projective limits $\{Ker\varphi_n\}$, $\{Coker\varphi_n\}$, $\{A_n\}$, $\{B_n\}$ has dense image, and the same is true for every map from the projective limit to the spaces which form it. All the hypotheses required to prove the duality theorems for inductive and projective limits ([MO-SCH], [MOH]) are verified for $(3.20)'_n$ and $(3.20)_n$. We get a perfect duality between (3.15) (which is the projective limit of $(3.20)_n$) and (3.18) (which is the inductive limit of $(3.20)'_n$) i.e. each space in (3.15) is reflexive and has its strong dual in (3.18) (this last sequence is exact by the exactness of the direct limit).

Remark. For $(3.15)_i$ and $(3.18)_i$ the arguments are the same.
But the topologies we found are the induced from $\mathcal{A}$, and $\mathcal{O}^q_{\overset{\circ}{B}_{\epsilon^i}}$, and we don't know if they coincide with the canonical ones which can be defined in $H^\bullet_c(\overset{\circ}{B}_{\epsilon^i}, \mathcal{M})$.

β) We will prove that the maps ψ^t ,δ^t in (3.18) are homeomorphisms to the image using the canonical topologies: this will give the proof that the topology which gave the perfect duality in α) (i.e.induced from the canonical topologies in $H^n_c(\overset{\circ}{B}_{\epsilon^i}, \mathcal{A})$ and $H^n_c(\overset{\circ}{B}_{\epsilon^i}, \mathcal{O}^q_{\overset{\circ}{B}_{\epsilon^i}})$,by induction) coincides with the canonical topology on $H^n_c(\overset{\circ}{B}_{\epsilon^i}, \mathcal{M})$ and $H^{n-1}_c(\overset{\circ}{B}_{\epsilon^i}, \mathcal{M})$. For β) one can study the following exact sequence for every $\eta < \epsilon^i$

$$
\begin{aligned}
0 \to H^{n-2}(\overset{\circ}{B}_{\epsilon^i} \setminus B_\eta, \mathcal{M}) &\xrightarrow{\delta^t} H^{n-1}(\overset{\circ}{B}_{\epsilon^i} \setminus B_\eta, \mathcal{A}) \xrightarrow{\varphi^t} \\
\to H^{n-1}(\overset{\circ}{B}_{\epsilon^i} \setminus B_\eta, \mathcal{O}^q_{\overset{\circ}{B}_{\epsilon^i}}) &\xrightarrow{\psi^t} H^{n-1}(\overset{\circ}{B}_{\epsilon^i} \setminus B_\eta, \mathcal{M}) \to 0.
\end{aligned}
\tag{3.21}
$$

The limit of (3.21) for the various $\eta < \epsilon^i$ gives (3.18) and the canonical topologies on it. It is enough to show that the maps in (3.21) are homeomorphisms for the canonical topologies. Taking ,then, a countable affinoid covering $\mathcal{U}$ of $\overset{\circ}{B}_{\epsilon^i} \setminus B_\eta$ one can calculate the previous maps using the sequence of Fréchet spaces:

$$
0 \to C^\bullet(\mathcal{U}, \mathcal{A}) \longrightarrow C^\bullet(\mathcal{U}, \mathcal{O}^q_{\overset{\circ}{B}_{\epsilon^i}}) \longrightarrow C^\bullet(\mathcal{U}, \mathcal{M}) \to 0\,. \tag{3.22}
$$

Using the open mapping theorem [BGR], [BOU] and the facts that $H^{n-1}(\overset{\circ}{B}_{\epsilon^i} \setminus B_\eta, \mathcal{O}^q_{\overset{\circ}{B}_{\epsilon^i}})$ is a Hausdorff space and that all the maps are continuous we obtain the desired conclusion.

So, we know the theorem for

$$
H^\bullet_c(\overset{\circ}{B}_{\epsilon^i}, \mathcal{M}) \quad \text{and} \quad Ext^\bullet(\overset{\circ}{B}_{\epsilon^i}, \mathcal{M}, \Omega^n_{\mathbf{A}^n_K})
$$

where

$$
H^\bullet_c(\overset{\circ}{B}_{\epsilon^i}, \mathcal{M}) = \varinjlim_{\eta<\epsilon^i} H^\bullet_{B_\eta}(\overset{\circ}{B}_{\epsilon^i}, \mathcal{M})
$$

with the canonical topologies. One notes that:

$$
H^\bullet_c(\overset{\circ}{B}_{\epsilon^i}, \mathcal{M}) = \varinjlim_{\eta<\epsilon^i} H^\bullet_{B_\eta}(\mathbf{A}^n_K, \mathcal{M})
$$

and the isomorphism is, actually a homeomorphism (3.8). For (3.14) we now need to prove that

$$\begin{aligned}\varprojlim Ext^\bullet(\overset{\circ}{B}_{\epsilon^i},\mathcal{M},\Omega^n_{\mathbf{A}^n_K}) &= Ext^\bullet(\mathbf{A}^n_K,\mathcal{M},\Omega^n_{\mathbf{A}^n_K})\\ \varinjlim H^{n-\bullet}_c(\overset{\circ}{B}_{\epsilon^i},\mathcal{M}) &= \varinjlim(\varinjlim_{\eta<\epsilon^i} H^{n-\bullet}_{B_\eta}(\overset{\circ}{B}_{\epsilon^i},\mathcal{M})) = H^{n-\bullet}_c(\mathbf{A}^n_K,\mathcal{M}),\end{aligned} \tag{3.23}$$

endowed with the canonical topologies, are in perfect duality (each one the strong dual of the other) via the pairing (3.2) and that $Ext^\bullet(\mathbf{A}^n_K,\mathcal{M},\Omega^n_{\mathbf{A}^n_K})$ is reflexive. The canonical topologies coincide with the topology given by the two limits in (3.23), once we have imposed the canonical topologies for each "ϵ^i-step". We write

$$Ext^\bullet(\overset{\circ}{B}_{\epsilon^i},\mathcal{M},\Omega^n_{\mathbf{A}^n_K}) \longrightarrow Ext^\bullet(B_{\epsilon^{i-1}},\mathcal{M},\Omega^n_{\mathbf{A}^n_K}) \longrightarrow Ext^\bullet(\overset{\circ}{B}_{\epsilon^{i-1}},\mathcal{M},\Omega^n_{\mathbf{A}^n_K})$$

and the space in the middle, which we will denote by D_{i-1}, is a Banach space of countable type. All maps in the sequence $\{D_i\}$ are continuous and with dense image (as those in the inverse limit in (3.23). Use proof of (3.10)) : if the field K is spherically complete the maps are c-compact, if not each D_i is reflexive [SCH],[vRO].

Take the strong dual sequence

$$H^{n-\bullet}_c(\overset{\circ}{B}_{\epsilon^i},\mathcal{M}) \hookleftarrow D_{i-1}' \hookleftarrow H^{n-\bullet}_c(\overset{\circ}{B}_{\epsilon^{i-1}},\mathcal{M})$$

and get an injective limit (of injective maps) of Banach spaces.

For the two limits we have the hypotheses of the theorems of duality ([MO-SCH],[MOH]): it follows that the inverse limit of $\{D_i\}$, $Ext^\bullet(\mathbf{A}^n_K,\mathcal{M},\Omega^n_{\mathbf{A}^n_K})$, with inverse limit topology (which is the canonical) is reflexive and in perfect duality (via the pairing) with the direct limit of $\{D_i'\}$, $H^{n-\bullet}_c(\mathbf{A}^n_K,\mathcal{M})$, with the direct limit topology (which is the canonical topology) Q.E.D.

§4

We want to extend the Serre type duality to the case of a rigid Stein space. We recall the definition (note the difference from that for quasi-Stein [KI],[LIU]).

Definition 4.1.[LU] A rigid analytic variety X is a Stein space if it is separated and if there exists an admissible countable affinoid covering $\{U_i\}_{i\in\mathbf{N}}$, $U_i = SpmA_i$, such that

$$U_i \subset U_{i+1}$$

and

$$\mathcal{O}_X(U_{i+1}) \longrightarrow \mathcal{O}_X(U_i)$$

is dense, and such that the U_i are defined by

$$U_i = \{x \in U_{i+1} \quad | f^{i+1}_j(x) |\leq| a_{i+1} | \quad j = 1,\dots,n_{i+1}\}$$

with $a_{i+1}\in K^*$, $| a_{i+1} |< 1$ and (f^{i+1}_j) a topologically generating system of A_{i+1}.

As in the classical case it was shown by [LU] that there exists a closed immersion of X in an affine rigid space

$$i \quad : X \longrightarrow \mathbf{A}^n_K.$$

By means of this map we want to transfer the duality for X in $\mathbf{A}^n_K$. We first need some generalities about closed immersions and the relative adjoint functors.

Definition 4.2. A morphism

$$i \quad : X_1 \longrightarrow X_2$$

of analytic rigid varieties is a closed immersion if there exists an admissible affinoid covering $\{U_i\}_{i\in I}$ of X_2 such that the induced map, for each i,

$$i^{-1}(U_i) \xrightarrow{\bar{i}} U_i$$

is a morphism of affinoid spaces, $i^{-1}(U_i) = SpmB_i$, $U_i = SpmA_i$ and the inherent map

$$\bar{i}^* \quad : A_i \longrightarrow B_i$$

is surjective ([BGR]9.5.3).

Remark. In general it can be proved that in a closed immersion the inverse image of an affinoid is an affinoid.

We want to show how a generic closed immersion

$$i \quad : X_1 \longrightarrow X_2 \tag{4.3}$$

represents $Rig(X_1)$ as a closed subtopos of $Rig(X_2)$ [SGAIV]. By definition (4.2), $i(X_1)$ is a closed analytic subset of X_2 ([BGR],[FRE]). In the saturated topology of X_2, $U = X_2 \setminus i(X_1)$ is an open admissible:

$$j \quad : U = X_2 \setminus i(X_1) \longrightarrow X_2 \quad .$$

By the fact all the morphisms in the situs are injections, U will be an open object in $Rig(X_2)$ and will characterize an open subtopos. From i (4.3) we have

$$i_* \quad : Rig(X_1) \longrightarrow Rig(X_2)$$

Proposition 4.4. *The functor i_* is a fully faithfull functor. An object $\mathcal{F} \in obRig(X_2)$ is a sheaf of the form $i_*\mathcal{G}$, $\mathcal{G} \in obRig(X_1)$, if and only if $j^*\mathcal{F}$ (restriction) is the final object in $Rig(U)$.*

Proof. From i of (4.3) we get a morphism of topos which we write in the same way. By general theory i_* and i^* (inverse image as sheaves of sets) are adjoint. The functor i_* will be fully faithfull if ([SGAIV]VIII.6):

$$i^* i_* \mathcal{G} \longrightarrow \mathcal{G}$$

is an isomorphism for every $\mathcal{G} \in obRig(X_1)$. It is enough to prove this (for example) on each affinoid which defines the immersion (4.3): $(\{U_i\}_{i\in I})$, $i^{-1}(U_i)$ in X_1. On each such affinoid we show the isomorphism of sheaves for the weak rational topology [FRE] (The weak rational topology is a Grothendieck topology where the admissible opens are the rationals and admissible coverings are the finite rational ones. The relative strong topology is the usual one (§1. and [BGR]). The usual "unicity" theorems connecting sheaves for the weak rational topology and the relative extension to the strong still hold). In this case if we take $W \subseteq i^{-1}(U_i)$ rational, there exists $W' \subseteq U_i$ such that $W' \cap i^{-1}(U_i) = W$ ([FRE]pag.232). Then:

$$i^* i_* \mathcal{G}(W) = \mathcal{G}(W)$$

as presheaves. But, in particular, $\mathcal{G}$ is a sheaf for the weak rational topology, too: it follows that $i^*i_*\mathcal{G}$ and $\mathcal{G}$ are isomorphic as sheaves for the weak rational topology (see [BGR]). Since the extention to the strong topology is unique, they will be isomorphic for the strong topology, too.

For the second part of the proposition we will work as in ([SGAIV]VIII.6). It remains to show that if $\mathcal{F} \in obRig(X_2)$ has a restriction to U which is the final object, then there exists a $\mathcal{G} \in obRig(X_1)$ such that $i_*\mathcal{G} = \mathcal{F}$. This is equivalent to the fact that the canonical homomorphism

$$\mathcal{F} \longrightarrow i_*i^*\mathcal{F} \tag{4.5}$$

is an isomorphism on X_2. In order to prove this isomorphism we restrict , as before, to an affinoid U_i of a covering $\{U_i\}$ related to the closed immersion. On each such U_i it is enough to prove that there is an isomorphism in (4.5) at the level of presheaves for the weak topology ([BGR]. Here the weak topology on affinoid space is the topology in which the admissibles are the affinoids and the admissible coverings the finite affinoid ones). Take $W \subseteq U_i$ affinoid and consider at level of presheaves

$$i_*i^*\mathcal{F}(W) \longleftarrow \mathcal{F}(W).$$

What we need to prove is the following fact : take an admissible open T of U_i such that $T \cap i^{-1}(U_i) = W \cap i^{-1}(U_i)$ then, in the previous hypothesis on $\mathcal{F}$,

$$\mathcal{F}(T) = \mathcal{F}(W).$$

Of course one can suppose $T \subseteq W$ and $U_i = W$. The new situation is

$$i^{-1}(U_i) \subseteq T \subseteq W \qquad i^{-1}(U_i) \cap W \subseteq T.$$

For our purposes it is enough to show that there exists an admissible affinoid covering $\{X_\alpha\}$ of W such that

$$\mathcal{F}(X_\alpha) \longrightarrow \mathcal{F}(X_\alpha \cap T)$$

is an isomorphism for every α. But $W = SpmA$ and $i^{-1}(U_i) \cap W = V(\mathcal{A})$,$\mathcal{A} = (f_1, \ldots, f_k)$, $f_i \in A$ (A affinoid algebra is noetherian), by definition of closed immersion. Then there exists $\pi \in \mid K^* \mid$ such that $T \supseteq V_\pi(\mathcal{A})$ where

$$V_\pi(\mathcal{A}) = \{x \in SpmA \; ; \quad \mid f_i(x) \mid \leq \pi, \quad i \in [1,k]\}$$

Put $Z_i = \{x \in W = SpmA \quad \mid f_i(x) \mid \geq \pi\}$ $i = 1, \ldots, n$. Then

$$\widetilde{\mathcal{U}} = \{ V_\pi(\mathcal{A}), Z_1, \ldots, Z_n \}$$

is an admissible affinoid covering of W and $Z_i \cap i^{-1}(U_i) = \emptyset$ and $(Z_i \cap T) \cap i^{-1}(U_i) = \emptyset$. **Q.E.D.**

Remark. The idea of the covering is from Fresnel's book [FRE].

In particular (4.4) shows how to a closed immersion

$$i \quad : X \longrightarrow \mathbf{A}^n_K$$

there corresponds a closed subtopos of $Rig(\mathbf{A}^n_K)$ complementary of the open $Rig(U)$, $U = \mathbf{A}^n_K \setminus i(X)$. From the previous results , as in ([SGAIV]VIII) :

$$\begin{aligned} i_* \quad &: Rig(X)_{ab} \longrightarrow Rig(\mathbf{A}^n_K)_{ab} \\ i^* \quad &: Rig(\mathbf{A}^n_K)_{ab} \longrightarrow Rig(X)_{ab} \end{aligned} \tag{4.6}$$

induces an equivalence of the categories of the abelian sheaves on X and the abelian sheaves on $\mathbf{A}^n_K$ whose restriction to $U = \mathbf{A}^n_K \setminus i(X)$ is zero. In particular i_* is an exact functor [SGAIV].

Using the structural sheaf $\mathcal{O}_{\mathbf{A}^n_K}$ on $Rig(\mathbf{A}^n_K)$ and the induced one $i^{-1}\mathcal{O}_{\mathbf{A}^n_K}$ (this is the inverse image as set. For visual reasons we decided for this notation instead of $i^*\mathcal{O}_{\mathbf{A}^n_K}$: $i^{-1}\mathcal{O}_{\mathbf{A}^n_K}$ is the notation for topological spaces) on X ([SGAIV]IV.14), the functors of (4.6) specialize to

$$\begin{aligned} i_* &: Rig(X)_{i^{-1}\mathcal{O}_{\mathbf{A}^n_K}} \longrightarrow Rig(\mathbf{A}^n_K)_{\mathcal{O}_{\mathbf{A}^n_K}} \\ i^* &: Rig(\mathbf{A}^n_K)_{\mathcal{O}_{\mathbf{A}^n_K}} \longrightarrow Rig(X)_{i^{-1}\mathcal{O}_{\mathbf{A}^n_K}} \end{aligned} \tag{4.6}$$

which maintains the same properties as (4.6). In particular for i_* there exists a right adjoint ([SGAIV]IV.14):

$$i^! : Rig(\mathbf{A}^n_K)_{\mathcal{O}_{\mathbf{A}^n_K}} \longrightarrow Rig(X)_{i^{-1}\mathcal{O}_{\mathbf{A}^n_K}}$$

i.e. if $\mathcal{F} \in Rig(X)_{i^{-1}\mathcal{O}_{\mathbf{A}^n_K}}$ and $\mathcal{G} \in Rig(\mathbf{A}^n_K)_{\mathcal{O}_{\mathbf{A}^n_K}}$:

$$Hom_{\mathcal{O}_{\mathbf{A}^n_K}}(i_*\mathcal{F},\mathcal{G}) = Hom_{i^{-1}\mathcal{O}_{\mathbf{A}^n_K}}(\mathcal{F},i^!\mathcal{G}) \tag{4.7}$$

Using the definition of ([SGAIV]IV.10.5.1) one can show that (4.7) can be localized in

$$\mathcal{H}om_{\mathcal{O}_{\mathbf{A}^n_K}}(i_*\mathcal{F},\mathcal{G}) = i_*\mathcal{H}om_{i^{-1}\mathcal{O}_{\mathbf{A}^n_K}}(\mathcal{F},i^!\mathcal{G}) \tag{4.8}$$

($\mathcal{H}om(\mathcal{F},\mathcal{G})$ is the sheaf which represents $V \to Hom(\mathcal{F} \times V, \mathcal{G})$. The functors i_* and i^* are exact : $i_*(\mathcal{F} \times i^*V) = i_*\mathcal{F} \times i_*i^*V = i_*\mathcal{F} \times V$ because $i_*\mathcal{F}$ has restriction to the complement of $i(X)$ which is the final object ([SGAIV]IV.14.3)). In particular $i^!$ is left exact and it sends injective objects to injective ones. It should be noted that if we restrict i_* to

$$i_* : Rig(X)_{\mathcal{O}_X} \longrightarrow Rig(\mathbf{A}^n_K)_{\mathcal{O}_{\mathbf{A}^n_K}} \tag{4.9}$$

the adjoint $i^!$ can be defined

$$i^! : Rig(\mathbf{A}^n_K)_{\mathcal{O}_{\mathbf{A}^n_K}} \longrightarrow Rig(X)_{\mathcal{O}_X}$$

maintaining the same properties ([SGAIV]IV.14.5). In fact from the previous results (see [SGAIV]IV.12):

$$\begin{aligned} Hom_{\mathcal{O}_{\mathbf{A}^n_K}}(i_*\mathcal{F},\mathcal{G}) &= Hom_{i^{-1}\mathcal{O}_{\mathbf{A}^n_K}}(\mathcal{F},i^!\mathcal{G}) = Hom_{i^{-1}\mathcal{O}_{\mathbf{A}^n_K}}(\mathcal{F},\mathcal{H}om_{i^{-1}\mathcal{O}_{\mathbf{A}^n_K}}(\mathcal{O}_X,i^!\mathcal{G})) = \\ &= Hom_{i^{-1}\mathcal{O}_{\mathbf{A}^n_K}}(\mathcal{F},i^*\mathcal{H}om_{\mathcal{O}_{\mathbf{A}^n_K}}(i_*\mathcal{O}_X,\mathcal{G})) = Hom_{\mathcal{O}_X}(\mathcal{F},i^*\mathcal{H}om_{\mathcal{O}_{\mathbf{A}^n_K}}(i_*\mathcal{O}_X,\mathcal{G})). \end{aligned}$$

So, in the case (4.9),

$$i^!\mathcal{G} = i^*\mathcal{H}om_{\mathcal{O}_{\mathbf{A}^n_K}}(i_*\mathcal{O}_X,\mathcal{G}). \tag{4.10}$$

(i^*, image inverse as sheaves of sets : it is exact) This agrees with the notion of $i^\flat$ in ([HARD]II.6).

Remark. These arguments are valid for every closed immersion.

We want to specialize the closed immersion (4.3) to the case in which X is a Stein space, irreducible and regular of dimension q. In this case there exists a closed immersion

$$i : X \longrightarrow \mathbf{A}^n_K \tag{4.11}$$

and all the properties we found still hold. We study $i^!$ and its derived functors in this particular case.

If I is the coherent sheaf of ideals of $\mathbf{A}^n_K$ defining the closed immersion (4.11) we can write

$$\mathcal{I}/\mathcal{I}^2 \longrightarrow \Omega^1_{\mathbf{A}^n_K} \otimes \mathcal{O}_X \longrightarrow \Omega^1_{X/K} \longrightarrow 0 \tag{4.12}$$

which is an exact sequence of coherent sheaves ($\Omega^1_{\mathbf{A}^n_K}$ and $\Omega^1_{X/K}$ are locally free by hypothesis of regularity [F-vdP], [BKKN]). One can show exactly as in the algebraic case ([HAA]II.8), using some results obtained by Fresnel ([FRE]pag.222) on the dimension of irreducible rigid varieties, that (4.12) is an exact sequence of locally free sheaves on the left ,too :

$$0 \longrightarrow \mathcal{I}/\mathcal{I}^2 \longrightarrow \Omega^1_{\mathbf{A}^n_K} \otimes \mathcal{O}_X \longrightarrow \Omega^1_{X/K} \longrightarrow 0 \tag{4.13}$$

and $\mathcal{I}$ is locally generated by $n-q$ elements (locally in the rigid sense ...), if $q = dimX$. From (4.13) we can write [HARD]:

$$\wedge^n \Omega^1_{\mathbf{A}^n_K} \otimes \mathcal{O}_X \xrightarrow{\sim} \wedge^{n-q} \mathcal{I}/\mathcal{I}^2 \otimes \wedge^q \Omega^1_{X/K} \quad .$$

Putting

$$\wedge^n \Omega^1_{\mathbf{A}^n_K} = \Omega^n_{\mathbf{A}^n_K}$$

and

$$\wedge^q \Omega^1_{X/K} = \Omega^q_{X/K}$$

(invertible sheaves), it is possible to rewrite the previous isomorphism:

$$i^*\Omega^n_{\mathbf{A}^n_K} \otimes \omega_{\mathbf{A}^n_K/X} \simeq \Omega^q_{X/K}$$

where $\omega_{\mathbf{A}^n_K/X} \simeq \left(\wedge^{n-q}\mathcal{I}/\mathcal{I}^2\right)'$ and i^*, this time, indicates the inverse image as sheaves of modules. We will now study $i^!\Omega^n_{\mathbf{A}^n_K}$ and its derived functors. To this end we consider (4.10):

$$\mathcal{E}xt^{\bullet}_{\mathcal{O}_{\mathbf{A}^n_K}}(\mathcal{O}_X, \Omega^n_{\mathbf{A}^n_K}).$$

(i^* the inverse image as sheaves of sets is exact, $\mathcal{O}_X$ is understood in $\mathbf{A}^n_K$ i.e. $I_*\mathcal{O}_X$). By previous considerations (3.3) , these sheaves are coherent (i_* of a coherent sheaf is coherent [BGR]) : we can study them locally (using global sections on affinoid domains). By choosing an admissible affinoid covering $\{U_i\}$ of $\mathbf{A}^n_K$ such that on each $U_i = SpmA_i$

$$\mathcal{O}_{X_{|U_i}} = Spm \frac{A_i}{(f_1^{(i)}, \ldots, ,f_{n-q}^{(i)})}$$

with $\{f_k^{(i)}\}$ an A-sequence , $\mathcal{I}_{|U_i} \simeq (f_k^{(i)})$: one can work exactly as in [HARD] by mean of the Koszul resolution associated to $\{f_k^{(i)}\}$.

We conclude

$$\mathcal{E}xt^i_{\mathcal{O}_{X_{|U_i}}}(\mathcal{O}_{X_{|U_i}}, \Omega^n_{\mathbf{A}^n_K\,|U_i}) = 0 \qquad \text{if} \quad i \neq n-q,$$

and then by local arguments

$$\mathcal{E}xt^i_{\mathcal{O}_X}(\mathcal{O}_X, \Omega^n_{\mathbf{A}^n_K}) = 0 \qquad i \neq n-q.$$

For the case $i = n - q$, always from the complex, on each U_i there is an isomorphism, which will depend on the choice of $\{f_k^{(i)}\}$:

$$\varphi_i \quad : \mathcal{E}xt^{n-q}_{\mathcal{O}_{X|U_i}}(\mathcal{O}_{X|U_i}, \Omega^n_{\mathbf{A}^n_K|U_i}) \xrightarrow{\sim} \frac{\Omega^n_{\mathbf{A}^n_K|U_i}}{\mathcal{I}_{|U_i}\Omega^n_{\mathbf{A}^n_K|U_i}}$$

All these local (on U_i) isomorphisms can be glued using the morphism

$$\frac{\Omega^n_{\mathbf{A}^n_K|U_i}}{\mathcal{I}_{|U_i}\Omega^n_{\mathbf{A}^n_K|U_i}}{}_{|U_i\cap U_j} \xrightarrow{det(c_{ij})} \frac{\Omega^n_{\mathbf{A}^n_K|U_j}}{\mathcal{I}_{|U_j}\Omega^n_{\mathbf{A}^n_K|U_j}}{}_{|U_i\cap U_j}$$

in $U_i \cap U_j$, where $(c_{ij}) \in \mathcal{O}_{\mathbf{A}^n_K}(U_i \cap U_j)$ is the matrix of the map between: (f^i_k) and (f^j_k). We can then define a global map which is an isomorphism

$$\varphi \quad : \mathcal{E}xt^{n-q}_{\mathcal{O}_X}(\mathcal{O}_X, \Omega^n_{\mathbf{A}^n_K}) \quad \xrightarrow{\sim} \quad \mathcal{H}om_{\mathcal{O}_X}(\wedge^{n-q}\frac{\mathcal{I}}{\mathcal{I}^2}, \frac{\Omega^n_{\mathbf{A}^n_K}}{\mathcal{I}\Omega^n_{\mathbf{A}^n_K}})$$

(on each U_i, φ composed with the evaluation on $f_1^{(i)} \wedge \ldots\ldots \wedge f_{n-q}^{(i)}$ is φ_i). We conclude with the following isomorphism

$$\varphi \quad : \mathcal{E}xt^{n-q}_{\mathcal{O}_X}(\mathcal{O}_X, \Omega^n_{\mathbf{A}^n_K}) \quad \xrightarrow{\sim} \quad \Omega^n_{\mathbf{A}^n_K} \otimes \mathcal{O}_X \otimes (\wedge^{n-q}\frac{\mathcal{I}}{\mathcal{I}^2})' \tag{4.14}$$

Applying the inverse image (as sheaves of sets) which is an exact functor:

$$i^*\mathcal{E}xt^{n-q}_{\mathcal{O}_X}(\mathcal{O}_X, \Omega^n_{\mathbf{A}^n_K}) \simeq i^*\Omega^n_{\mathbf{A}^n_K} \otimes \omega_{\mathbf{A}^n_K/X} \simeq \Omega^q_{X/K}$$

and

$$i^*\mathcal{E}xt^i_{\mathcal{O}_X}(\mathcal{O}_X, \Omega^n_{\mathbf{A}^n_K}) = 0$$

if $i \neq n - q$. In the language of the derived categories we state

$$\mathbf{R}i^!(\Omega^n_{\mathbf{A}^n_K}) \simeq \Omega^q_{X/K}[q-n]. \tag{4.15}$$

⋆ ⋆

We recall our situation : X is Stein , irreducible, regular of $dimX = q$ and we have [LU] a closed immersion

$$i \quad : X \quad \longrightarrow \mathbf{A}^n_K \tag{4.11}$$

Proposition 4.16. *Under the previous hypothesis on X, if $\mathcal{M}$ is a coherent $\mathcal{O}_X$-module, there exist the isomorphisms:*

$$Ext^i_{\mathcal{O}_{\mathbf{A}^n_K}}(i_*\mathcal{M}, \Omega^n_{\mathbf{A}^n_K}) = Ext^{i-(n-q)}_{\mathcal{O}_X}(\mathcal{M}, \Omega^q_X)$$

$$\mathcal{E}xt^i_{\mathcal{O}_{\mathbf{A}^n_K}}(i_*\mathcal{M}, \Omega^n_{\mathbf{A}^n_K}) = i_*\mathcal{E}xt^{i-(n-q)}_{\mathcal{O}_X}(\mathcal{M}, \Omega^q_X)$$

Proof. The closed immersion i_* is, in particular, proper : $i_*\mathcal{M}$ is coherent [KIE]. To prove the proposition it is enough to take an injective resolution:

$$0 \longrightarrow \Omega^n_{\mathbf{A}^n_K} \longrightarrow \mathcal{I}^\bullet$$

and apply

$$Hom(i_*\mathcal{M}, \mathcal{I}^\bullet) = Hom(\mathcal{M}, i^!\mathcal{I}^\bullet)$$

using the fact that $i^!$ sends injective objects to injectives [SGAIV] (the same argument for $\mathcal{H}om$ being i_* exact). **Q.E.D.**

Proposition 4.17. *For every sheaf $\mathcal{M}$ of $\mathcal{O}_X$-modules:*

$$H_c^\bullet(\mathbf{A}_K^n, i_*\mathcal{M}) \simeq H_c^\bullet(X, \mathcal{M}).$$

Proof. We need to show that each affinoid $V \subseteq X$ is the intersection of a finite union of affinoids of $\mathbf{A}_K^n$ and X i.e. there exists $I_1, \ldots, \ldots I_m$ admissible affinoids in $\mathbf{A}_K^n$ such that

$$\cup_i I_i \cap X = V.$$

Take $\{U_i\}_{i \in \mathbf{N}}$ an affinoid countable covering of $\mathbf{A}_K^n$ and study $\{U_i \cap X\}_{i \in \mathbf{N}}$ which is an admissible affinoid covering of X and $\{U_i \cap V\}_{i \in \mathbf{N}}$ which is an affinoid covering of V (X is separated). This implies that there will be a finite subcover :

$$\bigcup_{j=1}^{k} (U_i \cap V) = V.$$

For each $j = 1, \ldots, \ldots, k$

$$U_j \cap V \subseteq U_j \cap X \subseteq U_j$$

where $U_J \cap V$ is a closed affinoid space of U_j . As in ([FRE],[BGR]) one can prove that there exists a finite number of rationals $\{I_{j,l}\}_{l=1}^{n_l}$ of U_j (and so admissible affinoids of $\mathbf{A}_K^n$) such that

$$(\bigcup_l I_{j,l}) \cap (U_j \cap X) = U_j \cap V.$$

By the fact that i_* (exact functor) sends injective objects into flasques which are acyclic for the functor "sections with support" ([SGAIV]V), we conclude. **Q.E.D.**

Remark 4.18. If $\mathcal{F}$ is an $\mathcal{O}_X$-module then

$$\mathcal{H}om_{\mathcal{O}_X}(\mathcal{F}, -)$$

of an injective object is a flasque sheaf ([SGAIV]V.4.10). In particular it is $\Gamma_c(X, -)$ acyclic. From (4.16) and (4.17) it follows that:

$$Ext^i_{\mathcal{O}_{\mathbf{A}_K^n},c}(\mathbf{A}_K^n, i_*\mathcal{M}, \Omega^n_{\mathbf{A}_K^n}) = Ext^{i-(n-q)}_{\mathcal{O}_X,c}(\mathcal{M}, \Omega^q_X).$$

Remark 4.19. Trivially one has

$$H^\bullet(\mathbf{A}_K^n, i_*\mathcal{M}) = H^\bullet(X, \mathcal{M})$$

(i_* is exact, and sends injective in flasque ([SGAIV]V.4)). It follows that the only group different from zero is H^0 . On the other hand the two canonical topologies coincide (§3. cf. remark (4.20)).

Remark 4.20. In X, for a coherent sheaf $\mathcal{M}$, it is possible to define a canonical topology on $H^0(X, \mathcal{M})$. By (4.1) X admits an admissible countable affinoid covering and the sections of a coherent sheaf on an affinoid form a Banach space (2.0). The same argument can be repeated for $H_c^\bullet(X, \mathcal{M})$ since it can be written as the direct limit of spaces of the type $H^{n-1}(X \setminus U, \mathcal{M})$ where U is an affinoid. For the relationship between the canonical topologies $H^0(\mathbf{A}_K^n, i_*\mathcal{M})$ and $H^0(X, \mathcal{M})$ it is enough to notice that $i^{-1}(U)$ of an affinoid $U \subseteq \mathbf{A}_K^n$ is an affinoid : the two topologies are the same.

Finally we can state

Theorem 4.21. *If X is a regular, irreducible Stein space (4.1) of dimension q, the spaces $Ext^{q-i}_{\mathcal{O}_X}(\mathcal{M},\Omega^q_X)$ and $H^i_c(X,\mathcal{M})$ endowed with the canonical topologies (4.20) are, respectively, a reflexive Fréchet space of countable type, and a complete, reflexive Hausdorff space*
There exists a pairing

$$H^i_c(X,\mathcal{M}) \times Ext^{q-i}_{\mathcal{O}_X}(\mathcal{M},\Omega^q_X) \longrightarrow K \tag{4.22}$$

which is perfect (making one the strong dual of the other). Again for the canonical topology, $H^0(X,\mathcal{M})$ is a reflexive, Fréchet space of countable type, and there exists a pairing

$$H^0(X,\mathcal{M}) \times Ext^q_{\mathcal{O}_X,c}(\mathcal{M},\Omega^q_X) \longrightarrow K \tag{4.23}$$

making $Ext^q_{\mathcal{O}_X,c}(\mathcal{M},\Omega^q_X)$ its dual.

Proof. By [LU] there exists a closed immersion

$$i \quad : X \longrightarrow \mathbf{A}^n_K. \tag{4.11}$$

Then by (4.17) we have

$$H^\bullet_c(\mathbf{A}^n_K, i_*\mathcal{M}) \simeq H^\bullet_c(X,\mathcal{M})$$

and this is a homeomorphism for the canonical topologies (4.20). On the other hand by (4.16)

$$\mathcal{E}xt^{n-i}_{\mathcal{O}_{\mathbf{A}^n_K}}(i_*\mathcal{M},\Omega^n_{\mathbf{A}^n_K}) = i_*\mathcal{E}xt^{q-i}_{\mathcal{O}_X}(\mathcal{M},\Omega^q_X)$$

and the isomorphism is a homeomorphism for the canonical topologies. In fact from (4.16) they are the global sections of two isomorphic coherent sheaves on $\mathbf{A}^n_K$ (4.20) (and a A-linear map between two finite A-modules is automatically continuous if A is an affinoid algebra ([BGR]3.7.3)). We use the duality in $\mathbf{A}^n_K$ to conclude. Same argument for (4.23).

By the way, one can note that, putting the strong dual topology, arising from the pairing, on $Ext^q_{\mathcal{O}_X,c}(\mathcal{M},\Omega^q_X)$ the pairing (4.23) becomes a perfect duality (one space is the strong dual of the other) **Q.E.D.**

§5

Implicitly we have found a notion of residue for a regular, irreducible Stein space. One fixes a closed immersion $i : X \longrightarrow \mathbf{A}^n_K$. It is, in fact, possible to define a map

$$\tilde{i} \quad : H^q_c(X,\Omega^q_{X/K}) \longrightarrow H^n_c(\mathbf{A}^n_K,\Omega^n_{\mathbf{A}^n_K})$$

in the following way:

$$\begin{aligned} H^q_c(X,\Omega^q_{X/K}) &\simeq H^q_c(\mathbf{A}^n_K, i_*\Omega^q_{X/K}) \simeq H^q_c(\mathbf{A}^n_K, i_*\mathbf{R}i^!\Omega^n_{\mathbf{A}^n_K}[n-q]) \simeq \\ \simeq H^q_c(\mathbf{A}^n_K, \mathcal{E}xt^{n-q}(i_*\mathcal{O}_X,\Omega^n_{\mathbf{A}^n_K})[n-q]) &\simeq Ext^n_{\mathcal{O}_{\mathbf{A}^n_K}}(\mathbf{A}^n_K, i_*\mathcal{O}_X,\Omega^n_{\mathbf{A}^n_K}) \longrightarrow H^n_c(\mathbf{A}^n_K,\Omega^n_{\mathbf{A}^n_K}) \end{aligned} \tag{5.1}$$

($\mathcal{E}xt^{n-q}(i_*\mathcal{O}_X,\Omega^n_{\mathbf{A}^n_K})$ is the only cohomological sheaf different than zero). And finally

$$H^q_c(X,\Omega^q_{X/K}) \xrightarrow{\tilde{i}} H^n_c(\mathbf{A}^n_K,\Omega^n_{\mathbf{A}^n_K}) \xrightarrow{res_n} K\,.$$

One can ask if this notion is independent of the closed immersion we are using. We will answer this question in the affirmative (up to a multiplication by a constant).

Consider another closed immersion for X

$$j \quad : X \longrightarrow \mathbf{A}^n_K$$

we obtain the commutative diagram

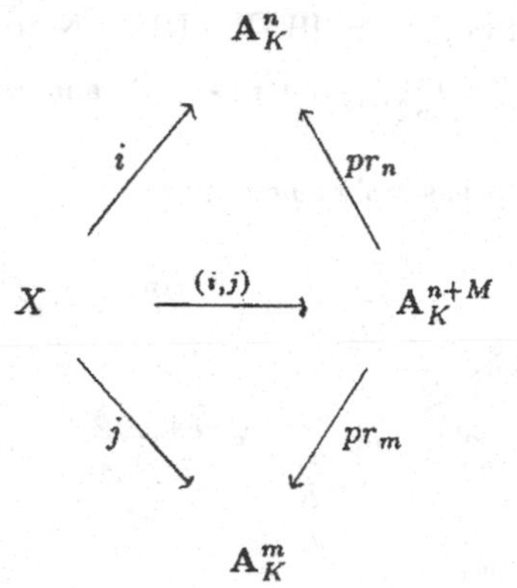

(5.2)

where (i,j) is a closed immersion. It is possible to construct k_n, k_m closed immersions such that the diagram

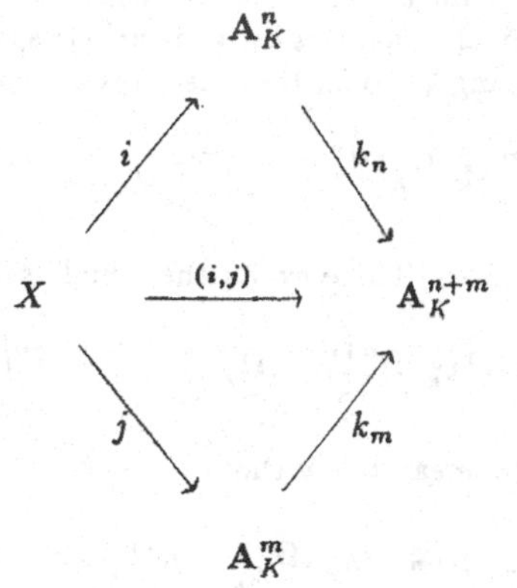

(5.3)

is commutative. In order to see the existence of k_n (the method for k_m is similar) consider the sheaf of ideals $\mathcal{I}$ defining the closed immersion i in $\mathbf{A}^n_K$. The facts that $\mathbf{A}^n_K$ is a Stein space and $\mathcal{I}$ is a coherent sheaf (the affinoid algebras are noetherian) imply that the sequence

$$\mathcal{O}_{\mathbf{A}^n_K}(\mathbf{A}^n_K) \longrightarrow \frac{\mathcal{O}_{\mathbf{A}^n_K}}{\mathcal{I}}(\mathbf{A}^n_K) \longrightarrow 0$$

is exact. If we take $\overline{j} \; : \; i(X) \rightarrow \mathbf{A}^m_K$ and compose with the various projections of $\mathbf{A}^m_K$ we get $\overline{j}_i \in \frac{\mathcal{O}_{\mathbf{A}^n_K}}{\mathcal{I}}(i(X)) = \frac{\mathcal{O}_{\mathbf{A}^n_K}}{\mathcal{I}}(\mathbf{A}^n_K)$. By the previous observation each $\overline{j}_i$ can be extented to a global section J_i. Finally we can define $k_n(x) = (x, J_1, \dots, J_m)$.

From the upper part of (5.3) we obtain the commutative diagram:

$$\begin{array}{ccc} & H_c^n(\mathbf{A}_K^n, \Omega^n_{\mathbf{A}_K^n}) & \\ \nearrow \tilde{i} & & \tilde{k}_n \searrow \\ H_c^q(X, \Omega^q_{X/K}) & \xrightarrow{\widetilde{(i,j)}} & H_c^{n+m}(\mathbf{A}_K^{n+m}, \Omega^{n+m}_{\mathbf{A}_K^{n+m}}) \end{array} \tag{5.4}$$

(For showing the commutativity one can use the fact that $i^!$, $(i,j)^!$, $k_n^!$ transform injective objects into injective objects §4, and work as in ([HARD]III.7). Note that $\mathbf{R}i^!(\mathbf{R}j^!(\Omega^{n+m}_{\mathbf{A}_K^{n+m}})[(n+m)-n])[n-p] \simeq \Omega^p_{X/K} \simeq \mathbf{R}(i,j)^!(\Omega^{n+m}_{\mathbf{A}_K^{n+m}})[n+m-p]$) (analogous situation for the lower part of (5.3))

Finally the problem of unicity is simply to compare

$$\begin{array}{ccc} H_c^n(\mathbf{A}_K^n, \Omega^n_{\mathbf{A}_K^n}) & \xrightarrow{\tilde{k}_n} & H_c^{n+m}(\mathbf{A}_K^{n+m}, \Omega^{n+m}_{\mathbf{A}_K^{n+m}}) \\ res_n \searrow & & \swarrow res_{n+m} \\ & K & \end{array} \tag{5.5}$$

and we expect that the two residues

$$res_n \quad ; \quad res_{n+m} \circ \tilde{k}_n$$

are equal up to a constant. The point is to show that $res_{n+m} \circ \tilde{k}_n$ satisfies a duality theory as res_n did with the same canonical topologies on the various spaces : it will then sufficient to apply proposition (2.8). Applying §4 to k_n there is an isomorphism (5.1):

$$H_c^n(\mathbf{A}_K^{n+m}, \mathcal{E}xt^m_{\mathcal{O}_{\mathbf{A}_K^{n+m}}}(k_{n*}\mathcal{O}_{\mathbf{A}_K^n}, \Omega^{n+m}_{\mathbf{A}_K^{n+m}})) \longrightarrow Ext^{n+m}_{\mathcal{O}_{\mathbf{A}_K^{n+m}},c}(k_{n*}\mathcal{O}_{\mathbf{A}_K^n}, \Omega^{n+m}_{\mathbf{A}_K^{n+m}}) \tag{5.6}$$

(degeneracy of the spectral sequence). There exists the "dual" sequence:

$$Ext^m_{\mathcal{O}_{\mathbf{A}_K^n}}(\mathcal{E}xt^m_{\mathcal{O}_{\mathbf{A}_K^{n+m}}}(k_{n*}\mathcal{O}_{\mathbf{A}_K^n}, \Omega^{n+m}_{\mathbf{A}_K^{n+m}}), \Omega^{n+m}_{\mathbf{A}_K^{n+m}}) \longleftarrow H^0(\mathbf{A}_K^{n+m}, k_{n*}\mathcal{O}_{\mathbf{A}_K^n}) \tag{5.7}$$

By mean of the Koszul complex one can prove that

$$\mathcal{E}xt^m_{\mathcal{O}_{\mathbf{A}_K^n}}(\mathcal{E}xt^m_{\mathcal{O}_{\mathbf{A}_K^{n+m}}}(k_{n*}\mathcal{O}_{\mathbf{A}_K^n}, \Omega^{n+m}_{\mathbf{A}_K^{n+m}}), \Omega^{n+m}_{\mathbf{A}_K^{n+m}}) \simeq i_*\mathcal{O}_{\mathbf{A}_K^n}$$

and so (5.7) is a homeomorphism for the canonical topologies, too. But (5.7) is in fact the transposed situation of (5.6) using the duality theory in $\Omega^{n+m}_{\mathbf{A}_K^{n+m}}$. From the reflexivity of the spaces which appear one concludes that (5.6) is a homeomorphism for the strong dual topology via the pairing in $\Omega^{n+m}_{\mathbf{A}_K^{n+m}}$. Note that the strong dual topology for

$$H_c^n(\mathbf{A}_K^{n+m}, \mathcal{E}xt^m_{\mathcal{O}_{\mathbf{A}_K^{n+m}}}(k_{n*}\mathcal{O}_{\mathbf{A}_K^n}, \Omega^{n+m}_{\mathbf{A}_K^{n+m}}))$$

is the canonical one §3. On the other hand

$$H_c^n(\mathbf{A}_K^n, \Omega^n_{\mathbf{A}_K^n}) \simeq H_c^n(\mathbf{A}_K^{n+m}, k_{n*}\Omega^n_{\mathbf{A}_K^n}) \simeq H_c^n(\mathbf{A}_K^{n+m}, \mathcal{E}xt^m_{\mathcal{O}_{\mathbf{A}_K^{n+m}}}(k_{n*}\mathcal{O}_{\mathbf{A}_K^n}, \Omega^{n+m}_{\mathbf{A}_K^{n+m}}))$$

are all homeomorphisms ((4.20) §3). it follows that

$$res_{n+m} \circ \widetilde{k}_n \quad : H^n_c(\mathbf{A}^n_K, \Omega^n_{\mathbf{A}^n_K}) \longrightarrow Ext^{n+m}_{\mathcal{O}_{\mathbf{A}^{n+m}_K},c}(k_{n*}\mathcal{O}_{\mathbf{A}^n_K}, \Omega^{n+m}_{\mathbf{A}^{n+m}_K}) \longrightarrow$$
$$\longrightarrow \quad H^{n+m}_c(\mathbf{A}^{n+m}_K, \Omega^{n+m}_{\mathbf{A}^{n+m}_K}) \stackrel{res_{n+m}}{\longrightarrow} K$$

is continuous for the canonical topology and it satisfies the duality for this topology. We can apply (2.8).

Note 5.8. Professor Berthelot pointed out to us the possibility of giving a "unicity statement" for our residue. Consider X a Stein space and fix (possibly after an extension of the ground field K) a K-rational point. Consider local coordinates and impose that in those coordinates the residue is defined as in §2. This notion is unique (note 2.12), at least, for projective automorphisms of the coordinates.

Bibliography

[AM] Amice Y. : "Les nombres p-adiques". Presses Universitaires de France, Paris (1975).

[AK] Altman A.; Kleiman S. : "Introduction to Grothendieck duality theory" . Lecture Notes in Mathematics 146, Springer–Verlag, Berlin (1970).

[AR] Artin M. : "Grothendieck topologies". Notes on a seminar by M. Artin, Harvard University (1962).

[BER] Berthelot P. : "Cohomologie rigide et cohomologie rigide à support propre". A paraitre dans Astérisque.

[BGR] Bosch S.; Güntzer U.; Remmert R. : "Non archimedean Analysis". Grundlehren der Math. Wissenschaften 261, Springer–Verlag, Berlin (1984).

[BKKN] Berger R.; Kiehl R.; Kunz E.; Nastold N.J. : "Differentialrechnung in der analytischen Geometrie". Lecture Notes in Mathematics 38, Springer–Verlag, Berlin (1967).

[BOU] Bourbaki N. : "Espaces vectoriels topologiques". Hermann, Paris (1958).

[CA] Cartier P. : "Les groupes Ext^s(A,B)". Séminaire Grothendieck t.1, 1957 n.3 Secrétariat Math., Paris (1958).

[FRE] Fresnel J. : "Cours de géométrie analytique rigide". Université de Bordeaux I (1984).

[F-vdP] Fresnel J.; Van der Put M. : "Géométrie analytique rigide et applications". Progress in Math. 18, Birkhauser (1981).

[GROT] Grothendieck A. : "Sur quelques points d'algèbre homologique". Tohoku Math. J. 9, 119–221 (1957).

[GRU] Gruson L. : "Fibrés vectoriels sur un polydisque ultramétrique". Ann. Scient. Ec. Norm. Sup. 4e série, 1, 45–89 (1968).

[HARA] Harthshorne R. : "Algebraic Geometry". Springer–Verlag , Berlin (1977).

[HARD] Harthshorne R. : "Residues and Duality". Lecture Notes in Mathematics 20, Springer–Verlag, Berlin (1966).

[IN] Ingleton A.W. : "The Hahn-Banach theorem for non-Archimedean valued fields". Proc. Cambridge Phil. Soc. 48, 41–45 (1952).

[KIE] Kiehl R. : "Der Endlichkeitssatz fur eigentliche Abbildungen in der nichtarchimedischen Funktionentheorie". Inventiones Math. 2, 191–214 (1967).

[KI] Kiehl R. : "Theorem A und Theorem B in der nichtarchimedischen Funktionentheorie". Inventiones Math. 2, 256–273 (1967).

[KO] Kopf U. : "Uber eigentliche Familien algebraischer Varietaten uber affinoiden Raumen". Schriftenreihe Math. Inst. Univ. Münster 2 Serie, Heft 7 (1974).

[LIU] Liu Q. : "Un contre example au critère cohomologique d'affinoidité". C.R. Acad. Sci. Paris t.307 Série 1, 83–86 (1988).

[LU] Lütkebohmert W. : "Steinsche Räume in der nichtarchimedischen Funktionentheorie". Schriftenreihe Math. Inst. Univ. Münster 2 Serie, Heft 6 (1973).

[LUT] Lütkebohmert W. : " Formal-algebraic and rigid-analytic geometry". Math. Ann. 286, 341-371 (1990).

[MOH] Morita Y. : "A p-adic theory of hyperfunctions I". Publ. Res. Inst. Math. Sci. 17, 1–24 (1981).

[MOS] Morita Y. : "A p-adic theory of hyperfunctions II". in "Algebraic Analysis", vol.I, 457–472, Academic Press (1988).

[MO-SCH] Morita Y.; Schikhof W.H. : "Duality of proiective limit spaces and inductive limit spaces over a non-spherically complete non archimedean field". Tohoku Math. J. 38, 387–397 (1986).

[R.R.] Ramis J.P.; Ruget G. : "Complexe dualisant et théorèmes de dualité en géométrie analytique complexe". Publ. Math. I.H. ES 38, 77–91 (1970)

[vRO] van ROOIJ A.C.M. : "Nonarchimedean functional analysis". Monographs and Textbooks in Pure and Applied Math. 51, Marcel Dekker, Inc. Basel (1978).

[SCH] Schikhof W.H. : "Locally convex spaces over non spherically complete valued field I, II". Tijdschrift van het Belgisch Wiskundig Genootschap Serie B, fasc. I, 28 187–224 (1986).

[SED] Serre J.P. : "Un théorème de dualité". Comm. Math. Helv. 29, 9–26 (1955).

[SGAIV] "Séminaire de Géométrie Algébrique. Théorie des Topos et Cohomologie étale des Schémas". M. Artin ; A. Grothendieck L.N. in Math., 269–270 (1972).

[SUO] Suominen K. : "Duality for coherent sheaves on analytic manifolds". Annales Ac. Scient. Fennicae, 424 (1968).

On the Frobenius Matrices of Fermat Curves

Robert F. Coleman
Department of Mathematics
University of California
Berkeley, CA 94720

Let p be a rational prime. If A is an abelian variety with good reduction over an unramified extension K of $\mathbf{Q}_p$, then there is a natural Frobenius-linear endomorphism of $H^1_{DR}(A,K)$. The Jacobian of a Fermat curve $F_m: x^m+y^m=1$ has good reduction over $\mathbf{Q}_p$, so long as $(m,p)=1$, and in [C-GK] using ideas of Katz, Gross, Koblitz and Dwork we computed the matrix of the corresponding endomorphism with respect to the the basis represented by the differentials of the second kind on F_m

$$\omega_{m,i,j} = x^i y^j (y/x) d(x/y)$$

where $\{0<i<m, 0<j<m, i+j \neq m\}$ in terms of special values of the p-adic gamma function at rational arguments. Here, as we shall do henceforth, we identify the first de Rham cohomology group of F_m with that of its jacobian.

In general, the jacobian of F_m has good reduction over an algebraic closure $\overline{\mathbf{Q}}_p$ of $\mathbf{Q}_p$ but not over an unramified extension of $\mathbf{Q}_p$ (see [CM]). As a result we no longer have a canonical Frobenius-linear endomorphism. Instead we have a canonical action of the crystalline Weil Group (see [BO] or [CM §2]) on $H^1_{DR}(F_m,K)$. We call the corresponding matrices Frobenius matrices.

In this article, we use the results of [CM] on the stable model of F_m as well as those of Ogus [O] on CM motives, in this volume, to compute these matrices when $p>3$. We were not able to complete the computation using only what we knew about the stable models. Gratifyingly, we were able to obtain just enough information about these matrices to be able to plug into Ogus' theory and finish the job. These matrices involve special values of an extension of the Morita gamma function to all of $\mathbf{Q}_p$ and bear a striking resemblance to the formula for the complex periods in terms of the classical gamma function. This is explained, in a large part, by an observation of Anderson [A] elucidated in this context by Ogus [O], which asserts roughly that the "Fermat motive" factors naturally into the product of three versions of the same "ulterior motive."

In Section I, the problem is precisely stated. We define the extended gamma function in Section II and describe our main results in Section III. In Section IV, we extend the methods of Katz [K] for computing Frobenius matrices to the case of bad reduction. The relevant results of [CM] are recalled in Section V and we complete the computation of the matrices by Sections VI and VII.

Finally we note that from the results proven in this paper, one may deduce the formula for the local components of Jacobi sum Hecke characters proven in [CM]. Thus this formula now has this Crystalline proof, an ℓ-adic étale proof given in [CM] and a p-adic étale proof sketched in [C-AI].

We are indebted to Ogus for suggesting the problem solved in this paper as well as for explaining to us how to apply his recent results [O].

I. The General Set Up.

Fix an algebraic closure $\overline{\mathbf{Q}}_p$ of $\mathbf{Q}_p$, a non-trivial non-archimedean absolute value $|\ |$ on $\overline{\mathbf{Q}}_p$ and let v be the valuation on $\overline{\mathbf{Q}}_p$ such that $v(p)=1$. Let $I=\{(r,s) \in \mathbf{Q}/\mathbf{Z}\times\mathbf{Q}/\mathbf{Z}, r\neq 0, s\neq 0, r+s\neq 0\}$. To each element (r,s) in I and each element σ in the Weil group of $\mathbf{Q}_p$ (see below) we shall associate an element $\beta_\sigma(r,s)$ in $\overline{\mathbf{Q}}_p$.

For $x \in \mathbf{Q}/\mathbf{Z}$ or $\mathbf{Q}$, we let $\langle x\rangle$ denote its unique representative mod $\mathbf{Z}$ in $(0,1]\cap\mathbf{Q}$ ($\langle x\rangle = x-[x]$, if $x \in \mathbf{Q}-\mathbf{Z}$

and 1 otherwise). For g, a function defined on $\mathbf{Q}^*$, and $r \in \mathbf{Q}/\mathbf{Z}$, we will let $g\langle r\rangle$ denote $g(\langle r\rangle)$. Let $(r,s) \in I$. Set $\varepsilon(r,s) = \langle r\rangle+\langle s\rangle-\langle r+s\rangle$ and $L(r,s) = \langle r+s\rangle^{\varepsilon(r,s)}$ (this equals $(-1)^{\varepsilon(r,s)}K(r,s)$ in the notation of [C-GK]). Let m be a positive integer such that $m\cdot(r,s) = 0$ and let $v_{m,r,s}$ denote the class of $mL(r,s)\omega_{m,i,j}$ in $H^1_{DR}(F_m,\overline{\mathbf{Q}}_p)$ where $i = m\cdot\langle r\rangle$ and $j = m\cdot\langle s\rangle$. If n is a positive integer divisible by m and $f: F_n \rightarrow F_m$ is the obvious map, then

$$f^*v_{m,r,s} = v_{n,r,s}. \tag{1.1}$$

Remark. Note that

$$\omega_{m,i,j} = -x^{i-m}y^j dy/y = m^{-1}x^{i-m}y^{j-m}dx^m. \tag{1.2}$$

Suppose K is a finite extension of $\mathbf{Q}_p$. The Weil group of K is the subgroup W(K) of $\mathrm{Gal}(\overline{\mathbf{Q}}_p/K)$ consisting of elements σ whose restriction to the maximal unramified extension of K is an integral power of Frobenius. This integer will be called the degree of σ and denoted $\deg(\sigma)$. We set $W = W(\mathbf{Q}_p)$. Now if X is a curve over K with stable model over a Galois extension K' of K in $\overline{\mathbf{Q}}_p$ and $\sigma \in W(K)$ such that $\deg(\sigma) \geq 0$ there is a canonical element $\Phi(\sigma) \in \mathrm{End}(\tilde{\ })$ (see [CM] §2). If $\tilde{\ }$ is arboreal (which is equivalent to the Jacobian of X having good reduction), this endomorphism induces an endomorphism $\Phi^*(\sigma)$ of $H^1_{DR}(X,K')$ and this map extends to an anti-homomorphism from W(K) into $\mathrm{End}_{K'}(H^1_{DR}(X,K'))$. Moreover, if Y is another curve over K with arboreal reduction over K' and $f: X \rightarrow Y$ is a morphism defined over K', then as morphisms from $H^1_{DR}(Y,K')$ to $H^1_{DR}(X,K')$,

$$\Phi^*(\sigma)\circ f^{\sigma *} = f^*\circ\Phi^*(\sigma). \tag{1.3}$$

In [BO], a semi-linear action ρ_{cris} of the crystalline Weil group $W_{cris}(K)$ on $H^1_{DR}(X,K')$ is defined. When $K = \mathbf{Q}_p$, $W_{cris}(K)$ is just W. Moreover, $\mathrm{Gal}(\overline{\mathbf{Q}}_p/K)$ acts on $H^1_{DR}(X,K') \cong H^1_{DR}(X,K)\otimes K'$ and if $\sigma \in W(K)$, $\rho_{cris}(\sigma) = \Phi^*(\sigma)\circ\sigma$ (see Theorem 4.8 and the remark following its proof in [BO]).

We let $\mathrm{Gal}(\overline{\mathbf{Q}}_p/\mathbf{Q}_p)$ act on $(\mathbf{Q}/\mathbf{Z})^u$, $u \geq 0 \in \mathbf{Z}$, by setting $\sigma r = \rho r$, for $\sigma \in \mathrm{Gal}(\overline{\mathbf{Q}}_p/\mathbf{Q}_p)$ and $r \in (\mathbf{Q}/\mathbf{Z})^u$, where ρ is the finite idèle of $\mathbf{Q}$ which describes the action of σ on roots of unity, (i.e., the inverse image of $\sigma|_{\mathbf{Q}^{ab}/\mathbf{Q}}$ under the Artin map).

Proposition 1.4. Let $\sigma \in W$. Suppose $q \in I$ and m is a positive integer such that $mq = 0$. Then

$$\Phi^*(\sigma)v_{m,\sigma^{-1}q} = \beta_\sigma(q)v_{m,q}$$

for some $\beta_\sigma(q) \in \mathbf{Q}_p$ which does not depend on m (because of (1.1)).

We now fix m and drop it from the notation unless there is danger of confusion.

Suppose $q \in I$. Using the fact that Φ^* is an anti-homomorphism we deduce,

$$\beta_{\sigma\tau}(q) = \beta_\sigma(\tau^{-1}q)\beta_\tau(q). \tag{1.5.1}$$

Using the fact that ρ_{cris} is a homomorphism, we deduce

$$\beta_{\sigma\tau}(q) = \beta_\tau(\sigma^{-1}q)^\sigma\beta_\sigma(q). \tag{1.5.2}$$

Combining (1.5.1) and (1.5.2) we see that $\beta_\sigma(q)^\sigma = \beta_\sigma(q)$.

For $q \in I$, $\mu(\sigma,r) = \langle\sigma^{-1}r\rangle\sigma\text{-}\langle r\rangle \in \mathbf{Q}[W]$. If $\alpha \in \mathbf{Q}_p$, $\psi \in \mathbf{Q}[W]$ we let $\alpha^\psi = (\alpha^{1/n})^{n\psi}$ for any positive integer n such that $n\cdot\psi \in \mathbf{Z}[W]$ and any n-th root in $\overline{\mathbf{Q}}_p$, $\alpha^{1/n}$, of α. It will be implicit in use of such an expression, without further qualification, that the resulting element of $\overline{\mathbf{Q}}_p$ is independent of the choices of n and $\alpha^{1/n}$.

Lemma 1.6. Let $(r,s) \in I$, $t = r+s$ and $\sigma \in W$. Then $\beta_\sigma(s,r) = \beta_\sigma(r,s)$ and $\beta_\sigma(r,-t) = (-1)^{\mu(\sigma,s)-\mu(\sigma,t)}\beta_\sigma(r,s)$.

<u>Remark.</u> $(-1)^{\mu(\sigma,s)-\mu(\sigma,t)} = \pm 1$ and if p is odd, $u \in \mathbf{Q}/\mathbf{Z}$ and $\deg(\sigma) = 1$, $(-1)^{\mu(\sigma,u)} = (-1)^{p\langle\sigma^{-1}u\rangle-\langle u\rangle}$ regarding $p\langle\sigma^{-1}u\rangle-\langle u\rangle$ as a 2-adic integer and this latter equals $[p\langle\sigma^{-1}u\rangle]$ when $\langle u\rangle \in \mathbf{Z}_p$.

<u>Proof.</u> This follows from (1.4) using the automorphisms defined over $\overline{\mathbf{Q}}_p$, $f: (x,y) \to (y,x)$ and $g: (x,y) \to (\zeta x/y, 1/y)$ of F_m where ζ is an m-th root of -1 and the facts that $f^*v_{r,s} = v_{s,r}$ and $g^*v_{r,s} = (-1)^{\varepsilon(r,s)}\zeta^{m\langle r\rangle}v_{r,-(r+s)}$. □

The following result is Theorem 19 of [C-GK].

Theorem 1.7. Suppose $q = (r,s) \in I$. If p does not divide the order of (r,s), and $\deg(\sigma) = 1$, then

$$\beta_\sigma(q) = (-1)^{\varepsilon(q)}p^{\varepsilon(-\sigma^{-1}q)}\Gamma_p\langle r+s\rangle/\Gamma_p\langle r\rangle\Gamma_p\langle s\rangle.$$

<u>Note.</u> The sign here is different from that found in [C-GK] when p = 2. This discrepancy is explained in the section at the end of this paper, "Errata to [C-GK]."

Both proofs of this theorem presented in [C-GK] use the fact that F_m has good reduction over $\mathbf{Q}_p$ when p does not divide m. In this paper we will apply the results of [CM] on the stable reduction of F_m to investigate the case in which p does divides m.

It is also worthwhile to record the relationship between $\beta_\sigma(r,s)$ and the analogous numbers corresponding to the curve F_m : $u^m+v^m+1 = 0$. For $q \in I$, let η_q denote the cohomology class of the differential $L(q)u^iv^j(v/u)d(u/v)$ where $(i,j) = m\langle q\rangle$. Define $\gamma_\sigma(q)$ by the formula

$$\Phi^*(\sigma)\eta_{\sigma^{-1}q} = \gamma_\sigma(q)\eta_q$$

Lemma 1.8. If $q = (r,s) \in I$, $\gamma_\sigma(q) = (-1)^{\mu(\sigma,r)+\mu(\sigma,s)}\beta_\sigma(q)$.

<u>Remark.</u> $(-1)^{\mu(\sigma,r)+\mu(\sigma,s)} = \pm 1$. If m is prime to p and ζ is an m-th root of -1 then $\zeta^\sigma = \zeta^p$ when p is odd, and $(-1)^m = -1$ when p = 2. In these cases, $\gamma_\sigma(r,s) = (-1)^{k(\sigma,r,s)}\beta_\sigma(r,s)$, where $k(\sigma,r,s) = [p\langle\sigma^{-1}r\rangle] + [p\langle\sigma^{-1}s\rangle]$ when p is odd and $\langle\sigma^{-1}r\rangle-\langle r\rangle + \langle\sigma^{-1}s\rangle-\langle s\rangle$ when p = 2.

<u>Proof.</u> This follows from (1.4) using the morphism $f: (u,v) \to (\zeta u,\zeta v)$ from F_m to F_m and the fact that $mf^*v_{r,s} = \zeta^{m(\langle r\rangle+\langle s\rangle)}\eta_{r,s}$. □

Proposition 1.9. Suppose m is prime to p and $q = (r,s) \in I$, $mq = 0$. Then if $\deg(\sigma) = 1$,

$$\gamma_\sigma(q) = (-p)^{\varepsilon(-\sigma^{-1}q)}\Gamma_p\langle r\rangle^{-1}\Gamma_p\langle s\rangle^{-1}\Gamma_p\langle-(r+s)\rangle^{-1}.$$

Proof. This follows from Lemma 1.8 and the remark following it, using: If $t \in \mathbf{Q}/\mathbf{Z}$, $mt = 0$ and $t \neq 0$, $\ell\langle t\rangle = p-[p\langle\sigma^{-1}t\rangle]$ and $[\langle u+v\rangle]+[\langle u\rangle]+[\langle v\rangle] \equiv \varepsilon(\sigma^{-1}(u,v))+\varepsilon(u,v)$ modulo 2 when p is odd as well as $\ell\langle t\rangle = 1+\langle\sigma^{-1}t\rangle-[2\langle\sigma^{-1}t\rangle] \equiv 1+\langle\sigma^{-1}t\rangle-\langle t\rangle$ modulo 2 when $p = 2$. (Here ℓ is as defined in [C-GK] as well as in the next section.) □

We will also find it convenient to define $\beta'_\sigma(q) = (L(q)/L(\sigma^{-1}q))\beta_\sigma(q)$ so that in cohomology $\Phi^*(\sigma)\omega_{i',j'} = \beta'_\sigma(q)\omega_{i,j}$ where $(i,j) = m\langle q\rangle$ and $(i',j') = m\langle\sigma^{-1}q\rangle$.

II. An extension of the Morita Gamma Function to $\mathbf{Q}_p$.

If $x \in \mathbf{Q}_p$, we let $\langle x\rangle_p = \langle x \bmod \mathbf{Z}_p\rangle$ identifying $\mathbf{Q}_p/\mathbf{Z}_p$ with the subgroup $\mathbf{Z}[1/p]/\mathbf{Z}$ of $\mathbf{Q}/\mathbf{Z}$ in the natural way and if $x \in \mathbf{Q}_p$, we set $[x]_p = x$ if $x \in \mathbf{Z}_p$ and $x-\langle x\rangle_p$ otherwise. In addition, if $x \in \mathbf{Q}/\mathbf{Z}$, we let x_p denote the image of x in $\mathbf{Q}_p/\mathbf{Z}_p$ under the natural projection. If a is a non-zero element of $\overline{\mathbf{Q}}_p$ with integral valuation, $\omega(a)$ will denote the unique root of unity of order prime to p in $\overline{\mathbf{Q}}_p$ such that $|\omega(a)-a/p^{v(a)}| < 1$. In this case, we also set $\kappa(a) = p^{v(a)}\omega(a)$ and $a^* = a/\kappa(a)$. Note that $\kappa\langle r\rangle = \kappa\langle r_p\rangle$ when $r \in \mathbf{Q}/\mathbf{Z}$, $r_p \neq 0$.

The natural ring homomorphism from $\mathbf{Z}$ into $R_o =: \mathbf{Z}/(p-1)\times\mathbf{Z}_p$ extends naturally to an injective ring homomorphism from $\mathbf{Z}[1/p]$ into $R =: \mathbf{Z}/(p-1)\times\mathbf{Q}_p$ and we identify $\mathbf{Z}[1/p]$ with its image. For $\alpha \in \mathbf{Q}_p$, we set $\{\alpha\} = 1$ if $\alpha \in p\mathbf{Z}_p$ and $p^{-v(\alpha)}\alpha$, otherwise. For $\alpha \in \mathbf{Z}[1/p]_{>0}$ we set

$$\hat{\Gamma}_p(\alpha,\alpha) = \prod\{\langle\alpha\rangle+k\} \tag{2.1}$$

where the product runs over integers k, $0 \leq k \leq [\alpha]-1$. If $z = (x,a) \in R$, we set $s(z) = a$. Then $\hat{\Gamma}_p$ extends uniquely to a continuous function from R into $\mathbf{Z}_p^*$. It is determined by the functional equation

$$\hat{\Gamma}_p(z+1) = \{s(z)\}\hat{\Gamma}_p(z) \tag{2.2}$$

and the normalization $\hat{\Gamma}_p(z) = 1$ for $z \in \mathbf{Z}[1/p]$, $0 < z \leq 1$. For $z = (x,s) \in R$, set $[z]_o = (x+1-\langle -s\rangle_p,[s]_p)$ which lies in R_o. (Note: For $\alpha \in \mathbf{Z}[1/p]$, $[(\alpha,\alpha)]_o = ([\alpha],[\alpha])$.) For $z = (x,s) \in R_o$ and $a \in \mathbf{Z}_p^*$ we set $a^z = \omega(a)^x a^{*s}$. Then $\hat{\Gamma}_p$ also satisfies

$$\hat{\Gamma}_p(z)\hat{\Gamma}_p(1-z) = (-1)^{i(z)} \tag{2.3}$$

where $i(z) = (z-1)-[(z-1)/p]$ for z in R_o and $i(z) = [z]_o$ otherwise. Moreover $\hat{\Gamma}_p(z) \equiv \{s(z)\}^{[z]_o}$ modulo $s^{-1}\mathbf{Z}_p$.

For $s \in \mathbf{Q}_p$ we set $\Gamma_p(s) = \hat{\Gamma}_p(\langle s\rangle,s)$ when $p \neq 2$ and $\Gamma_2(z) = \varepsilon(z)\hat{\Gamma}_2(0,z)$ where $\varepsilon(z) = -1$ when $|z| \geq 1$ and 1 when $|z| < 1$. The restriction of Γ_p to $\mathbf{Z}_p$ is the Morita gamma function. Then when $p \neq 2$, and $|z| > 0$,

$$\hat{\Gamma}_p(z) = \omega(s(z))^{[z]_o}\Gamma_p(s(z)). \tag{2.4}$$

It follows that,

$$\Gamma_p(s) \equiv (-1)^{p-1}s_*^{[s]_p} \bmod s^{-1}\mathbf{Z}_p, \tag{2.5}$$

$$(2.6)\qquad \Gamma_p(s+1) = \begin{cases} s^*\Gamma_p(s) & \text{when } |s| > 1 \\ -\{s\}\Gamma_p(s) & \text{when } |s| \le 1. \end{cases}$$

Also,

$$(2.7)\qquad \Gamma_p(s)\Gamma_p(1-s) = (-1)^{\ell(s)}$$

where

$$\ell(s) = \begin{cases} 0 & |s| > 1, p > 2 \\ [s]_2 & |s| > 1\ p = 2 \\ \langle s/p \rangle_p & |s| \le 1, p > 2 \\ 1+[s/2]_2 & |s| \le 1, p = 2 \end{cases}$$

In the above and the following, we regard an element a of $\mathbf{Z}[1/p]$ as contained in $\mathbf{Z}_2$ when p is odd and a appears in the exponent of -1. We have the following Gauss multiplication formulas: If $n > 0$, $(n,p) = 1$ then,

$$(2.8)\qquad \begin{aligned} \prod_{i=0}^{n-1}\Gamma_p((s+i)/n) &= \Gamma_p(s){n^*}^{-[s]_p}\prod_{i=0}^{n-1}\Gamma_p((\langle s\rangle_p+i)/n) \quad \text{if } |s| > 1 \\ \prod_{i=0}^{n-1}\Gamma_p((s+i)/n) &= \Gamma_p(s)n^{-B(s-1)}\prod_{i=0}^{n-1}\Gamma_p(i/n) \quad \text{if } |s| \le 1. \end{aligned}$$

where $B(s) = z-[z/p]_o$ for any $z \in R$ such that $s(z) = s$. Finally,

$$(2.9)\qquad \begin{aligned} \prod_{i=0}^{p-1}\Gamma_p((s+i)/p) &= \Gamma_p(s) \quad \text{if } |s| > 1 \text{ and} \\ \prod_{i=0}^{p-1}\Gamma_p((s+i)/p) &= \Gamma_p(s)\Gamma_p([s/p]_p)^* \quad \text{if } |s| \le 1. \end{aligned}$$

We state the duplication formula when $p > 2$ and $|s| > 1$ which is a special case of (2.8):

$$(2.10)\qquad \Gamma_p(s)\Gamma_p(s+1/2) = \Gamma_p(2s)\Gamma_p(\langle s\rangle_p+(-1)^{\langle 2s\rangle_p}/2)(1/2^*)^{[2s]_p}.$$

(Note that $\Gamma_p(\langle 2s\rangle_p/2)\Gamma_p(\langle 2s\rangle_p/2+1/2) = \Gamma_p(\langle s\rangle_p+(-1)^{\langle 2s\rangle_p}/2) = \Gamma_p(\langle s\rangle_p+1/2)\{\langle s\rangle_p -1/2\}^{*\,-\varepsilon(s_p,s_p)}$.) The proofs of the above formulas follow the same lines as the proofs of the corresponding formulas in [C-GK].

For future reference, we record: Suppose $(r,s) \in I$ and $(r+s)_p \neq 0$. Then

$$(2.11)\qquad \Gamma_p(\langle r\rangle+\langle s\rangle) = \Gamma_p(\langle r+s\rangle)L(r,s)\kappa\langle r+s\rangle^{-\varepsilon(r,s)}.$$

III. Statement of the Main Results.

If A is a group acting on an abelian group M on both the right and the left, we say these actions commute if $(\sigma m)\tau = \sigma(m\tau)$ for $\sigma, \tau \in A$ and $m \in M$. If g_σ is a one-cocycle on A with values in M for the left action, we say that g_σ is a left cocycle and if for the right action, we say g_σ is a right cocycle. If it is a cocycle for both these actions, we call g_σ a bi-cocycle. The unmodified expression "cocycle" will mean left cocycle.

Suppose S is a set on which W acts through its abelian quotient (like $\mathbf{Q}/\mathbf{Z}$) and T is a left W module like $\overline{\mathbf{Q}}_p$. Let M(S,T) denote the group of functions from S into T. We let W act on M(S,T) on the left via $(\sigma H)(r) = H(\sigma^{-1}r)^\sigma$ and also on the right via $(H\sigma)(r) = H(\sigma^{-1}r)$ for $H \in M(S)$. We set $M(S) = M(S,\overline{\mathbf{Q}}_p)$. These two actions commute and by formulas (1.5.1) and (1.5.2), β_σ is a bi-cocycle with values in M(I).

We now reformulate some results of [O] in our context. Let μ_∞ denote the group of roots of unity in $\overline{\mathbf{Q}}_p$. Let $\Gamma_1(\sigma)(r) = p\Gamma_\pi(\sigma)(-r)$ in the notation of [O] (π is a prime of $\overline{\mathbf{Q}}$ above p in [O] but its choice turns out not to affect $\Gamma_\pi(\sigma)(r)$). Then $\Gamma_1(\sigma)$ is a left cocycle on W with values in $M(\mathbf{Q}/\mathbf{Z},\overline{\mathbf{Q}}_p/\mu_\infty)$ and it follows from Lemma 1.8, [O; Theorem 2.5] and [O; Theorem 3.13] that

(3.1) $$\Gamma_1(\sigma)(r)\Gamma_1(\sigma)(s)/\Gamma_1(\sigma)(t) = \beta_\sigma(r,s) \mod \mu_\infty.$$

for $(r,s) \in I$, $t = r+s$. Remark 3.7 of [O], after applying a Tate twist, asserts in our language,

Lemma 3.2. Let S be a subset of $\mathbf{Q}/\mathbf{Z}$ stable under the action of the finite idèles and multiplication by 2. Let $I(S) = I \cap S^2$. Suppose T_σ is a cocycle on W with values in $M(S,\overline{\mathbf{Q}}_p/\mu_\infty)$ such that $T_\sigma(0) = 1$ if $0 \in S$ and $T_\sigma(1/2) = p^{\deg(\sigma)/2}$ if $1/2 \in S$. Define a cocycle B_σ on W with values in $M(I(S),\overline{\mathbf{Q}}_p/\mu_\infty)$ by

$$B_\sigma(q) = T_\sigma(r)T_\sigma(s)/T_\sigma(t)$$

for $q = (r,s) \in I(S)$, $t = r+s$. Suppose for every $r \in S-\{0,1/2\}$ and $\sigma \in W$, $B_\sigma(r,r) = \beta_\sigma(r,r) \mod \mu_\infty$. Then, $B_\sigma(q) = \beta_\sigma(q) \mod \mu_\infty$ for all $\sigma \in W$, $q \in I(S)$. Moreover, $T_\sigma(r) = \Gamma_1(\sigma)(r)$.

Lemma 3.3 Let A be a group and d a homomorphism A onto $\mathbf{Z}$. Suppose M is a left A module and for each $\sigma \in A$ such that $d(\sigma) = 1$ we have a $g_\sigma \in M$ such that $\sigma^{-1}(g_{\sigma\tau}-g_\sigma) = g_{\tau\sigma}-\tau g_\sigma$ for $\sigma, \tau \in W^u(\mathbf{Q}_p)$ such that $\deg(\sigma) = 1$ and $\deg(\tau) = 0$, is independent of σ defines a cocycle on the kernel of d. Then g_σ extends uniquely to a left cocycle on A with values in M.

<u>Proof</u>. Uniqueness is clear. We will define g_τ for $\tau \in A$, $d(\tau) \geq 0$ by induction on $d(\tau)$. For $d(\tau) = 1$, it is already defined. For $d(\tau) = 0$, we set $g_\tau = \sigma^{-1}(g_{\sigma\tau}-g_\sigma)$. Suppose $n \geq 1$ and g_τ is defined for $0 \leq d(\tau) \leq n$ and satisfies

P(n) : $$g_{\sigma\tau} = \sigma g_\tau + g_\sigma, \text{ for } \sigma, \tau \in A,\ d(\sigma), d(\tau) \geq 0,\ d(\sigma\tau) \leq n.$$

This is true for n = 1 by hypothesis. We will define g_τ for $d(\tau) \leq n+1$ so that P(n+1) is satisfied. Suppose $\alpha, \beta \in A$ such that $\alpha\beta = \tau$. If $\deg(\alpha), \deg(\beta) > 0$ set $g_\tau = \alpha g_\beta + g_\alpha$. We must check that this is independent of the choice of α and β. Suppose $0 \leq d(\rho) < d(\beta)$. Let $\gamma = \alpha\rho$ and $\delta = \rho^{-1}\beta$. Then $d(\delta) \geq 0$ and $\tau = \gamma\delta$ and

$$\alpha g_\beta + g_\alpha = \alpha(\rho g_\delta + g_\rho) + g_\alpha = \alpha\rho g_\delta + \alpha g_\rho + g_\alpha = \gamma g_\delta + g_\gamma,$$

by induction. By reversing the roles of α and β in this argument we deduce that g_τ is well defined and satisfies P(n+1) for $\sigma, \tau \in A$, $d(\sigma), d(\tau) > 0$, $d(\sigma\tau) \leq n$. Suppose therefore that $d(\tau) = n+1$ and $\tau = \alpha\beta$, with $d(\beta) = 0$. Write $\alpha = \sigma\rho$, $\gamma = \rho\beta$ with $d(\rho) = 1$. Then, $g_\tau = \sigma g_\gamma + g_\sigma = \sigma(\rho g_\beta + g_\rho) + g_\sigma = \sigma\rho g_\beta + \sigma g_\rho + g_\sigma = \alpha g_\beta + g_\alpha$ by induction. Now for $d(\tau) < 0$, we set $g_\tau = -\tau g_{\tau^{-1}}$ and one can easily check that g_σ is a cocycle. □

Lemma 3.4. Let A, M and d be as in the previous lemma but also suppose that A acts on M on the right and these two actions commute. Suppose that g_σ is a left cocycle which satisfies the following property

$$P'(n): \qquad g_{\sigma\tau} = g_\sigma\tau + g_\tau \text{ and } \sigma g_\sigma = g_\sigma\sigma \text{ for } d(\sigma), d(\tau) \geq 0 \text{ and } d(\sigma\tau) \leq n$$

for $n = 1$. Then g_τ is a right cocycle as well.

Proof. Suppose g_σ satisfies $P'(n)$ for $n \geq 1$. We will show it satisfies $P'(n)$ for $n+1$. Suppose $d(\tau) \geq d(\sigma) > 0$ and $d(\sigma\tau) = n+1$. Let $\tau = \sigma\rho$. Then, as $d(\sigma\rho) \leq n$ and $d(\rho) \geq 0$, $g_\rho = g_{\sigma\rho} - g_\sigma\rho = \sigma g_\rho + g_\sigma - g_\sigma\rho$ so $\sigma g_\rho = g_\rho + g_\sigma(\rho - 1)$ and

$$g_{\sigma\tau} = \sigma^2 g_\rho + \sigma g_\sigma + g_\sigma = \sigma g_\rho + \sigma(g_\sigma(\rho-1)) + \sigma g_\sigma + g_\sigma = \sigma g_\rho + g_\sigma + g_\sigma\sigma\rho = g_\tau + g_\sigma\tau,$$

as $\sigma g_\sigma = g_\sigma\sigma$. One can make similar arguments when $d(\sigma) \geq d(\tau) > 0$ or either is in the kernel of d. Finally, $\sigma\tau g_{\sigma\tau} = \sigma\tau(g_\sigma\tau + g_\tau) = \sigma(\tau g_\sigma + g_\tau)\tau = \sigma g_{\tau\sigma}\tau = \sigma(g_\tau\sigma + g_\sigma)\tau = (\sigma g_\tau + g_\sigma)\sigma\tau = g_{\sigma\tau}\sigma\tau$.

It is now easy to show that g_σ is a cocycle for the right action. □

Let G_σ denote the left cocycle on W with values in $M(\mathbf{Q}_p/\mathbf{Z}_p)$ such that $G_\sigma(r) = \Gamma_p(r)^{-1}$ if $r_p = 0$ and $\deg(\sigma) = 1$ and $G_\sigma(r) =: \Gamma_p(\sigma^{-1}r)/\Gamma_p(r)$ if $r_p \neq 0$. That this exists follows from Lemma 3.3 and one can show it is a bi-cocycle using Lemma 3.4. Let $\chi : W \to \mathbf{Z}_p^*$ denote the cyclotomic character, W_χ the kernel of χ and W_χ^0 the subgroup of W_χ consisting of elements degree zero. For $\sigma \in W$, we let $\omega(\sigma) = \omega(\chi(\sigma))$ and W_ω the kernel of ω in W. Let π be a fixed (p-1)-st root of -p in $\overline{\mathbf{Q}}_p$. For σ of degree one, let

$$\Gamma_\sigma(r) = \begin{cases} (-1)^{\mu(\sigma,r)} p\pi^{\langle\sigma^{-1}r\rangle(\sigma-p)} G_\sigma(r) & \text{if } r_p = 0, \\ -\Gamma_p(1/2)p\pi^{(\sigma-p)/2}(2\langle r\rangle)^{(\sigma-1)/2}\kappa(r)^{\mu(\sigma,r)} G_\sigma(r) & \text{if } r_p \neq 0. \end{cases}$$

Remark. $\Gamma_\sigma(r)$ is of the form $\zeta p^{f(\sigma,r)} G_\sigma(r)$, where ζ is a (p-1)m-th root of unity and $f(\sigma,r) = \langle -\sigma^{-1}r\rangle$ if $\deg(\sigma) = 1$, $r_p = 0$ and $f(\sigma,r) = \deg(\sigma)/2 + v(r)(\langle\sigma^{-1}r\rangle - \langle r\rangle)$ if $r_p \neq 0$. Also, $\Gamma_\sigma(0) = \omega(\sigma)$ and $\Gamma_\sigma(1/2) = \iota\omega(\sigma)\Gamma_p(1/2)\pi^{(p-\sigma)/2}$ where $\iota = 1$ when p is odd and $2^{(\sigma-1)/2}$ when $p = 2$.

Proposition 3.5. Suppose p is odd. There exists a unique extension of Γ_σ to a left cocycle on W_ω with values in $M(\mathbf{Q}/\mathbf{Z})$ and if $B_\sigma(r,s) = \omega(\sigma)^{-\varepsilon(\sigma^{-1}q)}\Gamma_\sigma(r)\Gamma_\sigma(s)/\Gamma_\sigma(r+s)$ for $(r,s) \in I$, $\sigma \in W$ of degree 1, then B_σ extends uniquely to a left cocycle on W with values in M(I). Moreover, the restriction of this extension of Γ_σ to W_χ is a bi-cocycle.

Proof. Let $\sigma \in W$ such that $\deg(\sigma) = 1$, $\tau \in W^0$, $r \in \mathbf{Q}/\mathbf{Z}$, $r \neq 0$. Note that $\tau r = r$, if $r_p = 0$. Then

$$(\Gamma_{\sigma\tau}(\sigma r)/\Gamma_\sigma(\sigma r))^{\sigma^{-1}} = \begin{cases} \pi^{\langle r\rangle(\tau-1)}, & \text{if } r_p = 0, \\ \omega(\sigma)^{\mu(\tau,r)}((2\pi\langle r\rangle))^{(\tau-1)/2}\kappa(r)^{\mu(\tau,r)}, & \text{if } r_p \neq 0, \end{cases}$$

from which it is obvious that $H_\tau(r) =: (\Gamma_{\rho\tau}(\rho r)/\Gamma_\rho(\rho r))^{\rho^{-1}}$ for $\rho \in W_\omega$, $\deg(\rho) = 1$, is independent of the choice of ρ. Moreover, H_τ is a cocycle on W^0_ω with values in $M(\mathbf{Q}/\mathbf{Z})$ (in fact, a coboundary) whose restriction to W^0_χ is a bi-cocycle. Now, since p is odd,

$$\Gamma_{\tau\sigma}(r)/\Gamma_\sigma(\tau^{-1}r)^\tau = \begin{cases} \omega(\tau)^{\mu(\sigma,r)}\pi^{\langle r\rangle(\tau-1)}, & \text{if } r_p = 0. \\ \omega(\tau)^{\tau\mu(\sigma,\tau^{-1}r)}((2\pi\langle r\rangle))^{(\tau-1)/2}\kappa\langle r\rangle^{\mu(\tau,r)}, & \text{if } r_p \neq 0. \end{cases}$$

It follows from Lemma 3.3 that Γ_σ extends to a left cocycle on W_ω. It also follows easily from this and Lemma 3.3 that B_σ extends to a left cocycle on W with values in M(I).

Finally, we must show that the restriction of Γ_σ to W_χ is a right cocycle. First it is clear that for σ of degree 1 in W_χ, $\sigma\Gamma_\sigma = \Gamma_\sigma\sigma$. For degree zero this same identity follows from the fact that Γ_σ is a bi-cocycle on W^0_χ. Next, if $\deg(\sigma) = 1$, $\deg(\tau) = 0$ and $r_p = 0$, $\Gamma_{\tau\sigma}(\sigma r)/\Gamma_\sigma(\sigma r) = (\pi^{\langle r\rangle\sigma})^{(\tau-1)} = \Gamma_\tau(r)$ since τ fixes all roots of unity and $\Gamma_{\sigma\tau}(r)/\Gamma_\sigma(r) = \pi^{\langle\sigma^{-1}r\rangle(\sigma\tau-\sigma)} = \Gamma_\tau(r)$. As we can also prove similar formulas when $r_p \neq 0$, the proposition follows from Lemma 3.4. □

Lemma 3.6. Suppose, $\sigma \in W$, $\deg(\sigma) = 1$, $t \in \mathbf{Q}/\mathbf{Z}$ and $t \neq 0$. Then if p is odd or $t_p = 0$

$$\Gamma_\sigma(t)\Gamma_\sigma(-t) = (-1)^{\mu(\sigma,t)}\omega(\sigma)p.$$

Proof. Suppose now that $t \neq 0$ and $\deg(\sigma) = 1$. First, if $t_p = 0$, $\Gamma_\sigma(-t) =$

$$(-1)^{\mu(\sigma,-t)}p\pi^{(1-\langle\sigma^{-1}t\rangle)(\sigma-p)}\Gamma_p(1-\langle t\rangle)^{-1} = (-1)^{\ell\langle t\rangle-\mu(\sigma,t)}p\pi^{\sigma-p}\pi^{-\langle\sigma^{-1}t\rangle(\sigma-p)}\Gamma_p\langle t\rangle.$$

From this the lemma follows, in this case (see the proof of Proposition 1.9).

Second if $t_p \neq 0$, p is odd and $\sigma \in W_\chi$ then $(2\pi(1-\langle t\rangle))^{(\sigma-1)/2} = (-2\pi\langle t\rangle)^{(\sigma-1)/2}$, $\mu(\sigma,-t) = (\sigma-1)-\mu(\sigma,t)$, $\kappa\langle -t\rangle = -\kappa\langle t\rangle$, $\Gamma_p(1-\langle\sigma^{-1}t\rangle)/\Gamma_p(1-\langle t\rangle) = G_\sigma(t)^{-1}$, $\pi^{(\sigma-p)/2} = -\omega(\sigma)/p\pi^{(\sigma-p)/2}$ and $\Gamma_p(1/2) = (-1)^{(p+1)/2}\Gamma_p(1/2)^{-1}$. The lemma follows. □

Let $(\mathbf{Q}/\mathbf{Z})_{p'} = \{r \in \mathbf{Q}/\mathbf{Z} : r_p = 0\} = \mathbf{Q}/\mathbf{Z} \cap \mathbf{Z}_p/\mathbf{Z}$ and let $I_0 = \{(r,s) \in I : \{r, s, r+s\} \subseteq (\mathbf{Q}/\mathbf{Z})_{p'}\}$, $I_2 = \{(r,s) \in I : \{r, s, r+s\} \cap (\mathbf{Q}/\mathbf{Z})_{p'} = \emptyset\}$, $I_1 = I-(I_0 \cup I_2)$. Also let $I'_2 = \{(r,s) \in I_2 : v\langle r\rangle = v\langle s\rangle = v\langle r+s\rangle\}$ and $I' = I_0 \cup I_1 \cup I'_2$.

Lemma 3.7. There exists a bi-cocycle H_σ on W_χ with values in M(I) such that if $\sigma \in W_\chi$ and $q = (r,s) \in I$, $r+s = t$ then $H_\sigma(q) = 1$ if $q \in I_0 \cup I_1$ and

$$H_\sigma(\sigma q) = (\{\langle r\rangle\}^{\langle r\rangle}\{\langle s\rangle\}^{\langle s\rangle}/\{\langle r\rangle+\langle s\rangle\}^{\langle r\rangle+\langle s\rangle})_*^{\sigma-1}$$

if $q \in I_2$ and $\deg(\sigma) = 1$.

Proof. Since I_i, $i \in \{0,1,2\}$, is stable under the action of W we may assume $q \in I_2$. We will use Lemmas 3.3 and 3.4. Then, for $\tau \in W^0_\chi$,

$$(H_{\sigma\tau}(\sigma q)/H_\sigma(\sigma q))^{\sigma^{-1}} = (\{\langle r\rangle\}^{\langle r\rangle}\{\langle s\rangle\}^{\langle s\rangle}/\{\langle r\rangle+\langle s\rangle\}^{\langle r\rangle+\langle s\rangle})^{\tau-1}$$

which depends only on τ and is a cocycle on W^0_χ. On the other hand, if we let $\langle\sigma^{-1}r\rangle-\langle r\rangle = a$, $\langle\sigma^{-1}s\rangle-\langle s\rangle = b$, then a and b are elements of $\mathbf{Z}_p$ and so as τ is inertial

$$H_{\tau\sigma}(q)/H_\sigma(\tau^{-1}q) = (\{\langle\sigma^{-1}r\rangle\}^{\langle r\rangle}\{\langle\sigma^{-1}s\rangle\}^{\langle s\rangle}/\{\langle\sigma^{-1}r\rangle+\langle\sigma^{-1}s\rangle\}^{\langle r\rangle+\langle s\rangle})^{\tau-1}.$$

Now, since none of r, s or t lie in $(\mathbf{Q}/\mathbf{Z})_{p'}$,

$$(\{\langle\sigma^{-1}r\rangle\}^{\langle r\rangle})^{\tau-1} = (\langle r\rangle(1+pa)^{1/p})^{\tau-1},\quad (\{\langle\sigma^{-1}s\rangle\}^{\langle s\rangle})^{\tau-1} = (\langle s\rangle(1+pb)^{1/p})^{\tau-1}$$

and

$$(\{\langle\sigma^{-1}r\rangle+\langle\sigma^{-1}s\rangle\}^{\langle r\rangle+\langle s\rangle})^{\tau-1} = ((\langle r\rangle+\langle s\rangle)(1+p(a+b))^{1/p})^{\tau-1}$$

Hence, $(H_{\sigma\tau}(\sigma q)/H_\sigma(\sigma q))^{\sigma^{-1}} = H_{\tau\sigma}(q)/H_\sigma(\tau^{-1}q)$.

The computations necessary to show H_σ is a bi-cocycle are similar. □

We note that $H_\sigma(q) = H_\sigma(q_p)$ and $H_\sigma(\sigma q) = (((1-p\varepsilon(q))^{1/p}\{\langle r\rangle\}^{\langle r\rangle}\{\langle s\rangle\}^{\langle s\rangle}/\{\langle t\rangle\}^{\langle t\rangle})^{\sigma-1})^*$ when $q = (r,s) \in I_2$. Using this one can show that the restriction of H_σ to the fixer of $(1+p)^{1/p}$ in W_χ factors naturally into the product of three bi-cocycles with values in $M(\mathbf{Q}/\mathbf{Z})$. We may now state,

Proposition 3.8. Let $q = (r,s) \in I$, $t = r+s$. Suppose $p > 3$ or $p = 3$ and $q \in I'$. Then for $\sigma \in W_\chi$,

$$\beta_\sigma(q) = H_\sigma(q)\Gamma_\sigma(r)\Gamma_\sigma(s)/\Gamma_\sigma(t). \tag{3.9}$$

One can show that this formula, in the case in which $\sigma q = q$, is equivalent to the formula for the local components of the Jacobi sum Hecke characters, Theorem 5.3 of [CM]. In fact, this theorem implies that (3.9) is false when $p = 3$ in the excluded cases (although, it true up to roots of unity.) (See [P] for the formula corresponding to [CM; Theorem 5.3] in the case $p = 2$.) We deduce from Theorem 1.7 and Proposition 3.5, the following theorem when $q \in I_0$.

Proposition 3.10. Suppose $\sigma \in W$, $\deg(\sigma) = 1$, $q = (r,s) \in I_0 \cup I_1$ and $t = r+s$. Then if p is odd or $t_p \neq 0$

$$\beta_\sigma(q) = \omega(\sigma)^{-\varepsilon(\sigma^{-1}q)}\Gamma_\sigma(r)\Gamma_\sigma(s)/\Gamma_\sigma(t).$$

When $q \in I_1$, this will be proven in Section VI. In the excluded case, it is true up to a sign. The conclusion of Proposition 3.8, in the case $q \in I_0 \cup I_1$ is an immediate corollary.

We also state the corresponding result for the symmetric Fermat curve. Let $\Gamma'_\sigma(r) = (-1)^{\mu(\sigma,r)}\Gamma_\sigma(r)$ if $r \in \mathbf{Q}/\mathbf{Z}$ and $\sigma \in W$. Then, it follows from Proposition 3.5 and Proposition 3.10 that under the same hypotheses as in Proposition 3.10,

$$\gamma_\sigma(q) = p^{-\deg(\sigma)}\omega(\sigma)^{\varepsilon(-\sigma^{-1}q)}\Gamma'_\sigma(r)\Gamma_\sigma(s)\Gamma_\sigma(-t). \tag{3.11}$$

As was pointed out in [O], it follows immediately from this and Lemma 3.2 that $\Gamma_1(\sigma)(r) = \Gamma_\sigma(r) \bmod \mu_\infty$ when $r_p = 0$. We also obtain a formula for $\Gamma_1(\sigma)(r)$ when $r_p \neq 0$ as Corollary 6.5 below.

Suppose now that p is odd. To completely determine $\beta_\sigma(q)$, at least when $p > 3$, we define

$$A_\sigma(r) = (2^{-\mu(\sigma,2r)}) * \Gamma_p(\langle r\rangle + (-1)^{\langle 2r\rangle}/2)/\Gamma_p(\langle \sigma^{-1}r\rangle + (-1)^{\langle 2\sigma^{-1}r\rangle}/2).$$

for $r \in \mathbf{Q}_p/\mathbf{Z}_p$ and, if f is the order of 2 mod $p^{-v(r)}$, we define an element $D_\sigma(r)$ in $\overline{\mathbf{Q}}_p^*/\mu_\infty$ by

$$D_\sigma(r) = \prod_{i=1}^{f}(A_\sigma(r/2^i))^{a_i} \text{ mod } \mu_\infty$$

where $a_i = 2^{i-1}/(2^f-1)$. (Here $r/2^i$ denotes the element s of $\mathbf{Q}_p/\mathbf{Z}_p$ such that $2^i s = r$.) Then D_σ is the unique cocycle, on W with values in $M(\mathbf{Q}_p/\mathbf{Z}_p; \overline{\mathbf{Q}}_p^*/\mu_\infty)$ such that, if $D_\sigma(r)^2/D_\sigma(2r) = A_\sigma(r)$ mod μ_∞. (See [O; Lemma 3.3].) Also note that when $\sigma \in W_\chi$, $A_\sigma = 1$ and $D_\sigma = 1$ mod μ_∞. We extend H_σ to all of W and

Proposition 3.12. There exists a unique cocycle H_σ on W with values in M(I), if $p > 3$, and on M(I'), if $p = 3$, with the following properties: Suppose $q \in I$, $q \in I'$, if $p = 3$ and $q \in I_0 \cup I_1$, if $p = 2$. Let $q_p = (r,s)$, $t = r+s$ and $\sigma \in W$. If $q \in I_2$,

$$H_\sigma(q) = D_\sigma(r)D_\sigma(s)/D_\sigma(t) \text{ mod } \mu_\infty$$

and

$$H_\sigma(q) \equiv (\chi(\sigma)^{1/2-\varepsilon(\sigma^{-1}q_p)}) * (\{\langle -r\rangle\}^{\mu(\sigma,r)}\{\langle -s\rangle\}^{\mu(\sigma,s)}/\{\langle -r\rangle + \langle -s\rangle\}^{\mu(\sigma,t)}) *$$

modulo $(1+p^{n-e}\mathbf{Z}_p)$, and $H(q) = 1$, otherwise. Moreover, when $r \in \mathbf{Q}_p/\mathbf{Z}_p$, $H_\sigma(r,r) = A_\sigma(r)$.

It is clear that H_σ must be the same cocycle as that defined above when $\sigma \in W_\chi$. This proposition and the following theorem will be proved in Section VII. The determination of β_σ, when $p > 3$, is completed by,

Theorem 3.13. Let $q = (r,s) \in I$, $t = r+s$. Suppose $p > 3$ or, $p = 3$ and $q \in I'$, or $p = 2$, $q \in I_0 \cup I_1$ and $t_p \neq 0$. Then if $\sigma \in W$ and $\deg(\sigma) = 1$,

$$\beta_\sigma(q) = \omega(\sigma)^{-\varepsilon(\sigma^{-1}q)}H_\sigma(q)\Gamma_\sigma(r)\Gamma_\sigma(s)/\Gamma_\sigma(t).$$

Propositions 3.8 and 3.10 are immediate corollaries of this theorem.

For $\sigma \in W$ and $r \in \mathbf{Q}/\mathbf{Z}$ set $F_\sigma(r) = \kappa(r_p)^{\mu(\sigma,r)-\mu(\sigma,r_p)}G_\sigma(r)$. Since $H_\sigma(q)$ only depends on q_p and $G_\sigma(u) = 1$ for $u \in \mathbf{Q}_p/\mathbf{Z}_p$, $u \neq 0$, the following is a corollary of Theorem 3.13, although we actually prove it first.

Proposition 3.14. Let $q = (r,s) \in I_2$, $t = r+s$. Suppose $p > 3$ or $p = 3$ and $q \in I'$. Then if $\sigma \in W$,

$$\beta_\sigma(q)/\beta_\sigma(q_p) = \omega(\sigma)^{\varepsilon(\sigma^{-1}q_p)-\varepsilon(\sigma^{-1}q)}F_\sigma(r)F_\sigma(s)/F_\sigma(t).$$

It is interesting to observe that the restriction of $\sigma \to F_\sigma$ to W_ω is a one-cocycle and, using the Gauss multiplication formula (2.9), that

$$\prod_{i=0}^{p-1}F_\sigma((r+i)/p) = F_\sigma(r) \quad \text{if } r_p \neq 0. \tag{3.15}$$

Also, using the above results one can prove:

(3.16) $$\beta_\sigma(s,t)\beta_\sigma(r+s,t) = \beta_\sigma(r,s+t)\beta_\sigma(r,s)$$

when all the terms in this equation are defined. (One should also be able to give a geometric proof of this equation using the ideas in [A]. See [O; Remark 3.5].)

IV. Computing Using Formal Expansions

In this section we generalize [C-GK; Theorem 17] which was a generalization of [K; Theorem 6.2].

Let K be a finite extension of $\mathbf{Q}_p$, with ring of integers A and residue field k. Suppose X and Y are complete curves over A with arboreal reductions $\bar{X}$ and $\bar{Y}$. Suppose $f: \bar{X} \to \bar{Y}$ is a morphism of curves over k. Because the Jacobians of X and Y have good reduction, f induces a homomorphism f^* from $H^1_{DR}(Y,K)$ to $H^1_{DR}(X,K)$. Suppose P and Q are smooth points on $\bar{X}$ and $\bar{Y}$ over k such that $f(P) = Q$. Let s and t be rational functions on X and Y which are uniformizing parameters on the residue disks above P and Q respectively (i.e. map these disks onto the open unit disk), ε an element of A^* and q is a positive integer such that $f^*t = \varepsilon s^q$. Suppose ω is a differential of the second kind on Y regular on the disk Q and v is a differential of the second kind on X regular on the disk P.

Let

$$\omega = \Sigma_{n=1}^{\infty} a(n)t^n dt/t \quad \text{and} \quad v = \Sigma_{n=1}^{\infty} b(n)s^n ds/s$$

be the expansions of ω in t and v in s.

Theorem 4.1. Suppose $f^*\omega = \alpha v$ in cohomology and $\{n_i\}$ is a sequence of positive integers such that $|a(n_i)/n_i| \to \infty$. Then

$$\alpha = \mathrm{Lim}_{i\to\infty}\ q\varepsilon^{n_i} a(n_i)/b(qn_i).$$

<u>Proof</u>. Let U be the dagger lifting of $\tilde{X}^{ns}$ in X and V the dagger lifting on $\tilde{Y}^{ns}$ in Y. Let S be the connected component of U above the component of $\tilde{X}^{ns}$ containing P and T the connected component of V above the component of $\tilde{Y}^{ns}$ containing Q. The main thing to observe is that the de Rham cohomology of X (resp. Y) injects into the dagger cohomology of U (resp. V) and the morphism f may be lifted to a morphism F from U to V such that on S, $(F^*t)|_T = \varepsilon s^q$. The rest of the proof follows the same lines as the proof of [C-GK; Theorem 17]. □

Unfortunately, one does not know too much, a priori, about the sequence $\{n_i\}$ in the above theorem. Since the set $\{a(n) : n \geq 0\}$ is bounded, it follows that $n_i \to 0$ as $i \to \infty$. The following result guided in us during our computations.

Proposition 4.2. With notation as in the above theorem, suppose there exists an automorphism α of $\tilde{T}$ of order e such that $\alpha(\tilde{Q}) = \tilde{Q}$, the quotient $\tilde{T}/\langle\alpha\rangle$ has genus zero and $\alpha^*t = \tau t$ for some $\tau \in k^*$ of order e. Then n_i is not congruent to zero mod e for large i.

<u>Proof</u>. Let $\hat{\tau}$ be the Teichmüller lifting of τ in K. The automorphism α lifts to an automorphism $\hat{\alpha}$ of order e of T, such that $\hat{\alpha}(Q) = Q$ and $\hat{\alpha}^*t = \hat{\tau}t$. It follows that $T/\langle\hat{\alpha}\rangle$ is a dagger space of genus zero. From this and the

fact that ω is of the second kind we see that $\eta = \sum_0^{qf-1} (\alpha^*)^i\omega$ is exact on T. As

$$\eta = qf\sum_{n=0}^{\infty} a(qn)t^{qn}dt/t,$$

we see that the set $\{a(qn)/qn\}$ is bounded. □

V. Recollections of the Stable Reduction of Fermat Quotients

Fix $m = p^n d$, $(p,d) = 1$ and let a and b be integers such that $(m,a,b) = 1$. Let $F_{m,a,b}$ denote the quotient of F_m whose function field is generated by $w = x^a y^b$ and $u = x^m$. Then $F_{m,a,b}$ has affine equation $w^m = u^a(1-u)^b$. Let $c = a+b$. In [CM], we calculated the stable model of the curve with affine equation $w^m = u^a(1-u)^b(-1)^c$ over $\overline{\mathbf{Q}}_p$ when p was greater than 3 or when p = 3 and either 3^n divides a (or b or c) or $(3,abc) = 1$. This curve is isomorphic to $F_{m,a,b}$ over an unramified extension of $\mathbf{Q}_p$ so long as $p > 2$ and therefore has the same theory of stable reduction. We will recall and slightly generalize some of the results of [CM]. Suppose $n > 1$.

If p^n divides b, then the map from F_m to $F_{m,a,b}$ factors through the curve with affine equation

$$x^m + t^d = 1. \tag{5.1}$$

It turns out that the computation of the stable model of this curve is very similar to that of $F_{m,a,b}$.

Suppose either (i) X_m is the curve with affine equation (5.1) or (ii) $X_m = F_{m,a,b}$, p^n does not divide a or b, $(p,c) = 1$ and either $p > 3$ or $p = 3$ and $(3,abc) = 1$. Because of the similarities between these two cases we will describe what we need to know about the stable models of these curves simultaneously.

Let ${}_m$ denote the stable model for X_m over $\overline{\mathbf{Q}}_p$. Then by [CM Theorem 3.4], $\tilde{}_m$ may be defined over $\mathbf{F}_p$, $\tilde{}_m = A \cup B$, where A and B are each closed curves contained in $\tilde{}_m$ such that $A \cap B$ is finite and the natural morphism from $\tilde{}_m$ to $\tilde{}_{m/p}$ collapses each component of B to a smooth point. The curve A is isomorphic to $\tilde{}_{m/p}$ in such a way that the map from A to it induced by the natural map from X_m to $X_{m/p}$ is the absolute Frobenius morphism. Moreover B consists of the m/p disjoint irreducible components each isomorphic to the Artin-Schreier curve $z^p - z = v^q$ (where q = d in case (i) and 2 in case (ii)) and each is attached to A at its "point at infinity." We will call the components of B the new components. We will now describe the affinoid subdomains of X_m corresponding to new components (i.e., if Y is an irreducible component of $\tilde{}_m$, the set of points of X_m which reduce to a point in the interior of Y is an affinoid subdomain).

Recall, $\pi^{p-1} = -p$. For a positive integer k let μ_k denote the group of k-th roots of unity in $\overline{\mathbf{Q}}_p$.

<u>Case</u> (i). Fix a τ in $\overline{\mathbf{Q}}_p$ such that $\tau^d = -p^n\pi$ and suppose $\gamma \in \mu_m$. Define v and z_γ by the equations $t = \tau v$ and $x = \gamma(1-\pi d^{-1}z_\gamma)$. Then

$$(p^n\pi)^{-1}(x^m - (1-t^d)) \equiv z_\gamma^p - z_\gamma - v^d \mod \pi$$

and for each $\gamma \in \mu_m$, $X_\gamma = X_{m,\gamma} =: \{(x,y) \in X : |y| \le |p^n\pi|^{1/d}$ and $|x-\gamma| \le |\pi|\}$ is an affinoid subdomain of X_m which corresponds to a new component B_γ with affine equation $z_\gamma^p - z_\gamma = v^2$. Moreover, $B_\gamma = B_{\gamma'}$ iff $\gamma'/\gamma \in \mu_p$ (these are the m/p new components). In this case, let $\Delta = \mu_m/\mu_p$ and for $\delta \in \Delta$, set $X_\delta = X_\gamma$ and $B_\delta = B_\gamma$ for any representative $\gamma \in \mu_m$ of δ. In this case, $z_{\zeta\gamma} = d(1-\zeta^{-1})/\pi + z_\gamma$ and $d(1-\zeta^{-1})/\pi$ is an integer when ζ is a p-th root of unity.

Case (ii). Let $C(a,b) = a^a b^b/c^c$ and $\Gamma = \{\gamma \in \bar{\mathbb{Q}}_p : \gamma^m = C(a,b)\}$ and let $e = v(ab)$. Set $\rho = \rho(a,b) =: -c^3/2ab$. Fix a τ in $\bar{\mathbb{Q}}_p$ such that $\tau^2 = -p^{n+e}\pi$. Define v by $u = a/c + \tau v$ and z_γ for $\gamma \in \Gamma$ by $w = \gamma(1+\pi p^e \rho d^{-1} z_\gamma)$. Then

$$(-p^{n+e}\pi\rho C(a,b))^{-1}(w^m - u^a(1-u)^b) \equiv z_\gamma^p - z_\gamma - v^2 \mod \pi .$$

In this case, the affinoid subdomain

$$X_\gamma = X_{m,\gamma} =: \{(u,v) \in F_{a,b} : |u-a/c| \le |p^{n+e}\pi|^{1/2},\ |w-\gamma| \le |C(a,b)|^{1/m}|\pi|\},$$

$\gamma \in \Gamma$ of $F_{m,a,b}$ corresponds to a new component B_γ with affine equation $z_\gamma^p - z_\gamma = v^2$; $B_\gamma = B_{\gamma'}$ iff $\gamma'/\gamma \in \mu_p$. In this case , let $\Delta = \Gamma/\mu_p$ and for $\delta \in \Delta$, set $X_\delta = X_\gamma$ and $B_\delta = B_\gamma$ for any representative $\gamma \in \Gamma$ of δ. In this case, $z_{\zeta\gamma} = d(1-\zeta^{-1})/p^e\rho\pi + z_\gamma$ and $d(1-\zeta^{-1})/p^e\rho\pi$ is an integer when ζ is a p-th root of unity.

By functoriality (see (1.3)) the elements of σ of W of positive degree act on B. By [CM; Theorem 4.3] this action is described by

$$\text{(5.2)} \qquad \Phi(\sigma)^* : \quad \begin{aligned} z_\gamma\sigma &\mapsto \pi^{\sigma-1}(z_\gamma)^{p^{\deg(\sigma)}} \\ v &\mapsto \tau^{\sigma-1} v^{p^{\deg(\sigma)}} . \end{aligned}$$

In particular, $\Phi(\sigma)$ takes B_δ to $B_{\delta\sigma}$.

Suppose $\zeta \in \mu_m$. Let α_ζ denote the automorphism of X such that $\alpha_\zeta(x,w) = (\zeta x,w)$, in case (i), and the automorphism such that $\alpha_\zeta(u,w) = (u,\zeta w)$, in case (ii), then

$$\text{(5.3)} \qquad \alpha_\zeta^* : \quad \begin{aligned} z_{\zeta\gamma} &\mapsto z_\gamma \\ v &\mapsto v. \end{aligned}$$

The following proposition will also follow from our explicit computations, nevertheless we include a conceptual proof by way of motivation.

Proposition 5.4. Suppose $h = (r,s) \in I$ and $\omega_{r,s}$ is the pullback of a differential η on X. Suppose $\delta \in \Delta$. Then the restriction of η to X_δ is exact iff $(m/p)w = 0$.

Note: The hypotheses imply $s_p = 0$, in case (i), and $mh \equiv k(a,b)$ mod m, for integer k, in case (ii).

Proof. First, $(m/p)h = 0$ iff η is the pullback of a differential η' on $X_{m/p}$. Let P denote the smooth point on $\tilde{\ }_{m/p}$ to which B_δ maps.

Suppose $(m/p)h = 0$. Then as η' is regular on the residue class above P and hence exact on this residue class since every differential on a disk is exact it follows that η is exact on X_δ.

Now suppose that η is exact on X_δ. Let V_δ denote the wide open above B_δ. It follows that η is exact on V_δ. It follows from (5.3) that $\alpha_\zeta(V_\delta) = V_{\zeta\delta}$. Since η is an eigenvector for the action of $\{\alpha_\zeta : \zeta \in \mu_m\}$, we see that η is exact on $\cup_{\delta\in\Delta} V_\delta = V$ the wide open above B.

Let W denote the wide open space in X_m above A. Since the components of B are each attached at only one point to A, it follows that $H^1_{DR}(X_m,\bar{\mathbb{Q}}_p)$ is isomorphic to $H^1_{DR}(W,\bar{\mathbb{Q}}_p) \oplus H^1_{DR}(V,\bar{\mathbb{Q}}_p)$ via the natural map (see [C-RLC §IV]). Next, the statement that the map from A to $\tilde{\ }_{m/p}$ is finite, radicial and surjective implies that the composition

$$H^1_{DR}(X_{m/p},\bar{\mathbb{Q}}_p) \to H^1_{DR}(X_m,\bar{\mathbb{Q}}_p) \to H^1_{DR}(W,\bar{\mathbb{Q}}_p)$$

is an isomorphism. This implies that the class of η in $H^1_{DR}(X_m,\bar{\mathbb{Q}}_p)$ must lie in the image of $H^1_{DR}(X_{m/p},\bar{\mathbb{Q}}_p)$ and so $(m/p)w = 0$. □

Corollary 5.5. Let notation be as in the proposition. Suppose $\delta \in \Delta$ and $(m/p)h \neq 0$. Let U be a residue class of X_δ and F a rigid analytic function on U such that $dF = \eta$. Then F is unbounded on U.

Proof. By [K], if ω is an overconvergent differential on X_δ, then it is the exterior derivative of a bounded function on U iff its de Rham cohomology class lies in the unit root subspace. Hence the corollary follows immediately from the proposition and the fact that B_δ is an Artin-Schreier curve (as the unit root subspace of the first crystalline cohomology group of such a curve is zero). □

VI. A Computation.

In this section we will prove Proposition 3.10 when $q \in I_1$ and deduce from this a formula for $\beta_\sigma(r,s)$ when $r = s$. Let notation be as in §V, case (i). In particular, $\tau^d = -p^n\pi$. In this section, if $j \in \mathbb{Z}$, we let $(-p^n\pi)^{j/d}$ denote τ^j.

Suppose $q = (r,s) \in I$, $mr = ds = 0$, $m\langle r\rangle = i$ and $d\langle s\rangle = j$. Let $\eta_{r,s}$ denote the differential $x^{i-m}v^j dv/v$ on X_m. Using (1.2), we see that the pullback of $\eta_{r,s}$ to F_m via $(x,y) \to (x,y^{p^n})$ is $-mp^n(-p^n\pi)^{-\langle s\rangle}\omega_{i,j}$.

We may expand $\eta_{r,s}$ at the point $w = 0$, $x = 1$ (which is a point on the affinoid subdomain $X_{m,1}$) in v (which is a parameter on the residue class of this point on $X_{m,1}$). We find

$$\eta_{r,s} = \sum_{\ell=0}^{\infty} (p^n\pi)^\ell \binom{\langle r\rangle-1}{\ell} v^{d\ell+j} dv/v := \sum_{k=0}^{\infty} a_{r,s}(k) v^k dv/v$$

Suppose that $(m/p)q \neq 0$, which means p does not divide i. Now $\Phi^*(\sigma)\eta_{r,s} = \alpha\eta_{\sigma r,\sigma s}$ in cohomology for some α in $\bar{\mathbb{Q}}_p$ and $\beta'_\sigma(\sigma q) = (-p^n\pi)^{\langle s\rangle-\langle\sigma s\rangle}\alpha$. It is easy to see that $|a_{r,s}(k)/k| \to \infty$ as $|k| \to 0$, $k \equiv j \bmod d$, $k > 0$. (This gives another proof of Proposition 5.4 in this case.) It then follows from Theorem 4.1 and (5.2) that

$$\alpha = \mathrm{Lim}_{k\equiv j(d),k\to 0}\, p(\tau^{\sigma-1})^k a_{r,s}(k)/a_{\langle\sigma r\rangle,\langle\sigma s\rangle}(pk)$$

$$= (-p^n\pi)^{\langle s\rangle(\sigma-1)}\mathrm{Lim}_{\ell\to-\langle s\rangle}\, p(\pi^{\sigma-1})^\ell (p^n\pi)^\ell \binom{\langle r\rangle-1}{\ell} / (p^n\pi)^{p\ell+t}\binom{\langle\sigma r\rangle-1}{p\ell+t}$$

where $t = p\langle s\rangle - \langle ps\rangle = [p\langle s\rangle]$ which is less than p. If $\ell \equiv 0 \bmod (p-1)$, the ℓ-th term of this limit equals

$$p\pi^{-t}(-1)^\ell((p\ell+t)!/p^n\ell!)\prod_{k=1}^{\ell}(i/d-p^nk)/\prod_{k=1}^{p\ell+t}(i'/d-p^nk)$$

$$= -p\pi^{-t}\Gamma_p(1+p\ell+t)\prod_{k=1}^{\ell}\{-\langle r\rangle+p^nk\}/\prod_{k=1}^{p\ell+t}\{-\langle\sigma r\rangle+p^nk\}$$

Taking limits in R, we have

$$\mathrm{Lim}_{\ell\to(0,-\langle s\rangle)}\prod_{k=1}^{\ell}\{-\langle r\rangle+k\} = \mathrm{Lim}_{\ell\to(0,-\langle s\rangle)}\mathrm{Lim}_{h\to((-r)_p,\langle -r\rangle)}\prod_{k=1}^{\ell}\{h-1+k\}$$

where $\ell \in \mathbf{Z}_{>0}$ and $h \in \mathbf{Z}[1/p]_{>0}$. As

$$\prod_{k=1}^{\ell}\{h-1+k\} = \prod_{g=0}^{[h+\ell]-1}\{\langle h\rangle+g\}/\prod_{g=0}^{[h]-1}\{\langle h\rangle+g\}$$

$$= \hat{\Gamma}_p(h+\ell)/\hat{\Gamma}_p(h) = \Gamma_p(h+\ell)/\Gamma_p(h)$$

for h close to $(\langle -r\rangle_p,\langle -r\rangle)$ by (2.4), $\mathrm{Lim}_{\ell\to(0,-\langle s\rangle)}\prod_{k=1}^{\ell}\{-\langle r\rangle+k\} =$

$$\mathrm{Lim}_{\ell\to-\langle s\rangle}\mathrm{Lim}_{h\to1-\langle r\rangle}\Gamma_p(h+\ell)/\Gamma_p(h) = \Gamma_p(1-\langle r\rangle-\langle s\rangle)/\Gamma_p(1-\langle r\rangle)$$

$$= (-1)^f\Gamma_p\langle r\rangle/\Gamma_p(\langle r\rangle+\langle s\rangle)$$

where f equals $\langle s\rangle$ when $p = 2$ and 0 otherwise, since $\langle r\rangle_p = \langle r+s\rangle_p$. Similarly, $\mathrm{Lim}_{\ell\to(0,-\langle s\rangle)}\prod_{k=1}^{p\ell+t}\{-\langle\sigma r\rangle+k\} =$

$$\mathrm{Lim}_{\ell\to(0,-\langle s\rangle)}\mathrm{Lim}_{h\to(\langle-\sigma r\rangle_p,\langle-\sigma r\rangle)}\hat{\Gamma}_p(h+p\ell+t)/\hat{\Gamma}_p(h)$$

$$= (-1)^{f'}\omega(-\langle\sigma r\rangle)^t\Gamma_p(\langle\sigma r\rangle)/\Gamma_p(\langle\sigma r\rangle+\langle\sigma s\rangle),$$

by (2.4) where f' equals $\langle\sigma s\rangle$ when $p = 2$ and 0 otherwise. Noting that $\sigma s = ps$, $[p\langle s\rangle] = p\langle s\rangle-\langle ps\rangle$, $\langle-\langle ps\rangle/p\rangle_p = [p\langle s\rangle]/p$ and $[\langle 2s\rangle/2]_2 \equiv \langle s\rangle-\langle 2s\rangle \bmod 2$ when $p = 2$, we see that

$$-(-1)^{f-f'}(-1)^{\ell\langle\sigma s\rangle}(\pi\omega(-\langle\sigma r\rangle))^{-[p\langle s\rangle]} = \pi^{-[p\langle s\rangle]}\omega\langle\sigma r\rangle^{\langle\sigma s\rangle-\langle s\rangle\sigma}$$

and so $\beta'_\sigma(\sigma q)$ equals

$$p\pi^{\langle s\rangle(\sigma-p)}(-\kappa\langle\sigma r\rangle)^{\langle\sigma s\rangle-\langle s\rangle\sigma}G_\sigma(\sigma r)G_\sigma(\sigma s)\Gamma_p(\langle\sigma r\rangle+\langle\sigma s\rangle)\Gamma_p(\langle r\rangle+\langle s\rangle)^{-1}. \tag{6.1}$$

Hence using (2.11) we conclude that $\beta_\sigma(q)$ equals $G_\sigma(r)G_\sigma(s)/G_\sigma(r+s)$ times

$$(-1)^{\mu(\sigma,s)}\omega(\sigma)^{-\epsilon(\sigma^{-1}q)}p\pi^{\langle\sigma^{-1}s\rangle(\sigma-p)}\kappa\langle r\rangle^{\mu(\sigma,r)-\mu(\sigma,t)}.$$

This combined with Lemmas 1.6 and 3.6 implies Proposition 3.10, in the case $q \in I_1$. In particular,

Proposition 6.2. Suppose $q = (r,s) \in I_1$, $t = r+s$ and $s_p = 0$. Then $v(\beta_\sigma(\sigma q)) =$

$$\deg(\sigma)\langle -s\rangle + v(r)(\langle r\rangle-\langle\sigma r\rangle-(\langle t\rangle-\langle\sigma t\rangle)).$$

Observe now that the quotient of F_m, $Y = F_{m,1,1}$: $w^m = u(1-u)$ is isomorphic to the curve $X : x^m+y^2 = 1$, via $f^*(x,y) = (4^{1/m}w, 1-2u)$ (where $4^{1/m}$ will be a fixed m-th root of 4 for this section and by $4^{x/m}$ we will mean $(4^{1/m})^x$.) Hence, we can determine the Frobenius matrix of Y from that of X.

More precisely, if $\omega_i = x^{i-m}dy$ ("=" $-m^{-1}\omega_{2m,2i,m}$), $v_i = w^{i-m}du$ ("=" $m^{-1}\omega_{m,i,i}$), then $f^*\omega_i = -2^{-1}\cdot 4^{\langle r\rangle}v_i$. Let σ be an element W of degree one. Let $r = i/m$ and $s = 1/2$. Then in cohomology

$$\Phi^*(\sigma)f^{\sigma*}\omega_i = f^*\Phi^*(\sigma)\omega_i = f^*\beta'_\sigma(\sigma(r,s))\omega_{i'} = -2^{-1}\cdot 4^{\langle\sigma r\rangle}\beta'_\sigma(\sigma(r,s))v_{i'}$$

where $i' = m\cdot\langle\sigma r\rangle$ and using (1.3). Hence, if $q = (r,r)$,

$$\beta'_\sigma(\sigma q) = 4^{-\mu(\sigma,\sigma r)}\beta'_\sigma(\sigma(r,s))$$

By (6.1), $\beta'_\sigma(\sigma(r,s))$ equals $G_\sigma(\sigma r)G_\sigma(1/2)(\Gamma_p(\langle\sigma r\rangle+1/2)/\Gamma_p(\langle r\rangle+1/2))$ times $p\pi^{(\sigma-p)/2}(-\langle\sigma r_p\rangle)^{(1-\sigma)/2}$ and $\beta_\sigma(q) = (L(q)/L(\sigma q))\beta'_\sigma(q)$. Now $G_\sigma(1/2) = (-1)^{(p+1)/2}\Gamma_p(1/2)$, $G_\sigma(r)(\Gamma_p(\langle\sigma r\rangle+1/2)/\Gamma_p(\langle r\rangle+1/2) =$

$$(G_\sigma(r))^2\cdot\Gamma_p\langle\sigma r\rangle\Gamma_p(\langle\sigma r\rangle+1/2)\cdot(\Gamma_p\langle r\rangle\Gamma_p(\langle r\rangle+1/2))^{-1}$$

and by (2.10), $\Gamma_p\langle\sigma r\rangle\Gamma_p(\langle\sigma r\rangle+1/2)/\Gamma_p\langle r\rangle\Gamma_p(\langle r\rangle+1/2) =$

$$(2*)^{[2\langle r\rangle]_p-[2\langle\sigma r\rangle]_p}\cdot\Gamma_p(\langle\sigma r\rangle_p+(-1)^{\langle 2\sigma r\rangle}p/2)\Gamma_p(2\langle\sigma r\rangle)\cdot(\Gamma_p(\langle r\rangle_p+(-1)^{\langle 2r\rangle}p/2))\Gamma_p(2\langle r\rangle))^{-1}.$$

Using (2.11), we see that $(L(\sigma^{-1}q)/L(q))(\Gamma_p(2\langle r\rangle)/\Gamma_p(2\langle\sigma^{-1}r\rangle)) = \omega(\sigma)^{-\varepsilon(\sigma^{-1}q)}\kappa\langle 2r\rangle^{\varepsilon(\sigma^{-1}q)-\varepsilon(q)}$. Moreover, $(2*)^{[2\langle\sigma^{-1}r\rangle]_p-[2\langle r\rangle]_p} = (4^{\mu(\sigma,r)}2^{-\mu(\sigma,2r_p)})*$, and

$$\omega(4^{-\mu(\sigma,r)})\kappa\langle 2r\rangle^{\varepsilon(\sigma^{-1}q)-\varepsilon(q)} = \kappa(\langle r\rangle/\langle 2r\rangle)^{2\mu(\sigma,r)}\kappa\langle 2r\rangle^{\varepsilon(\sigma^{-1}q)-\varepsilon(q)} = \kappa\langle r\rangle^{2\mu(\sigma,r)}/\kappa\langle 2r\rangle^{\mu(\sigma,2r)}.$$

as $\varepsilon(\sigma^{-1}q)-\varepsilon(q) = 2\mu(\sigma,r)-\mu(\sigma,2r) + \varepsilon(\sigma^{-1}q)(1-\sigma)$. Hence, we have,

Proposition 6.3. Suppose $r \in \mathbf{Q/Z}$, $r_p \neq 0$, $q = (r,r)$, $\sigma \in W$, and $\deg(\sigma) = 1$. Then $\beta_\sigma(q)$ equals the product of $G_\sigma(r)^2/G_\sigma(2r)$, $\omega(\sigma)^{-\varepsilon(\sigma^{-1}q)}\kappa\langle r\rangle^{2\mu(\sigma,r)}/\kappa\langle 2r\rangle^{\mu(\sigma,2r)}$ and

$$-\Gamma_p(1/2)p\pi^{(\sigma-p)/2}\langle r_p\rangle^{(\sigma-1)/2}A_\sigma(q_p).$$

Corollary 6.4. The conclusion of Proposition 3.14 in is true when $r = s$ and true modulo μ_∞ in general.

<u>Proof</u>. The first part follows immediately from the proposition.

Let n be a positive integer. To prove the second part we will apply Lemma 3.2 with $S = \{u \in \mathbf{Q/Z}: v\langle u\rangle = -n\}$ and $T_\sigma(r) = \kappa(r_p)^{\mu(\sigma,r)-\mu(\sigma,r_p)}G_\sigma(r)\Gamma_1(\sigma)(r_p)$ for $\sigma \in W$ and $r \in S$. Then T_σ is clearly a cocycle on W with values in $M(S,\overline{Q}_p/\mu_\infty)$. The first part of the corollary combined with (3.1) implies that T_σ fulfils the other requirements of Lemma 3.2. It follows from that lemma that $T_\sigma(r)T_\sigma(s)/T_\sigma(t) = \beta_\sigma(r,s) \bmod \mu_\infty$ for $(r,s) \in I(S)$, $t = r+s$. Applying (3.1) again we deduce the corollary. □

Also from Lemma 3.2 we deduce,

Corollary 6.5. Suppose $\sigma \in W$, $r \in \mathbf{Q/Z}$ and $r_p \neq 0$. Then,

$$\Gamma_1(\sigma)(r) = p^{-\deg(\sigma)/2} p^{\langle\sigma^{-1}r\rangle - \langle r\rangle} D_\sigma(r_p) G_\sigma(r) \mod \mu_\infty.$$

VII. A Congruence.

In this section we will complete the proofs of Proposition 3.14 and Theorem 3.13.

Let p be an odd prime and suppose $q = (r,s) \in I_2$, $t = r+s$ and $v(s) = v(t)$. Let $(a,b) = m(\langle r\rangle-1,\langle s\rangle-1)$, $(a',b') = m(\langle \sigma r\rangle-1,\langle \sigma s\rangle-1)$, $c = a+b$, $c' = a'+b'$. Then $(p,bc) = 1$. Let $e = v(a)$. We will use the notation, where appropriate, of case (ii) §V. In particular, τ will be a square root of $-p^{n+e}\pi$ and γ an m-th root of $C(a,b)$ (γ "=" $C(\langle r\rangle-1,\langle s\rangle-1)$). Then

$$C(a',b') = C(a,b)^f (a^i b^j/c^k)^m (1+mi/fa)^{a'} (1+mj/fb)^{b'} /(1+mk/fc)^{c'}$$

where f, i, j and k are integers such that $(a',b') = f(a,b) + m(i,j)$ and $k = i+j$. As $(1+mi/fa)^{a'} \equiv 1+mi \mod (m^2/a)\mathbf{Z}_p$ etc., it follows that there exists a unique m-th root γ' of $C(a',b')$ in $\gamma^f(a^i b^j/c^k)(1+p^{n-e}\mathbf{Z}_p)$ and γ' is independent of the choices of f, i and j. We note that the functions u and $w' = w^f u^i (1-u)^j$ are also co-ordinates on $F_{m,a,b} = F_{m,a',b'}$. If Q is the point $(a/c,\gamma^\sigma)$ in (u,w)-coordinates and P is the point $(a'/b',\gamma')$ in (u, w')-coordinates, then Q lies on $X_{\gamma\sigma}$, P lies on $X_{\gamma'}$ and $\Phi(\sigma)(\tilde{P}) = \tilde{Q}$. Define v and v' by $u = a/c + \tau v = a'/c' + \tau v'$. Then $\omega_{a+m,b+m} = w du = -\tau w dv$ and $\omega_{a'+m,b'+m} = -\tau w' dv'$. Thus we may expand $\omega_{a+m,b+m}$ at Q in v (which is a parameter on the residue class $\tilde{Q}$),

$$\omega_{a+m,b+m} = -\tau\gamma^\sigma \sum_{\ell=0}^{\infty} A(\ell+1) v^\ell \, dv$$

where

$$\sum_{\ell=0}^{\infty} A(\ell+1) v^\ell = \left((1-(c\tau/a)v)^a (1+(c\tau/b)v)^b\right)^{1/m}.$$

Similarly, we may expand $\omega_{a'+m,b'+m}$ at P in v',

$$\omega_{a'+m,b'+m} = -\tau\gamma' \sum_{\ell=0}^{\infty} A'(\ell+1) v'^\ell \, dv'.$$

In fact,

$$A(\ell+1) = (c\tau)^\ell \sum_{u+v=\ell} (-1)^u a^{-u} b^{-v} \binom{a/m}{u}\binom{b/m}{v}$$

with a similar formula for $A'(\ell+1)$. The above formula for $A(\ell+1)$ was applied by David Cruz-Uribe, using Maxsyma, to obtain some computational evidence supporting Proposition 3.8 (before it was proved). However, it is difficult to use this formula to get congruences. Therefore, define the coefficients B(h) by the equation:

$$(1-(c\tau/a)v)^a (1+(c\tau/b)v)^b = 1+\sum_{h=1}^{\infty} B(h)\tau^h v^h.$$

Then it follows that $B(1) = 0$ and $B(2) = -c^3/2ab = \rho(a,b)$ and

$$\text{(7.1)} \qquad v(B(h)) \geq \mathrm{Min}\{v\tbinom{a}{i}-ie : 0\leq i\leq h\} \geq -he + (e - \mathrm{Max}(e,v(h))).$$

In particular, $v(B(h)\tau^h) \geq n + 1/(p-1)$ since $p > 3$ or $p = 3$ and $e = 0$. Moreover,

$$\left((1-(c\tau/a)v)^a(1+(c\tau/b)v)^b\right)^{1/m} = \sum_{j=0}^{\infty}\binom{1/m}{j}\Big(\sum_{h=1}^{\infty} B(h)\tau^h v^h\Big)^j.$$

In the following the expression $x \equiv y$ modulo $(1+p^u\mathbf{Z}_p)$ will mean $y \neq 0$ and $x/y \in (1+p^u\mathbf{Z}_p)$.

Lemma 7.2. Suppose $\deg(\sigma) = 1$ and u is a positive integer. Then

$$A(p^u) \equiv \binom{1/m}{(p^u-1)/2}(-p^{n+e}\pi\rho(a,b))^{(p^u-1)/2} \text{ modulo } (1+p^{n-e}\mathbf{Z}_p) \tag{7.3}$$

and $v(A(p^u)) = u/2$. Moreover the same is true for $A'(p^u)$ with (a,b) replaced by (a',b').

<u>Proof</u>. Fix a positive integer u. Now $A(p^u) = \sum_{j=1}^{\infty} \alpha_j$ where

$$\alpha_j = \binom{1/m}{j}\sum\prod_{h=2}^{\infty}(B(h)\tau^h)^{i_h}$$

and the sum runs over sequences of positive integers $(i_2,i_3,\ldots)$ such that $\sum i_h = j$ and $\sum h i_h = p^u-1$.

For a positive integer x, let S(x) denote the sum of its p-adic digits. Let $K = (p^u-1)/2$. Now $\alpha_j = 0$ for $j > K$. Moreover, $\alpha_{(p^u-1)/2}$ equals the right hand side of the congruence (7.3) which has valuation u/2, since $S(K) = u(p-1)/2$. Hence all we have to do is show $v(\alpha_j) > (n-e) + u/2$ for $j < K$. Suppose $j < K$ and let $i = K - j$. Then $v(\tau^{p^u-1}\binom{1/m}{j}) =$

$$Kv(\tau) - jn - v(j!) = (n+e+1/(p-1))K - jn - (j-S(j))/(p-1)$$

$$= in + eK + S(j)/(p-1) + i/(p-1).$$

Now $S(x-i) > S(x) - i$. Therefore,

$$v(\tau^{p^u-1}\binom{1/m}{j}) \geq in + eK + u/2.$$

When $p = 3$, $e = 0$ and so $v(\alpha_j) \geq v(\tau^{p^u-1}\binom{1/m}{j})$ which concludes the proof in this case. Assume now that $p > 3$. In general we have,

$$v(\alpha_j) = v(\tau^{p^u-1}\binom{1/m}{j}) + \sum v(B(h))i_h$$

and

$$\sum v(B(h))i_h \geq \sum((1-h)e - v(h))i_h = (j-K)e - \sum v(h)i_h$$

(using (7.1) so $v(\alpha_j) \geq i(n-e) + u/2 - \sum v(h)i_h$. Since $n > e$, it is enough to show $i > \sum v(h)i_h$ or what is the same that $\sum(v(h)+1)i_h < K/2$. But $\sum(v(h)+1)i_h = \sum(t+1)D_t$ where $D_t = \sum_{v(h)=t} i_h$. Now since $\sum hi_h = 2K$ and $h \geq \mathrm{Max}(2,p^{v(h)})$, we see that $2D_0 + pD_1 + \ldots + p^tD_t + \ldots \leq 2K$. Thus,

$$\sum(t+1)D_t \leq K - \sum_{t\geq 1}(p^t/2 - (t+1))D_t < K,$$

since $p \geq 5$ unless $D_t = 0$ for $t > 0$. But in this case, $\sum(v(h)+1)i_h = D_0 = j < K$. This concludes the proof. □

As this lemma implies $|A_{a,b}(p^u)/p^u| \to \infty$ as $u \to \infty$, we may use Theorem 4.1, (5.2) and the fact that $\tilde{\nabla} = \tilde{\nabla}'$ to deduce:

Proposition 7.4. Suppose $\sigma \in W$, and $\deg(\sigma) = 1$.

$$\beta'_\sigma(\sigma h) = (\gamma^\sigma/\gamma) \operatorname{Lim} p(\tau^{\sigma-1})^{p^u} A(p^u)/A'(p^{u+1}).$$

Remark. That a limit formula of this shape must exist follows from Proposition 5.5. Also one can predict that the formula must involve $A(k)$ for k odd using Proposition 4.2, since the hyperelliptic involution of $\tilde{X}_\gamma$ fixes $\tilde{P}$ and takes $\tilde{\nabla}$ to $-\tilde{\nabla}$.

Proposition 7.5. Suppose $q = (r,s) \in I_2$, $t = r+s$ and $\sigma \in W$ such that $\deg(\sigma) = 1$. Then $\beta_\sigma(q)$ is congruent to $B\omega(\sigma)^{-\varepsilon(\sigma^{-1}q)}\Gamma_\sigma(r)\Gamma_\sigma(s)/\Gamma_\sigma(t)$ modulo $(1+p^{n-e}\mathbf{Z}_p)$, where

$$B = (\chi(\sigma)^{1/2-\varepsilon(\sigma^{-1}q_p)})*(\{\langle -r_p\rangle\}^{\mu(\sigma,r_p)}\{\langle -s_p\rangle\}^{\mu(\sigma,s_p)}/\{\langle -r_p\rangle+\langle -s_p\rangle\}^{\mu(\sigma,t_p)})*.$$

Proof. Using Lemma 1.6 we may suppose that $v(s) = v(t)$. Let notation be as above. First,

$$p(-p^n\pi)^{(p^u-p^{u+1})/2}\binom{1/m}{(p^u-1)/2}\binom{1/m}{(p^{u+1}-1)/2}^{-1}$$

$$= (-d)^{p^u(p-1)/2}p((p^{u+1}-1)/2)!/\pi^{p^u(p-1)/2}((p^u-1)/2)!)\cdot(1+mx)$$

for some m-adic integer x and $((p^{u+1}-1)/2)!/\pi^{p^u(p-1)/2}((p^u-1)/2)!) =$

$$\pi^{-(p-1)/2}((p^{u+1}-1)/2)!/\pi^{(p-1)(p^u-1)/2}((p^u-1)/2)!) = \pi^{(p-1)/2}((p^{u+1}-1)/2)!/(-p)^{(p^u-1)/2}((p^u-1)/2)!)$$

$$= -(-\pi)^{-(p-1)/2}\Gamma_p(1/2+p^{u+1}/2).$$

Since $(a,b,c) \equiv f(a',b',c') \bmod p^n$, $\rho(a',b')/\rho(a,b) \in f(1+p^{n-e}\mathbf{Z}_p)$. Hence setting $\rho = p^e\rho(a,b)$ and using Lemma 7.2 we observe that modulo $p^{n-e+1/2}$

$$pA(p^u)/A'(p^{u+1}) \equiv -p\pi^{-(p-1)/2}(d/\rho)^{p^u(p-1)/2}f^{(1-p^{u+1})/2}\Gamma_p(1/2+p^{u+1}/2).$$

Now $(\tau^{\sigma-1})^{p^u}f^{(1-p^{u+1})/2} \equiv (-\omega(\sigma)p^{n+e}\pi)^{(\sigma-1)/2}(\chi(\sigma)^{1/2})*$ modulo $(1+p^n\mathbf{Z}_p)$ and $(2\omega(\sigma)abc^{-3}m)^{(\sigma-1)/2} = (2\omega(\sigma)\langle r\rangle\langle s\rangle/\langle t\rangle)^{(\sigma-1)/2}$. Hence, if $u \geq n-e$, $(\tau^{\sigma-1})^{p^u}pA(p^u)/A'(p^{u+1})$ is congruent to

$$-p\pi^{(\sigma-p)/2}(2\omega(\sigma)\langle r\rangle\langle s\rangle/\langle t\rangle)^{(\sigma-1)/2}\Gamma_p(1/2)(\chi(\sigma)^{1/2})*$$

modulo $(1+p^{n-e}\mathbf{Z}_p)$.

On the other hand: Let $\bar{a} = p^{-e}a$, $\bar{a}' = p^{-e}a'$, $A = p^{-ea}C(a,b)$, $A' = p^{-ea'}C(a',b')$. Fix an m-th root λ of A and let λ' be the m-th root of A' in $\lambda^f(a^ib^j/c^k)(1+p^{n-e}\mathbf{Z}_p)$. of p. Then $\gamma^\sigma/\gamma = p^{e\mu(\sigma,\sigma r)}\lambda^\sigma/\lambda'$ and

$$\lambda^\sigma/\lambda' \in \lambda^{\sigma-f}(a^ib^j/c^k)^{-1}(1+p^{n-e}\mathbf{Z}_p) = B(1+p^{n-e}\mathbf{Z}_p)$$

where $B = \{\langle r\rangle-1\}^{\mu(\sigma,\sigma r)}\{\langle s\rangle-1\}^{\mu(\sigma,\sigma s)}/\{\langle r\rangle+\langle s\rangle-2\}^{\mu(\sigma,\sigma r)+\mu(\sigma,\sigma s)}$. Let $\delta = \varepsilon(q)-\varepsilon(\sigma q)$. Now, as $\mu(\sigma,\sigma r)+\mu(\sigma,\sigma s)-\mu(\sigma,\sigma t) = \varepsilon(q)\sigma-\varepsilon(\sigma q)$,

$$\omega(B) = \omega\langle t\rangle^{-\delta}\omega(\langle r\rangle^{\mu(\sigma,\sigma r)}\langle s\rangle^{\mu(\sigma,\sigma s)}/\langle t\rangle^{\mu(\sigma,\sigma t)})$$

and, $p^{e\mu(\sigma,\sigma r)} = p^{-v\langle t\rangle\delta}p^{v\langle r\rangle\mu(\sigma,\sigma r)}p^{v\langle s\rangle\mu(\sigma,\sigma s)}/p^{\langle t\rangle\mu(\sigma,\sigma t)}$. Also,

$$B^* = (\{\langle -r\rangle\}^{[\langle r\rangle]_p-[\langle\sigma r\rangle]_p}\{\langle -s\rangle\}^{[\langle s\rangle]_p-[\langle\sigma s\rangle]_p}/\{\langle -r\rangle+\langle -s\rangle\}^{[\langle t\rangle]_p-[\langle\sigma t\rangle]_p+\delta})^*$$

$$\cdot(\{\langle -r\rangle\}^{\mu(\sigma,\sigma r_p)}\{\langle -s\rangle\}^{\mu(\sigma,\sigma s_p)}/\{\langle -r\rangle+\langle -s\rangle\}^{\mu(\sigma,\sigma t_p)})^*.$$

Moreover,

$$(\{\langle -r\rangle\}^{[\langle r\rangle]_p-[\langle\sigma r\rangle]_p}\{\langle -s\rangle\}^{[\langle s\rangle]_p-[\langle\sigma s\rangle]_p}/\{\langle -r\rangle+\langle -s\rangle\}^{[\langle t\rangle]_p-[\langle\sigma t\rangle]_p+\delta})^* \equiv$$

$$(\{\langle t\rangle\}^{-\delta})^*(\{\langle r\rangle\}^{[\langle r\rangle]_p-[\langle\sigma r\rangle]_p}\{\langle s\rangle\}^{[\langle s\rangle]_p-[\langle\sigma s\rangle]_p}/\{\langle t\rangle\}^{[\langle t\rangle]_p-[\langle\sigma t\rangle]_p})^*$$

and

$$(\{\langle -r\rangle\}^{\mu(\sigma,\sigma r_p)}\{\langle -s\rangle\}^{\mu(\sigma,\sigma s_p)}/\{\langle -r\rangle+\langle -s\rangle\}^{\mu(\sigma,\sigma t_p)})^* \equiv$$

$$(\{\langle -r_p\rangle\}^{\mu(\sigma,\sigma r_p)}\{\langle -s_p\rangle\}^{\mu(\sigma,\sigma s_p)}/\{\langle -r_p\rangle+\langle -s_p\rangle\}^{\mu(\sigma,\sigma t_p)})^*$$

mod $(1+p^{n-e}\mathbf{Z}_p)$. (Note, here we used $(\{\langle -u\rangle+x\}^{\mu(\sigma,\sigma u)})^* \equiv \{\langle -u\rangle\}^{\mu(\sigma,\sigma u)}((1+p^nx)^{1/p^n})^{\sigma-\chi(\sigma)}$ mod $(1+p^f\mathbf{Z}_p)$, if $x \in \mathbf{Z}_p$, $u \in \mathbf{Q}_p/\mathbf{Z}_p$ and $n \le v\langle r\rangle = f$.) As $L(q)/L(\sigma q) = p^{v\langle t\rangle\delta}\{\langle t\rangle\}^{\varepsilon(q)}\{\langle\sigma t\rangle\}^{-\varepsilon(\sigma q)} \equiv p^{v\langle t\rangle\delta}\{\langle t\rangle\}^{\delta}\chi(\sigma)^{-\varepsilon(\sigma q)}$ modulo $(1+p^n\mathbf{Z}_p)$ and

$$G_\sigma(\sigma r)G_\sigma(\sigma s)/G_\sigma(\sigma t) \equiv$$

$$\chi(\sigma)^{*\varepsilon(\sigma q_p)-\varepsilon(\sigma q)}(\{\langle r\rangle\}^{[\langle r\rangle]_p-[\langle\sigma r\rangle]_p}\{\langle s\rangle\}^{[\langle s\rangle]_p-[\langle\sigma s\rangle]_p}/\{\langle t\rangle\}^{[\langle t\rangle]_p-[\langle\sigma t\rangle]_p})^*$$

modulo $(1+p^{n-e}\mathbf{Z}_p)$ by (2.5), the proposition follows. □

Corollary 7.6. With the above hypotheses, $v(\beta_\sigma(r,s)) =$

$$\deg(\sigma)/2 + v\langle r\rangle(\langle\sigma^{-1}r\rangle-\langle r\rangle)+v\langle s\rangle(\langle\sigma^{-1}s\rangle-\langle s\rangle)-v\langle t\rangle(\langle\sigma^{-1}t\rangle-\langle t\rangle).$$

We see that the two sides of the equation in Proposition 3.14 are congruent modulo $(1+p^{n-e}\mathbf{Z}_p)$ when $q \in I_2$. This combined with Corollary 6.4 implies that they are equal.

Let $J = I_2 \cap (\mathbf{Q}_p/\mathbf{Z}_p)^2$. By Lemma 3.5 the ratio, $d_\sigma(q)$, of $\beta_\sigma(q)$ and $\omega(\sigma)^{-\varepsilon(\sigma^{-1}q)}\Gamma_\sigma(r)\Gamma_\sigma(s)/\Gamma_\sigma(t)$ extends to cocycle on W with values in M(J). Moreover, it follows from the proposition that $d_\sigma(q)$ is congruent B modulo $(1+p^{n-e}\mathbf{Z}_p)$. Since, $\beta_\sigma(q) = \Gamma_1(\sigma)(r)\Gamma_1(\sigma)(s)/\Gamma_1(\sigma)(t) \bmod \mu_\infty$ by (3.1), it follows from Corollary 6.5 that $d_\sigma(r,s) = D_\sigma(r)D_\sigma(s)/D_\sigma(t) \bmod \mu_\infty$. Finally, from Proposition 6.3, we see that $d_\sigma(r,r) = A_q(r)$ when $r \in \mathbf{Q}_p/\mathbf{Z}_p$. By taking $H_\sigma = d_\sigma$, we complete the proof of Proposition 3.12 and Theorem 3.13 in the case $q \in J$. Theorem 3.13, in general, follows from this and Proposition 3.14. □

References.

[A] Anderson, G., cyclotomy and a Covering of the Taniyama group. Compositio Math., 57 (1985), 153-217.

[BO] Berthelot, P., and A. Ogus, F-isocrystals and de Rham cohomology. I, Invent. Math. 72, 2 (1983), 159-199.

[C-GK] Coleman, R., On the Gross Koblitz Formula, Advanced Studies in Pure Math. 12 (1987), 21-52.

[C-RL] ______, Reciprocity Laws on Curves, Compositio, 72, (1989) 205-235.

[C-AI] ______, Anderson-Ihara Theory: Gauss Sums and Circular Units, Advan. in Math., 17, (1989) 55-72.

[CM] ______ and W. McCallum, Stable Reduction of Fermat Curves and Local Components of Jacobi Sum Hecke Characters, J. reine angew. Math. **385** (1988) 41-101.

[K] Katz, N., Crystalline cohomology, Dieudonné modules and Jacobi sums, in Automorphic Forms, Representation Theory and Arithmetic, (Tata Institute of Fundamental Research, Bombay, India, 1979), 165-245.

[O] Ogus, A., A p-adic Analogue of the Chowla-Selberg Formula, this volume.

[P] Prapavessi, D., On Jacobi Sum Hecke Characters Ramified only at 2, to appear in The Journal of Algebra.

Errata for [C-GK]

On page 24, Proposition 2 requires the hypothesis $r \geq |p|$.

On page 44, line 12, $I = \{(r,s) \in (1/m)\mathbf{Z}/\mathbf{Z}\times(1/m)\mathbf{Z}/\mathbf{Z},\ r \neq 0,\ s \neq 0 \text{ and } r+s \neq 0\}$.

On page 45, line -7, the term "(p-1)⟨pa/m⟩" should be replaced by "(p-1)⟨a/m⟩".

On pages 46-47, in lines 4 and 7 on page 46 and line 12 on page 47, h and k should be switched

On page 47, lines 1, 2, and 10 on page 47, the first h in the exponent should be replaced by a k; lines 4 and 6 switch h and k; in the corollary, ι should be -1 for all p; in Theorem 19, $\iota_{r,s} = (-1)^{\varepsilon(r,s)}$ for all p.

On page 48, line 10, "1+K(pr,ps)" should be replaced by "1-K(pr,ps)" (note that $\varepsilon(pr,ps)$ must be 1 in this case); lines -10, -12 and -13, the values of Γ_p on the left hand side of the equations should be replaced by their inverses.

On page 49, line 1, insert "let $x \in \mathbf{Z}_p$"; line 2, replace with "(i) $\ell(1-x) \equiv \ell(x) \bmod 2$".

Supercongruences

Matthijs J. Coster
Centre for Mathematics and Computer Science
P.O. Box 4079, 1009 AB Amsterdam, the Netherlands

Abstract. *In this report we will discuss special congruences. We explain how the congruences arise from formal groups and then we give some examples.*

Key Words & phrases: (Super-)Congruences, Formal Groups, Conjecture of Atkin and Swinnerton-Dyer.

1. Introduction.

This paper deals with so called "supercongruences". Before we will explain this term, we give some definitions which we need in the explanation. Let K be an algebraic extension of $\mathbf{Q}$. Let p be a prime which splits in K as $p = \pi\bar{\pi}$. Let $|\cdot|_p$ be the valuation on $\mathbf{Q}$ in such a way that $|\bar{\pi}|_p = 1$ and $|\pi|_p = p^{-1}$. We will consider π as an element of $\mathbf{Z}_p$. Let $\{u_n\}_{n=1}^{\infty}$ be a sequence of rational or p-adic integers. In this paper we will consider the congruences

$$u(mp^r) \equiv a \cdot u(mp^{r-1}) \bmod p^{\lambda r}, \tag{1A}$$

and

$$u(mp^r) \equiv \alpha \cdot u(mp^{r-1}) \bmod \pi^{\lambda r}, \tag{1B}$$

where λ, m and r are positive integers and a is an integer and π is an p-adic integer. Therefore (1A) is a congruence in $\mathbf{Z}$ and (1B) is a congruence in $\mathbf{Z}_p$. In Section 2 we will give an introduction in formal groups. We will show that congruences (1A) and (1B) with $\lambda = 1$ arise in a natural way from formal groups. Especially, we will give a sketch of the Conjecture of Atkin and Swinnerton-Dyer. In Section 3, 4 and 5 we give some examples of congruences (1A) and (1B) with coefficient $\lambda > 1$. In such cases we call the congruence *supercongruence*. At the moment supercongruences cannot be proved by use of formal groups. In each case a separated proof has to be given. A lot of proofs will be omitted in this paper. For these proofs we refer to [12]. In Section 6 some conjectures are given.

2. The conjecture of Atkin and Swinnerton-Dyer.

Let K be a commutative field with char$(K) = 0$ and let R be a subring. (In our case we will choose $R = \mathbf{Z}_p$). We denote by $R[[T]]$ the set of power series in the variable T with coefficients in R.

Let $F(X,Y) \in R[[X,Y]]$. We call $F(X,Y)$ a commutative formal group law if $F(X,Y)$ satisfies the following properties.

$$\begin{aligned} F(X,Y) &= X + Y + (\text{terms of degree} \geq 2), \\ F(X, F(Y,Z)) &= F(F(X,Y),Z), \\ F(X,Y) &= F(Y,X). \end{aligned} \tag{2}$$

We derive from (2) that $F(X,Y)$ satisfies moreover the following properties.

$$F(X,0) = X,$$
$$\text{there is a unique } i(T) \in R[[T]] \text{ such that } F(T, i(T)) = 0.$$

Let $\mathcal{R}_T = \{X(T) \in R[[T]] : X(0) = 0\}$. We define a formal addition $+_{\mathcal{F}}$ on $\mathcal{R}_T$ by

$$X(T) +_{\mathcal{F}} Y(T) = F(X(T), Y(T)).$$

It turns out that $\mathcal{R}_T$ with $+_{\mathcal{F}}$ is a group. This group is called a formal commutative group in one variable over R. From now on let $\mathcal{F}$ be a formal group over R (i.e. $\mathcal{F} = (\mathcal{R}_T, +_{\mathcal{F}})$).

We define the logarithm $f(T)$ of the formal group $\mathcal{F}$ by

$$\begin{aligned} f(T) &\in K[[T]], \\ f(T) &= T + (\text{terms of degree} \geq 2), \\ f(F(X,Y)) &= f(X) + f(Y). \end{aligned} \tag{3}$$

The last condition can be replaced by

$$F(X,Y) = f^{-1}(f(X) + f(Y)),$$

where $f^{-1}(T) \in K[[T]]$ is the power series such that $f^{-1}(f(T)) = T$. We find that $f(T)$ satisfies the property

$$f(T) = \sum_{n=1}^{\infty} u(n) \cdot T^n/n \text{ with } u(n) \in R. \tag{4}$$

We call

$$\omega = f(T)\, \mathrm{d}T \tag{5}$$

the differential form related to the formal group $\mathcal{F}$. We consider the formal Dirichlet series

$$L(s, \mathcal{F}) = \sum_{n=1}^{\infty} u_n/n^s \tag{6}$$

where $\mathcal{F}(T) = \sum_{n=1}^{\infty} u_n T^n/n$ is the logarithm of the formal group $\mathcal{F}$.

Two formal groups $\mathcal{F}$ (with f and $L(s, \mathcal{F})$) and $\mathcal{G}$ (with g and $L(s, \mathcal{G})$) are isomorphic over R if there is a formal group homomorphism $h: \mathcal{F} \to \mathcal{G}$ with $h(T) \in R[[T]]$ and $h(F(X,Y)) = G(h(X),h(Y))$. In our case (that char$(K) = 0$) we have

$$h(T) = g^{-1}(f(T)) \text{ and } L(s, \mathcal{F})/L(s, \mathcal{G}) = \sum_{n=1}^{\infty} \frac{v(n)}{n^s}, \text{ with } |v(n)|_p \leq |n|_p .$$

We have a theorem due to Honda which says that for each formal group $\mathcal{F}$ there exists a formal group $\mathcal{G}$ such that the Dirichlet series related to $\mathcal{G}$ has p-adic numbers as coefficients. In formula:

Theorem 1. (Honda). *Let $\mathcal{F}$ be a formal group over $\mathbf{Z}_p$. Then there exists a formal group $\mathcal{G}$, isomorphic to $\mathcal{F}$ over $\mathbf{Z}_p$ such that*

$$L(s, \mathcal{G}) = \left(1 - \sum_{j=1}^{\infty} p^{j-1-js} b(j)\right)^{-1} \tag{7}$$

where $b(j)$ are p-adic integers.

Proof. See [20, pp. 441-445] or [12, pp. 18-23].■

Corollary 2. *Let $\mathcal{F}$ be a formal group over $\mathbf{Z}_p$. Let $f(T) = \sum_{n=1}^{\infty} u(n) \cdot T^n/n$ be the related formal group. Then there exists p-adic numbers $b(j)$ such that*

$$u(mp^r) - b(1) \cdot u(mp^{r-1}) - \ldots - p^{r-1} \cdot b(r) \cdot u(m) \equiv 0 \bmod p^r, \tag{8}$$

for m,r positive integers and p is not a divisor of m.

Proof. See [20, pp. 441-445].■

We will apply Theorem 1 and Corollary 2 on formal groups related to elliptic curves. Let $\mathcal{E}$ be an elliptic curve over $\mathbf{Z}$ and let ω be a holomorphic differential form on $\mathcal{E}$. Let $\mathcal{F}_\omega$ be the formal group related to ω. Suppose that $\mathcal{F}_\omega$ is a formal group over $\mathbf{Z}$. Let

$$L_p(s) = \left(1 - a_p \cdot p^{-s} + p \cdot b_p \cdot p^{-2s}\right)^{-1}$$

be the Dirichlet series related to the Hasse-Weil zeta-function of the elliptic curve. Then a theorem of Honda Cartier and Hill says that the formal series related to the Dirichlet series is isomorphic to the formal group $\mathcal{F}_\omega$.

Corollary 3. (Conjecture of Atkin and Swinnerton-Dyer). *Let $\mathcal{F}_\omega$ be the formal group as defined above.*

(i) *We have*

$$u(mp^r) - a_p \cdot u(mp^{r-1}) + p \cdot b_p \cdot u(mp^{r-2}) \equiv 0 \bmod p^r. \tag{9}$$

(ii) *If $\mathcal{E}$ is ordinary over $\mathbf{Z}_p$, (i.e. $b_p = 1$ and $a_p \neq 0$) then we have*

$$u(mp^r) \equiv \bar{\pi} \cdot u(mp^{r-1}) \bmod p^r. \tag{10}$$

where $\bar{\pi}$ *such that* $|\bar{\pi}|_p = 1$ *and* $\bar{\pi}$ *is a root of* $X^2 - a_p X + 1 = 0$.

Proof. (*i*) Corollary 2 says that (9) must be a congruence of form (8). Then we use the theorem of Honda Cartier and Hill, which was mentioned above. This theorem says that the coefficients $b(1)$ and $b(2)$ coincide with the integers a_p and b_p of the Dirichlet series and that the other coefficients $b(n)$ equal zero.

(*ii*) Notice that

$$L_p(s) = \left(1 - \pi \cdot p^{-s}\right)^{-1} \cdot \left(1 - \bar{\pi} \cdot p^{-s}\right)^{-1}.$$

It is not difficult to see that the formal group related to $L_p(s)$ is isomorphic to the formal group which is related to the Dirichlet series $\left(1 - \bar{\pi} \cdot p^{-s}\right)^{-1}$. (See [20]).■

3. Generalized Apéry numbers.

The numbers $b(n) = \sum_{k=1}^{n} \binom{n}{k}^2 \cdot \binom{n+k}{k}$ and $d(n) = \sum_{k=1}^{n} \binom{n}{k}^2 \cdot \binom{n+k}{k}^2$ were introduced by Roger Apéry and played a role in the proof of the irrationality of $\zeta(2)$ and $\zeta(3)$ respectively. Many papers deal with congruences on these numbers. We mention Chowla, cowles and Cowles [9] and Gessel [16]. For these numbers Mimura [21] proved some congruences of the form $u_{p-1} \equiv 1 \bmod p^3$, where p is a prime, $p \geq 5$. F. Beukers [4] generalized these congruences to

$$u(mp^r - 1) \equiv u(mp^{r-1} - 1) \bmod p^{3r},$$

where m and r are any positive integers. Now we consider the so called *generalized Apéry numbers*, which are defined by

$$w_{AB\varepsilon}(n) = \sum_{k=1}^{n} \binom{n}{k}^A \cdot \binom{n+k}{k}^B \cdot \varepsilon^k \tag{11}$$

where $A, B \in \mathbf{Z}_{\geq 0}$ and $\varepsilon = \pm 1$.

We have for the generalized Apéry numbers the following theorem.

Theorem 4. *Let* $w(n)$ *be as defined above. Let* $p \geq 5$ *be a prime. Then for any* $m, r \in \mathbf{Z}_{\geq 1}$ *we have*

$$w(mp^r) \equiv w(mp^{r-1}) \bmod p^{3r} \, for \begin{cases} A \geq 2 \\ A = 1 \text{ and } B \geq 1, \varepsilon = -1 \end{cases}$$

and

$$w(mp^r - 1) \equiv w(mp^{r-1} - 1) \bmod p^{3r} \, for \begin{cases} B \geq 2 \\ B = 1 \text{ and } A \geq 1, \varepsilon = (-1)^A. \end{cases}$$

Proof. The proof is very technical. See [12, pp. 49-55].■

4. Binomial coefficients.

Since the work of Fermat it is known that every prime $p \equiv 1 \bmod 4$ can be written as $p = a^2+b^2$ for integers a and b in an essentially unique way. Without loss of generality we may assume that $a \equiv 1 \bmod 4$. Gauss proved by counting the number of solutions of the elliptic curve $\mathscr{E}$: $Y^2 = X^4+1 \bmod p$ in two, essentially different ways, that

$$\binom{\frac{p-1}{2}}{\frac{p-1}{4}} \equiv 2a \bmod p \tag{12}$$

By applying Corollary 3 on the elliptic curve $\mathscr{E}$, congruence (12) can be generalized to

$$\binom{\frac{mp^r-1}{2}}{\frac{mp^r-1}{4}} \equiv (a+bi) \cdot \binom{\frac{mp^{r-1}-1}{2}}{\frac{mp^{r-1}-1}{4}} \bmod p^r \tag{13}$$

where m,r are positive integers and $m \equiv 1 \bmod 4$. Here i denotes a p-adic integer such that $i^2 = -1$ and $bi \equiv -a \bmod p$. Beukers conjectured in [4] the congruence

$$\binom{-\frac{1}{2}}{\frac{p-1}{4}} \equiv (a+bi) \bmod p^2 \tag{14}$$

This was proved by Chowla, Dwork and Evans [10]. Van Hamme [18] generalized (14) to

$$\binom{-\frac{1}{2}}{\frac{p^r-1}{4}} \equiv (a+bi) \cdot \binom{-\frac{1}{2}}{\frac{p^{r-1}-1}{4}} \bmod p^r \tag{15}$$

for any positive integer r. This congruence can be generalized to the supercongruence

$$\binom{\frac{mp^r-1}{2}}{\frac{mp^r-1}{4}} \equiv (a+bi) \cdot \binom{\frac{mp^{r-1}-1}{2}}{\frac{mp^{r-1}-1}{4}} \bmod p^{2r} \tag{16}$$

We get another example by considering primes $p \equiv 1 \bmod 3$. Then $4p = e^2+3f^2$ for certain values e and f. Without loss of generality we may assume that $e \equiv -1 \bmod 3$. Choose the p-adic number $\bar{\pi} = (e+3f\sqrt{3})/2$ such that $|\bar{\pi}|_p = 1$. Starting from the elliptic curve $\mathscr{E}$: $Y^2 = 1-4X^3$, Corollary 3 implies the congruence

$$\binom{\frac{2}{3}(mp^r-1)}{\frac{1}{3}(mp^r-1)} \equiv \bar{\pi} \cdot \binom{\frac{2}{3}(mp^{r-1}-1)}{\frac{1}{3}(mp^{r-1}-1)} \bmod p^r \tag{17}$$

for any positive integers m,r with $m \equiv 1 \bmod 3$. However congruence (17) can be improved to the supercongruence

$$\binom{\frac{2}{3}(mp^r-1)}{\frac{1}{3}(mp^r-1)} \equiv \bar{\pi} \cdot \binom{\frac{2}{3}(mp^{r-1}-1)}{\frac{1}{3}(mp^{r-1}-1)} \bmod p^{2r} \tag{18}$$

In the general case we define for α, β positive integers with $\alpha+\beta \leq d$ the binomial coefficient

$$v(n) = \begin{cases} \binom{\frac{(\alpha+\beta)(n-1)}{d}}{\frac{\alpha(n-1)}{d}} & \text{if } n \equiv 1 \bmod d \\ 0 & \text{else.} \end{cases} \tag{19}$$

We have for these coefficients the congruence

$$v(mp^r) \equiv \bar{\pi} \cdot v(mp^{r-1}) \bmod p^r \tag{20}$$

where $\bar{\pi} = \dfrac{\Gamma_p(\frac{\alpha}{d})\Gamma_p(\frac{\beta}{d})}{\Gamma_p(\frac{\alpha+\beta}{d})}$.

This result can be found using formal group theory (namely $f(T) = \sum_{n=1}^{\infty} \frac{v(n)}{n} \cdot T^n$ is a formal logarithm over $\mathbf{Z}_p$ for $p \equiv 1 \bmod d$) or the p-adic Γ-function (cf. [22, pp. 111-114]). In the case that $d = 2, 3, 4$ or 6 we can improve congruence (20). The following theorem deals with the improvement.

Theorem 5. *Let d be 2, 3, 4 or 6. Let p be a prime with* $p \equiv 1 \bmod d$. *Let m and r be positive integers with* $m \equiv 1 \bmod d$. *Let* $\alpha, \beta \in \mathbf{Z}_{\geq 1}$ *with* $\alpha+\beta \leq d$. *Then the binominal coefficient v(n) satisfies the supercongruence*

$$v(mp^r) \equiv g(p)^{mp^{r-1}} \cdot \bar{\pi} \cdot v(mp^{r-1}) \bmod p^{2r} \tag{21}$$

where $g(p) \in \mathbf{Z}_p$ *with* $g(p) \equiv 1 \bmod p$ *and* $\bar{\pi} = \dfrac{\Gamma_p(\frac{\alpha}{d})\Gamma_p(\frac{\beta}{d})}{\Gamma_p(\frac{\alpha+\beta}{d})}$.

Proof. We prove congruence (21) using the p-adic Γ-function. The proof is based on a formula of Gross and Koblitz [17] which expresses the p-adic Γ-function in terms of Gauss sums and on a formula of Diamond [14] which expresses the logarithmic derivative in terms of the p-adic logarithm. See [11]. ∎

5. Values of the Legendre polynomials.

This section contains joined work with L. van Hamme. Nice supercongruences exist for the values of some Legendre polynomials. These polynomials can be defined by

$$P_n(t) = \sum_{k=0}^{n} \binom{n}{k} \cdot \binom{n+k}{k} \cdot \left(\frac{t-1}{2}\right)^k \tag{22}$$

and they satisfy

$$F(X) = \frac{1}{\sqrt{1-2tX+X^2}} = \sum_{n=0}^{\infty} P_n(t) X^n \tag{23}$$

Let K be an algebraic extension of $\mathbf{Q}$. Let p be a prime which splits in K as $p = \pi\bar{\pi}$. Let $t \in K$ with $|t|_p \leq 1$ and consider the differential form

$$\frac{\mathrm{d}X}{\sqrt{1-2tX^2+X^4}} = \sum_{n=0}^{\infty} P_n(t) X^n \mathrm{d}X \tag{24}$$

on the elliptic curve $\mathscr{E} : y^2 = x(x^2 + Ax + B)$. The theory of formal groups predicts a congruence of the form as described in Corollary 3

$$P_{\frac{1}{2}(mp^r-1)}(t) \equiv \bar{\pi} \cdot P_{\frac{1}{2}(mp^{r-1}-1)}(t) \bmod p^r \tag{25}$$

for any positive integer r and positive odd integer m.

It turns out that if $\mathscr{E}$ has complex multiplication, congruence (25) can be changed into a congruence mod π^{2r}. We have the following theorem.

Theorem 6. *Let $K = \mathbf{Q}(\sqrt{-d}, \sqrt{d})$ with d a square-free positive integer. Consider the elliptic curve*

$$\mathscr{E} : y^2 = x(x^2 + Ax + B) \text{ with } A, B \in K \tag{26}$$

Let $\Delta = A^2 - 4B$. Let ω and ω' be a basis of periods of $\mathscr{E}$ and suppose that $\tau = \omega'/\omega \in \mathbf{Q}(\sqrt{-d})$ (which implies that the curve has complex multiplication), τ has positive imaginary part and $A = 3\wp(\omega/2)$, $\sqrt{\Delta} = \wp(\omega/2 + \omega'/2) - \wp(\omega'/2)$, where $\wp(z)$ is the Weierstrass $\wp$-function. Let p be an odd prime which does not divide d and $p = \bar{\pi}\pi$, where π, $\bar{\pi} \in \mathbf{Q}(\sqrt{-d})$. Suppose that $\pi = u + v\tau$ and $\pi\tau = x + y\tau$ with u, v, x, y integers and v even. Then we have

$$P_{\frac{1}{2}(mp^r-1)}\left(\frac{A}{\sqrt{\Delta}}\right) \equiv \varepsilon^{mp^{r-1}} \cdot \bar{\pi} \cdot P_{\frac{1}{2}(mp^{r-1}-1)}\left(\frac{A}{\sqrt{\Delta}}\right) \bmod \pi^{2r} \tag{27}$$

where $\varepsilon = i^{y(1-x)+p-2}$. Here $i = \sqrt{-1}$.

We first give an example in which Theorem 6 can be applied. Let $\mathscr{E}: y^2 = x(x^2+3x+2)$. We can choose periods ω and ω' in such a way that $\wp(\omega/2) = 1$ and $\omega'/\omega = \tau = i$. Let $p \equiv 1 \bmod 4$ be a prime. Let i be a p-adic number such that $i^2 = -1$. Fix the sign of bi such that $a \equiv bi \bmod p$. Let $\pi = a-bi$. Then we have $\pi\tau = \pi i = b+ai$. Hence $\varepsilon = i^{\,y(1-x)+p-2} = i^{-b} = (-1)^{(p-1)/4}$. We denote $a(n) = \sum_{k=1}^{n} \binom{n}{k} \cdot \binom{n+k}{k}$. The numbers $a(n)$ have been used for proving that log 2 is irrational with measure of irrationality 4.622 [1]. Carlitz proved that the numbers $a(n)$ satisfy for $p \equiv 1 \bmod 4$ the congruence

$$a\left(\frac{p-1}{2}\right) \equiv (-1)^{\frac{1}{4}(p-1)} \cdot 2a \bmod p. \tag{28}$$

Since $a(n) = P_n(3)$, we have for those primes the supercongruence

$$a\left(\frac{mp^r-1}{2}\right) \equiv (-1)^{\frac{1}{4}(p-1)} \cdot \bar{\pi} \cdot a\left(\frac{mp^{r-1}-1}{2}\right) \bmod p^{2r}. \tag{29}$$

Another proof of this supercongruence in the case $m = r = 1$ has been given by van Hamme in [18].

Sketch of the proof of Theorem 6. Let $L = \mathbf{Q}(\sqrt{-d}, \sqrt{d})$ and $R = \{\alpha \in L : \mathrm{ord}_\pi(\alpha) \geq 0\}$. In this proof we will denote

$$c(n) = \sqrt{\Delta}^{\,n} \cdot P_n\left(\frac{A}{\sqrt{\Delta}}\right). \tag{30}$$

We consider the holomorphic differential form $\omega = -\frac{\mathrm{d}x}{2y}$. Let $t = \frac{x}{y}$ be a local parameter at infinity. We express ω in terms of t and we get

$$\omega = \frac{\mathrm{d}t}{\sqrt{1-2At^2+\Delta t^4}} = \sum_{n=0}^{\infty} c(n) \cdot t^{2n}\mathrm{d}t. \tag{31}$$

Then we define the local parameter z at infinity by

$$\mathrm{d}z = \omega \tag{32}$$

Hence z can be expressed as a function of t by

$$z = \sum_{k=0}^{\infty} \frac{c(k)}{2k+1} t^{2k+1} \tag{33}$$

and t can be expressed as a function of z by

$$t = z + \ldots = -2 \cdot \frac{\wp(z) - \wp(\frac{\omega}{2})}{\wp'(z)}. \tag{34}$$

Notice that $t(z)$ is an elliptic function. Since $\mathcal{E}$ has complex multiplication we have $\pi \in \mathrm{End}(\mathcal{E})$. More specified we have

$$\begin{aligned} t(\pi z) &= F(t(z)) \\ &= \eta t^{p}(z) \cdot \frac{1 + \pi a_2 t^{-2}(z) + \pi a_4 t^{-4}(z) + \ldots + \pi a_{p-1} t^{1-p}(z)}{1 - \pi d_2 t^{-2}(z) - \pi d_4 t^{-4}(z) + \ldots - \pi d_{p-1} t^{1-p}(z)} \end{aligned} \tag{35}$$

where $\eta, a_i, d_j \in \mathbf{R}$. This formula is due to Weber (cf. [23]). Formula (33) imply the formulas

$$\pi z = \sum_{l=0}^{\infty} \frac{c(l)}{2l+1} t^{2l+1}(\pi z) \tag{36}$$

and

$$\pi z = \sum_{k=0}^{\infty} \pi \cdot \frac{c(k)}{2k+1} t^{2k+1}. \tag{37}$$

Substitute (35) for $t(\pi z)$ in (36). Consider in equations (36) and (37) the coefficient of $t^{\,mp^r} \bmod \frac{\pi p^{2r}}{mp^r}$. We get the coefficients

$$\frac{1}{mp^{r-1}} \cdot c\left(\frac{1}{2}(mp^{r-1} - 1)\right) \cdot \eta^{mp^{r-1}} \tag{38}$$

and

$$\frac{\pi}{mp^r} \cdot c\left(\frac{1}{2}(mp^r - 1)\right) \tag{39}$$

respectively from formulas (36) and (37) respectively. This implies the congruence of the theorem. We can calculate that $\eta = i^{y(1-x)+p-2} \cdot (\sqrt{\Delta})^{\frac{1}{2}(p-1)}$. See a more detailed proof in [13].■

There are only 8 values t with these nice supercongruences over $\mathbf{Z}$ (cf. [12, pp. 87-89]).

6. Conclusion.

In Section 5 we introduced the numbers $a(n)$, which are generalised Apéry numbers. They satisfy supercongruence (29). The numbers $a(n)$ are related to an elliptic curve. The Apéry numbers $b(n)$ and $d(n)$ as defined in Section 4 are related to $K3$ surfaces (cf. [7]). They satisfy other congruences which are comparable to congruence (29), namely

$$b\left(\frac{mp^r-1}{2}\right) \equiv (a+bi)^2 \cdot b\left(\frac{mp^{r-1}-1}{2}\right) \bmod p^r \tag{40}$$

and

$$d\left(\frac{mp^r-1}{2}\right) \equiv \bar{\pi} \cdot d\left(\frac{mp^{r-1}-1}{2}\right) \bmod p^r \tag{41}$$

where $a+bi$ is as defined in section 4 and $\bar{\pi}$ is a root of some polynomial of degree 3 (see [6] or [24]). Beukers and Stienstra conjectured in [6] and [7] the supercongruences

$$b\left(\frac{mp^r-1}{2}\right) \equiv (a+bi)^2 \cdot b\left(\frac{mp^{r-1}-1}{2}\right) \bmod p^{2r} \tag{42}$$

and

$$d\left(\frac{mp^r-1}{2}\right) \equiv \bar{\pi} \cdot d\left(\frac{mp^{r-1}-1}{2}\right) \bmod p^{2r}. \tag{43}$$

Van Hamme [19] proved (42) in the case that $m = r = 1$. Recently Young [24] proved (43) in the case that $m = r = 1$. The rest of the conjectures is at the moment unproved. Perhaps, the proof of Theorem 4 gives a good possibility to prove the rest of the conjectures.

7. References.

[1] K. Alladi and M.L. Robinson: *On certain values of the logarithm*, Lecture Notes **751**, 1-9.

[2] R. Apéry: Irrationalité de ζ(2) et ζ(3), Astérisque 61 (1979), 11-13.

[3] A.O.L. Atkin and H.P.F. Swinnerton-Dyer: *Modular forms on noncongruence subgroups*, Proc. of Symposia in Pure Math., A.M.S. **19** (1971), 1-25.

[4] F.Beukers: *Arithmetical properties of Picard-Fuchs equations, Séminaire de théorie des nombres*, Paris 82-83, Birkhäuser Boston, 1984, 33-38.

[5] F. Beukers: *Some congruences for the Apéry numbers*, J. Number Theory **21** (1985), 141-150.

[6] F. Beukers: *Another congruence for the Apéry numbers*, J. Number Theory **25** (1987), 201-210.

[7] F. Beukers and J. Stienstra: *On the Picard-Fuchs equation and the formal Brauer group of certain elliptic K3-surfaces*, Math. Annalen **271** (1985), 293-304.

[8] L. Carlitz: *Advanced problem 4268*, A.M.M. **62** (1965) p. 186 and A.M.M. **63** (1956) 348-350.

[9] S. Chowla, J. Cowles and M. Cowles: *Congruence properties of Apéry numbers*, J. Number theory **12** (1980), 188-190.

[10] S. Chowla, B. Dwork and R.J. Evans: *On the* mod p^2 *determination of* $\binom{\frac{p-1}{2}}{\frac{p-1}{4}}$, J. Number theory **24** (1986), 188-196.

[11] M.J. Coster: *Generalisation of a congruence of Gauss*, J. Number theory **29** (1988), 300-310.

[12] M.J. Coster: *Supercongruences*, [Thesis] Univ. of Leiden, the Netherlands, 1988.

[13] M.J. Coster and L. van Hamme: *Supercongruences of Atkin and Swinnerton-Dyer type for Legendre polynomials*, to appear in J. of Number Theory in 1990.

[14] J. Diamond: *The p-adic log gamma function and p-adic Euler constants*, Trans. Amer. Math. Soc. **233** (1977), 321-337.

[15] C.F. Gauss: *Arithmetische Untersuchungen (Disquisitiones arithmeticae)*, [Book] Chelsea Publishing Company Bronx, New York, reprinted 1965.

[16] I. Gessel: *Some congruences for Apéry numbers*, J. Number theory **14** (1982), 362-368.

[17] B. Gross and M. Koblitz: *Gauss sums and the p-adic Γ-function*, Ann. Math. **109** (1979), 569-581.

[18] L. van Hamme: *The p-adic gamma function and congruences of Atkin and Swinnerton-Dyer*, Groupe d'étude d'analyse ultramétrique, 9e année 81/82, Fasc. 3 no. J17-6p.

[19] L. van Hamme: *Proof of a conjecture of Beukers on Apéry numbers*, Proceedings of the conference of p-adic analysis, Hengelhoef, Belgium (1986), 189-195.

[20] M. Hazewinkel: *Formal groups and applications*, [Book] Academic Press, New York, 1978.

[21] Y. Mimura: *Congruence properties of Apéry numbers*, J. Number theory **16** (1983), 138-146.

[22] W.H. Schikhof: *Ultrametric calculus*, [Book] Cambridge University Press, Cambridge, 1984.

[23] H. Weber: *Lehrbuch der Algebra*, [Book] dritter dand, Friedrich Vieweg und Sohn, Braunschweig, 1908.

[24] P.T. Young: *Further congruences for the Apéry numbers*, to appear, 1989.

Witt realization of the p-adic Barsotti-Tate groups; some applications

Valentino Cristante

Dipartimento di Matematica Pura e Applicata dell'Università di Padova

via Belzoni , 7 - 35131 Padova , Italia.

0. - Introduction .

It is now some twenty years since Barsotti's paper "Varietà abeliane su corpi p-adici; parte prima" appeared (cf. [VP]). In it the Barsotti-Tate groups over p-adic rings are studied by introducing an embedding of their affine algebra R in a ring of Witt vectors with components over an algebra $\mathcal{R}_0$, of characteristic p, depending functorially on the reduction mod p of R.

In my opinion, this embedding is a good tool for handling a great many problems; so I would like first to explain some of the ideas leading to this embedding, and then show how it can be used in situations involving the local deformation of abelian varieties.

In order to say something a little bit more precise, let us consider a Barsotti-Tate (B-T) group G over the ring $W = W(k)$ of Witt vectors with components in a perfect field k of characteristic $p > 0$, let us denote by R the affine algebra of G and by R_0 the affine algebra of its special fibre; R_0 is endowed with the usual topology as a profinite ring and we define $\mathcal{R}'_0$ as the direct limit

$$\mathcal{R}'_0 = \varinjlim (R_0 \xrightarrow{[p]} R_0 \xrightarrow{[p]} \cdots) ,$$

where $[p]$ denotes the endomorphism corresponding to the multiplication by p in G_0, with the limit topology. The algebra depending functorially on R_0, to which we alluded above, is exactly the completion $\mathcal{R}_0$ of $\mathcal{R}'_0$. In what follows the R_0-algebra structure of $\mathcal{R}_0$ is fixed once and for all by an embedding i: $R_0 \longrightarrow \mathcal{R}_0$, which is a bialgebra homomorphism, so that each bialgebra extension of R_0 gives an analogous extension of $\mathcal{R}_0$. The main point in our construction of the embedding of R in $W(\mathcal{R}_0)$ depends on the following remark: one can functorially associate to R an extension $E = E(R)$ of R_0 by a Witt bialgebra, and the extension of $\mathcal{R}_0$ corresponding to E splits. As a consequence there exists a bialgebra homomorphism j sitting in the following commutative diagram:

$$\begin{array}{ccc} R & \xrightarrow{\quad j \quad} & W(\mathcal{R}_0) \\ \Big\downarrow{\scriptstyle can} & & \Big\downarrow{\scriptstyle \rho} \\ R\otimes k = R_0 & \xrightarrow{\quad i \quad} & \mathcal{R}_0 , \end{array}$$

where *can* and ρ denote the reduction mod p of R and of $W(\mathcal{R}_0)$, respectively.

Now let us start with a fixed R_0, and consider the set of the liftings of R_0 to W: we mean the set of the pairs (R,σ), where R is the affine algebra of a B-T group over W and σ is a bialgebra homomorphism which can be factored in the following way:

$$\begin{array}{ccc} & R & \\ {}^{can}\swarrow & & \searrow{}^{\sigma} \\ R\otimes k & \xrightarrow{\;\sim\;} & R_0 \end{array} .$$

On this set there is a natural notion of equivalence: the liftings (R,σ) and (R',σ') are equivalent if and only if there exists an isomorphism τ of R in R' which makes the following diagram commutative:

$$\begin{array}{ccc} & R & \\ {}^{\tau}\swarrow & & \searrow{}^{\sigma} \\ R' & \xrightarrow[\sigma']{} & R_0 \end{array} .$$

This shows that $W(\mathcal{R}_0)$ is the natural place where all the liftings of R_0 to W live; but there is something more here, as well. In fact, for each class of equivalent liftings there exists only one representative embedded in $W(\mathcal{R}_0)$, so that after our construction, coordinatizing the classes of equivalent liftings becomes quite natural: it amounts to giving the position inside $W(\mathcal{R}_0)$ of the corresponding embedded representatives. With these coordinates, the usual structures associated with deformation theory, e.g. the Kodaira-Spencer mapping and the Gauss-Manin connection on the de Rham cohomology of the universal lifting, can be described in a very natural and simple way.

In fact, in the second part of the present paper we will describe the situation which arises when R_0 comes from an ordinary abelian variety; our results are similar,to, and in fact inspired by, those obtained by N. Katz in [KA], so that in the end we can conclude that our coordinates are essentially the same of those of Serre and Tate.

The first part is devoted to giving the definitions and the results about the canonical embedding necessary for intelligibility of the second part: it is in fact an abstract of a work in progress in collaboration with M. Candilera, where the canonical embedding is used in order to discuss the reduction mod p of the theta functions. In the meantime, Candilera's thesis (cf. [CA]), where part of our results were first published with proofs, may be useful.

Most of the main ideas used here, are - in a more or less explicit way - contained in the works of Barsotti, and it is impossible for me to think about these arguments without a deep feeling of gratitude for his heritage.

1. - Witt extensions of B-T groups; their trivialization and related embeddings.

The notations in the introduction maintain their meaning; moreover we will use the symbol $BW(S)$ to denote the ring of infinite (Witt) bivectors with components in any k-bialgebra S. General results about bivectors can be found in [MA] or in [FO]. The coproduct, inversion and coidentity of S, will be denoted by $\mathbb{P}$, τ, and ε, respectively.

Now we will describe the Witt extension of the affine algebra R_0 of a *B-T* group G_0 over k. A *Witt factor system* of R_0 is an element $\gamma \in W(R_0 \otimes R_0)$ with the following properties:

(i) $\gamma = sc(\gamma)$;

(ii) $(\varepsilon \otimes id)\gamma = (id \otimes \varepsilon)\gamma = 0$;

(iii) $(\mathbb{P} \otimes id)\gamma + (\gamma \otimes 1) = (id \otimes \mathbb{P})\gamma + (1 \otimes \gamma)$,

where sc and id are the twist of $W(R_0 \otimes R_0)$ and the identity of $W(R_0)$, respectively.

Using γ we will construct a bialgebra $E_0(\gamma)$ in the following way:

$$E_0(\gamma) = R_0[[U_0, U_1, \dots]],$$

where $(U_i)_{i \in \mathbb{N}}$ is a family of indeterminates over R_0 and the bialgebra structure is given by extending the coproduct, coidentity and inversion of R_0, in the following way: $\mathbb{P}U_i$, εU_i, τU_i are the components of place i of the following Witt vectors:

$$\mathbb{P}U = U \otimes 1 + 1 \otimes U + \gamma,\ \varepsilon U = 0,\ \tau U = -U,$$

where U denotes the vector $(U_0, U_1, \dots)$.

Observe that when $R_0 = k$ and $\gamma = 0$, then $E_0(\gamma)$ becomes the usual Witt formal k-bialgebra $\mathcal{W}$; the bialgebra $E_0(\gamma)$ just described is an extension of R_0 by $\mathcal{W}$.

As the embedding $i: R_0 \longrightarrow \mathcal{R}_0$ is fixed, we can speak of bivectors of $BW(\mathcal{R}_0)$ whose components of negative index are in R_0; among these the subset of the *canonical bivectors*,

$$\{X: \mathbb{P}X = X \otimes 1 + 1 \otimes X\},$$

has a W-module structure: it will be denoted by $M(R_0)$ and called the *canonical* or the *Dieudonné module* of R_0. The Frobenius and the Verschiebung maps operating on R_0, $\mathcal{R}_0$, $BW(\mathcal{R}_0)$, $M(R_0)$, etc. will be denoted by F and V. Usually R_0^{et}, R_0^{π}, R_0^{r} denote the étale, multiplicative and radical components of R_0.

THEOREM 1.1. *Let γ be a Witt factor system of R_0; let us denote by $\mathcal{E}_0(\gamma)$ the bialgebra $i_*(E_0(\gamma))$, i.e. the extension coming from $E_0(\gamma)$ by means of the embedding $i: R_0 \longrightarrow \mathcal{R}_0$. Then $\mathcal{E}_0(\gamma) = \mathcal{R}_0 \otimes \mathcal{W}$; more precisely, there exist two elements $\lambda \in W(\mathcal{R}_0)$ and $\delta \in BW(R_0)$, such that $\gamma = \Delta\lambda = \Delta\delta$.*

Here, and in what follows, Δ denotes the operator $\mathbb{P} - id \otimes 1 - 1 \otimes id$ both on $W(\mathcal{R}_0)$ and on $BW(R_0)$. The vector λ is unique, its components λ_i are in $V^{-(i+1)}R_0$; in particular if $R_0 = R_0^{\pi}$, each extension of R_0 by $\mathcal{W}$ is trivial. On the contrary, δ is unique if and only if $R_0^{\pi} = k$.

REMARK 1.2. *If the Witt vectors are identified with the Witt bivectors having components of negative indices equal to 0, then $\eta = \lambda - \delta$ is a canonical bivector of $\mathcal{R}_0$ whose components of negative indices are in R_0, i.e. it is in $M(R_0)$. The bialgebra $E_0(\gamma)$ is essentially determined by the bivector η: in fact if η' is an element of $BW(R_0)$ whose components with negative indices coincide with the corresponding components of η, then the factor system $\Delta\eta'$ gives an extension equivalent to $E_0(\gamma)$.*

If $\mathcal{W}^r$ denotes the product of r copies of $\mathcal{W}$, we can consider extensions of R_0 by $\mathcal{W}^r$. Everything works as in the previous case: if $\delta = (\delta_i)_{1 \le i \le r}$ is a trivialization of the factor system $\gamma = (\gamma_i)_{1 \le i \le r}$ of E in

$BW(R_0)$, we will consider the submodule $N(\delta)$ of $M(R_0)$ generated by the canonical bivectors η_i whose components with negative indices coincide with the corresponding ones of δ_i. We will identify two trivialized extension if they give the same submodule $N(\delta)$.

Now we will try to explain the relation between the Witt extensions of R_0 and its liftings to W.

Let us consider a lifting (R,σ) of R_0 to W. When the parameters $X = (X_i)_{1\leq i\leq g}$ are chosen, we have $R = R^{et}[[X_1, \ldots, X_g]]$, where R^{et}, which is the lifting of R_0^{et} to W will be identified with $W(R_0^{et})$ and $\sigma|_{R^{et}}(a_0, a_1, \ldots) = a_0$. Let us denote by F_j and g_j the power series which give the coproduct and the inversion, respectively:

$$\mathbb{P}X_j = F_j(X_1\otimes 1, \ldots, X_g\otimes 1, 1\otimes X_1, \ldots, 1\otimes X_g), \quad \tau X_j = g_j(X_1, \ldots, X_g).$$

Observe that the coefficients of F_j are in $W(R_0^{et}\otimes R_0^{et})$, and those of g_j are in $W(R_0^{et})$. By means of these series we will construct a k-bialgebra: let (X_{ij}) be a family of indeterminates over R_0^{et}, where $1\leq i\leq g$, and $0\leq j\leq\infty$, put $S = R_0^{et}[[(X_{ij})]]$ and denote by $\tilde{X}_j$ the Witt vector $(X_{j0}, X_{j1}, \ldots)$. A bialgebra structure on S extending the structure of R_0^{et} is given by defining $\mathbb{P}X_{ij}$ as the j-th component of

$$F_i(\tilde{X}_1\otimes 1, \ldots, \tilde{X}_g\otimes 1, 1\otimes\tilde{X}_1, \ldots, 1\otimes\tilde{X}_g), \text{ and } \tau X_{ij} = g_j(\tilde{X}_1, \ldots, \tilde{X}_g).$$

At this point it is quite easy to verify that the subring $R_0^{et}[[X_{10}, \ldots, X_{g0}]]$ is canonically isomorphic to R_0, so that S can be canonically identified with $R_0[[(X_{ij})]]$, where $1\leq i\leq g$, and $1\leq j\leq\infty$, but the complete result is the following:

THEOREM 1.3. *Let R_0 be a B-T group of dimension g over k and (R,σ) a lifting. Then the bialgebra S corresponding to the lifting (R,σ) just constructed is an extension of R_0 by $\mathcal{W}^g$. Let S and S' the extensions corresponding to the liftings (R,σ) and (R',σ'), then they can be trivialized in such a way that the W-module $N(\delta)$ and $N(\delta')$ coincide (i.e. they are equivalent in the sense of* 1.2) *if and only if (R,σ) and (R',σ') are equivalent in the sense of the introduction.*

Here we will give an idea of how the module $N(\delta)$ arises, the same argument will explain the existence of the embedding of R in $W(\mathcal{R}_0)$. Let us denote by $j'\colon R \longrightarrow W(S)$ the map which extends the identity of R^{et} and sends X_i in $\tilde{X}_i$. The map j' is a bialgebra embedding such that $\rho j' = \sigma$. It can be extended to a sub-bialgebra of R_K (the general fibre) which contains the W-module $I_1(R)$ of the integrals of the first kind. The j'-images of the integrals of $I_1(R)$ are canonical bivectors whose components with non-positive indices are in R_0.Thus that to each $h\in I_1(R)$ there corresponds an element jh in $M(R_0)$: the set of such images is just $N(\delta)$. Now, we have an embedding $j\colon I_1(R) \longrightarrow BW(\mathcal{R}_0)$, and it can be proved that if (h_i) is a basis of $I_1(R)$, there exist g power series, G_i, such that

$$R' = R^{et}[[G_1(jh_1, \ldots, jh_g), \ldots, G_g(jh_1, \ldots, jh_g)]],$$

with the structure induced by means of the components, is a sub-bialgebra of $W(\mathcal{R}_0)$. If, as in the introduction, ρ denotes the usual reduction mod p of $W(\mathcal{R}_0)$, then (R',ρ) is the unique lifting of R_0 equivalent to (R,σ) embedded in $W(\mathcal{R}_0)$. It will be called *the embedded lifting* equivalent to (R,σ).

The sub-module $N(\delta)$ of $M(R_0)$ in the previous discussion is the usual filtration appearing in several points of the literature (for instance cf. [BIV], [VP] or [MM]), so that the following result is well known:

THEOREM 1.4. *Let R_0 be the affine algebra of a B-T group of dimension g over k; then a W-submodule N of $M(R_0)$ comes from a lifting (R,σ) if and only if $rk(N) = g$ and $N + pM(R_0) = VM(R_0)$.*

REMARK 1.5. *Let N be a W-submodule of $M(R_0)$ coming from a lifting of R_0 to W. Then there exists a submodule N' such that $M(R_0) = N \oplus N'$. Moreover, for each $f \in Hom(N, pN')$ the submodule*

$$N_f = \{x + fx : x \in N\}$$

comes from a lifting, and all the modules coming from liftings of R_0 to W are of this form.

The pair (N, N') is a *frame* for the set of the classes of equivalent liftings of R_0 and f is the *coordinate* of the lifting corresponding to N_f.

REMARK 1.6. *If $R_0^{loc} = R_0^{\pi}$, i.e. the local component of R_0 is multiplicative, then $M(R_0^{loc})$ satisfies the conditions of 1.4., and $M(R_0)$ has the canonical direct decomposition $M(R_0) = M(R_0^{loc}) \oplus M(R_0^{et})$. Hence in this case we can speak of canonical frame and canonical coordinate. The lifting which has coordinate 0 is called the canonical lifting. It is the direct product of the canonical liftings of R_0^{π} and of R_0^{et}.*

We assume R_0 as in 1.6.; let g and c be the dimension and codimension of R_0, respectively. Assume $1 \le r \le g$ and $1 \le s \le c$, denote by (t_{rs}) a family of cg indeterminates over W, fix a basis (η_i) of $M(R_0^{loc})$ and a basis (λ_i) of $M(R_0^{et})$, then the g elements

$$\eta_i + \sum_{j=1}^{c} t_{ij}\lambda_j$$

of $M(R_0) \otimes W[[(t_{rs})]]$ give a basis for the module of the integrals of the first kind of a B-T group over $W[[(t_{rs})]]$: this group is the *universal lifting of R_0 to $W[[(t_{rs})]]$*, its coordinate f is the homomorphism $M(R_0^{loc}) \otimes W[[(t_{rs})]] \longrightarrow M(R_0^{et}) \otimes W[[(t_{rs})]]$, given by

$$\eta_i \longrightarrow \sum_{j=1}^{c} t_{ij}\lambda_j .$$

Observe that for each point $P \in Hom(W[[(t_{rs})]], W)$ the map given by

$$\eta_i \longrightarrow \sum_{j=1}^{c} P(t_{ij})\lambda_j$$

is an element in $Hom(M(R_0^{loc}), pM(R_0^{et}))$ and that each such element arises by specializing f by means of a suitable P.

DEFINITION 1.7. *Let $R = R^{et}[[X_1, \ldots, X_g]]$ be the affine algebra of a B-T group over W; an* *integral of the second kind of R is a power series*

$$h = \sum_{\nu} a_{\nu} X^{\nu}$$

with coefficients in $BW(\mathcal{R}_0^{et})$, *such that* :

$$a_{(0,\ldots,0)} \in BW(R_0^{et}), \qquad \nu! a_{\nu} \in R^{et}(=W(R_0^{et})), \quad and \quad \mathbb{P}h - h\otimes 1 - 1\otimes h \in R\hat{\otimes}R.$$

The W-module of the integrals of the second kind will be denoted by $I_2(R)$. *The quotient* $I_2(R)/R$ *is the usual* $H^1_{DR}(R)$.

Let (R,σ) be an embedded lifiting of R_0. Then the embedding j can be extended to the module $I_2(R)$; the images of the integrals are in $BW(\mathcal{R}_0)$; more precisely, as a consequence of the relation

$$\mathbb{P}h - h\otimes 1 - 1\otimes h \in R\hat{\otimes}R,$$

we have $\mathbb{P}jh - jh\otimes 1 - 1\otimes jh \in W(\mathcal{R}_0\hat{\otimes}\mathcal{R}_0)$, so that we can conclude that jh has the same components with negative indices as a canonical bivector η_h. The map $h \longrightarrow \eta_h$ induces the classical isomorphism between $H^1_{DR}(R)$ and $M(R_0)$.

The most important result for applications to deformation theory is the following

THEOREM 1.9. *Let* (R,σ) *be an embedded lifiting of* R_0; *to each integral of the second kind jh, by means of the factor system* $\mathbb{P}jh - jh\otimes 1 - 1\otimes jh \in R\hat{\otimes}R$, *there corresponds an embedded extension E of R by the additive bialgebra* $\mathcal{D}$. *The elements of the module* $I_1(R) + <\eta_h>$ *are the integrals of the first kind of E. More generally, there exist a natural bijective correspondence between the set of the classes of additive extensions of R, and the set of W-submodules of* $M(R_0)$ *containing the module* $I_1(R)$: *it is the correspondence induced by* $E \longrightarrow I_1(E)$; *in fact for each class there exists one and only one embedded representative. In particular, for each lifting* (R,σ) *there exists a class of extensions corresponding to the module* $M(R_0)$. *Its embedded representative is an extension of R by* $\mathcal{D}^c$ *which is independent of* (R,σ): *it is the usal universal extension,* $U(R)$, *of R.*

For a definition of $U(R)$ one can look at [BIV] or [MM].

2. - Local deformations.

Let A be an abelian variety over k and $\hat{A}$ the corresponding B-T group; let us denote by $R_0 = R_0(A)$ the affine algebra of $\hat{A}$, by $M(A)$ and $M^*(A)$ the Dieudonné modules $M(R_0(A))$ and $M(R_0^*(A))$, respectively: here $R_0^*(A)$ denotes the dual of R_0. The components of the elements of $M(R_0^*(A))$ are in $\mathcal{R}_0^*(A)$ (which is constructed as $\mathcal{R}_0(A)$, but starting from $R_0^*(A)$). The duality between $\mathcal{R}_0(A)$ and $\mathcal{R}_0^*(A)$ gives an identification of $M^*(A)$ with $Hom\,(M(A), W)$.

If $\mathcal{L}$ is a line bundle on A, then in [MA] a Dieudonné module homomorphism

$$\varphi_{\mathcal{L}} : M^*(A) \longrightarrow M(A)$$

is defined. This map, which can also be obtained from a theta function associated to a section of $\mathcal{L}$ cf. [CC],

[BR] or [CR]), is a Riemann form; i.e. the map $\mathcal{L} \longrightarrow \varphi_{\mathcal{L}}$ induces an injective group homomorphism from the Neron-Severi group, $NS(A)$, into the group of alternating 2-forms

$$M^*(A) \times M^*(A) \longrightarrow W.$$

Now, let us consider the following particular situation: A is replaced by $A \times A^*$, where A^* denotes the dual of A, and $\mathcal{L}$ by the Poincaré bundle $\mathcal{P}$ on $A \times A^*$. If we identify $M^*(A \times A^*)$ with $M^*(A) \oplus M^*(A^*)$ and analogously $M(A \times A^*)$ with $M(A) \oplus M(A^*)$, then $\varphi_{\mathcal{P}}$ can be represented by a matrix $\begin{pmatrix} \alpha & \varphi_A \\ \gamma & \beta \end{pmatrix}$ which works in the following way:

$$\begin{pmatrix} \varphi_{\mathcal{P}}\eta \\ \varphi_{\mathcal{P}}\eta' \end{pmatrix} = \begin{pmatrix} \alpha & \varphi_A \\ \gamma & \beta \end{pmatrix} \begin{pmatrix} \eta \\ \eta' \end{pmatrix},$$

for each $\eta \in M^*(A)$, $\eta' \in M^*(A^*)$, $\alpha \in Hom(M^*(A), M(A))$, $\varphi_A \in Hom(M^*(A^*), M(A))$, etc..

The results about φ_A used later on are described in the following

THEOREM 2.1. (i) *φ_A is a Dieudonné module isomorphism.* (ii) *if B is another abelian variety, α a homomorphism of A in B and α^* its dual, then we have the following commutative diagram:*

$$\begin{array}{ccc} M^*(B^*) & \xrightarrow{\ \varphi_B\ } & M(B) \\ \downarrow{\scriptstyle M(\alpha^*)} & & \downarrow{\scriptstyle M(\alpha)} \\ M^*(A^*) & \xrightarrow{\ \varphi_A\ } & M(A) \end{array};$$

(iii) *φ_A and φ_{A^*} are related by the following duality relation: $\varphi_{A^*} = -\varphi_A^*$.*

All this was announced in '62, cf [BRX], then proved with all details in chap. 7 of [MA].

COROLLARY 2.2. *Let us denote by e_A the map of $M(A) \times M(A^*)$ with values in W, defined for each $\eta \in M(A)$, $\eta^* \in M(A^*)$ by*

$$e_A(\eta, \eta^*) = \langle \varphi_A^{-1}\eta, \eta^* \rangle,$$

where $\langle -, - \rangle$ means the usual duality between $M^*(A^*)$ and $M(A^*)$; then

$$e_A(\eta, \eta^*) = -e_{A^*}(\eta^*, \eta) \tag{2.3}$$

PROOF. The duality relation of Th. 2.1., i.e. $\varphi_{A^*} = -\varphi_A^*$, means

$$\langle \varphi_{A^*} d, d^* \rangle = -\langle \varphi_A d^*, d \rangle \tag{2.4}$$

for each $d \in M^*(A)$, $d^* \in M^*(A^*)$.

Since φ_A is isomorphism, (2.4) holds if and only if

(2.5) $$<\varphi^{-1}_A\eta, \eta^*> = - <\varphi^{-1}_{A^*}\eta^*, \eta>$$
for each $\eta \in M(A)$, $\eta^* \in M(A^*)$; but (2.5) is exactly (2.3), QED.

REMARK 2.6. *Let us consider the universal lifting $\mathcal{A}$ of $\hat{A}$ and its dual $\mathcal{A}^*$, which is canonically isomorphic to the universal lifting of the dual of $\hat{A}$; then there is the following relation between the modules of the integrals of the first kind :*

$$\varphi_A(I_1(\mathcal{A}^*)^{\perp}) = I_1(\mathcal{A}).$$

This can be checked by recalling the commutativity of the following diagram:

$$\begin{array}{ccccccccc}
0 & \longrightarrow & I_1(R) & \longrightarrow & M(A) & \longrightarrow & M(A)/I_1(R) & \longrightarrow & 0 \\
 & & \uparrow r & & \uparrow \varphi_A & & \uparrow \delta & & \\
0 & \longrightarrow & I_1(R^*)^{\perp} & \longrightarrow & M^*(A^*) & \longrightarrow & Lie(R^*) & \longrightarrow & 0
\end{array},$$

where R is the affine algebra of any lifting of $\hat{A}$ to W, r is the restriction of φ_A and δ the map induced by φ_P.

REMARK 2.7. *If $M^{\pi}(A)$ and $M^{et}(A)$ denote the multiplicative and the étale component of $M(A)$, and if A is ordinary, i.e. $M(A) = M^{\pi}(A) \oplus M^{et}(A)$, then with respect to the bilinear form*

$$e_A : M(A) \times M(A^*) \longrightarrow W$$

we have the following orthogonality relations :

$$(M^{\pi}(A))^{\perp} = M^{\pi}(A^*),\ (M^{et}(A))^{\perp} = M^{et}(A^*).$$

In fact, since φ_A is a Dieudonné module isomorphism,

$$\varphi^{-1}_A(M^{\pi}(A)) = (M^*(A^*))^{\pi} = \{\, f \in Hom(M(A^*),W) \mid fM^{\pi}(A) = 0 \,\} = Hom\,(M^{et}(A^*),W);$$

the analogous relation for $M^{et}(A^*)$ is

$$\varphi^{-1}_A(M^{et}(A)) = Hom\,(M^{\pi}(A^*),W).$$

As we said in part 1, if t denotes the family of indeterminates (t_{ij}), $1 \leq i,j \leq g$ (=dimA), the universal lifting $\mathcal{A}$ of $\hat{A}$ is given by means of a $W[[t]]$-module homorphism

$$f_{\mathcal{A}} : M^{\pi}(A) \otimes W[[t]] \longrightarrow M^{et}(A) \otimes W[[t]],$$

which has been called the coordinate of $\mathcal{A}$; in what follows $f_{\mathcal{A}}$ will be replaced by the $W[[t]]$- bilinear form

$$q_{\mathcal{A}} : (M^{\pi}(A) \otimes W[[t]]) \times (M^{\pi}(A^*) \otimes W[[t]]) \longrightarrow W[[t]]$$

defined by

(2.8) $$q_{\mathcal{A}}(\eta, \eta^*) = e_A(f_{\mathcal{A}}\,\eta, \eta^*);$$

here, of course, e_A means the extension by $W[[t]]$-linearity of the map previously defined in (2.2).

Now we'll write down (2.8) more explicitly, in order to understand the way in which $q_{\mathcal{A}}$ too works like a coordinate.

Let us denote by $(\eta_i)_{1\le i\le g}$ and $(\eta_i^*)_{1\le i\le g}$ a basis of $M^{\pi}(A)$ and $M^{\pi}(A^*)$, respectively, by $({}^t\eta_i)_{1\le i\le g}$ the basis of $(M^*(A^*))^{et}$ dual to $(\eta_i^*)_{1\le i\le g}$ and by $(\delta_i)_{1\le i\le g}$ the basis $(\varphi_A{}^t\eta_i)_{1\le i\le g}$ of $M^{et}(A)$.

First we show the following

PROPOSITION 2.9 *The coordinates $f_{\mathcal{A}}$ and $q_{\mathcal{A}}$ are related as follows:*

$$f_{\mathcal{A}}(\eta) = \sum_j q_{\mathcal{A}}(\eta, \eta_j^*)\delta_j$$

for each $\eta \in M^{\pi}(A)\otimes W[[t]]$.

PROOF. It suffices to show that $f_{\mathcal{A}}(\eta_i) = \sum_j q_{\mathcal{A}}(\eta_i, \eta_j^*)\delta_j$, for $i = 1, \ldots, g$. Now by (2.8) and (2.2) we have

$$q_{\mathcal{A}}(\eta_i, \eta_j^*) = e_A(f_{\mathcal{A}}\,\eta_i, \eta_j^*) = -<f_{\mathcal{A}}\,\eta_i, \varphi_A^{-1}\eta_j^*>.$$

Then, if we can show that $(\varphi_{A^*}^{-1}\eta_i^*)_{1\le i\le g}$ is the basis of $(M^*)^{\pi}(A)$, dual to (δ_i), the proposition follows immediately. Now by (iii) of (2.1), we have

$$<-\varphi_{A^*}^{-1}\eta_j^*), \delta_i> = <-\varphi_{A^*}^{-1}\eta_j^*), \varphi_A{}^t\eta_i> = <(\varphi_A^*)^{-1}\eta_j^*), \varphi_A{}^t\eta_i> = <\eta_j^*, {}^t\eta_i> = \delta_{ij}, \text{ QED.}$$

The following theorem shows why $q_{\mathcal{A}}$ is a good coordinate; its point (*i*) is just a translation of 2.1 in terms of $q_{\mathcal{A}}$.

THEOREM 2.10. *Let us denote by A and B two ordinary abelian varieties over k, by $\mathcal{A}$ and $\mathcal{B}$ the universal liftings of the corresponding B-T groups, and by $q_{\mathcal{A}}$, $q_{\mathcal{B}}$ the coordinates; then:*

(*i*) $$q_{\mathcal{A}}(\eta, \eta^*) = q_{\mathcal{A}^*}(\eta^*, \eta),$$

for each $\eta \in M^{\pi}(A)\otimes W[[t]]$, $\eta^* \in M^{\pi}(A^*)\otimes W[[t]]$

(*ii*) *A homomorphism $G \in Hom(A, B)$ can be lifted to $\mathcal{G} \in Hom(\mathcal{A}, \mathcal{B})$ if and only if*

$$q_{\mathcal{A}}(M(G)\gamma, \eta^*) = q_{\mathcal{B}}(\gamma, M(G^*)\eta^*),$$

for each $\gamma \in M^{\pi}(B)\otimes W[[t]]$, $\eta^* \in M^{\pi}(A)\otimes W[[t]]$, where $M(G)$ denotes the extension by $W[[t]]$-linearity of the element of $Hom(M(B), M(A))$ induced by G.

PROOF. The relation between $f_{\mathcal{A}}$ and $f_{\mathcal{A}^*}$ is the following:

(2.11) $$e_A(\eta + f_{\mathcal{A}}\eta, \eta^* + f_{\mathcal{A}^*}\eta^*) = 0.$$

In fact (2.11) means

(2.12) $$<\varphi_A^{-1}(\eta + f_{\mathcal{A}}\eta), \eta^* + f_{\mathcal{A}^*}\eta^*> = 0,$$

i.e. $\varphi_A^{-1}(I_1(\mathcal{A})) = I_1(\mathcal{A}^*)^{\perp}$, as remarked in 2.6.

Now by (2.12), recalling the $e_{\mathcal{A}}$-orthogonality of $M^{\pi}(A)$ with $M^{\pi}(A^*)$ and of $M^{et}(A)$ with $M^{et}(A^*)$, cf. 2.7, we conclude:

$$<\varphi_A^{-1}\eta, f_{\mathcal{A}^*}\eta^*> + <\varphi_A^{-1}(f_{\mathcal{A}}\eta), \eta^*> = - <\varphi_{A^*}^{-1} f_{\mathcal{A}^*}\eta^*, \eta> + <\varphi_A^{-1} f_{\mathcal{A}}\eta, \eta^*> = 0$$

so that, using the definition 2.8, the last relation can be translated as follows:

$$q_{\mathcal{A}}(\eta, \eta^*) = q_{\mathcal{A}^*}(\eta^*, \eta),$$

this is just what we claimed in (i).

Now we will prove (*ii*): G can be lifted to $\mathcal{G}$ if and only if $M(G)(I_1(\mathcal{B})) \subseteq I_1(\mathcal{A})$; therefore for each $\gamma \in M^{\pi}(B)$ there must exist $\eta' \in M^{\pi}(A)$ such that

(2.13) $$M(G)(\gamma + f_{\mathcal{B}}\gamma) = \eta' + f_{\mathcal{A}}\eta'.$$

In view of the decomposition $M(B) = M(B)^{\pi} \oplus M(B)^{et}$ and of the analogue for $M(A)$, (2.13) implies

(2.14) $$M(G)(f_{\mathcal{B}}\gamma) = f_{\mathcal{A}}(M(G)(\gamma)).$$

Now, by (2.14), and recalling the commutativity of the diagram in (ii) of 2.1, we can write:

$$q_{\mathcal{A}}(M(G)(\gamma), \eta^*) = <\varphi_A^{-1} f_{\mathcal{A}}((M(G)(\gamma)), \eta^*> = <\varphi_A^{-1} M(G)(f_{\mathcal{B}}\gamma), \eta^*> =$$

$$<M^*(G^*)(\varphi_B^{-1}(f_{\mathcal{B}}\gamma)), \eta^*> = <\varphi_B^{-1}(f_{\mathcal{B}}\gamma), M(G^*)\eta^*> = q_{\mathcal{B}}(\gamma, M(G^*)(\eta^*)),$$ QED.

Our next goal is to compute the differential of $q_{\mathcal{A}}$ with respect to the parameters.

By the results described in the first part, cf. in particular 1.9., the universal extension of $\mathcal{A}$, $U(\mathcal{A})$, can be canonically embedded in $biv(\mathcal{R}(\mathring{A}) \otimes W[[t]])$. And the well known constance of $U(\mathcal{A})$ implies that, for each homomorphism P *of* $W[[t]]$ in W, the universal extension $U(\mathring{A}_P)$ of the lifting $\mathring{A}_P$ corresponding to P, doesn't depend on P; so that, once $U(\mathcal{A})$ is identified with its embedded image, there is the decomposition

(2.15) $$U(\mathcal{A}) = U \otimes W[[t]],$$
where U denotes $U(\hat{A}_P)$ for each $P \in Hom(W[[t]],W)$; of course, the same holds for the integrals of the first kind:

(2.16) $$I_1(U(\mathcal{A})) = M(A) \otimes W[[t]].$$

REMARK 2.17 *Let us denote by* ∇ *the connection on* $I_1(U(\mathcal{A}))$ *whose module of the horizontal sections is* $M(A)$. *By means of the canonical identification*

(2.18) $$\psi : H^1_{DR}(\mathcal{A}) \longrightarrow I_1(U(\mathcal{A})),$$

∇: $H^1_{DR}(\mathcal{A}) \longrightarrow H^1_{DR}(\mathcal{A}) \otimes_{W[[t]]} \Omega_{W[[t]]/W}$ *becomes the Gauss-Manin connection on* $H^1_{DR}(\mathcal{A})$; *the corresponding module of the horizontal sections,* $\psi^{-1}(M(A))$, *will be denoted by* M.

In fact, when all things are embedded, it is easy to check that $M = H^1_{DR}(\hat{A}_P)$ for each $P \in Hom(W[[t]],W)$.

Now, let us consider the decomposition of $H^1_{DR}(\mathcal{A})$ induced by the identification (2.18):

(2.19) $$H^1_{DR}(\mathcal{A}) = I_1(\mathcal{A}) \oplus (M^{et} \otimes W[[t]]),$$
where $M^{et} = M^{et}(A)$.

In this situation, the natural isomorphism ι: $M^{et} \otimes W[[t]] \longrightarrow Lie(\mathcal{A}^*)$ can be described explicitly :

PROPOSITION 2.20. *The map* ι *is determined by* e_A *in the following way: for each* $\eta \in M^{et} \otimes W[[t]]$ *and* $h^* \in I_1(\mathcal{A}^*)$, *we have:*

$$[\iota(\eta)](h^*) = e_A(\eta, \eta^*),$$

where η^* *is given by*

$$\eta^* + f_{\mathcal{A}^*}\eta^* = h^*.$$

This result can be deduced from theorem 8.1 of [VP].

REMARK 2.21. *The map*

$$Kod: I_1(\mathcal{A}) \longrightarrow Lie(\mathcal{A}^*) \otimes_{W[[t]]} \Omega_{W[[t]]/W}$$

defined by $h \longmapsto \iota(pr_2(\nabla(h)))$, *where* pr_2 *is the projection relative to the decomposition* (2.19), *is the usual Kodaira-Spencer map.*

For our next application we prefer a bilinear variation of *Kod* :

$$[-,-] : I_1(\mathcal{A}^*) \times I_1(\mathcal{A}) \longrightarrow \Omega_{W[[t]]/W}$$

defined as follow : $[h^*, h] = (\langle -,-\rangle \otimes id)((id \times Kod)(h^*, h))$, where id denotes the identity map and $\langle -,-\rangle$ denotes the usual pairing between $I_1(\mathcal{A}^*)$ and $Lie(\mathcal{A}^*)$.

The following is the main result:

THEOREM 2.22. *For each* $(\eta,\eta^*) \in (M^{\pi}(A)\otimes W[[t]]) \times (M^{\pi}(A^*)\otimes W[[t]])$, *we have:*

(2.23) $$\partial q_{\mathcal{A}}(\eta,\eta^*) = [\eta^* + f_{\mathcal{A}^*}\eta^*, \eta + f_{\mathcal{A}}\,\eta].$$

PROOF. First, by 2.21., recalling 2.9., we can compute Kod :

$$Kod(\eta + f_{\mathcal{A}}\,\eta) = \iota[pr_2(\nabla(\eta + \sum_j q_{\mathcal{A}}(\eta, \eta_j^*)\,\delta_i))] = \sum_j \iota(\delta_i) \otimes \partial q_{\mathcal{A}}(\eta, \eta_j).$$

Then we observe that η_j^*

$$<\eta_j^* + f_{\mathcal{A}^*}\eta_j^*, \iota(\delta_i)> = e_A(\delta_i, \eta_j^*) = \delta_{ij};$$

in fact,

$$e_A(\delta_i, \eta_j^*) = e_A(\varphi_A\,{}^t\eta_i, \eta_j^*) = <{}^t\eta_i, \eta_j^*>\ ,$$

and $({}^t\eta_i)$ is the basis dual to (η_j^*).

As a consequence,

(2.24) $$(<-,->\otimes id)\,(\eta_j^* + f_{\mathcal{A}^*}\eta_j^*, Kod(\eta + f_{\mathcal{A}}\eta)) = \partial q_{\mathcal{A}}(\eta, \eta_j^*);$$

finally (2.23) comes from (2.24) by linearity,QED.

REFERENCES

[MA] I. Barsotti, *Metodi analitici per le varietà abeliane in caratteristica positiva*, seven chapters divided into five publications in Ann. Scuola Norm. Sup. Pisa, from **18** (1964) to **20** (1966).

[VP] - *Varietà abeliane su corpi p-adici*, Symp Math. **1** (1968), 109-173.

[BIV]- *Bivettori*, Symp. Math. **24** (1981), 23-63.

[BRX] - *Analytical methods for abelian varieties in positive characteristic*, Colloque sur la théorie des groupes algébriques, Bruxelles, 1962, 78-85.

[FO] J-M Fontaine, *Modules galoisiens, modules filtrés et anneaux de Barsotti-Tate*, Asterisque **65** III (1979), 3-80.

[KA] N. Katz, *Serre-Tate local moduli*, in "Surfaces Algébriques", Lecture Notes in Math., vol. **868**, Springer-Verlag, 1981, 138-202.

[BR] L. Breen, *Function thêta et théorème du cube*, Lecture Notes in Math., vol. **980**, Springer-Verlag, 1983.

[CC] M. Candilera and V. Cristante, *Biextension associated to divisors on abelian varieties and theta functions*, Ann. Scuola Norm. Sup. Pisa **10** (1983), 437-491.

[CA] M. Candilera, *Riduzione di funzioni theta*, Thesis, University of Padova, 1986-87.

[CR] V. Cristante, *Theta functions and Barsotti-Tate groups*, Ann. Scuola Norm. Sup.Pisa **7** (1980), 181-215.

[MM] B. Mazur and W. Messing, *Universal extensions and one dimensional crystalline cohomology*, Lecture Notes in Math., vol. **370**, Springer-Verlag, 1974.

POLYEDRES DE NEWTON ET POIDS DE SOMMES EXPONENTIELLES

J. Denef
Department of Mathematics
University of Leuven
Celestijnenlaan 200 B
B-3030 Leuven
Belgium

et

F. Loeser
Université Paris 6 et Ecole Polytechnique
Adresse :
Centre de Mathématiques
Ecole Polytechnique
F-91128 Palaiseau Cedex
France

1. Introduction et énoncé des résultats

1.1 - On note k le corps fini $\mathbf{F}_q$ à q éléments, et ℓ un nombre premier ne divisant pas q. Si K est un corps, on note $\bar{K}$ une clôture algébrique de K. On fixe un caractère additif non trivial $\psi : k \to \mathbf{C}^\times$ et on note $\mathcal{L}_\psi$ le $\bar{\mathbf{Q}}_\ell$ faisceau sur $\mathbf{A}^1_k$ associé à ψ et au revêtement d'Artin-Schreier $t^q - t = x$. Soit X un schéma de type fini sur k, si f est un morphisme $f : X \to \mathbf{A}^1_k$, on considère la somme exponentielle $S(f) = \sum\limits_{x \in X(k)} \psi(f(x))$.

D'après la formule des traces de Grothendieck, on a

$$S(f) = \sum_i (-1)^i \operatorname{Tr}(F, H^i_c(X \otimes \bar{k}, f^*\mathcal{L}_\psi)),$$

F désignant le morphisme de Frobenius (géométrique) de k.

Dans cette note nous expliquons comment on peut déterminer explicitement le module des valeurs propres de F quand X est un tore et f est non dégénéré pour son polyèdre de Newton à l'infini.

1.2 - Si A est un anneau commutatif, on note $\mathbf{T}^n_A = \operatorname{Spec} A[x_1, \ldots, x_n, x_1^{-1}, \ldots, x_n^{-1}]$ le tore de dimension n sur A. Etant donné $f : \mathbf{T}^n_A \to \mathbf{A}^1_A$ un A-morphisme, on écrit f comme un polynôme de Laurent $f = \sum\limits_{i \in \mathbf{Z}^n} c_i x^i$.

Le polyèdre de Newton $\Delta_\infty(f)$ de f à l'infini est l'enveloppe convexe dans $\mathbf{Q}^n$ de $\{i \in \mathbf{Z}^n \, ; \, c_i \neq 0\} \cup \{0\}$. Pour toute face τ de $\Delta_\infty(f)$ on écrit $f_\tau = \sum\limits_{i \in \tau} c_i x^i$. On dit que f est non dégénéré pour $\Delta_\infty(f)$ si, pour toute face τ de $\Delta_\infty(f)$ ne contenant pas l'origine, le sous-schéma de $\mathbf{T}^n_A$ défini par

$$\frac{\partial f_\tau}{\partial x_1} = \ldots = \frac{\partial f_\tau}{\partial x_n} = 0$$

est vide. On note Vol $(\Delta_\infty(f)$) le volume de $\Delta_\infty(f)$.

1.3 - Notre premier résultat est le suivant.

Théorème 1.— *Soit* $f : \mathbf{T}_k^n \to \mathbf{A}_k^1$ *un morphisme non dégénéré pour* $\Delta_\infty(f)$. *On suppose que* $\Delta_\infty(f) = n$.
Alors

a) $H_c^i(\mathbf{T}_{\bar{k}}^n, f^*\mathcal{L}_\psi) = 0$ *pour* $i \neq n$,

b) $\dim H_c^n(\mathbf{T}_{\bar{k}}^n, f^*\mathcal{L}_\psi) = n!$ *Vol* $(\Delta_\infty(f))$.

Si de plus l'origine appartient à l'intérieur de $\Delta_\infty(f)$, *alors*

c) $H_c^n(\mathbf{T}_{\bar{k}}^n, f^*\mathcal{L}_\psi)$ *est pur de poids* n *(i.e. toutes les valeurs propres de* F *sont de module* $\sqrt{q}^n$ *).*

Ce théorème a été démontré par Adolphson et Sperber ([A-S 1], [A-S 2]) pour presque tout p et ils ont conjecturé que le résultat était vrai pour tout p.

Contrairement à leur approche, notre preuve est purement ℓ-adique et utilise des compactifications toroïdales.

1.4 - D'après les résultats fondamentaux de Deligne ([De]), les valeurs propres de F sur $H_c^n(\mathbf{T}_{\bar{k}}^n, f^*\mathcal{L}_\psi)$ sont de module $\sqrt{q}^w$ avec $w \in \mathbf{N}$, $w \leq n$. Dans [D-L] nous utilisons la cohomologie d'intersection pour déterminer le nombre e_w de valeurs propres de module $\sqrt{q}^w$ (comptées avec multiplicités). Autrement dit, on détermine le polynôme

$$E(\mathbf{T}_k^n, f) = \sum_{w=0}^{n} e_w T^w .$$

Ce problème a été posé par Adolphson et Sperber ([A-S 3]) qui ont également traité certains cas particuliers. Pour énoncer nos formules, il nous faut introduire encore quelques notations.

1.5 - Pour tout cône convexe polyédral σ, ayant l'origine pour sommet, on définit poly (σ) comme le polytope convexe obtenu en coupant σ avec un hyperplan dans $\mathbf{Q}^n$ ne passant pas par l'origine et intersectant toutes les faces de dimension 1 de σ. Ce polytope est bien défini à équivalence combinatoire près.

Si Δ est un polytope convexe dans $\mathbf{Q}^n$ et τ une face de Δ, on note $\mathrm{cone}_\Delta(\tau)$ le cône convexe polyédral dans $\mathbf{Q}^n$ engendré par $\Delta - \tau = \{x - y\,;\, x \in \Delta\,,\, y \in \tau\}$. On définit $\mathrm{cone}_\Delta^0(\tau)$ comme le cône convexe polyédral de sommet l'origine obtenu en coupant $\mathrm{cone}_\Delta(\tau)$ par un sous-espace affine passant par l'origine complémentaire du sous-espace engendré par $\tau - \tau$. Ce cône est bien défini à équivalence affine près.

Avec les notations précédentes on peut maintenant définir par récurrence sur la dimension de σ et de Δ des polynômes $\alpha(\sigma)$ et $\beta(\Delta)$ en la variable T par

(1.5.1) $$\alpha(\sigma) = \mathop{\mathrm{tronc}}_{\leq \dim \sigma - 1}\big((1 - T^2)\beta(\mathrm{poly}(\sigma))\big) \quad \text{si} \quad \dim \sigma > 0$$

(1.5.2) $$\beta(\Delta) = (T^2 - 1)^{\dim \Delta} + \sum_{\substack{\tau \text{ face de } \Delta \\ \tau \neq \Delta}} (T^2 - 1)^{\dim \tau} \alpha(\mathrm{cone}_\Delta^0(\tau))$$

(1.5.3) $$\alpha(\{0\}) = 1.$$

(Ici $\mathop{\mathrm{tronc}}_{\leq k}(\ \ )$ désigne le polynôme obtenu par troncation en ne gardant que les monômes de degré $\leq k$.) Les polynômes $\alpha(\sigma)$ et $\beta(\Delta)$ ne dépendent que du type combinatoire de σ et Δ, et ont été introduits par Stanley ([S]). On vérifie que $\alpha(\sigma) = 1$ si σ est un cône simplicial. On note $\alpha(\sigma)(1)$ la valeur de $\alpha(\sigma)$ en $T = 1$.

1.6 - Soit $f : \mathbf{T}_k^n \to \mathbf{A}_k^1$ un k-morphisme et τ une face de dimension d de $\Delta_\infty(f)$. Si $0 \in \tau$ on peut écrire $f_\tau = \widetilde{f}_\tau(x^{e_1}, \ldots, x^{e_d})$, avec $\widetilde{f}_\tau$ un polynôme de Laurent en d variables et $e_1, \ldots, e_d$ une base du réseau obtenu en intersectant $\mathbf{Z}^n$ et l'espace affine engendré par τ. Remarquons que $\widetilde{f}_\tau$ est défini à un automorphisme de $\mathbf{T}^d$ près, et que si f est non dégénéré pour $\Delta_\infty(f)$, $\widetilde{f}_\tau$ est non dégénéré pour $\Delta_\infty(\widetilde{f}_\tau)$. Le résultat suivant permet de calculer explicitement, de façon récursive, les e_w.

Théorème 2.— *Soit $f : \mathbf{T}_k^n \to \mathbf{A}_k^1$ un morphisme non dégénéré pour $\Delta = \Delta_\infty(f)$. On suppose que* $\dim \Delta = n$. *Alors*

a) $e_n = n!\,\mathrm{Vol}(\Delta) + \sum\limits_{0\in\tau} (-1)^{n-\dim\tau}(\dim\tau)!\,\mathrm{Vol}(\tau)\alpha(\mathrm{cone}^0_\Delta(\tau))(1)$

b) $E(\mathbf{T}_k^n, f) = e_n T^n - \sum\limits_{0\in\tau} (-1)^{n-\dim\tau} E(\mathbf{T}_k^{\dim\tau}, \widetilde{f}_\tau)\alpha(\mathrm{cone}^0_\Delta(\tau))$.

(On somme sur les faces propres τ de Δ contenant l'origine ; Vol(τ) est le volume de τ dans le sous-espace engendré par τ, normalisé pour prendre la valeur 1 sur un domaine fondamental du réseau induit par $\mathbf{Z}^n$.)

En fait Adolphson et Sperber ([A-S 3]) ont conjecturé une formule explicite pour les e_w. Notre résultat entraîne que cette conjecture est fausse, déjà pour $w = n = 5$ et un polyèdre $\Delta_\infty(f)$ dont toutes les faces sont des simplexes ([D-L]).

2. Nous allons donner dans ce paragraphe quelques indications sur les principales étapes et les principaux ingrédients de la démonstration des théorèmes 1 et 2. Le lecteur trouvera tous les détails dans [D-L].

2.1 - Schémas toriques

(2.1.1) Soit A un anneau commutatif. A tout cône convexe polyédral σ dans $\mathbf{Q}^n$ on associe le A-schéma torique affine $X_A(\sigma) := \mathrm{Spec}\ A[\sigma \cap \mathbf{Z}^n]$. De plus, à tout éventail (fini) Σ dans $\mathbf{Q}^n$ on associe un A-schéma torique $X_A(\Sigma)$ obtenu en recollant les schémas $X_A(\check\sigma)$, $\sigma \in \Sigma$, $\check\sigma$ désignant le cône dual de σ. Cette construction est donnée dans ([Da, § 5]) lorsque A est un corps, mais elle se généralise de façon évidente. Le schéma $X_A(\Sigma)$ est lisse sur A si et seulement si l'éventail Σ est régulier, il est propre sur A si et seulement si le support de Σ est égal à $\mathbf{Q}^n$ (cf. [Da]). A tout cône σ de dimension r de Σ on associe un A-tore $X_A^\sigma(\Sigma) \subset X_A(\check\sigma) \subset X_A(\Sigma)$ de dimension $n-r$. Ces tores forment une partition de $X_A(\Sigma)$. En prenant $\sigma = \{0\}$, on voit que $\mathbf{T}_A^n$ est un ouvert dense de $X_A(\Sigma)$. On note $\bar X_A^\sigma(\Sigma)$ l'adhérence de $X_A^\sigma(\Sigma)$ dans $X_A(\Sigma)$. Si σ est un cône polyédral convexe et τ une face de σ, on pose $X_A^\tau(\sigma) := X_A(\tau - \tau) \subset X_A(\sigma)$.

(2.1.2) **Définition.-** *Soit Y un schéma sur A et $y \in Y$. On dit que Y est toroïdal sur A en y si y a un voisinage étale isomorphe sur A à un voisinage étale d'un point dans un A-schéma torique affine $X_A(\sigma)$.*

Le lemme suivant est bien connu quand A est un corps.

(2.1.3) **Lemme.-** *Soit Z un diviseur de Cartier effectif de $X_A(\Sigma)$ et $z \in Z$. Soit σ l'unique cône de Σ avec $z \in X_A^\sigma(\Sigma)$. Supposons que l'intersection schématique $Z \cap X_A^\sigma(\Sigma)$ soit lisse sur A en z et qu'elle ne soit pas égale à $X_A^\sigma(\Sigma)$ en z. Alors Z est toroïdal sur A en z. Si de plus l'éventail Σ est régulier, alors Z est lisse sur A en z.*

(2.1.4) Soit Δ un polytope convexe de dimension n dans $\mathbf{Q}^n$. Pour $b \in \mathbf{Q}^n$ on définit $F_\Delta(b)$ comme l'ensemble des $x \in \Delta$ tels que le produit scalaire $b.x$ soit minimal. C'est une face de Δ. Pour $\sigma \subset \mathbf{Q}^n$ on pose $F_\Delta(\sigma) = \bigcap\limits_{b\in\sigma} F(b)$. On définit la relation d'équivalence suivante sur $\mathbf{Q}^n$: $b \sim b'$ si et seulement si $F_\Delta(b) = F_\Delta(b')$. Les adhérences des classes d'équivalence définissent un éventail dans $\mathbf{Q}^n$, l'éventail associé à Δ. On le note $\Sigma(\Delta)$. L'application $\sigma \to F_\Delta(\sigma)$ établit une bijection entre les cônes de $\Sigma(\Delta)$ et les faces de Δ. On note $X_A(\Delta)$ le A-schéma torique $X_A(\Sigma(\Delta))$. Pour $\sigma \in \Sigma(\Delta)$ on écrit $X_A^\sigma(\Delta)$ au lieu de $X_A^\sigma(\Sigma(\Delta))$. Si τ est une face de Δ on écrira $X_A^\tau(\Delta)$ pour l'unique $X_A^\sigma(\Delta)$ avec $F_\Delta(\sigma) = \tau$.

2.2 - Compactifications toroïdales

Soit $f : \mathbf{T}_k^n \to \mathbf{A}_k^1$ un morphisme non-dégénéré pour $\Delta = \Delta_\infty(f)$. On suppose que $\dim \Delta = n$. Si Σ est un éventail plus fin que $\Sigma(\Delta)$ on peut construire une compactification g_Σ de f de la façon suivante.

On pose $A = k[T]$. Le tore $\mathbf{T}_A^n = \mathbf{T}_k^n \times \mathbf{A}_k^1$ est un ouvert dense de la variété torique $X_A(\Sigma) = X_k(\Sigma) \times \mathbf{A}_k^1$. On pose $G = f - T$ considéré comme polynôme de Laurent sur A. On note Y_Σ l'adhérence dans $X_A(\Sigma)$ de l'hypersurface $G = 0$ dans $\mathbf{T}_A^n$, et g_Σ le morphisme propre $g_\Sigma : Y_\Sigma \to \mathbf{A}_k^1$ induit par la projection $X_A(\Sigma) \to \operatorname{Spec} A = \mathbf{A}_k^1$. Comme le graphe de f est le lieu de $G = 0$ dans $\mathbf{T}_A^n$, on voit que $\mathbf{T}_k^n$ est un ouvert de Y_Σ et que la restriction de g_Σ à $\mathbf{T}_k^n$ coïncide avec f. Pour $\sigma \in \Sigma$ on pose

$$Y_\Sigma^\sigma = Y_\Sigma \cap X_A^\sigma(\Sigma) \quad \text{et} \quad \bar{Y}_\Sigma^\sigma = Y_\Sigma \cap \bar{X}_A^\sigma(\Sigma).$$

L'énoncé suivant est crucial.

Lemme-clé [D-L].— *Supposons que pour tout $\sigma \in \Sigma$ tel que $0 \in F_\Delta(\sigma)$ on ait $\sigma \in \Sigma(\Delta)$. Alors Y_Σ est toroïdal sur $k[T]$ en dehors d'un ensemble fini de points. En particulier, g_Σ est localement acyclique en dehors d'un ensemble fini de points.*

2.3 - Ramification modérée

On démontre le résultat suivant dans [D-L].

Théorème.— *Soit $f : \mathbf{T}_k^n \to \mathbf{A}_k^1$ un morphisme non dégénéré pour $\Delta_\infty(f)$. Alors $R^i f_! \mathbf{Q}_\ell$ est modérément ramifié à l'infini, quel que soit i.*

Pour démontrer ce résultat, on utilise en particulier que si R est un anneau de valuation discrète de caractéristique nulle ayant k pour corps résiduel, on peut relever f en un polynôme de Laurent $\widehat{f}$ à coefficients dans R avec $\Delta_\infty(\widehat{f}) = \Delta_\infty(f)$ qui est non dégénéré pour $\Delta_\infty(\widehat{f})$.

2.4 - Cohomologie d'intersection

(2.4.1) Soit X un schéma séparé de type fini sur $k = \mathbf{F}_q$ purement de dimension n. Soit $\mathcal{F}$ un $\bar{\mathbf{Q}}_\ell$-faisceau constructible localement constant sur un ouvert de Zariski dense U de X. Alors, d'après ([B-B-D], 1.4.14), il existe un unique objet $K^\bullet$ de la catégorie dérivée $D_c^b(X, \bar{\mathbf{Q}}_\ell)$ vérifiant $K^\bullet{}_{|U} = \mathcal{F}[n]$ tel que $K^\bullet$ et son dual de Verdier $\mathcal{D}K^\bullet$ satisfassent

(2.4.1.1) $H^i K^\bullet = 0$ pour $i < -n$.

(2.4.1.2) $\dim \operatorname{Sup} H^i K^\bullet < -i$ pour $i > -n$.

Cet objet unique est le complexe d'intersection sur X associé à $\mathcal{F}$, on le note $I^\bullet{}_X(\mathcal{F})$. On note $I^\bullet{}_X := I^\bullet{}_X(\bar{\mathbf{Q}}_\ell)$. La cohomologie d'intersection de X est l'hypercohomologie $\mathbf{H}^i(X \otimes \bar{k}, I^\bullet{}_X[-n])$.

(2.4.2) Le résultat suivant permet de calculer la cohomologie d'intersection des variétés toriques et fournit une interprétation cohomologique des polynômes $\alpha(\sigma)$ et $\beta(\Delta)$.

Théorème.— *Soit σ un cône polyédral dans $\mathbf{Q}^n$, ayant l'origine pour sommet, et soit Δ un polytope convexe dans $\mathbf{Q}^n$. On suppose que σ et Δ sont de dimension n. On pose $k = \mathbf{F}_q$. Pour tout entier i, on a*

(2.4.2.1) $\dim H^i\left((I^\bullet_{X_k(\sigma)}[-n])_0\right) =$ *coefficient de T^i dans $\alpha(\sigma)$*

(2.4.2.2) $\dim \mathbf{H}^i\left(X_k(\Delta), I^\bullet_{X_k(\Delta)}[-n]\right) =$ *coefficient de T^i dans $\beta(\Delta)$*

(l'indice 0 désignant la fibre à l'origine). De plus on a

(2.4.2.3) $\left(I^\bullet_{X_k(\sigma)}\right)_0$ *est pur de poids n.*

Ce résultat est démontré dans [D-L]. L'énoncé (2.4.2.2) pour $k = \mathbf{C}$ est énoncé sans preuve par Stanley dans [S] où il est attribué à I.N. Bernstein, A.G. Khovanskii et R.D. MacPherson. Après avoir rédigé une version préliminaire de [D-L] nous avons reçu un preprint de K.H. Fieseler ([F]) dans lequel (2.4.2.1) et (2.4.2.2) sont démontrés pour $k = \mathbf{C}$. La preuve de [F] utilise la théorie de Morse équivariante et le théorème de décomposition. Pour finir, notons que (2.4.2.1) et (2.4.2.2) pour $k = \mathbf{C}$ sont une conséquence formelle du cas où k est fini.

2.5 - Pour démontrer le théorème 1, on utilise la proposition suivante, jointe à (2.2) et (2.3), ainsi qu'un théorème de D. Bernstein, A.G. Kushnirenko et A.G. Khovanskii ([B-K-Kh]).

Proposition [D-L].— *Soit Y un schéma de pure dimension n sur k et $g : Y \to \mathbf{A}^1_k$ un k-morphisme propre. On suppose que g est localement acyclique en dehors d'un ensemble fini de points et que $R^i g_* \mathbf{Q}_\ell$ est modérément ramifié à l'infini pour tout i. Alors*

(2.5.1) $H^i_c(Y \otimes \bar{k}, g^*\mathcal{L}_\psi) = 0$ *pour $i > n$,*

(2.5.2) les applications naturelles $H^i_c(Y \otimes \bar{k}, g^*\mathcal{L}_\psi) \to H^i(Y \otimes \bar{k}, g^*L_\psi)$ *sont des isomorphismes pour tout entier i.*

Si de plus Y est lisse sur k, alors

(2.5.3) $H^i_c(Y \otimes \bar{k}, g^*\mathcal{L}_\psi) = 0$ *pour $i \neq n$,*

(2.5.4) $H^n_c(Y \otimes \bar{k}, g^*\mathcal{L}_\psi)$ *est pur de poids n.*

2.6 - Pour démontrer le théorème 2, on utilise (2.4) et la proposition suivante, que l'on applique à $Y = Y_{\Sigma(\Delta)}$ et $g = g_{\Sigma(\Delta)}$. (On peut déduire de (2.2) et (2.3) que les hypothèses de la proposition sont alors vérifiées.)

Proposition [D-L].— *Soit Y un schéma purement de dimension n sur k et soit $g : Y \to \mathbf{A}^1_k$ un k-morphisme propre. On suppose qu'en dehors d'un ensemble fini de points g est localement acyclique relativement à $I^\bullet{}_Y$ et que tous les $R^i g_* I^\bullet{}_Y$ sont modérément ramifiés à l'infini.*

Alors

(2.6.1) $\mathbf{H}^i_c(Y \otimes \bar{k}, I^\bullet{}_Y[-n] \otimes g^*\mathcal{L}_\psi) = 0$ *pour $i \neq n$,*

(2.6.2) $\mathbf{H}^n_c(Y \otimes \bar{k}, I^\bullet{}_Y[-n] \otimes g^*\mathcal{L}_\psi)$ *est pur de poids n.*

Références

[A-S 1] A. Adolphson, S. Sperber : *Exponential sums and Newton polyhedra.* Bull. Amer. Math. Soc. 16 (1987), 282-286.

[A-S 2] A. Adolphson, S. Sperber : *Exponential sums and Newton polyhedra : cohomology and estimates.* Annals of Math. 130 (1989), 367-406.

[A-S 3] A. Adolphson, S. Sperber : *Exponential sums on* $\mathbf{G}_m^n$. Inventiones Math. (A paraître.)

[B-B-D] A.A. Beilinson, J. Bernstein, P. Deligne : *Faisceaux pervers.* Société Mathématique de France, Astérisque 100 (1983).

[B-K-Kh] D. Bernstein, A.G. Kushnirenko, A.G. Khovanskii : *Newton polyhedra.* Usp. Mat. Nauk., 31 n° 3 (1976), 201-202.

[Da] V.I. Danilov : *The geometry of toric varieties.* Russ. Math. Surveys 33 (1978), 97-154.

[De] P. Deligne : *La conjecture de Weil II.* Publ. Math. IHES 52 (1980), 137-252.

[D-L] J. Denef, F. Loeser : *Weights of exponential sums, intersection cohomology and Newton polyhedra.* (A paraître.)

[F] K.-H. Fieseler : *Rational intersection cohomology of projective toric varieties.* (Preprint

[S] R. Stanley : *Generalized H-vectors, intersection cohomology of toric varieties, and related results.* Advanced Studies in Pure Math. 11 (1987), Commutative Algebra and Combinatorics, 187-213.

DE RHAM COHOMOLOGY AND THE GAUSS-MANIN CONNECTION FOR DRINFELD MODULES

Ernst-Ulrich Gekeler
Institut des Hautes Etudes Scientifiques
35, route de Chartres
91440 Bures-sur-Yvette, FRANCE

Introduction. The purpose of the present article is to give a survey on new developments in the theory of Drinfeld modules, in particular on those points of the theory that are related to non-archimedean analysis. Since 1974, when Drinfeld introduced the notion of "elliptic module" [2], now commonly called "Drinfeld module", this theory showed increasing importance in the arithmetic of function fields.

Drinfeld modules are "motives" over global function fields K that may be considered as analogues of "abelian varieties without polarization" over number fields. Their schemes of torsion points and the associated modular schemes have properties close to those of abelian varieties. By means of rank one Drinfeld modules, one can construct all the abelian class fields of K . For higher ranks, there exist (proved or conjectured) reciprocity laws relating Galois representations on the ℓ-adic cohomologies of modular schemes with automorphic representations of adele-valued groups over K ([2],[1],[3]). But there is an important difference : Whereas abelian varieties are related to the representation theory of Sp(2n) , Drinfeld modules correspond to GL(n) .

For abelian varieties A , the interplay of the different cohomology theories (Betti, ℓ-adic, de Rham) is crucial for the arithmetic of A . Thus it is natural to look for Drinfeld module substitutes of these theories. Concerning Betti and ℓ-adic cohomologies, the constructions are evident and known since Drinfeld's original paper. From a seminar held at the Institut for Advanced Study in Princeton 1987/88, a definition H_{DR}^* for "de Rham cohomology for Drinfeld modules" evolved, due to the efforts of several people (e.g. G. Anderson, P. Deligne, J. Yu, the author).

Strictly speaking, H_{DR}^* is not a cohomology theory but rather a functor corresponding to H_{DR}^1 , but with essentially all the properties one would expect from a first de Rham cohomology of an abelian variety (except there is no canonical polarization). Among them, let us mention :

(a) H^*_{DR} commutes with base extension ;

(b) over the K-analogue C of the complex numbers, there is a natural de Rham isomorphism of H^*_{DR} with H^*_{Betti} ;

(c) over special fields L (including L = C), there are GAGA-type isomorphisms ;

(d) one disposes of a formalism of "vanishing cycles" that relates H^*_{DR} of a degenerate Drinfeld module with that of its stable reduction ;

(e) there is a Kodaira-Spencer isomorphism that expresses H^*_{DR} through the tangent sheaf of certain modular schemes.

We will survey these properties (and prove some of them), focusing on the analytical aspects, and then apply them to the study of modular forms.

The plan of the paper is as follows : In sections 1 and 2, we fix the notation, give general definitions (section 1) and the well-known analytical constructions of Drinfeld modules (section 2). In section 3, we introduce the functors H^*_{DR} and discuss the base extension problem. Due to the base change property stated in Thm. 3.5, a Drinfeld module Φ of rank r over a scheme S will define a locally free $\mathcal{O}_S$-sheaf $H^*_{DR}(\Phi)$ provided with a canonical decomposition into a one-dimensional part $H^*_1(\Phi)$ and an (r-1)-dimensional part $H^*_2(\Phi)$ (3.12). In section 4, we consider Drinfeld modules defined over fields complete w.r.t. a non-archimedean absolute value. Over such fields, there exists an obvious analytical version $H^{*,an}_{DR}$ of the de Rham modules. Theorems 4.3, 4.6, and 4.9 describe the relationship between H^*_{DR} and $H^{*,an}_{DR}$, H^*_{DR} and H^*_{Betti}, and the formalism of vanishing cycles for degenerating Drinfeld modules, respectively. In section 5, we give a new construction of the universal additive extension of a Drinfeld module, based on the decomposition of H^*_{DR}. Section 6 introduces the Gauss-Manin connection. Drinfeld modules Φ may be differentiated with respect to tangent vectors of their base schemes. There results a Kodaira-Spencer map (6.6) from the tangent bundle to *Hom* $(H^*_1(\Phi), H^*_2(\Phi))$, which for *modular* Drinfeld modules is an isomorphism (Thm. 6.11).

This fact corresponds to the isomorphy of the sheaf of modular forms of weight two with the sheaf of differentials on elliptic modular schemes. In the last section, restricting to the rank two case, we apply the machinery developed to the study of modular forms. Drinfeld modules of rank two are very close to elliptic curves, and both the analytic and the algebraic part of elliptic modular forms theory may be translated to the Drinfeld case ([11] [12] [5]). Modular forms for an arithmetic subgroup Γ of GL(2,K) are rigid-analytic functions defined on an "upper half-plane" $\Omega \subset C$, transforming nicely under elements of Γ, and satisfying "cuspidal conditions". For an elliptic curve E in Weierstrass normal form, the differentials $\omega = dX/Y$ and $\eta = X\,dX/Y$ constitute a canonical basis of $H^1_{DR}(E)$, which is essential for the discussion given in [13], A1. In our context, "ω" exists a priori, but "η" has to be constructed, using the Gauss-Manin connection (Prop. 7.7). These forms transform under Γ like modular forms of a certain weight and type. This leads to a description of the ring $M(\Gamma)$ of modular forms for the group Γ and the "Serre derivation" $\partial : M(\Gamma) \to M(\Gamma)$ in terms of H^*_{DR} of the "generic" rank 2 Drinfeld module over Ω. As a by-product, we obtain a nice formula for the analogue of the classical Δ'/Δ as a conditionally convergent lattice sum (7.10).

As mentioned above, the point of view is analytical, so we avoid all the complications that result e.g. from choosing the proper "arithmetic" uniformizers at the cusps ([5], p. 38-39). We do *not* study the nature of the t-expansion coefficients. Further, we limit our efforts to considering modular forms for *maximal* arithmetic subgroups Γ of GL(2,K). Thus our approach is merely a first step in investigating $M(\Gamma)$, in that it introduces the technical tools for an arithmetical study.

The following terminology is used without further reference. For a ring R and $r \in R$, R^*, R/r is the multiplicative group, the factor ring R/rR, respectively. $\bar{L}$ is the algebraic closure of the field L, and #(S) the cardinality of the set S. "Locally free sheaves" and "vector bundles" over a scheme are used synonymously. In general, we do not distinguish between polynomials (or convergent power series) and the mappings they induce.

This article was written during a stay at the "Institut des Hautes Etudes Scientifiques" in Bures-sur-Yvette. The author takes the opportunity to thank the IHES and its staff for their hospitality. Thanks are also due to Deutsche Forschungsgemeinschaft for its support through a Heisenberg grant.

1. The basic set-up. Throughout the paper, we use the following notation :

$\mathbb{F}_q$ = finite field with q elements, $q = p^f$ with a prime number p ;
K = function field in one variable over $\mathbb{F}_q$ (we suppose $\mathbb{F}_q$ algebraically closed in K) ;
A = subring of elements of K regular away from
∞ = place of K, fixed once for all, of degree δ over $\mathbb{F}_q$;
|?| = normalized absolute value on K associated with ∞ (i.e., $|a| = \#(A/a)$ for $0 \neq a \in A$) ;
$\deg : K^* \to \delta\mathbb{Z}$ the corresponding degree function $\deg x = \log_q |x|$ (we also put $\deg 0 = -\infty$) ;
K_∞ = completion of K w.r.t. |?| ;

$C = \hat{\bar{K}}$ = completion of $\bar{K}_\infty$ w.r.t. the canonical extension of $|?|$ to K_∞
= smallest field extension of K_∞ that is complete and algebraically closed.

1.1. EXAMPLE : If K is the rational function field $\mathbb{F}_q(T)$ and "∞" the usual infinite valuation, we have $A = \mathbb{F}_q[T]$, "deg" is the usual degree of polynomials, and K_∞ the field of formal Laurent series $\mathbb{F}_q((T^{-1}))$.

The finite places of K (i.e., those different from ∞) correspond bijectively to the (non-zero) prime ideals of A. We shall not distinguish between the "valuation" and the "ideal" point of view, thus using "place", "valuation", and "prime ideal" as synonyms.

(1.2) Recall that for any field L of characteristic p, the endomorphism ring $\mathrm{End}_L(G_a)$ of the additive group scheme $G_a|L$ is the set of additive polynomials, i.e., polynomials of the form $\Sigma\, a_i X^{p^i}$, where the multiplication is defined by insertion of polynomials. Denoting by τ_p the polynomial X^p, $\mathrm{End}_L(G_a)$ is the twisted polynomial ring $L\{\tau_p\}$ subject to the commutation rule $\tau_p\, a = a^p\, \tau_p$ for constants $a \in L$. We let $\tau = \tau_q = \tau_p^f$ $(q = p^f)$ be the operator corresponding to X^q and $L\{\tau\}$ the subalgebra generated by τ. According to $\tau\, a = a^q\, \tau$, we always write "polynomials" in τ with left coefficients, and let $\deg_\tau$ the degree in τ.

(1.3) Next, assume that L comes equipped with a structure $\gamma : A \to L$ as A-algebra. Let $S = \mathrm{Spec}\, L$. A *Drinfeld A-module of rank* r over L (or over S) is given by a group scheme $G|L$, isomorphic with $G_a|L$, and a structure of scheme in A-modules

$$\Phi : A \to \mathrm{End}_L(G)$$

$$a \mapsto \Phi_a$$

on G such that the following hold :

(i) The two structures of scheme in A-modules on $\mathrm{Lie}(G)$ induced by Φ and by γ agree ;

(ii) For each non-zero element a of A, the group scheme $\mathrm{Ker}(\Phi_a)$ is finite of degree $|a|^r$.

Choose an isomorphism $\alpha : G \overset{\cong}{\to} G_a$, and let $\Phi^{(\alpha)} : A \to \mathrm{End}_L(G_a)$ be given by

$a \mapsto \Phi^{(\alpha)}_a = \alpha \circ \Phi_a \circ \alpha^{-1}$. Writing for $a \neq 0$

$$\Phi_a^{(\alpha)} = \sum_{i \le N} a_i^{(\alpha)} \tau^i \qquad (a_i^{(\alpha)} \in L,\ a_N^{(\alpha)} \neq 0),$$

we have $a_i^{(c\alpha)} = c^{1-q^i} a_i^{(\alpha)}$ $(c \in L^*)$, and our conditions (i), (ii) boil down to

(i') $a_i^{(\alpha)} = \gamma(a)$;

(ii') $N = r \cdot \deg a$.

Thus e.g. in the situation of (1.1), a Drinfeld module of rank r is given by $\Phi_T^{(\alpha)} = \sum_{i \le r} a_i^{(\alpha)} \tau^i$, where $a_i^{(\alpha)} \in L$, $a_0^{(\alpha)} = \gamma(T)$, and $a_r^{(\alpha)} \neq 0$.

Let now $S \to \operatorname{Spec} A$ be an arbitrary A-scheme and $\gamma : A \to \mathcal{O}_S(S)$ the corresponding ring homomorphism. A *Drinfeld module of rank* r *over* S is given by a line bundle $\mathcal{G}|S$ and a structure of scheme in A-modules

$$\Phi : A \to \operatorname{End}_S(G)$$

$$a \mapsto \Phi_a$$

on the additive group scheme G underlying $\mathcal{G}$ such that for any local isomorphism $\alpha : \mathcal{G} \xrightarrow{\cong} \mathcal{O}_S$, (i') and (ii') hold, where the "leading coefficients" $a_{r \cdot \deg a}^{(\alpha)}$ are units. A *morphism* $u : (G,\Phi) \to (G',\Phi')$ of Drinfeld modules is a morphism $u : G \to G'$ of group schemes that commutes with the A-actions on G and G'. Non-vanishing morphisms have finite kernels, they are called *isogenies* . Being isogeneous (connected by an isogeny) is an equivalence relation on Drinfeld modules ; it implies equality of ranks. The scheme of *a-division points* $(a \in A)$, or more generally, the scheme of $\mathfrak{a}$*-division points* of (G,Φ) $(\mathfrak{a} \subset A$ an ideal) is ${}_a\Phi = \operatorname{Ker} \Phi_a$ or ${}_{\mathfrak{a}}\Phi = \bigcap_{a \in \mathfrak{a}} \operatorname{Ker} \Phi_a$, respectively. If (G,Φ) is defined over the algebraically closed A-field L and $\mathfrak{a}$ is prime to $\operatorname{Ker} \gamma$, ${}_{\mathfrak{a}}\Phi(L)$ is isomorphic with $(\mathfrak{a}^{-1}/A)^r$ as an A-module. This follows easily from (i') and (ii').

(1.4) From now on, we abbreviate "Drinfeld module over S" by "Dmod|S" . An "r-Dmod" will be a Drinfeld module of rank r . Given a Dmod (G,Φ) over S and a base extension $S' \to S$, there is an induced Dmod (G',Φ') over S' . If there is no danger of confusion, we will write (G,Φ) or even Φ for (G',Φ') .

(1.5) Given any locally free sheaf $\mathcal{G}|S$ with underlying group scheme $G|S$, we will write $\mathcal{G} = \mathrm{Lie}(G)$. If $\mathcal{G}$ is a line bundle, we will not distinguish between (a) isomorphisms of $\mathcal{G}$ with $\mathcal{O}_S$; (b) isomorphisms of G with $G_a|S$; (c) nowhere vanishing sections of $\mathcal{G}^\wedge$ = dual bundle of $\mathcal{G}$. Considering Dmod's over a field L, we usually choose without reference an isomorphism $\alpha : G \xrightarrow{\cong} G_a$, or, what is the same, assume that $G = G_a$.

2. Analytical constructions. The contents of this section is entirely due to Drinfeld [2]. The constructions given should be compared with the Weierstrass parametrization of elliptic curves over $\mathbb{C}$ (2.4), or with Tate's parametrization of elliptic curves over local fields (2.10).

An *A-lattice in* C is a finitely generated (thus projective) discrete A-submodule Λ of C. The discreteness of Λ means that its intersection with each bounded subset of C is finite. For an A-lattice Λ, put

(2.1) $$e_\Lambda(z) = z \prod_{0 \neq \lambda \in \Lambda} (1 - z/\lambda).$$

It is easy to verify that

(2.2) (i) the product converges for any z in C, uniformly on bounded sets, thereby defining an entire, surjective, Λ-periodic map e_Λ from C to C;

(ii) e_Λ is $\mathbb{F}_q$-linear, thus defined by a convergent power series $e_\Lambda(z) = \sum_{i \geq 0} \alpha_i z^{q^i}$;

(iii) the zeroes of e_Λ are the points of $\Lambda \subset C$, and are simple.

In what follows, r is the projective rank of the A-lattice Λ. Fix some non-zero $a \in A$, and let $\Phi_a^\Lambda \in C\{\tau\}$ be defined by the commutative diagram (whose rows are exact by the above):

(2.3)
$$\begin{array}{ccccccccc} 0 & \to & \Lambda & \to & C & \xrightarrow{e_\Lambda} & C & \to & 0 \\ & & \downarrow a & & \downarrow a & & \downarrow \Phi_a^\Lambda & & \\ 0 & \to & \Lambda & \to & C & \xrightarrow{e_\Lambda} & C & \to & 0 \end{array}$$

By diagram chasing, we see that

(2.4) (i) $\Phi_a^\Lambda = a + \ldots + a_N \tau^N$, where $N = r \cdot \deg a$, $a_N \neq 0$, and

(ii) $a \mapsto \Phi^\Lambda_a$ is a ring homomorphism $\Phi^\Lambda : A \to C\{\tau\}$,

i.e., Φ^Λ defines a structure of Drinfeld module of rank r on $G_a|C$. The construction may be inverted :

2.5. THEOREM [2] : *The association* $\Lambda \mapsto \Phi^\Lambda$ *defines a bijection of the set of lattices of rank* r *in* C *with the set of structures of Drinfeld module of rank* r *on* $G_a|C$.

2.6. REMARK : If Λ, Λ' are rank r lattices and $c \in C^*$ is such that $c\Lambda \subset \Lambda'$, there exists a unique $u \in C\{\tau\}$, $u = c + o(\tau)$ that for $a \in A$ satisfies $u \circ \Phi^\Lambda_a = \Phi^{\Lambda'}_a \circ u$, i.e., a morphism $u : \Phi^\Lambda \to \Phi^{\Lambda'}$. Thus if we define morphisms of lattices by $\mathrm{Hom}(\Lambda,\Lambda') = \{c \in C \mid c\Lambda \subset \Lambda'\}$, we obtain in (2.5) an equivalence of categories. Also, we could easily give a coordinate-free version of the above, i.e., without assuming Φ to be defined on $G_a|C$.

(2.7) Next, we try to imitate the construction for local fields other than C. Let $\gamma : A \to L$ be an A-field with a rank one valuation v, valuation ring B and residue class field $L(v)$. Reduction mod v and everything derived from it will be denoted by a bar "$\bar{?}$". We assume that γ takes values in B, and, for simplicity, that L is complete and algebraically closed (e.g. L = completed algebraic closure of $K((t))$, or of $K_{\mathfrak{p}}$ with some $\mathfrak{p} \in \mathrm{Spec}\ A$). Let $\Phi : A \to L\{\tau\}$ be a Drinfeld module of rank r over L. Replacing Φ by the isomorphic module $\Phi' = c \circ \Phi \circ c^{-1}$ with $c \in L^*$ if necessary, we may assume that Φ takes values in $B\{\tau\}$, and that $\bar{\Phi} : A \to L(v)\{\tau\}$ is different from $\bar{\gamma} : A \to L(v)$. In that case, $\bar{\Phi}$ is a Dmod over $L(v)$ of some rank $0 < r_1 \le r$, whose isomorphism class depends only on that of Φ. We call $\bar{\Phi}$ the *stable reduction* of Φ. If $r_1 = r$, we say that Φ has *good reduction*.

(2.8) Now assume that the Dmod Φ over L, of rank r_1, has good reduction. A *lattice in* Φ is a discrete, finitely generated projective A-submodule of (L,Φ). As in (2.1), we form the lattice function e_Λ for the lattice Λ in Φ of rank r_2. It has properties similar to those described in (2.2). Defining $\psi^\Lambda_a \in L\{\tau\}$ by the commutative diagram

$$(2.9)\qquad \begin{array}{ccccccccc} 0 & \to & \Lambda & \to & L & \overset{e_\Lambda}{\to} & L & \to & 0 \\ & & \downarrow\Phi_a & & \downarrow\Phi_a & & \downarrow\psi^\Lambda_a & & \\ 0 & \to & \Lambda & \to & L & \overset{e_\Lambda}{\to} & L & \to & 0, \end{array}$$

we get

(2.10) (i) $\psi^\Lambda : a \mapsto \psi^\Lambda_a$ defines a Dmod of rank $r = r_1 + r_2$ over L;

(ii) ψ^Λ has stable reduction of rank r_1.

Furthermore,

2.11. THEOREM [2] : *The association* $(\Phi,\Lambda) \mapsto \psi^\Lambda$ *defines a bijection of the set of isomorphism classes of pairs* (Φ,Λ), *where* Φ *is an* r_1-Dmod *over* L *with good reduction and* Λ *is a lattice of rank* r_2 *in* Φ, *with the set of isomorphism classes of* r-Dmod's *over* L *with stable reduction rank* r_1.

We call (Φ,Λ) the *Tate data* of $\psi = \psi^\Lambda$.

2.12. REMARK : Although we will not need this fact, let us mention that the above bijection also comes from an equivalence of categories, as follows from [4], §3.

3. De Rham cohomology. In this section, we associate an L-vector space $H^*_{DR}(\Phi)$ of dimension r with any r-Dmod Φ over the A-field L. In the analogy of Drinfeld modules with elliptic curves, it plays the role of the first de Rham cohomology module. The construction is functorial and gives rise to a locally free sheaf $\mathcal{H}^*_{DR}(\Phi)$ over S, whenever Φ is defined over the A-scheme S. Perhaps, some of the definitions to follow seem to fall out of the blue. They are motivated in [9], section 2.3 (see also [14]).

We first assume that $S = \mathrm{Spec}\, B$ is an affine A-scheme and "Φ" = (G,Φ) a Dmod|S. "G_a" denotes the additive group scheme over S with the tautological A-action induced by the structural map $\gamma : A \to B$. In what follows, "Hom" and "$\otimes$" without subscript are relative to $\mathbb{F}_q$.

(3.1) We put

$M(\Phi) = \mathrm{Hom}((G,\Phi),G_a)$ = A-bimodule of $\mathbb{F}_q$-linear morphisms of S-group schemes from G to G_a;

$N(\Phi) = \{m \in M(\Phi) \mid \mathrm{Lie}(m) = 0\}$.

Note that the left and right actions of $\mathbb{F}_q$ on $M(\Phi)$ agree. We therefore may regard $M(\Phi)$ and $N(\Phi)$ as $A \otimes A$-modules.

3.2. EXAMPLE : Suppose that $G = G_a|S = G_a|B$. Then $M(\Phi) = \mathrm{Hom}_{B,\mathbb{F}_q\text{-lin}}(G_a,G_a) = B\{t\}$. The left action of $a \in A$ is left multiplication by $\gamma(a)$, its right action is right multiplication by Φ_a , and $N(\Phi)$ is the sub-A-bimodule $B\{\tau\}\tau$.

(3.3) An $\mathbb{F}_q$-linear map $\eta : a \mapsto \eta_a$ from A to $N(\Phi)$ that satisfies the derivation rule

$$\eta_{ab} = \gamma(a)\,\eta_b + \eta_a \circ \Phi_b$$

will be called a *"derivation"* . Each $m \in M(\Phi)$ defines through $\eta_a^{(m)} = \gamma(a)\,m - m \circ \Phi_a$ a derivation $\eta^{(m)}$, which is called *inner (strictly inner* if $m \in N(\Phi)$). We put $D(\Phi)$, $D_i(\Phi)$, $D_{si}(\Phi)$ for the A-bimodule of derivations, inner derivations, strictly inner derivations, respectively. Note that $M(\Phi), \ldots, D_{si}(\Phi)$ have natural B-module structures compatible with their left A-actions, thus may be considered as $B \otimes A$-modules.

3.4. DEFINITION : The de Rham module of Φ is the B-module

$$H^*_{DR}(\Phi) = D(\Phi)/D_{si}(\Phi) .$$

Obviously, our modules $M(\Phi)$, $N(\Phi)$, $D(\Phi)$, $D_i(\Phi)$, $D_{si}(\Phi)$, $H^*_{DR}(\Phi)$ are contravariant functors in Φ , i.e., for $u : \Phi \to \Phi'$ a morphism of Dmod's over S , there results a map $u^* : M(\Phi') \to M(\Phi)$, etc. If the need arises, we label $M(\Phi) = M(\Phi,B), \ldots, H^*_{DR}(F) = H^*_{DR}(\Phi,B)$. All these modules are also covariant functors in B (or contravariant functors in S), where for a base change $S' = \mathrm{Spec}\ B' \to \mathrm{Spec}\ B = S$, $M(\Phi,B')$ is short for $M((G \times S',\Phi'), B')$ (see (1.4), similar notation for $N, \ldots, H^*_{DR}$). We have canonical *base change morphisms*

$$B' \underset{B}{\otimes} M(\Phi,B) \to M(\Phi,B')$$
$$\vdots$$
$$B' \underset{B}{\otimes} H^*_{DR}(\Phi,B) \to H^*_{DR}(\Phi,B')\ .$$

The following results are proved in [9], section 4.

3.5. THEOREM : *The functors* $M(\Phi,?)$, $N(\Phi,?)$, $D(\Phi,?)$, $D_i(\Phi,?)$, $D_{si}(\Phi,?)$, $H^*_{DR}(\Phi,?)$ *on* B-*algebras commute with arbitrary base changes, i.e., the above base change morphisms are isomorphisms.*

3.6. PROPOSITION : *Let the* r-Dmod Φ *be defined over the* A-*field* B . *Then* $H^*_{DR}(\Phi,B)$ *is an r-dimensional* B-*vector space.*

Now allow Φ to be defined over a not necessarily affine A-scheme S . It follows from (3.5) that, up to unique isomorphism, there exist unique quasi-coherent S-sheaves $\mathcal{M}(\Phi), \ldots, \mathcal{H}^*_{DR}(\Phi)$ whose sections on open affine subschemes $S' = \mathrm{Spec}\ B' \subset S$ are given by $\mathcal{M}(\Phi)(S') = M(\Phi,B'), \ldots, \mathcal{H}^*_{DR}(\Phi)(S') = H^*_{DR}(\Phi,B')$. Evaluating on residue class fields of S yields

3.7. COROLLARY : *Let* Φ *be an* r-Dmod *over the* A-*scheme* S . *The sheaf* $\mathcal{H}^*_{DR}(\Phi)$ *is locally free of rank* r *on* S . *Its formation commutes with arbitrary (not necessarily affine) base changes.*

We conclude the section in describing $H^*_{DR}(\Phi)$ as a direct summand of $D(\Phi)$. Suppose for the moment that Φ is defined on $G_a|B$, where B is an A-algebra. Recall that in this case $M(\Phi) = B\{\tau\}$ and $N(\Phi) = B\{\tau\}\tau$ with the $B \otimes A$-structure determined by $\Phi : A \to B\{\tau\}$. As usual, we write $m = \sum m_i \tau^i \in M(\Phi)$ with left coefficients m_i , so $\deg_\tau m = \max\{i \mid m_i \neq 0\}$ is well-defined. Let a be an arbitrary non-constant element of A , and r the rank of Φ .

(3.8) A derivation $\eta \in D(\Phi)$ is *reduced (strictly reduced)*, if

$$\deg_\tau \eta_a \leq r \cdot \deg a \quad (\text{resp. } \deg_\tau \eta_a < r \cdot \deg a)\ .$$

For η reduced, we let $\mathrm{def}\,\eta = r \cdot \deg a - \deg_\tau \eta_a$ be the *defect* of η . Let $D_{sr}(\Phi) \subset D_r(\Phi) \subset D(\Phi)$ be the B-submodules of strictly reduced, reduced derivations, respectively.

3.9. PROPOSITION ([9], sect. 5) : *(i) The notions of reducedness and of defect do not depend on the choice of* a .

(ii) For each $\eta \in D(\Phi)$, *there exists a unique* $n \in N(\Phi)$ *such that* $\eta' = \eta - \eta^{(n)}$ *is reduced.*

Therefore, we have a canonical decomposition $D(\Phi) = D_r(\Phi) \oplus D_{si}(\Phi)$, and $D_r(\Phi)$ maps bijectively to $H^*_{DR}(\Phi)$. The derivation $\eta^{(1)}$ given by $\eta^{(1)}_a = a - \Phi_a$ is reduced but not strictly reduced. Since the leading coefficient of $\eta^{(1)}_a$ is a unit, subtracting a suitable multiple of $\eta^{(1)}$ from $\eta \in D_r(\Phi)$ defines a decomposition $D_r(\Phi) = B\,\eta^{(1)} \oplus D_{sr}(\Phi)$. We let

(3.10)
$$H^*_{DR}(\Phi) = H^*_1(\Phi) \oplus H^*_2(\Phi)$$

be the corresponding decomposition of H^*_{DR} .

(3.11) Finally, we allow $\Phi = (G,\Phi)$ to be defined over an arbitrary A-scheme S , where G belongs to the line bundle $\mathfrak{g} = \mathrm{Lie}(G)$. Locally choosing isomorphisms $\alpha : \mathfrak{g} \to \mathcal{O}_S$, we may apply the above considerations to define (strict) reducedness of local sections of the sheaf $D(\Phi)$. There result subsheaves (in fact, locally direct summands) $D_{sr}(\Phi) \subset D_r(\Phi) \subset D(\Phi)$, whose sections are the (strictly) reduced sections of $D(\Phi)$. Clearly, $D_r(\Phi) \xrightarrow{\cong} H^*_{DR}(\Phi)$. An isomorphism $\alpha : \mathfrak{g} \to \mathcal{O}_S$ defines a section $\eta^{(\alpha)}$ of $D_r(\Phi)$. Viewing α as a nowhere vanishing section of the dual bundle $\mathfrak{g}^\wedge = \mathrm{Lie}(G)^\wedge$, $\alpha \mapsto \eta^{(\alpha)}$ defines an injection of $\mathrm{Lie}(G)^\wedge$ into $D_r(\Phi)$. (Note that $\eta^{(c\alpha)} = c\,\eta^{(\alpha)}$, c a unit). The local considerations show that

(3.12)
$$D(\Phi) = D_1(\Phi) \oplus D_{sr}(\Phi) \oplus D_{si}(\Phi)$$

with $D_1(\Phi)$ = image of $\mathrm{Lie}(G)^\wedge$ = subsheaf generated by the sections $\eta^{(\alpha)}$, and

$$D_1(\Phi) \oplus D_{sr}(\Phi) = D_r(\Phi) \xrightarrow{\cong} H^*_{DR}(\Phi) = H^*_1(\Phi) \oplus H^*_2(\Phi) .$$

4. Analytical properties of H^*_{DR} . Throughout the section, we assume Drinfeld modules $\Phi : A \to L\{\tau\}$ to be defined on the additive group G_a over the A-field L , where either

(a) $L = C$, or

(b) L is as in the end of section 2, and $\gamma : A \to L$ takes values in the ring B of integers in L.

In any case, L will be an algebraically closed A-field complete with respect to a non-archimedean absolute value.

(4.1) Let $L\{\{\tau\}\}$ be the L-algebra of non-commutative power series in τ with coefficients in τ (subject to the usual rule $\tau a = a^q \tau$ for constants $a \in L$). Further, let $L_{ent}\{\{\tau\}\}$ be the subalgebra of those power series that define entire functions on L (i.e., power series $f = \sum a_i \tau^i$ such that for every $z \in L$, the series $f(z) = \sum a_i z^{q^i}$ converges). For example, the functions e_Λ lie in $L_{ent}\{\{\tau\}\}$. First, we define analytical versions of the de Rham modules $H^*_{DR}(\Phi)$. Put

(4.2) $M^{an}(\Phi) = L\{\{\tau\}\}$ (with the A-bimodule structure described by (3.2)) ;

$N^{an}(\Phi) = L\{\{\tau\}\}\tau$;

$D^{an}(\Phi) = \{\text{derivations } \eta : a \mapsto \eta_a \text{ from } A \text{ to } N^{an}(\Phi)\}$

(i.e., η is $\mathbb{F}_q$-linear and satisfies $\eta_{ab} = \gamma(a)\,\eta_b + \eta_a \circ \Phi_b$) ;

$D_i^{an}(\Phi)$ (resp. $D_{si}^{an}(\Phi)$)

$= \{\text{derivations of the form } \eta^{(m)},\ m \in M^{an}(\Phi) \quad (\text{resp. } N^{an}(\Phi))\}$

(recall that $\eta_a^{(m)} = \gamma(a)m - m \circ \Phi_a$) ;

$H^{*,an}_{DR}(\Phi) = D^{an}(\Phi)/D^{an}_{si}(\Phi)$.

The relationship between "algebraic" and "analytic" de Rham modules is given by the following GAGA-type result.

4.3. THEOREM : *Let Φ be a Drinfeld module over L, and suppose that (a) or (b) hold. The canonical map $c_\Phi : H^*_{DR}(\Phi) \to H^{*,an}_{DR}(\Phi)$ is an isomorphism .*

For the proofs in the two cases, which are quite different, see [9], sections 6 and 7.

Let us now give, in case a), a description of $H^*_{DR}(\Phi)$ related to "path integration on Φ ". By Thm. 2.5, Φ corresponds to some A-lattice Λ . Comparing with the Weierstrass parametrization of elliptic curves, we regard $H_{Betti}(\Phi) := \Lambda$ as a *Betti homology* group of $\Phi|C$. Hence it is natural to expect some kind of cycle integral pairing $H^*_{DR}(\Phi) \times \Lambda \to C$, or, equivalently, a de Rham map

$$DR : H^*_{DR}(\Phi,C) \to \mathrm{Hom}_A(\Lambda,C) =: H^*_{Betti}(\Phi,C)$$

with reasonable properties.

Let $e = e_\Lambda \in C_{ent}\{\{\tau\}\}$ be the function associated with Λ. The next lemma is taken from [8], section 2.

4.4. LEMMA : *(i) Given* $\eta \in D(\Phi)$ *and* $a \in A$ *non-constant, there exists a unique solution* $F_\eta \in C_{ent}\{\{\tau\}\}$ *of the functional equation*

$$(*) \qquad F_\eta(az) - a\,F_\eta(z) = \eta_a(e(z)) .$$

(ii) F_η *is independent of the choice of* a .

As we read off from (*), the restriction χ^η of F_η to Λ is A-linear, so we have a C-morphism $\eta \mapsto \chi^\eta$ from $D(\Phi)$ to $\mathrm{Hom}_A(\Lambda,C)$. It factors through $H^*_{DR}(\Phi,C)$ since for $\eta = \eta^{(n)}$ strictly inner, $F_\eta(z) = -n(e(z))$ vanishes on Λ . The induced map

$$DR : H^*_{DR}(\Phi) \to \mathrm{Hom}_A(\Lambda,C)$$

is called the *de Rham map* . If $\eta \in D(\Phi)$ and $v \in \Lambda$, we formally write $\int_v \eta$ for $DR(\eta)(v)$. The pairing is functorial. If $u : \Phi \to \Phi'$ is a morphism, $\eta \in D(\Phi')$, and $v \in H_{Betti}(\Phi)$,

$$(4.5) \qquad \int_v u^*(\eta) = \int_{u(v)} \eta ,$$

as is immediate from definitions.

4.6. THEOREM ([8], Thm. 5.14) : *For any Drinfeld module* Φ *over* C , DR *is an isomorphism* .

Actually, part (a) of Thm. 4.3 follows easily from (the proof of) the present theorem. Also, we may describe the effect of DR on the decomposition (3.10) :

4.7. COROLLARY ([8], Thm. 6.10) : *(i) The class of* $\eta^{(1)}$ *is mapped under* DR *to* -id , *where* id : $\Lambda \subsetneq C$ *is the canonical inclusion.*

(ii) An A-*character* $\chi : \Lambda \to C$ *corresponds under* DR *to an element of* $H_2^*(\Phi)$ *if and only if the limit*

$$S(1,\chi) = \lim_{s\to\infty} \Sigma\, \chi(\lambda)/\lambda \quad (0 \neq \lambda \in \Lambda\,, |\lambda| \leq s)$$

vanishes.

Functions F_η of the type considered above are similar to quasi-periodic functions for complex lattices.

(4.8) Now consider case b), i.e., the situation of (2.7)-(2.12). We relate the de Rham modules of ψ and of Φ, where ψ has Tate data (Φ,Λ) (see (2.11)). Some function $F : L \to L$ is *quasi-periodic* for the lattice Λ in Φ if

(i) F is $\mathbb{F}_q$-linear and entire (given by a power series in $L_{ent}\{\{\tau\}\}$) and

(ii) for each $a \in A$, F satisfies a functional equation

$$F(\Phi_a(z)) - \gamma(a)\, F(z) = \eta_a(e(z))$$

with some $\eta_a \in L\{\tau\}\,\tau$.

Here, $e = e_\Lambda : L \to L$ is the lattice function of Λ . One easily verifies that for F given, the map $a \mapsto \eta_a$ is a derivation $\eta \in D(\psi)$, and that F restricted to Λ is A-linear. We let $QP(\Lambda)$ be the L-vector space of quasi-periodic functions for Λ . The next results are proved in [9].

4.9. THEOREM ([9], Thm. 7.7) : *The sequence of L-vector spaces*

$$0 \to N(\psi) \xrightarrow{s} QP(\Lambda) \xrightarrow{t} \mathrm{Hom}_A(\Lambda,L) \to 0$$

is exact, where $s(n) = n \circ e$ *and* $t(F) = F|\Lambda$.

Therefore, there exists a canonical map $i : \mathrm{Hom}_A(\Lambda,L) \to H^*_{DR}(\psi)$, defined by the commutative diagram

(4.10)
$$\begin{array}{ccccccccc} 0 & \to & N(\psi) & \to & QP(\Lambda) & \to & \mathrm{Hom}_A(\Lambda,C) & \to & 0 \\ & & \wr\!\!\parallel \downarrow & & \downarrow & & \downarrow i & & \\ 0 & \to & D_{si}(\psi) & \to & D(\psi) & \to & H^*_{DR}(\psi) & \to & 0\,. \end{array}$$

Next, we define $j : H^*_{DR}(\psi) \to H^*_{DR}(\Phi)$ by composing the map $[\eta] \mapsto [\eta \circ e] : H^*_{DR}(\psi) \to H^{*,an}_{DR}(\Phi)$ with the inverse of c_Φ (see Thm. 4.3). Here, $\eta \circ e \in D^{an}(\Phi)$ is the derivation $(\eta \circ e)_a := \eta_a \circ e$. The wanted relationship between $H^*_{DR}(\psi)$ and the Tate data (Φ,Λ) of ψ is given by

4.11. THEOREM ([9], Thm. 7.12) : *The sequence of vanishing cycles*

$$0 \to \mathrm{Hom}_A(\Lambda,L) \xrightarrow{i} H^*_{DR}(\psi) \xrightarrow{j} H^*_{DR}(\Phi) \to 0$$

is exact .

4.12. REMARK : For ease of presentation, we assumed ψ to be defined on $G_a|L$. However, it is purely formal to write all the results of this section in a coordinate-free language. The sequence of vanishing cycles is functorial in ψ , since isogenies $u_\psi : \psi \to \psi'$ induce isogenies $u_\Phi : \Phi \to \Phi'$ of the stable reduction parts ([4],§3). Also, the assumption on L to be algebraically closed is not essential. Statements correponding to (2.11) and to (4.11) hold true for Drinfeld modules ψ defined over arbitrary A-fields L complete with respect to a non-archimedean absolute value, but are more complicated to formulate.

5. The universal additive extension. Here, we give a more geometrical description of $H^*_{DR}(\Phi)$, proposed by Deligne. Let (G,Φ) be defined over the affine A-scheme $S = \mathrm{Spec}\, B$. For $\eta \in D(\Phi)$ and $a \in A$, consider the matrix

(5.1)
$$\Phi^\eta_a = \begin{pmatrix} \gamma(a) & \eta_a \\ 0 & \Phi_a \end{pmatrix}.$$

The derivation rule translates to $\Phi^\eta_{ab} = \Phi^\eta_a \circ \Phi^\eta_b$. With η , we associate the extension of schemes in A-modules

$$[\eta] \qquad 0 \to G_a \to G_a \oplus G \to (G,\Phi) \to 0,$$

where A acts upon the middle term via the matrices Φ^η_a. Since $\mathrm{Lie}(\Phi^\eta_a) = \begin{pmatrix} \gamma(a) & 0 \\ 0 & \gamma(a) \end{pmatrix}$, the sequence of Lie algebras of $[\eta]$ splits canonically. It is easy to see that $[\eta]$ (provided with its Lie splitting) is trivial if and only if η belongs to $D_i(\Phi)$ (resp. $D_{si}(\Phi)$). Associating $[\eta]$ with its canonical Lie-splitting to $\eta \in D(\Phi)$ yields a commutative diagram

$$(5.2) \qquad \begin{array}{ccccccc} H^*_1(\Phi) & & \overset{\cong}{\leftarrow} & & D_i(\Phi)/D_{si}(\Phi) & \overset{\cong}{\to} & \mathrm{Lie}(G)^\wedge \\ \downarrow & & & & \downarrow & & \downarrow s \\ H^*_{DR}(\Phi) & & = & & D(\Phi)/D_{si}(\Phi) & \overset{\cong}{\to} & \mathrm{Ext}^\#((G,\Phi),G_a) \\ \downarrow & & & & \downarrow & & \downarrow \\ H^*_2(\Phi) & \overset{\cong}{\leftarrow} & D_{sr}(\Phi) & \overset{\cong}{\to} & D(\Phi)/D_i(\Phi) & \overset{\cong}{\to} & \mathrm{Ext}((G,\Phi),G_a), \end{array}$$

where all the horizontal arrows are isomorphisms. Here, Ext (resp. $\mathrm{Ext}^\#$) is the B-module of extensions of schemes in A-modules over S (provided with a Lie splitting), and s is described as follows : A local section α of $\mathrm{Lie}(G)^\wedge$ is a map from $\mathrm{Lie}(G)$ to $\mathrm{Lie}(G_a)$, thus a Lie splitting for the trivial extension of (G,Φ) by G_a, whose class is $s(\alpha)$.

(5.3) Next, let $V|S$ be the additive group scheme underlying the dual of the locally free B-module $D_{sr}(\Phi) \overset{\cong}{\to} \mathrm{Ext}((G,\Phi),G_a)$. We construct an extension (#) of schemes in A-modules of (G,Φ) by V, the latter being equipped with the tautological A-action. According to (5.1), we define (#) by

$$(\#) \qquad 0 \to V \to V \oplus G \to (G,\Phi) \to 0,$$

where $a \in A$ acts on $V \oplus G$ through the matrix

$$\Phi^\#_a = \begin{pmatrix} \gamma(a) & \eta^\#_a \\ 0 & \Phi_a \end{pmatrix}.$$

Here, $\gamma(a)$ is the scalar action of a on V, Φ_a the Drinfeld operator on G, and

$\eta_a^\# \in \mathrm{Hom}(G,V)$ is defined by

$$\eta_a^\#(x)(\eta) = \eta_a(x) \qquad (x \in G, \eta \in D_{sr}(\Phi)).$$

In fact

(5.4) (i) $\eta_a^\#(x)(\eta)$ is B-linear in η, hence $\eta_a^\#(x) \in V$;

(ii) $\eta_a^\#(x)$ is $\mathbb{F}_q$-linear in x;

(iii) $\eta^\#$ satisfies the derivation rule $\eta_{ab}^\# = \gamma(a)\,\eta_b^\# + \eta_a^\# \circ \Phi_b$, so $\Phi_{ab}^\# = \Phi_a^\# \circ \Phi_b^\#$.

5.5. PROPOSITION : (#) *is the universal additive extension of* (G,Φ) (see [14]).

This means : Let X be the additive group scheme underlying a line bundle $\mathfrak{X}$ over S, provided with the tautological A-action. For any extension of schemes in A-modules

$$(*) \qquad 0 \to X \to Y \to (G,\Phi) \to 0,$$

there exists a unique morphism $f : \mathrm{Lie}(V) \to \mathfrak{X}$ such that (*) is induced by (#) through f, and "$(*) \mapsto f$" defines an isomorphism $\mathrm{Ext}((G,\Phi),X) \xrightarrow{\cong} \mathrm{Hom}_S(\mathrm{Lie}(V),\mathfrak{X})$.

PROOF : Since $D_{sr}(\Phi)$ is canonically isomorphic with $\mathrm{Ext}((G,\Phi),G_a)$, we have

$$\mathrm{Hom}_S(\mathrm{Lie}(V),\mathfrak{X}) = \mathrm{Hom}_S(D_{sr}(\Phi)^\wedge,\mathfrak{X}) \xrightarrow{\cong} D_{sr}(\Phi) \otimes_B \mathfrak{X} \xrightarrow{\cong}$$

$$\mathrm{Ext}((G,\Phi),G_a) \otimes_B \mathfrak{X} \xrightarrow{\cong} \mathrm{Ext}((G,\Phi),X),$$

which is the inverse of the stated isomorphism.

(5.6) Finally, we drop the assumption that S be affine. Everything remains valid except that we have to replace locally free B-modules by locally free S-sheaves. The universal additive extension of $(G,\Phi)|S$ is an extension by the additive group V of the sheaf $\mathcal{E}xt((G,\Phi),G_a)^\wedge$, which is locally free of rank $r-1$ (r = rank of Φ). We have canonical isomorphisms of sheaves

$$H^*_{DR}(\Phi) = D(\Phi)/D_{si}(\Phi) \xrightarrow{\cong} \mathcal{E}xt^\#((G,\Phi),G_a) \xrightarrow{\cong} \mathrm{Lie}(V \oplus G)^\wedge,$$

where $(V \oplus G, \Phi^{\#})$ is the maximal additive extension. Applying Lie(?)^ to (#) yields the canonically split short exact sequence $0 \to H_1^*(\Phi) \to H_{DR}^*(\Phi) \to H_2^*(\Phi) \to 0$ of (3.12).

6. The Gauss-Manin connection. We show how tangent vectors of the base scheme S of a Drinfeld module (G,Φ) act on $H^*_{DR}(\Phi)$. Assume first that $S = \operatorname{Spec} B$ is affine. If $D \in \operatorname{Der}_A(B)$ is an A-derivation of B and $f = \Sigma f_i \tau^i \in B\{\tau\}$, we put $D(f) = \Sigma D(f_i) \tau^i$.

6.1. LEMMA : *Let* $f,g \in B\{\tau\}$. *Then* $D(f \circ g) = D(f) \circ g + f_0 D(g)$.

PROOF : $D(f \circ g) = D(\sum_k \sum_{i+j=k} f_i g_j^{q^i} \tau^k) = \sum_k \sum_{i+j=k} (D(f_i) g_j^{q^i} + f_i D(g_j^{q^i})) \tau^k$

$$= \sum_k \sum_{i+j=k} D(f_i) g_j^{q^k} \tau^k + f_0 \Sigma D(g_k) \tau^k = D(f) \circ g + f_0 D(g).$$

Next, we define an operator ∇_D on $N(\Phi)$. Locally, chose an isomorphism $\alpha : \operatorname{Lie}(G) \xrightarrow{\cong} \mathcal{O}_S$. Then $n \in N(\Phi)$ may be written $n = f \circ \alpha$ with $f \in B\{\tau\}\tau$. Define

(6.2) $$\nabla_D(n) = D(f) \circ \alpha.$$

6.3. LEMMA : *(i)* $\nabla_D(n)$ *is independent of the choice of* α *(in particular, is globally defined)*.

(ii) For $b \in B$, $\nabla_D(bn) = D(b)n + b\,\nabla_D(n)$ *holds*.

PROOF : (i) Let $\alpha' = b^{-1}\alpha$ with some unit $b \in B^*$, and $f = \Sigma f_i \tau^i$. We have $n = f \circ \alpha = f \circ b\alpha' = \Sigma b^{q^i} f_i \tau^i \circ \alpha'$. Since $f_0 = 0$, $D(b^{q^i} f_i) = b^{q^i} D(f_i)$, thus calculating $\nabla_D(n)$ with respect to α' gives the same result $\Sigma b^{q^i} D(f_i) \tau^i \circ \alpha' = \Sigma D(f_i) \tau^i \circ \alpha$ as with respect to α.

(ii) Product rule for D !

For $\eta \in D(\Phi)$ and $a \in A$, put

(6.4) $$(\nabla_D(\eta))_a = \nabla_D(\eta_a).$$

6.5. LEMMA : *(i)* $\nabla_D(\eta) \in D(\Phi)$ *(i.e., it is* $\mathbb{F}_q$*-linear in* a *and satisfies (3.3))*.

(ii) If $\eta = \eta^{(n)}$ *is strictly inner,* $\nabla_D(\eta^{(n)}) = \eta^{(\nabla_D(n))}$.

(iii) For $b \in B$, *we have* $\nabla_D(b\eta) = D(b)\eta + b\,\nabla_D(\eta)$.

PROOF : (i) Let $a,b \in A$, and write locally $\eta_a = f \circ \alpha$.

$$\nabla_D(\eta)_{ab} = \nabla_D(\eta_{ab}) = \nabla_D(\gamma(a)\,\eta_b + \eta_a \circ \Phi_b) = \gamma(a)\,\nabla_D(\eta_b) + \nabla_D(f \circ \alpha \circ \Phi_b \circ \alpha^{-1} \circ \alpha)$$

$$= \gamma(a)\,\nabla_D(\eta)_b + D(f \circ \alpha \circ \Phi_b \circ \alpha^{-1}) \circ \alpha = \gamma(a)\,\nabla_D(\eta)_b + D(f) \circ \alpha \circ \Phi_b$$

(by (6.1), since the "constant" coefficient of f vanishes)

$= \gamma(a)\,\nabla_D(\eta)_b + \nabla_D(\eta)_a \circ \Phi_b$. (ii) and (iii) follow the same lines.

By abuse of notation, we also call

$$\nabla_D : H^*_{DR}(\Phi) \to H^*_{DR}(\Phi)$$

the mapping induced on the de Rham module. The *Gauss-Manin connection for* Φ is the collection of operators $\{\nabla_D \mid D \in \mathrm{Der}_A(B)\}$. Note that the deviation from B-linearity of ∇_D is annihilated in the composite mapping

$$\pi_D : H^*_1(\Phi) \subsetneq H^*_{DR}(\Phi) \xrightarrow{\nabla_D} H^*_{DR}(\Phi) \xrightarrow{\text{projection}} H^*_2(\Phi).$$

Therefore, the *Kodaira-Spencer map*

$$\text{(6.6)} \qquad KS : \mathrm{Der}_A(B) \to \mathrm{Hom}_B(H^*_1(\Phi), H^*_2(\Phi))$$

$$D \mapsto \pi_D$$

is well-defined. Let now $a \in A$ have degree $d > 0$, and write locally $\Phi^{(\alpha)}_a = \sum_{i \le rd} a^{(\alpha)}_i \tau^i$ (recall that r = rank of Φ, so $a^{(\alpha)}_{rd}$ is a unit).

6.7. LEMMA : *The quotient* $E^{(\alpha)}_D := D(a^{(\alpha)}_{rd})/a^{(\alpha)}_{rd}$ *does not depend on* a.

PROOF : Let $b \in A$ have degree $e > 0$, $b^{(\alpha)}_{re}$ and $c^{(\alpha)}_{rd+re}$ the leading coefficients of $\Phi^{(\alpha)}_b$,

$\Phi^{(\alpha)}_{ab}$, respectively. Then

$$c^{(\alpha)}_{rd+re} = a^{(\alpha)}_{rd}\,(b^{(\alpha)}_{re})^{q^{rd}},$$

thus its logarithmic derivative w.r.t. D equals that of $a^{(\alpha)}_{rd}$.

Now define $D(\Phi)_a \in \mathrm{End}_S(G)$ by

(6.8) $$D(\Phi)_a = \alpha^{-1} \circ (\nabla_D(\eta^{(\alpha)})_a - E^{(\alpha)}_D\,\eta^{(\alpha)}_a).$$

6.9. LEMMA : *(i) This definition does not depend on the choice of* $\alpha : \mathrm{Lie}(G) \xrightarrow{\cong} \mathcal{O}_S$.

(ii) For $a,b \in A$, *the derivation rule holds* : $D(\Phi)_{ab} = \gamma(a)\, D(\Phi)_b + D(\Phi_a) \circ \Phi_b$.

PROOF : Straightforward, using $\eta^{(b\alpha)} = b\eta^{(\alpha)}$ $(b \in B^*)$ and (6.5).

Hence, applying the procedure in the last section, D defines an extension of (G,Φ) by the tautological A-module G. An element a of A acts on $G \oplus G$ through the matrix $\begin{pmatrix} \gamma(a) & D(\Phi)_a \\ 0 & \Phi_a \end{pmatrix}$. Mapping D to the corresponding extension class in $\mathrm{Ext}((G,\Phi),G) \xrightarrow{\cong} \mathrm{Ext}((G,\Phi),G_a) \otimes \mathrm{Lie}(G)$ yields a B-morphism

$$\mathrm{Der}_A(B) \to \mathrm{Ext}((G,\Phi),G_a) \otimes \mathrm{Lie}(G),$$

which, tracing back our canonical isomorphisms, is nothing else than the Kodaira-Spencer map (6.6).

Everything said so far extends immediately to non-affine base schemes S (compare (3.11) and (5.6)), replacing $\mathrm{Der}_A(B)$ by the relative tangent sheaf $\mathcal{T}_{S|A}$, and Hom_B, Ext by their sheaf versions $\mathcal{H}om_S$, $\mathcal{E}xt$.

(6.10) We now investigate the Kodaira-Spencer map in the case of *modular* Drinfeld modules. Let $\mathfrak{a}$ be a non-trivial ideal of A with support $\mathrm{supp}(\mathfrak{a}) \subset \mathrm{Spec}\, A$. We consider $M = M^r(\mathfrak{a}) \times (\mathrm{Spec}\, A - \mathrm{supp}(\mathfrak{a}))$, where $M^r(\mathfrak{a})$ is the modular scheme for r-Dmod's with a level $\mathfrak{a}$ structure (see e.g. [2], [5], [1]). In this context, a level $\mathfrak{a}$ structure on an r-Dmod Φ is an isomorphism of schemes in A-modules of $(\mathfrak{a}^{-1}/A)^r$ with the scheme ${}_{\mathfrak{a}}\Phi$ of $\mathfrak{a}$-division points. It is known ([2], 5.4 Cor.) that $M|A$ is smooth of relative dimension $r - 1$, hence the tangent sheaf $\mathcal{T} = \mathcal{T}_{M|A}$ is locally free of rank $r - 1$. Let (G,Φ) be the universal r-Dmod over M.

6.11. THEOREM : *In the above situation, the Kodaira-Spencer map*

$$KS : \mathfrak{T}_{M|A} \to Hom_M(H_1^*(\Phi), H_2^*(\Phi))$$

is an isomorphism.

Before giving the proof, let us state some lemmata. Let $\gamma : A \to L$ be an A-field, $L[\varepsilon]$ the ring of dual numbers over L (i.e., $\varepsilon^2 = 0$), and $\Phi : A \to L\{\tau\}$ an r-Dmod over L.

6.12. LEMMA : *The set of extensions of* Φ *to an* r-Dmod *over* $L[\varepsilon]$ *corresponds bijectively to* $D_r(\Phi)$.

PROOF : Let $a \mapsto \tilde{\Phi}_a$ be an extension $\tilde{\Phi} : A \to L[\varepsilon]\{\tau\}$ of Φ to $L[\varepsilon]$. Then $\tilde{\Phi}_a = \Phi_a + \varepsilon\eta_a$, where $\eta_a \in L\{\tau\}$, and (i) $\eta_a \in L\{\tau\}\tau$; (ii) $\deg_\tau \eta_a \le r \cdot \deg a$; (iii) $\eta_{ab} = \gamma(a)\,\eta_b + \eta_a \circ \Phi_b$. Conditions (i) and (ii) result from (i'), (ii') of (1.3), and (iii) from $\tilde{\Phi}_{ab} = \tilde{\Phi}_a \circ \tilde{\Phi}_b$ and $\varepsilon^2 = 0$. Therefore, $\eta \in D_r(\Phi)$, and conversely, any $\eta \in D_r(\Phi)$ defines an extension $\tilde{\Phi}$ as above.

6.13. LEMMA : *Suppose that* $\mathrm{Ker}(\gamma)$ *does not divide* $\mathfrak{a}$. *Let* $\tilde{\Phi}$ *be an extension of the* r-Dmod Φ *on* L *to* $L[\varepsilon]$. *Each structure of level* $\mathfrak{a}$ *on* Φ *extends uniquely to* $\tilde{\Phi}$.

PROOF : It suffices to verify this for $\mathfrak{a} = (a)$ principal. Let $\tilde{\Phi}_a = \Phi_a + \varepsilon\eta_a$. For $u + \varepsilon v \in L[\varepsilon]$, $\tilde{\Phi}_a(u + \varepsilon v) = \Phi_a(u) + \varepsilon(\gamma(a)\, v + \eta_a \circ u)$, so $u \mapsto u + \varepsilon v$ with $v = -\gamma(a)^{-1}\,\eta_a \circ u$ defines a canonical isomorphism of $\mathrm{Ker}(\Phi_a)$ with $\mathrm{Ker}(\tilde{\Phi}_a)$, which gives the assertion.

6.14. LEMMA : *Let* Φ^η, Φ^φ *be two extensions of* Φ *to* $L[\varepsilon]$ *associated with* $\eta, \varphi \in D_r(\Phi)$. *They are isomorphic if and only if modulo* $D_i(\Phi)$, η *and* φ *are conjugate under the automorphism group* $\mathrm{Aut}(\Phi)$ *of* Φ.

PROOF : Note first that the invertible elements of $L\{\tau\}$ are those of the form $u + \varepsilon v$, where $u \in L^*$ and $v \in L\{\tau\}$. Now

$$(u + \varepsilon v) \circ \Phi^\eta_a = \Phi^\varphi_a \circ (u + \varepsilon v)$$

is equivalent with the system of equations in $L\{\tau\}$:

(i) $u\Phi_a = \Phi_a \circ u$

(ii) $u\eta_a + v \circ \Phi_a = \gamma(a)v + \varphi_a \circ u$.

Equation (i) says that u is an automorphism of Φ. Writing (ii) in the form

$$u\eta_a - \varphi_a \circ u = \gamma(a)v - v \circ \Phi_a \quad (= \eta^{(v)})$$

and counting τ-degrees, this forces $\deg_\tau v \le 0$, i.e., $v \in L$, and $\eta \equiv u^{-1} \varphi \circ u$ modulo $D_i(\Phi) \cap D_r(\Phi)$.

PROOF OF THE THEOREM : It suffices to show the bijectivity of KS in maximal points $x \in M$. Let L be the residue field at x, $\tilde{x} = \operatorname{Spec} L[\varepsilon]$, and $\Phi|L$ the corresponding Drinfeld module. We may suppose that Φ lives on $G_a|L$, thus $H_1^*(\Phi) \xrightarrow{\cong} L$. For the fiber $\mathcal{T}(x)$ of $\mathcal{T}_{M|A}$ in x, we have

$\mathcal{T}(x) \xrightarrow{\cong}$ {morphisms of $\tilde{x}$ to M centered at x }

$\xrightarrow{\cong}$ {isomorphism classes of extensions $\tilde{\Phi}$ of Φ, provided with its level structure, to $\tilde{x}$ }

by the modular property of M. But Φ with its level structure is rigid, i.e., has no non-trivial automorphisms, so by the preceding lemmata, this set is in canonical bijection with $D_r(\Phi)/D_r(\Phi) \cap D_i(\Phi) \xrightarrow{\cong} D_{sr}(\Phi) \xrightarrow{\cong} H_2^*(\Phi)$. The composite is easily verified to agree with KS(x).

6.15. REMARKS : (i) Theorem 6.11 is most naturally formulated in terms of the *modular stack* M^r for rank r Drinfeld modules. Here, no need arises to introduce rigidifying level structures. Essentially the same deformation argument works to give a Kodaira-Spencer isomorphism over M^r.

(ii) Let R be an arbitrary ring extension of A. Since both $\mathcal{T}_{M|A}$ and the $H_i^*(\Phi,?)$ (i = 1,2) commute with base changes, there result Kodaira-Spencer isomorphisms $\mathcal{T}_{M\times R|R} \xrightarrow{\cong} \mathit{Hom}_{M\times R}(H_1^*(\Phi), H_2^*(\Phi))$ between the extended data. In the next section, we will use this to investigate modular forms.

7. Applications to modular forms. In this final section, we only consider Dmod's of rank 2. We first briefly review the description of modular schemes and modular forms which, in this special case, is quite similar to that of elliptic modular schemes and forms (see [13], A1). The standard reference on the subject is [5], labelled DMC, in particular chapters V,VI.

(7.1) Let Y be an A-submodule of K^2, projective, of rank 2, and $\Gamma = GL(Y)$ the subgroup $\{\gamma | Y\gamma = Y\}$ of $GL(2,K)$. Without restriction (see discussion pp. 71 of DMC), we may assume that $Y = \mathfrak{a}(1,0) + \mathfrak{b}(0,1)$ with ideals $\mathfrak{a}, \mathfrak{b}$ of A. In this case, Γ is the group

$$\Gamma = \left\{ \begin{pmatrix} a & b \\ c & d \end{pmatrix} \middle| \begin{array}{l} a,d \in A,\ ad - bc \in \mathbb{F}_q^* \\ b \in \mathfrak{a}^{-1}\mathfrak{b},\ c \in \mathfrak{a}\mathfrak{b}^{-1} \end{array} \right\}.$$

Each z in the "Drinfeld upper half-plane" $\Omega = C - K_\infty$ defines by $(1,0) \mapsto z$, $(0,1) \mapsto 1$ an injection i_z of K^2 into C. We let $Y_z = i_z(Y)$, which is a 2-lattice in C. Recall that Ω is a rigid analytic space over C, on which $GL(2,K_\infty)$ acts through fractional linear transformations: $\begin{pmatrix} a & b \\ c & d \end{pmatrix}(z) = (az+b)/(cz+d)$. The corresponding action of Γ on Ω has finite stabilizers, so the quotient $\Gamma\backslash\Omega$ of Ω by Γ exists as an analytic space. Furthermore,

$$z \mapsto \Phi^{(z)} = \text{2-Dmod corresponding to } Y_z$$

induces a bijection of $\Gamma\backslash\Omega$ with

$$M_\Gamma(C) = \text{set of isomorphism classes of 2-Dmod's } \Phi \text{ of type } Y$$

$$(\text{i.e., } H_{\text{Betti}}(\Phi)\ A\text{-isomorphic with } Y).$$

M_Γ is a component of the base extension $M^2 \times C$ of the coarse modular scheme for 2-Dmod's, and the above bijection is an isomorphism of smooth one-dimensional analytic spaces over C.

7.2. DEFINITION: A (holomorphic) *modular form of weight* k *and type* m (k a non-negative integer, m a class in $\mathbb{Z}/(q-1)$) for Γ is a C-valued function f on Ω that satisfies

(i) f transforms under $\gamma = \begin{pmatrix} a & b \\ c & d \end{pmatrix} \in \Gamma$ according to

$$f(\gamma z) = (\det \gamma)^{-m} (cz+d)^k f(z);$$

(ii) f is holomorphic on Ω;

(iii) f is holomorphic at the cusps of Γ.

Condition (iii) is explained in detail in DMC, pp. 44. For example, the holomorphy of f at the cusp "∞" of Γ means that for large values of $\inf\{|z - x| \mid x \in K_\infty\}$ (= small values of

$|t(z)|$), $f(z)$ has a convergent power series expansion in

$$t(z) = e_{\mathcal{C}}^{-1}(z),$$

where $e_{\mathcal{C}}$ is the function associated with the one-lattice $\mathcal{C} = \mathfrak{a}^{-1}\mathfrak{b}$ in C. Thus t is similar to the uniformizer $q(z) = \exp(2\pi iz)$ in the classical case. We put $M_k^m(\Gamma)$ for the finite-dimensional C-vector space of modular forms of weight k and type m for Γ, and

$$M(\Gamma) = \bigoplus_{k,m} M_k^m(\Gamma).$$

It is easy to write down some examples (see e.g. [10], DMC, [6]) :

7.3. EXAMPLES : (i) For $a \in A$, let $\Phi_a^{(z)} = \sum_{i \leq 2\deg a} \ell_i(a,z)\, \tau^i$. Then $(z \mapsto \ell_i(a,z)) \in M_k^0(\Gamma)$, where $k = q^i - 1$. We define the (nowhere vanishing) modular form $\Delta_a(z)$ as $\ell_{2\deg a}(a,z)$.

(ii) The *Eisenstein series* (first studied in [11])

$$E^{(k)}(z) = \sum_{0 \neq \lambda \in Y_z} \lambda^{-k}$$

defines an element of M_k^0 (non-zero if $k \equiv 0(q-1)$).

(iii) Let H be the subgroup of elements $\begin{pmatrix} a & b \\ 0 & 1 \end{pmatrix}$ of Γ, and put $\alpha_{k,m}(\gamma,z) = (\det\gamma)^{-m}\,(cz+d)^k$ for $\gamma = \begin{pmatrix} a & b \\ c & d \end{pmatrix}$. The *Poincaré series*

$$P_{k,m}(z) = \sum_{\gamma \in H \backslash \Gamma} \alpha_{k,m}^{-1}(\gamma,z)\, t^m(\gamma z)$$

$(k,m \in \mathbb{N},\ k \equiv 2m\,(q-1),\ m \leq k/(q+1))$ defines an element of $M_k^{m \bmod q-1}$.

(iv) If $A = \mathbb{F}_q[T]$ as in (1.1), the C-algebra $M(\Gamma)$ is a polynomial ring $C[g,h]$ in $g = E^{(q-1)}$ and $h = P_{q+1,1}$ ([6] 5.13. We would like to know a similar presentation of $M(\Gamma)$ for base rings A other than $\mathbb{F}_q[T]$. Unfortunately, all we know about the C-algebra $M(\Gamma)$ is its Hilbert function. For the subalgebra $M^0(\Gamma) = \bigoplus_k M_k^0(\Gamma)$, it is given in DMC, p. 86, and it may be calculated for $M(\Gamma)$ following the same lines).

(7.4) In what follows, Φ is the "generic" Drinfeld module over Ω, defined by the lattice Y_z, where z varies. By (7.3) (i), we can also regard Φ as being defined over the ring R of holomorphic functions on Ω, which is the point of view of [13], A1. We want to describe a canonical basis of $H^*_{DR}(\Phi,R) = H^*_1(\Phi,R) \oplus H^*_2(\Phi,R)$. The class $[\omega]$ of $\omega = \eta^{(1)} \in D_r(\Phi,R)$ spans $H^*_1(\Phi,R)$. We use the Gauss-Manin connection to construct a basis vector $[\eta]$ of $H^*_2(\Phi,R)$. Henceforth, we identify D_r and H^*_{DR}, i.e., we omit the brackets "[ ]".

Let θ be the differential operator $\frac{d}{dz}$, $a \in A$ non-constant, and put

$$E(z) = \Delta'_a(z)/\Delta_a(z) = \theta(\Delta_a)/\Delta_a. \tag{7.5}$$

It follows from (6.7) that E is independent of the choice of a.

7.6. REMARK : There exists a canonical "discriminant function" Δ, which is a modular form of weight $q^{2\delta}-1$ and type 0 for Γ, and such that $E = \theta(\Delta)/\Delta$. That Δ has remarkable properties : It has a product expansion like the classical $\Delta = (2\pi i)^{12} q \prod (1-q^n)^{24}$, and up to constants, all the Δ_a are powers of Δ. Thus E is similar to the "false Eisenstein series of weight 2" in the theory of elliptic modular forms. From Δ, it inherits a simple t-expansion ([6], 8.2). See also (7.10) and (7.35) !

7.7. PROPOSITION : *The map* $(\nabla_\theta - E)$ *restricted to* $H^*_1(\Phi,R)$ *is an* R*-isomorphism with* $H^*_2(\Phi,R)$.

PROOF : Since $\nabla_\theta(\omega_a) = \sum_{i \le 2\deg a} \ell'_i(a,z)\,\tau^i = \Delta'_a(z)\,\tau^{2\deg a}$ + terms of lower degree in τ, the derivation $(\nabla_\theta - E)(\omega)$ is strictly reduced, i.e., $\nabla_\theta - E$ agrees on H^*_1 with the Kodaira-Spencer map, and in particular, is R-linear. Let $\hat{R}_z$ be the completed localization of R at z, which agrees with the completed localization of $M_{\Gamma(\mathfrak{a})}(C) = \Gamma(\mathfrak{a})\backslash\Omega$ in z, provided that $\mathfrak{a}$ is a non-trivial ideal of A. Here, $\Gamma(\mathfrak{a})$ is the principal congruence subgroup $\mathrm{Ker}(\Gamma = GL(Y) \to GL(Y/\mathfrak{a}Y))$, which contains no non-trivial elliptic elements, so $\Omega \to \Gamma(\mathfrak{a})\backslash\Omega$ is étale around z (DMC, p. 50). Now the stated bijectivity over R follows from that over $\hat{R}_z (z \in \Omega)$, which in turn results from Thm. 6.11.

We let $\eta = (\nabla_\theta - E)(\omega) \in D_{sr}(\Phi,R) \xrightarrow{\cong} H^*_2(\Phi,R)$ be the corresponding basis vector.

Denoting by

$$r_z : D(\Phi,R) \to D(\Phi^{(z)},C)$$

the restriction mappings, we have, for each point $z \in \Omega$, a basis $\{\omega^{(z)} = r_z(\omega), \eta^{(z)} = r_z(\eta)\}$ of $H^*_{DR}(\Phi^{(z)},C)$.

Next, we give an alternate description of ∇_θ. Using Thm. 4.6, we extend the cycle integral pairing to a perfect R-pairing

$$\text{(7.8)} \qquad H^*_{DR}(\Phi,R) \times Y \underset{A}{\otimes} R \to R$$

$$\varphi \qquad v \mapsto \int_v \varphi,$$

where the integral evaluated on z is $\int_{i_z(v)} r_z(\varphi)$. Equivalently, we have the bijective de Rham map over R

$$DR : H^*_{DR}(\Phi,R) \overset{\cong}{\to} \mathrm{Hom}_A(Y,R)$$

given by $\varphi \mapsto (v \mapsto \int_v \varphi)$. It is straightforward to show that under DR, ∇_θ corresponds to $\theta = \frac{d}{dz}$, i.e., that ∇_θ is the unique C-endomorphism of $H^*_{DR}(\Phi,R)$ that verifies

$$\text{(7.9)} \qquad \int_v \nabla_\theta(\varphi) = \frac{d}{dz}\int_v \varphi \quad (v \in Y, \varphi \in D(\Phi,R)).$$

7.10. COROLLARY : *The function* $E(z)$ *on* Ω *satisfies*

$$E(z) = -\lim_{s\to\infty} \Sigma \frac{y_1}{i_z(y)} \quad (0 \neq y \in Y, |i_z(y)| \leq s).$$

With our particular choice $Y = \mathfrak{a}(1,0) + \mathfrak{b}(0,1)$, *this may be written*

$$E(z) = -\lim_{s\to\infty} \Sigma \frac{a}{az+b} \quad ((0,0) \neq (a,b) \in \mathfrak{a} \times \mathfrak{b}, |az+b| \leq s).$$

PROOF : Let $\chi^{(z)} \in \mathrm{Hom}_A(Y_z,C)$ be the image of $\eta^{(z)}$ under DR. From the above, if $y = (y_1,y_2)$, $\chi^{(z)}(i_z(y)) = (\frac{d}{dz} - E(z))(y_1z + y_2) = y_1 - E(z)(y_1z + y_2)$. Due to (4.7), $\chi^{(z)}$ has the average property :

$$0 = \lim_{s\to\infty} \Sigma \frac{\chi^{(z)}(i_z(y))}{i_z(y)} \quad \text{(summing over } 0 \neq y \in Y,\ |i_z(y)| \leq s)$$

$$= \lim (\Sigma \frac{y_1}{i_z(y)} - \#\{y \neq 0,\ |i_z(y)| \leq s\}\ E(z)).$$

But the number # is congruent to -1 mod q if s is large enough, thus the result.

Let for the moment Φ be an arbitrary 2-Dmod over C with lattice $\Lambda = H_{Betti}(\Phi)$ and de Rham map $DR : H^*_{DR}(\Phi,C) \to Hom_A(\Lambda,C)$. Each pair (λ,μ) of K-linearly independent elements of $\Lambda \underset{A}{\otimes} K$ defines through

(7.11) $$<\varphi',\varphi''>_{\lambda,\mu} = \chi'(\lambda)\,\chi''(\mu) - \chi'(\mu)\,\chi''(\lambda)$$

$(\chi' = DR(\varphi'),\ \chi'' = DR(\varphi''))$ a non-degenerate (Thm. 4.6 !) alternating pairing on $H^*_{DR}(\Phi,C)$. The dependence on (λ,μ) is described by

(7.12) $$< >_{\lambda\gamma,\mu\gamma} = (\det \gamma) < >_{\lambda,\mu} \quad (\gamma \in GL(\Lambda \underset{A}{\otimes} K) \text{ acting from the right}).$$

7.13. REMARK : These pairings are a substitute for the cup-product (or the Legendre determinant) in the case of complex elliptic curves. Let e.g. $Y = A^2$ and $x = (1,0)$, $y = (0,1)$. If $DR(\eta^{(z)}) = \chi^{(z)} \in Hom_A(Y_z,C)$, then $\chi^{(z)}(z)$ and $\chi^{(z)}(1)$ are the *quasi-periods* of $\Phi^{(z)}$. The case of $A = \mathbb{F}_q[T]$ is studied in detail in [7]. Up to a constant, the function $z \mapsto \chi^{(z)}(1)$ agrees with $E(z)$!

Back now to our general situation. We let $x = (x_1,x_2)$, $y = (y_1,y_2)$ be K-linearly independent elements of Y, and define the perfect alternating R-pairing $< >_{x,y}$ on $H^*_{DR}(\Phi,R)$ by

(7.14) $$<\varphi',\varphi''>_{x,y}(z) = <r_z(\varphi'),\ r_z(\varphi'')>_{i_z(x),i_z(y)}.$$

Then $<\omega,\eta>_{x,y} = <\omega,(\nabla_\theta - E)\,\omega>_{x,y} = <\omega,\nabla_\theta\,\omega>_{x,y}$ evaluated at z gives

(7.15) $$<\omega,\eta>_{x,y}(z) = i_z(x)\frac{d}{dz}\,i_z(y) - i_z(y)\frac{d}{dz}\,i_z(x) = x_2y_1 - x_1y_2 = \text{const.} \neq 0.$$

Let $\gamma = \begin{pmatrix} a & b \\ c & d \end{pmatrix}$ be an element of $GL(Y)$ and $u = (cz+d)^{-1}$. Multiplication by u yields an isomorphism $Y_z \to Y_{\gamma(z)}$. By (2.6), u defines an isomorphism, also denoted by $u : \Phi^{(z)} \xrightarrow{\cong}$

$\Phi^{(\gamma z)}$, i.e., for each $a \in A$, $u \circ \Phi^{(z)}_a = \Phi^{(\gamma z)}_a \circ u$. Let $u^* : H^*_{DR}(\Phi^{(\gamma z)}) \to H^*_{DR}(\Phi^{(z)})$ be the induced arrow on H^*_{DR}. Then from (3.11),

(7.16) $$u^*(\omega^{(\gamma z)}) = u \cdot \omega^{(z)} = (cz+d)^{-1}\,\omega^{(z)}.$$

We will show at once the transformation formula for η:

(7.17) $$u^*(\eta^{(\gamma z)}) = (\det\gamma)^{-1}\,(cz+d)\,\eta^{(z)}.$$

If we define a C-linear operation $\varphi \mapsto \varphi_{[\gamma]}$ of Γ on $D_r(\Phi,R)$ by

(7.18) $$r_z(\varphi_{[\gamma]}) = u^*(r_{\gamma z}(\varphi)),$$

these formulae say that ω (resp. η) transform like modular forms of weight -1, type 0 (resp. weight 1, type 1).

PROOF OF (7.17) : $u^*(\eta^{(\gamma z)})$ and $\eta^{(z)}$ are proportional since $H^*_2(\Phi^{(z)})$ is one-dimensional. Let $z' = \gamma z$. An elementary calculation yields $u\, i_z(x) = i_{z'}(x)\,\gamma^{-1}$, so

$<u^*(\omega^{(z')}), u^*(\eta^{(z')})>_{i_z(x),i_z(y)} = <\omega^{(z')}, \eta^{(z')}>_{ui_z(x),ui_z(y)}$ (by definition of $<>$ and (4.5))

$= (\det\gamma)^{-1} <\omega^{(z')}, \eta^{(z')}>_{i_{z'}(x),i_{z'}(y)} = (\det\gamma)^{-1} <\omega,\eta>_{x,y}(z')$

$= (\det\gamma)^{-1} <\omega^{(z)}, \eta^{(z)}>_{x,y}$ (by (7.12) and (7.15)). Comparing with (7.16), we obtain the stated proportionality factor.

Next, we consider the action of θ on t-expansions. Recall that $t(z) = e^{-1}(z)$, where $e = e_C$ as in (7.2). Since $\frac{de(z)}{dz} = 1$,

(7.19) $$\theta = \frac{d}{dz} = -t^2\frac{d}{dt}.$$

The next theorem is proved, in the special case $A = \mathbb{F}_q[T]$, in ([6], 6.10, 3.10). The proof given there generalizes without complications to arbitrary A.

7.20. THEOREM : *Let the t-expansion of the Eisenstein series* $E^{(k)}$ *be given by* $E^{(k)}(z) = \sum a_i t^i$. *If* $k = q^j - 1$, $a_i \neq 0$ *implies* $i \equiv 0$ *or* $q-1 \bmod q(q-1)$.

In view of (7.19), this implies

7.21. COROLLARY : θ^2 *annihilates* $E^{(k)}(z)$.

7.22. COROLLARY : θ^2 *annihilates all the coefficient forms* $\ell_i(a,z)$.

PROOF : Define the Eisenstein series $E^{(0)}(z)$ as the constant -1. Then for $k \geq 0$, the relation

$$a\,E^{(q^k-1)} = \sum_{i+j=k} E^{(q^i-1)}\,\ell_j(a,z)^{q^i}$$

holds (DMC, p. 15), which together with (7.21) implies the assertion.

7.23. COROLLARY : $\theta(E) = -E^2$.

PROOF : Let $a \in A$ be non-constant, so $E = \theta(\Delta_a)/\Delta_a$ and $\theta(E) = -(\theta(\Delta_a)/\Delta_a)^2 = -E^2$ since $\theta^2(\Delta_a) = 0$.

We can now completely describe the Gauss-Manin connection with respect to the basis $\{\omega,\eta\}$ of $H^*_{DR}(\Phi,R)$. By definition of η,

$$\nabla_\theta(\omega) = E\omega + \eta.$$

7.24. COROLLARY : $\qquad \nabla_\theta(\eta) = -E\eta.$

PROOF : $\nabla_\theta(\eta) = \nabla_\theta(\nabla_\theta - E)(\omega) = \nabla^2_\theta(\omega) - \theta(E)\,\omega - E\nabla_\theta(\omega)$. The term $\nabla^2_\theta(\omega)$ vanishes by (7.22), and due to (7.23), the rest gives $-E\eta$.

We use these results to construct a "Serre derivation" ∂ on $M(\Gamma)$. Let

$$S^\ell(R) = \mathrm{Symm}^\ell\,(H^*_{DR}(\Phi,R)) \tag{7.25}$$

be the ℓ-th symmetric power. Since $H^*_{DR} = H^*_1 \oplus H^*_2$, it decomposes

$$S^\ell(R) = \bigoplus_{0\leq v\leq \ell} S^{\ell-v,v}(R), \tag{7.26}$$

where $S^{u,v}(R) = H_1^{*\otimes u}(\Phi,R) \otimes H_2^{*\otimes v}(\Phi,R)$ with basis $\omega^{\otimes u} \otimes \eta^{\otimes v}$. The same type of isomorphism holds over any ring R' containing the coefficients of Φ and stable under $\theta = \frac{d}{dz} = -t^2\frac{d}{dt}$.

If $i,j \geq 0$, $k = i - j$, $j \equiv -m(q-1)$, condition (i) of Definition 7.2 is equivalent with

$$(7.27) \qquad f\,\omega^{\otimes i} \otimes \eta^{\otimes j} \text{ is } \Gamma\text{-invariant}.$$

Correspondingly, we may express the "cusp condition" (iii) using H^*_{DR}. As in (7.2), we explain this only for the cusp "∞". Let R_t be the subring of R consisting of functions f on Ω invariant under transformations $z \mapsto z + a$ ($a \in \mathfrak{c} = \mathfrak{a}^{-1}\mathfrak{b}$) and possessing a finite-tailed Laurent expansion with respect to t, i.e., $R_t = \{$functions "meromorphic at ∞"$\}$. Let further $B_\infty = C[[t]]$ and $L_\infty = C((t))$. Then R_t embeds into L_∞. As stated in (7.3) (i), the generic Dmod Φ is already defined over R_t. In $H^*_{DR}(\Phi, L_\infty)$, consider the B_∞-lattice W = span of $\{\omega, \eta\}$. Its ℓ-th symmetric power is a lattice in $S^\ell(L_\infty)$, direct sum of lattices $W^{u,v} = B_\infty\,\omega^{\otimes u} \otimes \eta^{\otimes v}$ $(u + v = \ell)$. The holomorphy of f at the cusp ∞ (condition (iii) of (7.2)) is now equivalent with

$$(7.28) \qquad f\,\omega^{\otimes i} \otimes \eta^{\otimes j} \in W^{i,j}.$$

Since the module $\mathrm{Diff}_C(R)$ of C-differentials of R is free with basis dz, the Gauss-Manin connection defines an arrow

$$(7.29) \qquad \nabla : H^*_{DR}(\Phi,R) \to H^*_{DR}(\Phi,R) \otimes \mathrm{Diff}_C(R)$$

$$\varphi \mapsto \nabla_\theta(\varphi) \otimes dz$$

which is equivariant for the action of Γ (see (7.18)). The transpose of the R-isomorphism given by (7.7)

$$\mathrm{Der}_C(R) \xrightarrow{\cong} \mathrm{Hom}_R(H^*_1, H^*_2)$$

$$\theta \mapsto (\nabla_\theta - E)|H^*_1$$

is

$$\mathrm{Hom}_R(H^*_2, H^*_1) \xrightarrow{\cong} \mathrm{Diff}_C(R),$$

where "dz" corresponds to $(\eta \mapsto \omega)$. Thus, taking (7.26) into account, ∇ defines a connection, also denoted by ∇

$$\nabla : S^\ell = \bigoplus_{0 \leq v \leq \ell} S^{\ell-v,v} \to S^\ell \otimes \mathrm{Hom}_R(H^*_2, H^*_1) = \bigoplus S^{\ell-v+1,v-1}$$

on $S^{\ell} = \mathrm{Symm}^{\ell}(H^*_{DR}(\Phi,R))$. Let now $f \in M_k^m(\Gamma)$, $i,j \in \mathbf{N}$ satisfy $k = i - j$, $j \equiv -m\ (q-1)$. Then $f\,\omega^{\otimes i} \otimes \eta^{\otimes j}$ is Γ-invariant and contained in $W^{i,j}$, as is its image under ∇. The latter is

(7.30)
$$\nabla(f\,\omega^{\otimes i} \otimes \eta^{\otimes j}) = \theta(f)\,\omega^{\otimes(i+1)} \otimes \eta^{\otimes(j-1)} + if\,\omega^{\otimes i}(E\omega + \eta) \otimes \eta^{\otimes(j-1)}$$
$$+ jf\,\omega^{\otimes(i+1)} \otimes \eta^{\otimes(j-2)}(-E\eta)$$
$$= (\theta(f) + kfE)\,\omega^{\otimes(i+1)} \otimes \eta^{\otimes(j-1)} + if\,\omega^{\otimes i} \otimes \eta^{\otimes j}.$$

(Note that "$\otimes dz$" simply amounts to replacing a factor ω by η).

Thus the $S^{i+1,j-1}$-component

(7.31)
$$\partial_k(f) = \theta(f) + kfE$$

satisfies conditions (i), (ii), (iii) "at ∞" of (7.2) with (k,m) replaced by $(k+2, m+1)$. The holomorphy of $\partial_k(f)$ at cusps other than "∞" follows the same lines. Therefore, $\partial_k(f) \in M_{k+2}^{m+1}(\Gamma)$. Direct calculation shows :

(7.32) If $f_i \in M_{k_i}^{m_i}$ $(i = 1,2)$, $f = f_1 f_2$, and $k = k_1 + k_2$ then

$$\partial_k(f) = \partial_{k_1}(f_1)\,f_2 + f_1\,\partial_{k_2}(f_2).$$

Hence the

7.33. PROPOSITION : *The* C-*linear map* $\partial : M(\Gamma) \to M(\Gamma)$ *given on* M_k^m by $\partial_k = \theta + kE$ *is a derivation. It maps* $M_k^m(\Gamma)$ *to* $M_{k+2}^{m+1}(\Gamma)$.

There exists a very simple proof avoiding H^*_{DR}. Writing $E = \theta(\Delta_a)/\Delta_a$, and applying θ to the transformation equation

$$\Delta_a(\gamma z) = (cz+d)^{q^{2\deg a}-1}\,\Delta_a(z)$$

yields the transformation law for E :

(7.34)
$$E(\gamma z) = (\det\gamma)^{-1}(cz+d)^2 E(z) + c(cz+d)/\det\gamma$$

which, together with the law for $\theta(f)$ and some considerations at the cusps, implies that ∂_k

maps M_k^m to M_{k+2}^{m+1}. Nevertheless, it is important to have the "de Rham" interpretation of modular forms, and of ∂ (see below).

We finish with some concluding remarks.

7.35. REMARKS : (i) From the product expansion of Δ (DMC, p. 76), it is easy to see that the t-expansion of E is $t + O(t^2)$. Since $\theta = -t^2 \frac{d}{dt}$, ∂_k increases the order of f at ∞ by at least 1. The same holds for the other cusps.

(ii) The uniformizer t at the cusp "∞" does not properly correspond to $q(z) = \exp(2\pi iz)$. All the results presented so far remain true if we replace $t(z)$ by $c \cdot t(z)$ and θ by $c \cdot \theta$ with some $c \in \mathbf{C}^*$. Choosing c carefully (i.e., c some analogue of $(2\pi i)^{-1}$), the normalized Eisenstein series $c^k E^{(k)}$ and the $c^{(q^i-1)} \ell_i(a,z)$ will have t-expansions with algebraic coefficients (in fact, with bounded denominators). For the case $A = \mathbb{F}_q[T]$, see [6]. For general A, there are some still unsolved questions of "best choice" of c to get "canonical" t-expansions (DMC, p. 38-39). The operator ∂ should then act on the A-module of *algebraic modular forms* [11]. The definition given there has to be generalized to include non-trivial "types".

(iii) Having chosen the "correct" uniformizer t, the coefficients will have interesting arithmetical properties. One may apply the "de Rham" description of modular forms, together with the formalism of vanishing cycles and of the Gauss-Manin connection, to investigate their congruence properties. I hope to come back later to that problem.

REFERENCES

[1] P. Deligne - D. Husemöller : Survey of Drinfeld modules. Contemp. Math. 67, 25-91, 1987

[2] V.G. Drinfeld : Elliptic modules (Russian). Math. Sbornik 94, 594-627, 1974. English translation : Math. USSR-Sbornik 23, 561-592, 1974

[3] Y. Flicker - D. Kazdan : Drinfeld moduli schemes and automorphic forms. To appear

[4] E. - U. Gekeler : Zur Arithmetik von Drinfeld-Moduln. Math. Ann. 262, 167-182, 1983

[5] E. - U. Gekeler : Drinfeld modular curves. Lecture Notes in Mathematics 1231. Springer-Verlag. Berlin - Heidelberg - New York 1986

[6] E. - U. Gekeler : On the coefficients of Drinfeld modular forms. Inv. Math. 93, 667-700, 1988

[7] E. - U. Gekeler : Quasi-periodic functions and Drinfeld modular forms. Comp. Math. 69, 277-293, 1989

[8] E. - U. Gekeler : On the de Rham isomorphism for Drinfeld modules. J. reine angew. Math. 401, 188-208, 1989

[9] E. - U. Gekeler : De Rham cohomology for Drinfeld modules. To appear in Sem. Th. des Nombres Paris, vol. 1988/89

[10] L. Gerritzen - M. van der Put : Schottky groups and Mumford curves. Lecture Notes in Mathematics 817. Springer-Verlag. Berlin - Heidelberg - New York 1980

[11] D. Goss : π-adic Eisenstein series for function fields. Comp. Math 41, 3-38, 1980

[12] D. Goss : The algebraist's upper half-plane. Bull. AMS 2, 391-415, 1980

[13] N. Katz : P-adic properties of modular schemes and modular forms. Lecture Notes in Mathematics 350. Springer-Verlag. Berlin - Heidelberg - New York 1973.

[14] B. Mazur - W. Messing : Universal extensions and one dimensional crystalline cohomology. Lecture Notes in Mathematics 370. Springer-Verlag. Berlin - Heidelberg - New York 1974

THE NONARCHIMEDEAN EXTENDED TEICHMÜLLER SPACE

Frank Herrlich

Fak. u. Inst. f. Math., Ruhr-Universität

Postfach 1o2148, D-463o Bochum

Let k be an algebraically closed field which is complete with respect to a nonarchimedean valuation. For an integer $g \geq 2$ consider the moduli scheme $\overline{\mathfrak{M}}_g$ of stable curves of genus g and denote by $\overline{M}_g(k)$ the analytification of $\overline{\mathfrak{M}}_g \times k$. Finally let $\overline{\mathcal{A}}_g(k)$ be the k-analytically open subset of $\overline{M}_g(k)$ whose points correspond to the stable Mumford curves of genus g over k. In other words: a point $x \in \overline{M}_g(k)$ is in $\overline{\mathcal{A}}_g(k)$ if and only if the canonical stable reduction of the stable curve C_x represented by x, is totally degenerate, i.e. has only rational irreducible components.

In this note we present a "Teichmüller theory" for $\overline{\mathcal{A}}_g(k)$: this means that we construct a k-analytic manifold $\overline{S}_g(k)$ on which the group Out F_g of outer automorphisms of a free group of rank g acts discontinuously with orbit space analytically isomorphic to $\overline{\mathcal{A}}_g(k)$. The points of $\overline{S}_g(k)$ are stable Mumford curves of genus g together with a basis of the uniformizing Schottky group. The advantage here is that the construction includes, in contrast to classical complex Teichmüller theory, not only smooth Mumford curves but also those with nodes. On the other hand, of course, the space $\overline{\mathcal{A}}_g(k)$ is only a part of the whole moduli space $\overline{M}_g(k)$. A similar construction has been carried out over the complex numbers in [1]. Here $\overline{S}_g(\mathbb{C})$ maps surjectively onto $\overline{M}_g(\mathbb{C})$, but the map is not the quotient map for the action of Out F_g.

The result of this paper has been obtained by M. Piwek [5] in a slightly different form: he constructs an algebraic Teichmüller space for totally degenerated stable curves over k, then considers a formal neighbourhood of this space in a suitable larger scheme, and finally analytifies this formal scheme.

§ 1. Uniformization of stable Mumford curves

The uniformization of Mumford curves by Schottky groups is well known, see e.g. [3]: the k-analytic universal covering of a Mumford curve C is an unbounded Stein domain Ω in $\mathbb{P}^1(k)$; Ω is the set of ordinary points of a Schottky group, i.e. a free hyperbolic subgroup Γ of $PGL_2(k)$ of rank g = genus of C, and the orbit space Ω/Γ is k-analytically isomorphic to C.

As mentioned in the introduction, by a stable Mumford curve we mean a stable curve over k with totally degenerate stable reduction. The k-analytic universal covering of such a curve C is an open dense connected part Ω of a tree of projective lines, together with the action of a free group Γ of rank g = arithmetic genus of C. More precisely: each irreducible component L of Ω is an unbounded Stein domain in a projective line $\mathbb{P}^1(k)$, and the subgroup $\Gamma_L = \{\gamma \in \Gamma : \gamma(L) = L\}$ acts as a Schottky group (of some rank $\leq g$) on $\mathbb{P}^1(k)$, with L the set of ordinary points of Γ_L. In particular, Γ acts discontinuously on Ω, and Ω/Γ is isomorphic to C.

Conversely, given a tree of projective lines X and a discontinuous action of a free group Γ of rank g on X such that the region of discontinuity Ω of Γ is connected, the orbit space Ω/Γ is a curve of arithmetic genus g which becomes stable if all irreducible components are contracted that are isomorphic to $\mathbb{P}^1(k)$ and have only one or two intersection points with other irreducible components.

These considerations show that stable Mumford curves over k correspond in a unique way to "stable" discontinuous actions of a free group on a tree of projective lines. The notions of "tree of projective lines" and "stability" will be made precise in the next sections.

§ 2. Trees of projective lines

In this section k may be an arbitrary field. The letters TPL shall stand for "tree of projective lines".

In the following we give some definitions and properties of TPLs and sketch their classification. More details and proofs can be found in the forthcoming article [4].

A finite TPL over k is, as in [2], § 1, a projective k-variety C, such that

(i) all irreducible components of C are isomorphic to $\mathbb{P}^1_k$

(ii) irreducible components intersect, if at all, in k-rational ordinary double points

(iii) the intersection graph of C is a tree

A morphism $f: C \to C'$ of finite TPLs over k is called a contraction if

(i) for each irreducible component L of C, $f|L$ is either a constant map or an isomorphism onto f(L)

(ii) for each irreducible component L' of C' there is at most one component L of C with f(L) = L'

Obviously the composition of contractions is again a contraction. Hence the finite TPLs over k with contractions form a subcategory $\underline{TPL^f(k)}$ of the k-varieties. A projective system in this category has a limit in the category of locally ringed spaces over k, and any locally ringed space which is the projective limit of finite TPLs over k will be called a TPL over k. It turns out that TPLs over k can be characterized as follows:

Proposition 1: A connected locally ringed space $(C,\mathcal{O})$ over k is a TPL over k if and only if

(i) each irreducible component of C is either isomorphic to $\mathbb{P}^1_k$, or is a single point x with $\mathcal{O}_x \cong k$

(ii) the irreducible components of C that are isomorphic to $\mathbb{P}^1_k$ are a dense subset of C

(iii) for any closed point $x \in C$, the local ring $\mathcal{O}_x$ is isomorphic to k or to $\mathcal{O}_{\mathbb{A}^1_k,0}$ or to $\mathcal{O}_{V,(0,0)}$, where $V = V(t_1t_2) \subset \mathbb{A}^2_k$

(iv) for any $x \in C$, $C - \{x\}$ has at most two connected components

(v) if L_1, L_2, L_3 are irreducible components of C such that $L_1 \cap L_2$ and $L_1 \cap L_3$ is nonempty, then $L_2 \cap L_3 = \emptyset$.

(vi) For any two irreducible components L_1, L_2 of C there is a unique minimal closed connected subset $S(L_1,L_2)$ containing L_1 and L_2.

A contraction of TPLs over k is defined in the same way as for finite TPLs. In particular we have the following observation:

Proposition 2: Let $f: C \to C'$ be a contraction of TPLs over k, and let $(C_i)_{i\in I}$ and $(C'_j)_{j\in J}$ be projective systems of finite TPLs over k such that $C = \varprojlim C_i$, $C' = \varprojlim C'_j$. Then there exists for each $j \in J$ an index $i(j) \in I$ and a contraction $f_j: C_{i(j)} \to C'_j$ such that

$$\begin{array}{ccc} C & \xrightarrow{f} & C' \\ \downarrow & & \downarrow \\ C_{i(j)} & \xrightarrow{f_j} & C'_j \end{array} \quad \text{is commutative.}$$

If L is an irreducible component of a TPL C over k, there is a (unique) contraction $\pi_L: C \to L$. For any three distinct closed points x_1, x_2, x_3 of C there is a unique irreducible component $L = L(x_1, x_2, x_3)$, called the <u>median component of</u> x_1, x_2, x_3, such that $\pi_L(x_1), \pi_L(x_2)$ and $\pi_L(x_3)$ are three different points of L. This property is obvious for finite TPLs, hence also for projective limits of finite TPLs. Note in particular that the median component of three points can never be a one-point-component (see also (iv) of Prop. 1). Let M be a set and C a TPL over k. A <u>stable M-marking</u> of C is an injective map $\phi: M \to C(k)$ satisfying

(i) for any irreducible component L of C there are ν_1, ν_2, ν_3 in M such that $L = L(\phi(\nu_1), \phi(\nu_2), \phi(\nu_3))$.

(ii) for each $\nu \in M$, $\pi_{L_\nu}^{-1}(\phi(\nu)) = \{\phi(\nu)\}$, where L_ν denotes the irreducible component of C with $\phi(\nu) \in L_\nu$.

The second condition ensures in particular that no double points are marked. It is possible, however, that end points of C are marked. Note that for finite TPLs this definition agrees with the one in [2], § 1.

Let Γ be a group and M a set on which Γ acts. A stable M-marking ϕ of a TPL C over k is called <u>Γ-equivariant</u> if there is a homomorphism $\rho: \Gamma \to \operatorname{Aut} C$ such that

$$\rho(\gamma)(\phi(\nu)) = \phi(\gamma\nu)$$

for all $\nu \in M$, $\gamma \in \Gamma$.

We shall apply this concept in the following situation: Γ is a free group of rank g, and $M = \mathring{\Gamma}$ is the set of primitive elements of Γ, on which Γ acts by conjugation. If then $\phi: \mathring{\Gamma} \to C(k)$ is a stable

Γ-equivariant marking of a TPL C over k, it is clear that $\phi(\gamma)$ is a fixed point of $\rho(\gamma)$ for each $\gamma \in \Gamma$. Since γ^{-1} is in $\mathring{\Gamma}$ if and only if γ is we see that each $\rho(\gamma)$ has at least two fixed points on C.

§ 3. Classification of stable marked TPLs

In [2] we classified stable n-pointed TPLs by cross ratios. This approach can be extended to stable M-marked TPLs for arbitrary M as follows:

Let $q = (C,\phi)$ be a stable M-marked TPL over a field k. For any $\nu = (\nu_1,\nu_2,\nu_3,\nu_4) \in Q(M) := \{(\nu_1,\ldots,\nu_4) \in M^4 : \nu_i \neq \nu_j \text{ for } i \neq j\}$ let

$$\lambda_\nu(q) := \pi_{L(\nu_1,\nu_2,\nu_3)}(\phi(\nu_4))$$

Here we normalize $\pi_{L(\nu_1,\nu_2,\nu_3)}$ by an isomorphism $L \to \mathbb{P}^1_k$ that sends $\pi_L(\phi(\nu_i))$ to $0,\infty$ and 1 for $i = 1,2,3$, resp.

As for finite M (see [2], 1.3) these $\lambda_\nu = \lambda_\nu(q)$ satisfy the following <u>cross ration relations</u>:

(i) $\lambda_{\nu_1,\nu_2,\nu_3,\nu_4} = \lambda^{-1}_{\nu_2,\nu_1,\nu_3,\nu_4}$

(ii) $\lambda_{\nu_1,\nu_2,\nu_3,\nu_4} = 1 - \lambda_{\nu_2,\nu_3,\nu_4,\nu_1}$

(iii) $\lambda_{\nu_1,\nu_2,\nu_3,\nu_4}\,\lambda_{\nu_1,\nu_2,\nu_4,\nu_5} = \lambda_{\nu_1,\nu_2,\nu_3,\nu_5}$

for all distinct $\nu_1,\ldots,\nu_5$ in M.

If ϕ is Γ-equivariant for a group Γ that acts on M the $\lambda_\nu(q)$ satisfy in addition the <u>Γ-invariance equations</u>

(iv) $\lambda_{\gamma(\nu_1),\gamma(\nu_2),\gamma(\nu_3),\gamma(\nu_4)} = \lambda_{\nu_1,\nu_2,\nu_3,\nu_4}$

for all $(\nu_1,\ldots,\nu_4) \in Q(M)$ and $\gamma \in \Gamma$.

The moduli spaces B_n in [2] are in the general situation replaced by the following "provarieties" B_M:

For $Q := Q(M)$ let $\mathbb{P}^Q = \mathbb{P}^Q_{\mathbb{Z}}$ be the projective limit (in the category of locally ringed spaces) of the projective schemes $\mathbb{P}^{Q'}_{\mathbb{Z}} := \prod_{\nu' \in Q'} \mathbb{P}^1_{\mathbb{Z}}$ for all finite subsets Q' of Q. Thus for any $\nu \in Q(M)$ we have a well

defined projection $\lambda_\nu : \mathbb{P}^Q \to \mathbb{P}^1_{\mathbb{Z}}$. Now let B_M be the closed subset of $\mathbb{P}^Q$ defined by all equations (i) - (iii). It is clear that B_M is itself the projective limit of Zariski-closed, i.e. projective subsets of the $\mathbb{P}^{Q'}_{\mathbb{Z}}$, hence carries a natural structure of provariety over $\mathbb{Z}$ (i.e. locally ringed space that is limit of a projective system of $\mathbb{Z}$-varieties).
In the situation where Γ acts on M we define B^Γ_M in the same way using equations (i) - (iv). In particular, B^Γ_M is a subprovariety of B_M.

Proposition 3: For any field k, the k-valued points of B_M (resp. B^Γ_M) are in 1-1 correspondence with the isomorphy classes of stable M-marked TPLs over k (resp. stable Γ-equivariantly M-marked TPLs over k).

The construction of the universal family of stable M-marked TPLs over B_M is carried out in the same way as in [2], § 3 for finite M: choose an element $m' \notin M$, and let $\overline{M} := M \cup \{m'\}$. Now consider the natural projection $p_M : B_{\overline{M}} \to B_M$. Since p_M factors through the $B_{\overline{M'}} \to B_{M'}$ for finite $M' \subset M$ it is easy to see that for $q \in B_M$ the fibre $p_M^{-1}(q)$ is exactly the stable M-marked TPL over k(q) associated with q by Prop. 3. It is also possible to show that this family p_M is universal for a suitable definition of the notion of family of TPLs.

If Γ acts on M we extend this action to $\overline{M}$ by $\gamma(m') = m'$ for all $\gamma \in \Gamma$. Then Γ acts on B_M and on $B_{\overline{M}}$, and p_M is Γ-equivariant. Thus over B^Γ_M, which is the fixed point set of the Γ-action on B_M, Γ acts on the fibres. This shows that $p_M : F^\Gamma_M \to B^\Gamma_M$ is the universal family of stable Γ-equivariantly marked TPLs, where we denote by
$F^\Gamma_M = p_M^{-1}(B^\Gamma_M) \subset B_{\overline{M}}$.

§ 4. The p-adic extended Teichmüller space

To come back to the situation of § 1, i.e. Schottky uniformization of stable Mumford curves we have to consider several subspaces of B_M^Γ for Γ and M as at the end of § 2, i.e. $\Gamma = F_g$, free group of rank g, and $M = \mathring{\Gamma}$. To simplify notation we write B^g for $B_{\mathring{F}_g}^{F_g}$.

Let U^g be the subset of B^g defined by the inequalities

$$\lambda_{\gamma,\gamma^{-1},\alpha,\gamma\alpha\gamma^{-1}} \neq 1$$

for all $\gamma,\alpha \in \Gamma$. U^g is not open for the projective-limit-topology on B^g, but there is a reasonable topology on B^g for which U^g is open. Also, if U^g and B^g are considered over a field k as in the introduction, U^g is open in the k-analytic topology.

For $q \in B^g$, $\lambda_{\gamma,\gamma^{-1},\alpha,\gamma\alpha\gamma^{-1}}(q)$ is the multiplier of $\rho(\gamma)$ on the median component of $\phi(\gamma),\phi(\gamma^{-1}),\phi(\alpha)$. That it should not be equal to one means that $\rho(\gamma)$ may not act as a parabolic transformation on any component of the TPL C associated with q. In particular, $\rho(\gamma)$ is not the identity on any component. As a consequence, $\rho(\gamma)$ has exactly two fixed points that are not singular points of C. Thus given the Γ-action on C we have exactly two possibilities to put the markings $\phi(\gamma)$ and $\phi(\gamma^{-1})$ on C. For any Γ-orbit c on $\mathring{\Gamma}$, i.e. for any conjugacy class c of primitive elements of Γ, let σ_c be the bijection of Γ given by $\sigma_c(\gamma) = \gamma^{-1}$ for $\gamma \in c$, $\sigma_c(\gamma) = \gamma$ if $\gamma \notin c$. Let Σ^g be the group of permutations of Γ generated by all σ_c, c Γ-orbit of $\mathring{\Gamma}$. Then clearly Σ^g acts on B^g. If B^g is written as a limit of projective varieties $B^{(n)}$, Σ^g acts on $B^{(n)}$ through a finite quotient $\Sigma^{(n)}$. Thus $B^{(n)}/\Sigma^{(n)}$ is again a projective variety, and the limit of the $B^{(n)}/\Sigma^{(n)}$ is the ("geometric") quotient of B^g by Σ^g.
Obviously U^g is Σ^g-invariant, and the above remarks yield

Proposition 4: The points in $\overline{U}_g := U^g/\Sigma^g$ correspond one-one to conjugacy classes of stable nonparabolic Γ-actions on TPLs (here "stable" means that any irreducible component of C that is isomorphic to $\mathbb{P}^1$ is median component of fixed points of Γ, and "conjugacy classes" refers to conjugation in Aut C).

For a field k as in the introduction, i.e. algebraically closed and complete with respect to a nontrivial nonarchimedean valuation, let $\overline{S}_g(k)$ be the subset of $\overline{U}_g(k)$ where the Γ-action on the corresponding TPL is discontinuous with connected region of discontinuity. Note that the last condition implies in particular that all intersection points of irreducible components are ordinary points of Γ, and that all irreducible components of C that are reduced to a single point are end points of C.

It is clear that for $q = (C,\phi) \in \overline{S}_g(k)$ the orbit space C/Γ is a stable Mumford curve over k. Moreover Out F_g acts in an obvious manner on $\overline{S}_g(k)$ (in fact Out F_g acts on B^g, and the group generated by Out F_g and Σ^g is a semidirect product, as $\phi \circ \sigma_c \circ \phi^{-1} = \sigma_{\phi(c)}$ for $\phi \in$ Out F_g and a conjugacy class c in $\mathring{F}_g$). As for each $q = (C,\phi) \in \overline{S}_g(k)$, C is the analytic universal covering of the stable Mumford curve C/Γ (see § 1), we conclude that C/Γ and C'/Γ are isomorphic if and only if (C,ϕ) and (C',ϕ') are in the same Out F_g-orbit. In other words we have set theoretically

$$(*) \qquad \overline{S}_g(k)/\text{Out } F_g = \overline{\mathcal{M}}_g(k).$$

If $q = (C,\phi) \in \overline{S}_g(k)$ and $\gamma \in \Gamma$, than γ has exactly two fixed points on C, one of them attracting and the other repelling. If $\phi(\gamma)$ is the attracting fixed point of γ, the multiplier $t_\gamma = \lambda_{\gamma,\gamma^{-1},\alpha,\gamma\alpha\gamma^{-1}}$ (for an arbitrary $\alpha \in \mathring{\Gamma}$, $\alpha \notin \gamma^{\mathbb{Z}}$) has absolute value < 1. On the other hand, if $c(\gamma)$ denotes the conjugacy class of γ, we have $\sigma_{c(\gamma)}(t_\gamma) = t_\gamma^{-1}$. This shows that the inverse image of $\overline{S}_g(k)$ in U^g is a disjoint union of copies of $\overline{S}_g(k)$. As a consequence we can restrict the universal family of TPLs over B^g to $\overline{S}_g(k)$, and forming the quotient for the Γ-action we obtain a family of stable (Mumford) curves over $\overline{S}_g(k)$. Hence once we have shown that $\overline{S}_g(k)$ carries an analytic structure we can conclude from the coarse moduli space property of $\overline{M}_g(k)$ that the canonical quotient map $\overline{S}_g(k) \to \mathcal{M}_g(k)$ is analytic, i.e. that (*) is also an analytic isomorphism. Therefore we call $\overline{S}_g(k)$ the <u>k-analytic extended Teichmüller space</u>.

<u>Proposition 5</u>: $\overline{S}_g(k)$ is an analytic submanifold of $\overline{U}_g(k)$ of dimension $3g - 3$.

The proof is modelled on M. Piweks proof [5] that the Teichmüller space for totally degenerated curves is a locally noetherian scheme: he first fixes a basis $\varepsilon = (\varepsilon_1,\ldots,\varepsilon_g)$ of Γ and considers subspaces U^g_ε of U^g where ε is a "geometric" basis for the Γ-action on the corresponding TPL. These subspaces are given by inequalities of the form $\lambda_\nu \neq \infty$ for certain ν. Thus U^g_ε is open in U^g in the same sense as U^g itself is open in B^g. He then restricts to the subset $\overline{U}^g_\varepsilon$ where all multipliers are 0, and proves that $\overline{U}^g_\varepsilon$ is an affine noetherian scheme. But following his arguments (in particular lemma 1.8) it is not hard to see that the same holds for U^g_ε itself: in addition to [5] one only has to show that the multiplier of an arbitrary element of Γ can be expressed by the multipliers of the ε_i and the $\lambda_{\varepsilon_i,\varepsilon_j,\varepsilon_k,\varepsilon_l}$. Finally one glues, as in [5], the spaces U^g_ε for various ε and obtains the result.

References

[1] Gerritzen, L. and Herrlich, F.: The extended Schottky space. J. r. angew. Math. 389 (1988), 19o-2o8.

[2] Gerritzen, L., Herrlich, F. and van der Put, M.: Stable n-pointed trees of projective lines. Indag. math. 5o(1988), 131-163.

[3] Gerritzen, L. and van der Put, M.: Schottky groups and Mumford curve, LNM 817, Springer 198o.

[4] Herrlich, F.: Moduli for group actions on trees of projective lines. To appear.

[5] Piwek, M.: The formal Teichmüller space. Preprint Bochum, 1989.

SUR LES COEFFICIENTS DE DE RHAM-GROTHENDIECK DES VARIETES ALGEBRIQUES

Mebkhout Z., Narvaez-Macarro L.

M.Z., UFR de Mathématiques LA 212,
Université de Paris 7, 2 Place Jussieu 25175 Paris.

N.L. Departamento de Algebra C/ Tarfia s/n
Universidat de Sevilla, 41012 Sevilla (Espagne).

Sommaire

§ 0. Introduction.

Dans son article fondamental qui date de l'année 1966 [G_2] Grothendieck a proposé de construire une théorie des coefficients de type de de Rham pour les variétés algébriques définies sur un corps de caractéristique p. On se propose dans cet exposé de faire le point sur les résultats de la théorie des coefficients de Rham-Grothendieck et sur les méthodes de démonstrations.

Grothendieck a montré que la cohomologie de de Rham d'une variété algébrique non singulière sur un corps de caractéristique nulle fournissait les bons nombres de Betti [G_1], il a défini la notion de connexion de Gauss-Manin d'un morphisme lisse sur un schéma de base quelconque [G_2] et enfin il a souligné la nécessité d'avoir des coefficients généraux qui « *joueraient le rôle de* **C**-*vectoriels transcendants algébriquement constructibles, et seraient stables par les opérations habituelles (telles que les* $R^if_*(\mathcal{O}_{X/S})$) » ([$G_3$], p. 105). Il s'agit, bien entendu, du cristal unité et de l'image directe *cristalline.*

Deligne [D_1] a défini sur une variété algébrique non singulière sur un corps de caractéristique nulle une catégorie de coefficients lisses, les fibrés à connexion intégrable à singularités régulières, qui quand le corps de base est **C** est équivalente à la catégorie des systèmes locaux transcendants d'espaces vectoriels de dimension finie par le foncteur de de Rham transcendant. Dans [D_2] Deligne à défini une catégorie de coefficients sur une variété algébrique complexe, les cristaux discontinus, qui est équivalente à la catégorie des faisceaux d'espaces vectoriels complexes algébriquement constructibles.

La théorie des $\mathcal{D}_X$-modules a complété les résultats de Grothendieck-Deligne. Sur toute variété X non singulière sur un corps k de caractéristique nulle on dispose de la catégorie $D^b_h(\mathcal{D}_{X/k})$ des complexes holonomes stable par les six opérations de Grothendieck f^*, f_*, $f^!$, $f_!$, $\otimes$, hom et par la dualité. De plus $D^b_h(\mathcal{D}_{X/k})$ contient comme sous-catégorie pleine la catégorie des complexes holonomes réguliers $D^b_{hr}(\mathcal{D}_{X/k})$ qui est aussi stable par les six opérations de Grothendieck et qui est équivalente quand le corps de base est **C** par le foncteur de de Rham transcendant à la catégories des coefficients algébriquement constructibles $D^b_c(C_X)$. Ceci remplit le programme de Grothendieck en caractéristique nulle. Ce résultat de la théorie des $\mathcal{D}_X$-modules est indépendant du résultat de Deligne sur les cristaux discontinus. Le manuscrit de Deligne [D_2] n'a malheureusement pas été publié.

Du point de vue des démonstrations la stabilité de la catégories $D^b_h(\mathcal{D}_{X/k})$ par les opérations cohomologiques, en particulier le théorème de finitude de la cohomologie de de Rham, n'utilise que l'équation fonctionnelle de Bernstein-Sato et est purement algébrique alors que la stabilité de la catégorie $D^b_{hr}(\mathcal{D}_{X/C})$ par les opérations cohomologiques et de leur comportement par le foncteur de de Rham transcendant, en particulier le théorème de comparaison, fait intervenir la théorie des équations différentielles complexes et reposaient de façon essentielle sur le théorème de résolution des singularités de Hironaka [H].

Si en caractéristique nulle on avait depuis longtemps une théorie satisfaisante il ne va pas de même en caractéristique $p > 0$. La cohomologie cristalline ([G_2], [B_1]) fournit une bonne théorie pour les variétés propres et lisses mais présente des pathologies pour les autres variétés. Grothendieck avait signalé [G_2] que la cohomologie de Dwork-Monsky-Washnitzer devraient être considérée comme la base d'une théorie des coefficients de type de Rham en caractéristique $p > 0$. En effet Berthelot a défini une catégorie de coefficients lisses les F-isocristaux surconvergents sur une variété défini sur un corps de caractéristique $p > 0$ qui donnent naissance à la cohomologie rigide qui est de dimension finie pour le crystal unité si l'on dispose de la résolution des singularités en caractéristique $p > 0$ [B_2].

Si la théorie des $\mathcal{D}_{X/C}$-modules est une théorie de coefficients au sens de Grothendieck, à plus d'un titre, elle introduit deux nouvelles théories cohomologiques qui n'étaient pas prévues dans le programme de Grothendieck et permettent d'aller de l'avant. D'une part elle donne naissance à la théorie des faisceaux pervers complexes et de la cohomologie perverse. Cette théorie garde un sens pour les faisceaux l-adiques pour l différent de la caractérique p du corps de base (cf. [B-B-D]). D'autre part elle donne naissance à la théorie des coefficients complexes holonomes d'ordre *infini*. La catégorie des complexes holonomes d'ordre infini qui est aussi équivalente à la catégorie des complexes algébriquement constructibles rend compte des phénomènes qui sont de nature transcendante tout en gardant un sens dans le cas p-adique. En particulier cette catégorie est stable par *immersion ouverte*. Ce résultat est le dernier pas dans la théorie des $\mathcal{D}_{X/C}$-modules et peut être considéré comme le résultat le *plus profond*.

Grâce à la théorie des faisceaux pervers et à la théorie de la ramification sur le corps des complexes on a pu montrer que le théorème de la résolution des singularités n'était pas * au fond du problème * dans la théorie complexe [Me_7]. Le théorème de résolution des singularités était considédé jusqu'à alors comme le théorème de base dans l'étude de la cohomologie des variétés algébriques et on a toujours pas de démonstration en caractéristique $p > 0$. Ceci encourage, au moins spychologiquement, à chercher à démontrer directement si les résultats de la théorie complexe ont des analogues dans le cas p-adique. C'est ce que nous avons essayé de faire depuis quelques temps.Cependant il nous semble qu'on se heurte à un problème de fond : alors que dans le cas complexe le prolongement analytique, c'est à dire la monodromie, est au coeur du problème en dernière analyse, c'est précisemment son analogue qui pose problème dans le cas p-adique. Dans ce but nous avons rédigé cet exposé à l'intention du lecteur-chercheur en analyse p-adique pour contribuer au débat qui ne sera clos que le jour où on aura une théorie des $\mathcal{D}_X$-modules en caractéristique $p > 0$ qui a toute la souplesse de la théorie en caractéristique zéro.

Voici le contenu de cet exposé. Dans le § 1 nous avons rassemblé quelques propriétés algébriques des faisceaux opérateurs différentiels. En partant de la définition de ([E.G.A.IV], §16) du faisceau $\mathcal{D}iff_{X/S}(\mathcal{O}_X)$ des opérateurs différentiels sur un S-schéma X nous avons rappelé les opérations cohomologiques que l'on peut faire sur les catégories de modules sur le modèle de la caractéristique nulle. La différence avec la situation en caractéristique nulle est que $\mathcal{D}iff_{X/S}(\mathcal{O}_X)$ est un faisceau de $\mathcal{O}_X$-algébres qui ne sont pas de type fini même dans le cas d'un S-schéma lisse. Pour remédier à cet inconvéniant Chase [Ch] et Smith [Sm] ont utilisé dans le cas d'un corps de base k algébriquement clos de caractéristique $p > 0$ ce qu'ils appellent la p-filtration du faisceau $\mathcal{D}iff_{X/k}(\mathcal{O}_X)$ pour une variété non singulière X. C'est une filtration par des sous-faisceaux *d'anneaux* qui sont næthériens. Ils ont montré que la dimension homologique de $\mathcal{D}iff_{X/k}(\mathcal{O}_X)$ est égale à dim(X) et on peut déduire de leur résultats que le faisceau $\mathcal{D}iff_{X/k}(\mathcal{O}_X)$ est *cohérent!*. Haastert [Ha] a étudié les opérations cohomologiques sur les $\mathcal{D}_{X/k}$-modules par passage à la limite à partir de la p-filtration. De même dans le cas d'un $\mathbf{Z}_p$-schéma de base S, Berthelot [B_3] a défini la filtration de $\mathcal{D}iff_{X/S}(\mathcal{O}_X)$ par les sous-faisceaux d'opérateurs différentiels d'échelon m. En fait la reduction modulo p de la filtration par les échelons est la p-filtration décalée d'une unité. Dans le cas d'ordre fini les extensions d'un échelon au suivant n'étant pas plates on ne peut pas en déduire que le faisceau $\mathcal{D}iff_{X/S}(\mathcal{O}_X)$ est cohérent. Mais d'aprés Berthelot [B_3] ceci est vrai pour la limite inductive des complétés p-adiques des faisceaux d'opérateurs d'échelon fixe tensorisée par Q. Il y a tout lieu de croire que tout se passe bien du point de vue de la dimension homologique. Tout semble indiquer que l'on a de point de vue algébrique une théorie cohérente de

$\mathcal{D}_{X/S}$-modules.

Dans le § 2 nous rappelons la théorie des $\mathcal{D}_{X/k}$-modules en caractéristique nulle sans théorème de résolution des singularités, ce qui se fait surtout grâce à une bonne généralisation en dimension supérieure du nombre de Fuchs attaché à un point singulier d'une équation différentielle. Dans le § 3 nous rappelons la théorie des $\mathcal{D}_X^\infty$-modules complexes à partir du § 2. Enfin dans le § 4 nous décrivons la théorie des $\mathcal{D}_{X\dagger/W}\otimes_Z Q$-modules et ce que nous savons de la théorie des $\mathcal{D}^\dagger_{X\dagger/W}\otimes_Z Q$-modules. En particulier nous montrons en dimension un que la cohomologie locale analytique de Dwork-Monsky-Washnitzer d'un point à valeur dans un $\mathcal{D}^\dagger_{X\dagger/W}\otimes_Z Q$-module définie par une équation différentielle est $\mathcal{D}^\dagger_{X\dagger/W}\otimes_Z Q$-cohérente si et seulement si cette équation admet un indice dans l'espace des fonctions analytiques dans la boule de rayon un qui est le tube de ce point. On peut donc appliquer les théorèmes de Robba [Rb_1], [Rb_2]. Tout se passe comme dans le cas complexe et fait penser que les propriétés de finitudes des $\mathcal{D}^\dagger_{X\dagger/W}\otimes_Z Q$-modules sont liées à leurs propriétés de finitude de leurs solutions à valeur dans les espaces de fonctions analytiques dans des tubes convenables. Dans ce dernier § on se place dans le cadre des schémas formels faibles de Meridith [Mr] pour être un peu original par rapport à Berthelot [B_3] qui se place dans le cadre des schémas formels. Mais il est clair que les deux points de vue sont nécessaires.

Le premier auteur a commencé à étudier la théorie des $\mathcal{D}_X$-modules et en particulier la théorie des $\mathcal{D}_X^\infty$-modules en juin 1972. Il nous a semblé dés 1975 que la théorie des $\mathcal{D}_X$-modules est le cadre naturelle de la théorie cristalline. En mai 1983 et durant l'année 1984-85 nous avons eu des discusions avec Grothendieck sur la théorie générale qu'il appelle "des coefficients de de Rham". Grothendieck pense d'ailleurs qu'il doit exister une théorie des coefficients de de Rham sur **Z**. En 1985 [N-M_2] nous avons cherché à montrer le théorème de finitude de la cohomologie de Dwork-Monsky Washnitzer sur le modèle de la caractéristique nulle à l'aide de la théorie de polynôme de Bernstein-Sato. A partir de ce moment là ayant constaté que cela est insuffisant nous avons cherché à developper la théorie des $\mathcal{D}^\dagger_{X\dagger/W}\otimes_Z Q$-modules qui devait être l'analogue p-adique de la théorie complexes des $\mathcal{D}_X^\infty$-modules et en particulier montrer que l'image directe par une immersion ouverte du fibré trivial est un $\mathcal{D}^\dagger_{X\dagger/W}\otimes_Z Q$-module de présentation finie.

Nous aimerions remercier P. Berthelot, F. Baldassarri et G. Christol des discussions que l' on a eues ces dernières années qui nous a permi de s'introduire à la théorie p-adique.

Notations. Si $\mathcal{A}$ est un faisceau d'anneaux nous noterons par $\mathcal{A}$-mod la catégories des $\mathcal{A}$-modules *à gauche* et mod-$\mathcal{A}$ la catégorie des $\mathcal{A}$-modules *à droite.* Si A est une catégorie abélienne nous noterons par D(A) la catégorie dérivée de la catégorie A et $D^b(A)$ sa sous-catégorie des complexes à cohomologie bornée. On note par f* le foncteur image inverse pour un morphisme d'espaces annelés f et par dim(f) sa dimension relative quand elle a un sens.

Nous conseillons au lecteur pour lire cet article d'avoir à la main le cours [Me$_5$] pour tout ce qui concerne la théorie algébro-géométrique des $\mathcal{D}_X$-modules.

§1. Le formalisme des faisceaux d'opérateurs différentiels.

1.1. Les opérations sur les catégories de modules sur les faisceaux d'opérateurs différentiels.

1.1.1. Soit $h : X \to S$ un morphisme d'espaces annelés et m un entier positif ou nul. Le faisceau $\mathcal{D}iff^{(m)}_{X/S}(\mathcal{O}_X)$ ([E.G.A. IV], §16) des opérateurs différentiels d'ordre m est défini par récurrence sur m comme le sous-faisceau du faisceau des endomorphismes $h^{-1}\mathcal{O}_S$-linèaires de $\mathcal{O}_X$ en posant $\mathcal{D}iff^{(0)}_{X/S}(\mathcal{O}_X) := \mathcal{O}_X$ et un endomorphisme P est un opérateur différentiel d'ordre m si et seulement pour toute section locale a de $\mathcal{O}_X$ l'endomorphisme $[P,a] := Pa - aP$ est un opérateur différentiel d'ordre m - 1. On pose

$$\mathcal{D}iff_{X/S}(\mathcal{O}_X) := \cup_m \mathcal{D}iff^{(m)}_{X/S}(\mathcal{O}_X).$$

1.1.2. Si $h : X \to S$ est un morphisme de schémas notons par $X^{(m)}$ le m-ème voisinage infinitésimal de X pour le morphisme diagonal $\Delta_h : X \to X \times_S X$, par h_m le morphisme canonique $X^{(m)} \to X \times_S X$ et par $p_1^{(m)}$, $p_2^{(m)}$ les deux morphismes composés

$$p_1^{(m)}: X^{(m)} \to X \times_S X \to X \quad p_2^{(m)} : X^{(m)} \to X \times_S X \to X .$$

Notons par $\mathcal{P}^{(m)}_{X/S}(\mathcal{O}_X)$ le faisceau des parties principales d'ordre m du S-schéma X. Par définition on a

$$\mathcal{P}^{(m)}_{X/S}(\mathcal{O}_X) := (p_1^{(m)})_* (p_2^{(m)})^* (\mathcal{O}_X).$$

Le faisceau des parties principales d'ordre m est muni d'une structure de $\mathcal{O}_X$-algèbre à gauche et d'une structure de $\mathcal{O}_X$-algébre à droite induites par les deux projections $p_1^{(m)}$, $p_2^{(m)}$. On a alors l'isomorphisme ([E.G.A.IV], 16.8.4)

$$hom_{\mathcal{O}_X}(\mathcal{P}^{(m)}_{X/S}(\mathcal{O}_X), \mathcal{O}_X) \cong \mathcal{D}iff^{(m)}_{X/S}(\mathcal{O}_X)$$

et le faisceau $\mathcal{D}iff_{X/S}(\mathcal{O}_X)$ des opérateurs différentiels est alors un faisceau *d'anneaux* filtré par ses sous-faisceaux $\mathcal{D}iff^{(m)}_{X/S}(\mathcal{O}_X)$.

1.1.3. Supposons que le morphisme h est lisse et soient $x_1,\ldots,x_n$ des sections du faisceau $\mathcal{O}_X$ au dessus d'un ouvert U telles que $dx_1,\ldots,dx_n$ forment une $\mathcal{O}_U$-base du faisceau $\Omega_{X/S}$. Alors les opérateurs différentiels $\Delta^q := \Delta_1^{q_1}\ldots\Delta_n^{q_n} := (\partial_{x_1}^{q_1}/ q_1!)\ldots(\partial_{x_n}^{q_n}/ q_n!)$ forment une $\mathcal{O}_U$-base du faisceau $\mathcal{D}iff^{(m)}_{X/S}(\mathcal{O}_X)$ pour $q = (q_1,\ldots,q_n)$ tels que $|q| \leq m$ ([E.G.A. IV], 16.11.2).

1.1.4. Supposons que le morphisme h est lisse et que le schéma S est localement nœthérien. Notons par ω_h le complexe dualisant relatif pour le morphisme h. Le théorème de dualité [R.D] pour le morphisme fini $p_1^{(m)}$ s'écrit :

$$(p_1^{(m)})_*(p_1^{(m)})^!(\mathcal{O}_X) \cong hom_{\mathcal{O}_X}(\mathcal{P}^{(m)}_{X/S}(\mathcal{O}_X),\mathcal{O}_X).$$

Mais on a les isomorphismes de dualité

$$(p_1^{(m)})^!(\mathcal{O}_X) \cong (h_m)^!((p_1)^!(\mathcal{O}_X)) \cong \mathbf{R}hom_{\mathcal{O}_{X\times_S X}}(\mathcal{O}_{X^{(m)}}, (p_1)^!(\mathcal{O}_X)),$$

$$(p_1)^!(\mathcal{O}_X) \cong \omega_{p_1}[\dim(p_1)] \cong (p_2)^*(\omega_h)[\dim(h)].$$

D'où l'isomorphisme

$$\mathbf{R}hom_{\mathcal{O}_{X\times_S X}}(\mathcal{O}_{X^{(m)}}, (p_2)^*(\omega_h))[\dim(h)] \cong \mathcal{D}iff^{(m)}_{X/S}(\mathcal{O}_X).$$

Soit

$$\mathcal{E}xt^{\dim(h)}_{\mathcal{O}_{X\times_S X}}(\mathcal{O}_{X^{(m)}}, (p_2)^*(\omega_h)) \cong \mathcal{D}iff^{(m)}_{X/S}(\mathcal{O}_X).$$

Prenons la limite avec m on trouve

$$\lim_{\substack{\rightarrow \\ m}} \mathcal{E}xt^{\dim(h)}_{\mathcal{O}_{X\times_S X}}(\mathcal{O}_{X^{(m)}}, (p_2)^*(\omega_h)) \cong \mathcal{D}iff_{X/S}(\mathcal{O}_X).$$

Soit

$$\mathcal{H}_X^{\dim(h)}((p_2)^*(\omega_h)) \cong \mathcal{D}iff_{X/S}(\mathcal{O}_X)$$

qui est la définition de Mikio Sato des opérateurs différentiels ([S], [S.K.K]), incarnation en théorie des hyperfonctions du théorème des noyaux de Schwartz.

1.1.5. Notons $\mathcal{D}_{X/S}$ le faisceau des opérateurs différentiels $\mathcal{D}iff_{X/S}(\mathcal{O}_X)$. Si le morphisme h est lisse la donnée d'une structure de $\mathcal{D}_{X/S}$-modules *à gauche* sur un un $\mathcal{O}_X$-module quasi-cohérent $\mathcal{M}$ est équivalente à la donnée d'une *-stratification relativement à S sur $\mathcal{M}$ ([G$_2$], [B$_1$]) c'est à dire d'un isomorphisme pour tout m

$$p_1^{(m)*}\mathcal{M} \cong p_2^{(m)*}\mathcal{M}$$

satisfaisant à la condition de cocycle sur les images *-inverses sur les m-ème voisinages infinitésimaux de la diagonale du produit triple $X\times_S X\times_S X$ et compatibles quand m varie. Le faisceau structural $\mathcal{O}_X$ est donc un $\mathcal{D}_{X/S}$-module *à gauche*.

De même si le morphisme h est lisse la donnée d'une structure de $\mathcal{D}_{X/S}$-module *à droite* sur un un $\mathcal{O}_X$-module quasi-cohérent $\mathcal{M}$ est équivalente à la donnée d'une !-stratification relativement à S sur $\mathcal{M}$ c'est à dire d'un isomorphisme pour tout m

$$p_1^{(m)!}\mathcal{M} \cong p_2^{(m)!}\mathcal{M}$$

satisfaisant à la condition de cocycle sur les images !-inverses sur les m-ème voisinages infinitésimaux de la diagonale du produit triple $X\times_S X\times_S X$ et compatibles quand m varie. Cette description des $\mathcal{D}_{X/S}$-modules *à droite* est due à Grothendieck et à Berthelot. Il résulte du théorème de dualité pour les morphismes $p_1^{(m)}, p_2^{(m)}$ que le faisceau dualisant relatif ω_h est un $\mathcal{D}_{X/S}$-module *à droite*.

De même par dualité il résulte que le foncteur $\mathcal{M} \to \omega_h \otimes_{\mathcal{O}_X} \mathcal{M}$ est une équivalence de catégories entre la catégorie des $\mathcal{D}_{X/S}$-modules *à gauche* quasi-cohérents et la catégorie des $\mathcal{D}_{X/S}$-modules *à droite* quasi-cohérents. Le foncteur $\mathcal{N} \to \mathit{hom}_{\mathcal{O}_X}(\omega_h, \mathcal{N})$ étant un quasi-inverse.

Dans [Ha] Haastert décrit, sur un corps de base de caractéristique p, les $\mathcal{D}_{X/S}$-modules *à gauche* en terme de limite projective et les $\mathcal{D}_{X/S}$-modules *à droite* en terme de limite inductive et en déduit que ces deux catégories sont équivalentes.

1.1.6. Les $\mathcal{D}_{X/S}$-modules à gauche peuvent se décrire comme les *cristaux* de modules, objets du topos critallin ([G$_2$], [B$_1$]). De même, d'aprés Grothendieck et Berthelot les $\mathcal{D}_{X/S}$-modules à droite peuvent se décrire en terme de *co-cristaux* de modules, objets du topos co-cristallin. Le topos co -cristallin garde un sens quand le S-schéma X est singulier et fournit un substitut intrinséque à la catégorie des $\mathcal{D}_{X/S}$-modules à droite. En caractéristique nulle on dispose de tout ce qu'il faut pour développer une théorie des coefficients de de Rham-Grothendieck, les co-cristaux justement, pour un morphisme f de variétés algébriques.

1.1.7. Soient $f : X \to Y$ un morphisme de S-schémas lisses et $\Delta_f : X \to X \times_S Y$ le morphisme graphe de f. Nous supposons la base S localement nœtherienne. Notons par q_1 et q_2 les projections de $X \times_S Y$ sur X et Y respectivement. Posons

$$\mathcal{D}_{X\to Y/S} := q_{1*}\mathcal{H}_X^{\dim(Y/S)}(\mathcal{O}_X \otimes_S \omega_{Y/S})$$
$$\mathcal{D}_{S/Y\leftarrow X} := q_{1*}\mathcal{H}_X^{\dim(Y/S)}(\omega_{X/S} \otimes_S \mathcal{O}_Y).$$

La structure de $\mathcal{D}_{X/S}$-module *à gauche* de $\mathcal{O}_X$ induit une structure de $\mathcal{D}_{X/S}$-module *à gauche* sur le faisceau $\mathcal{D}_{X\to Y/S}$ et la structure de $\mathcal{D}_{Y/S}$-module *à droite* sur $\omega_{Y/S}$ induit une structure de $f^{-1}\mathcal{D}_{Y/S}$-module *à droite* sur le faisceau $\mathcal{D}_{X\to Y/S}$ qui est donc un $(\mathcal{D}_{X/S}, f^{-1}\mathcal{D}_{Y/S})$-bimodule. De même la structure de $\mathcal{D}_{X/S}$-module *à droite* de $\omega_{X/S}$ induit une structure de $\mathcal{D}_{X/S}$-module *à droite* sur le faisceau $\mathcal{D}_{S/Y\leftarrow X}$ et la structure de $\mathcal{D}_{Y/S}$-module *à gauche* sur $\mathcal{O}_Y$ induit une structure de $f^{-1}\mathcal{D}_{Y/S}$-module *à gauche* sur le faisceau $\mathcal{D}_{S/Y\leftarrow X}$ qui est donc un $(f^{-1}\mathcal{D}_{Y/S}\text{-}\mathcal{D}_{X/S})$-bimodule.

En vertu de 1.1.4 le faisceau d'anneaux $\mathcal{D}_{X\to X/S}$ pour le morphisme identique de X est isomorphe au faisceau d'anneaux $\mathcal{D}_{X/S}$. Et l'involution naturelle de $X \times_S X$ induit un isomorphisme de faisceau d'anneaux de $\mathcal{D}_{X\to X/S}$ sur $\mathcal{D}_{S/X\leftarrow X}$.

1.1.8. Toujours pour un morphisme $f : X \to Y$ de S-schémas lisses considérons le diagramme naturel :

$$\begin{array}{ccc} X & \to & X \times_S Y \\ \downarrow & & \downarrow \\ Y & \to & Y \times_S Y. \end{array}$$

On a les isomorphismes, avec un abus de notations évident,

$$\mathbf{L}f^*\mathcal{D}_{Y/S} \cong f^*\mathcal{D}_{Y/S} \cong \mathbf{L}f^*\mathbf{R}\Gamma_Y(\mathcal{O}_Y \otimes_S \omega_{Y/S}))[\dim(Y/S)] \cong \mathbf{R}\Gamma_X(\mathcal{O}_X \otimes_S \omega_{Y/S}))[\dim(Y/S)] \cong \mathcal{D}_{X\to Y/S}.$$

Il en résulte que le faisceau $f^*\mathcal{D}_{Y/S}$ est un $(\mathcal{D}_{X/S}, g^{-1}\mathcal{D}_{Y/S})$-bimodule.

Si $\mathcal{M}$ est un $\mathcal{O}_Y$-module muni d'une *-stratification relativement à S son image *-inverse par f est muni d'une *-stratification relativement à S de façon naturelle. C'est donc un $\mathcal{D}_{X/S}$-module à gauche. L'isomorphisme naturel de $\mathcal{O}_X$-modules

$$f^*\mathcal{M} \cong f^*\mathcal{D}_{Y/S} \otimes_{f^{-1}\mathcal{D}_{Y/S}} f^{-1}\mathcal{M} \cong \mathcal{D}_{X\to Y/S} \otimes_{f^{-1}\mathcal{D}_{Y/S}} f^{-1}\mathcal{M}$$

est en fait un isomorphisme de $\mathcal{D}_{X/S}$-modules à gauche. Ce foncteur d'image *-inverse est exact à droite et se dérive à gauche pour donner naissance à un foncteur exact de catégories triangulées

$$D^b(\mathcal{D}_{Y/S}\text{-mod}) \to D^b(\mathcal{D}_{X/S}\text{-mod})$$

$$\mathcal{M} \to Lf^*\mathcal{M} \cong \mathcal{D}_{X\to Y/S} \overset{L}{\otimes}_{f^{-1}\mathcal{D}_{Y/S}} f^{-1}\mathcal{M}.$$

Le faisceau $\mathcal{D}_{Y/S}$ étant $\mathcal{O}_Y$ localement libre tout $\mathcal{D}_{Y/S}$-module plat reste plat en tant que $\mathcal{O}_Y$-module. Le foncteur Lf^* de la catégorie $D^b(\mathcal{D}_{Y/S}\text{-mod})$ est donc la restriction du foncteur Lf^* de la catégorie $D^b(\mathcal{O}_Y\text{-mod})$ ce qui justifit les notations.

1.1.9. Pour un morphisme $f : X \to Y$ de S-schémas lisses le foncteur $f^!$ entre les catégories $D^b(\mathcal{O}_Y\text{-mod})$ et $D^b(\mathcal{O}_X\text{-mod})$ est défini en factorisant f par une immersion fermée $i : X \to X\times_S Y$ suivie d'un morphisme lisse $p : X\times_S Y \to Y$. Pour un morphisme lisse p le foncteur $p^!$ est défini comme $p^*(-)\otimes \omega_p[\dim(p)]$. Pour une immersion fermée i le foncteur $i^!$ est défini comme l'adjoint à droite du foncteur i_*. On pose alors $f^! := i^! p^!$. Si $\mathcal{M}$ est un $\mathcal{O}_Y$-module muni d'une !-stratification relativement à S les objets du complexe $f^!\mathcal{M}$ sont munis d'une !-stratification relativement à S. C'est donc un complexe de $\mathcal{D}_{X/S}$-modules à droite que nous allons décrire. En fait on a l'isomorphisme naturel de complexes de $\mathcal{D}_{X/S}$-modules à droite pour tout complexe $\mathcal{D}_{Y/S}$-modules à droite $\mathcal{M}$:

$$f^!\mathcal{M} \cong f^{-1}\mathcal{M} \overset{L}{\otimes}_{f^{-1}\mathcal{D}_{Y/S}} \mathcal{D}_{S/Y\leftarrow X}[\dim(X/S) - \dim(Y/S)].$$

On a ainsi défini un foncteur exact de catégories triangulées :

$$D^b(\text{mod-}\mathcal{D}_{Y/S}) \to D^b(\text{mod-}\mathcal{D}_{X/S})$$

$$\mathcal{M} \to f^!\mathcal{M}.$$

Le foncteur naturel de la catégorie des modules à gauche dans la catégorie des modules à droite transforme le foncteur *-image inverse en !- image inverse et réciproquement.

1.1.10. Pour un morphisme $f : X \to Y$ de S-schémas lisses on définit les foncteurs *-images directes (cristallines) f_*^c

$$f_*^c\ D^b(\mathcal{D}_{X/S}\text{-mod}) \to D^b(\mathcal{D}_{Y/S}\text{-mod}) \text{ par}$$

$$\mathcal{M} \to f_*^c\,\mathcal{M} :\cong Rf_*\mathcal{D}_{S/Y\leftarrow X} \overset{L}{\otimes}_{\mathcal{D}_{X/S}} \mathcal{M},$$

et

$$f_*^c\ D^b(\text{mod-}\mathcal{D}_{X/S}) \to D^b(\text{mod-}\mathcal{D}_{Y/S})$$

$$\mathcal{M} \to f_*^c\,\mathcal{M} := Rf_*\mathcal{M} \overset{L}{\otimes}_{\mathcal{D}_{X/S}} \mathcal{D}_{X\to Y/S}.$$

Il résulte de la formule de projection que le foncteur naturel de la catégorie des modules à gauche dans la catégorie des modules à droite et compatible aux foncteurs $*$-images directes.

Pour un morphisme lisse de variétés lisses le foncteur f_*^c coincide avec la définiton de la connextion de Gauss-Manin de Grothendieck quand la base est de caractéristique nulle. Mais bien entendu ce n'est pas le cas en caractéristique non nulle. Dans ([G$_2$], 3.5, p.330) Grothendieck donne un example où la connexion de Gauss-Manin d'un morphisme propre et lisse n'a pas une structure de $\mathcal{D}$-module. En fait la connexion de Gauss-Manin correspond à l'image directe des modules sur le faisceau des opérateurs différentiels d'échelon nul ou de p-filtration un cf. § 1.2 ci-dessous.

1.1.11. Si $\mathcal{M}$ est un complexe de la catégorie $D^b(\mathcal{D}_{X/S}\text{-mod})$ on définit son complexe dual $\mathcal{M}^*$ de la catégorie $D^b(\mathcal{D}_{X/S}\text{-mod})$ en posant

$$\mathcal{M}^* := \hom_{\mathcal{O}_X}(\omega_{X/S}, \mathrm{Rhom}_{\mathcal{D}_{X/S}}(\mathcal{M}, \mathcal{D}_{X/S}))[\dim(X/S)].$$

De même si $\mathcal{M}$ est un complexe de la catégorie $D^b(\text{mod-}\mathcal{D}_{X/S})$ on définit son complexe dual $\mathcal{M}^*$ de la catégorie $D^b(\text{mod-}\mathcal{D}_{X/S})$ en posant

$$\mathcal{M}^* := \omega_{X/S} \otimes_{\mathcal{O}_X} \mathrm{Rhom}_{\mathcal{D}_{X/S}}(\mathcal{M}, \mathcal{D}_{X/S}))[\dim(X/S)].$$

1.1.12. Si $f : X \to Y$ est un morphisme de S-schémas lisses on définit les !-images directes (cristallines)

$$f_!^c\ D^b(\mathcal{D}_{X/S}\text{-mod}) \to D^b(\mathcal{D}_{Y/S}\text{-mod}) \text{ par}$$

$$\mathcal{M} \to f_!^c\, \mathcal{M} := (f_*^c\, \mathcal{M}^*)^*$$

et

$$f_!^c\ D^b(\text{mod-}\mathcal{D}_{X/S}) \to D^b(\text{mod-}\mathcal{D}_{Y/S})$$

$$\mathcal{M} \to f_!^c\, \mathcal{M} := (f_*^c\, \mathcal{M}^*)^*.$$

Donc pour un morphisme $f : X \to Y$ de S-schémas lisses et pour les catégories $D^b(\mathcal{D}_{X/S}\text{-mod})$ et $D^b(\text{mod-}\mathcal{D}_{X/S})$ on dispose des opérations f^*, $f^!$, f_*, $f_!$ et de la dualité $(-)^*$.

1.2. Propriétés de finitude.

En général l'anneau $\Gamma(X; \mathcal{D}_{X/S})$ pour un S-schéma affine lisse n'est pas nœthérien. Mais si tous les nombres premiers sont inversibles sur S sauf un nombre p et si $x_1,\ldots,x_n$ sont des coordonnées locales la $\Gamma(X;\mathcal{O}_X)$-algèbre $\Gamma(X;\mathcal{D}_{X/S})$ est engendré par $\Delta_i^{p^m}$ pour m parcourant tous les entiers. En effet si q_i est un entier dont le développement p-adique s'écrit $q_i = a_0 + a_1p^1 + ... + a_lp^l$ alors le nombre $(p^1!)^{a_1}...(p^l!)^{a_l}/q_i!$ est une unité p-adique u et l'on a donc

$$\Delta_i^{q_i} := (\partial_{x_i}^{q_i}/q_i!) = u(\partial_{x_i})^{a_0}(\partial_{x_i}^{p^1}/p^1!)^{a_1}...(\partial_{x_i}^{p^l}/p^l!)^{a_l}.$$

Si on note par $\Gamma(X;\mathcal{D}_{X/S,m})$ pour tout entier m la $\Gamma(X;\mathcal{O}_X)$-algèbre est engendré par $\Delta^{p^{m'}}$ pour $m' \leq m$ on obtient pour m variable la p-filtration de $\Gamma(X;\mathcal{D}_{X/S})$ quand S est le spectre d'un corps algébriquement clos k ([Ch], [Sm], [Ha]) et la filtration de Berthelot [B_3] par les échelons quand S est un $\mathbf{Z}_p$-schéma. Nous allons voir que ces filtrations ne dépendent pas des coordonnées et permettent d'étudier l'anneau $\Gamma(X;\mathcal{D}_{X/S})$.

1.2.1. Cas où S est le spectre d'un corps k algébriquement clos de caractéristique p > 0.

Pour étudier le faisceau $\mathcal{D}_{X/k}$ on introduit , à coté de la filtration naturelle par l'ordre, la p-filtration $\mathcal{D}_{X/k,m}$ $(m \in \mathbb{N})$ ([Ch], [Sm], [Ha]). Pour tout "niveau " m on note par $F^m := X \to X^m$ le m-éme itéré du morphisme de Frobénius. La sous-algébre $\mathcal{D}_{X/k,m} := hom_{\mathcal{O}_{X^{m+1}}}(\mathcal{O}_X,\mathcal{O}_X)$ de $hom_k(\mathcal{O}_X,\mathcal{O}_X)$ est en fait une sous-algébre d'opérateurs différentiels et l'on a l'égalité ([Ch],[Sm])

$$\cup\mathcal{D}_{X/k,m} = \mathcal{D}_{X/k}.$$

On peut consider les catégories $D^b(\mathcal{D}_{X/S,m}\text{-mod})$,$D^b(\mathcal{D}_{X/S}\text{-mod})$,$D^b(\text{mod-}\mathcal{D}_{X/S,m})$ et $D^b(\text{mod-}\mathcal{D}_{X/S})$. Pour tout m et tout morphisme $X \to Y$ on a les faisceaux $\mathcal{D}_{X\to Y/S,m}$ et $\mathcal{D}_{S/Y\leftarrow X,m}$ de façon évidente. Pour tout m on les opérations f^*_m, $f^!_m$, $f_{*,m}$, $f_{!,m}$, $(-)^*$ pour les catégories $D^b(\mathcal{D}_{X/S,m}\text{-mod})$,$D^b(\text{mod-}\mathcal{D}_{X/S,m})$ qui donnent par passage à la limite quand m tend vers l'infini les opérations analogues pour les catégories $D^b(\mathcal{D}_{X/S}\text{-mod})$, $D^b(\text{mod-}\mathcal{D}_{X/S})$ à condition de se restreindre aux coefficients quasi-cohérents en tant que $\mathcal{O}_X$ -modules [Ha].

Proposition **(1.2.2).**— *Si* X *est k-schéma affine non singulier la* $\Gamma(X;\mathcal{O}_X)$*-algébre* $\Gamma(X;\mathcal{D}_{X/k,m})$ *est engendré par les opérateurs* $\Delta_i^{p^k}$, $k = 0,...,m$, $i = 1,...,n$, *où les* Δ *sont associés à un systéme de coordonnées locales* $x = (x_1,...,x_n)$ *au dessus de* X.

Preuve. Les opérateurs $\Delta_i^{p^k}$, $k = 0,...,m$ sont dans $\Gamma(X;\mathcal{D}_{X/k,m})$ puisque leur ordre est $< p^{m+1}$ ([Sm], Th., 2.7). Réciproquement soit P un opérateur de $\Gamma(X;\mathcal{D}_{X/k,m})$. On a $[P, x_i^{p^{m+1}}] = 0$ pour $i = 1,...,n$. Soit d son degré nous allons montrer que $d<p^{m+1}$. Supposons que $d \geq p^{m+1}$, écrivons $P = P_0 + ... P_d\Delta_n^d$ où les P_i commutent avec x_n; par hypothése $[P, x_n^{p^{m+1}}] = 0$. Donc $\Sigma P_i[\Delta_n^i, x_n^{p^{m+1}}] = 0$. Mais $[\Delta_n^d,$

$x_n^{p^{m+1}}] = \Delta_n^{d-p^{m+1}} + x_n^{p^{m+1}}\Delta_n^d$. Donc $0 = P_d x_n^{p^{m+1}}\Delta_n^d$ + termes de plus bas degré en Δ_n. Ceci entraine que $P_d = 0$. Donc si P est un opérateur de p-filtration m+1 la plus grande puissance de Δ_i; $i = 1\ldots n$, qui apparait dans sa décomposition est strictement plus petite que p^{m+1}. En vertu du développement p-adique d'un entier si $d < p^{m+1}$, Δ_i^d appartient à l'anneau engendré par $\Delta_i^{p^k}$, k = 0,..., m.

La p-filtration est croissante, pour $m \geq m'$ l'extension $\mathcal{D}_{X/k,m'} \to \mathcal{D}_{X/k,m}$ est plate à droite et à gauche ([Ch], [Sm]) et pour tout m l'anneau $\Gamma(U;\mathcal{D}_{X/k,m})$ est nœthérien a droite et à gauche pour tout ouvert affine U de X. On en déduit en particulier que le faisceau $\mathcal{D}_{X/k,m}$ d'anneaux est cohérent pour tout m. L'anneau $\Gamma(U;\mathcal{D}_{X/k})$ n'est pas nœthérien. Cependant de ces remarquables résultats on obtient le théorème :

Théorème **(1.2.3).**— *Pour toute variété* X *non singulière sur un corps* k *algébriquement clos de carastéristique* p > 0 *le faisceau* $\mathcal{D}_{X/k}$ *est un faisceau d'anneaux* cohérent.

Preuve. Il faut montrer que le noyau de tout morphisme $\mathcal{D}_{X/k}$-linéaire : $(\mathcal{D}_{X/k})^q \to \mathcal{D}_{X/k}$ (à droite ou à gauche) est un $\mathcal{D}_{X/k}$-module de type fini. Mais le faisceau $\mathcal{D}_{X/k}$ étant quasi-cohérent en tant que $\mathcal{O}_X$-module il suffit de montrer que le noyau du morphisme $\Gamma(U;(\mathcal{D}_{X/k})^q) \to \Gamma(U;\mathcal{D}_{X/k})$ pour tout ouvert affine U est de type fini. Mais pour U affine $\Gamma(U;\mathcal{D}_{X/k})$ est réunion des anneaux $\Gamma(U;\mathcal{D}_{X/k,m})$ pour $m \in \mathbb{N}$. Il en résulte que les opérateurs différentiels donnant le morphisme précédent sont dans l'anneau $\Gamma(U;\mathcal{D}_{X/k,m})$ pour m assez grand. Mais l'extention $\Gamma(U;\mathcal{D}_{X/k,m}) \to \Gamma(U;\mathcal{D}_{X/k})$ étant plate ([Ch],[Sm]) le noyau du morphisme $\Gamma(U;(\mathcal{D}_{X/k})^q) \to \Gamma(U;\mathcal{D}_{X/k})$ provient par extention des scalaires du noyau du morphisme $\Gamma(U;(\mathcal{D}_{X/k,m})^q) \to \Gamma(U;\mathcal{D}_{X/k,m})$ pour m assez grand. Mais l'anneau $\Gamma(U;\mathcal{D}_{X/k,m})$ étant nœthérien ce noyau est de type fini.

On peut donc comme en caractéristique nulle considérer les catégories abéliennes des $\mathcal{D}_{X/k}$-modules cohérents et les catégories triangulées $D_c^b(\mathcal{D}_{X/k}\text{-mod})$, $D_c^b(\text{mod-}\mathcal{D}_{X/k})$ des complexes de $\mathcal{D}_{X/k}$-modules à cohomologie bornée et cohérente. Remarquons cependant que le faisceau structural $\mathcal{O}_X$ et le faisceau dualisant $\omega_{X/k}$ ne sont pas $\mathcal{D}_{X/k}$-cohérents. On a alors le théorème de Chase-Smith ([Ch],[Sm]):

Théorème **(1.2.3).**— *Pour tout ouvert affine* U *d'une variété* X *non singulière sur un corps k algébrique clos de caractéristique* p > 0 *la dimension homologique de l'anneau* $\Gamma(U;\mathcal{D}_{X/k})$ *est égale à* la dimension de X.

On peut alors définir la catégorie des $\mathcal{D}_{X/k}$-modules holonomes:

Définition **(1.2.4).**— *On dit alors qu'un* $\mathcal{D}_{X/k}$*-module cohérent* $\mathcal{M}$ *est holonome si les faisceaux* $Ext^i_{\mathcal{D}_{X/k}}(\mathcal{M},\mathcal{D}_{X/k})$ sont nul *pour* $i \neq \dim(X)$.

Il n'est pas clair que la catégories des $\mathcal{D}_{X/k}$-modules holonomes soit abélienne bien que cela semble probable. On note par $D^b_h(\mathcal{D}_{X/k}\text{-mod})$, $D^b_h(\text{mod-}\mathcal{D}_{X/k})$ les catégories des complexes de $\mathcal{D}_{X/k}$-modules à cohomologie bornée et holonome. Il n'est pas clair que les catégories $D^b_h(\mathcal{D}_{X/k}\text{-mod})$, $D^b_h(\text{mod-}\mathcal{D}_{X/k})$ soient triangulées. De même pour tout m puisque le faisceau $\mathcal{D}_{X/k,m}$ est cohérent de dimension homologique égale à dim(X) on a les catégories $D^b_h(\mathcal{D}_{X/k,m}\text{-mod})$, $D^b_h(\text{mod-}\mathcal{D}_{X/k,m})$. On obtient des catégories de coefficients dont on a pas encore exorciser les propriétés de finitude si cela est raisonnable.

1.2.5. Cas d'un $\mathbf{Z}_p$-schéma de base S .

Soit X un S-schéma lisse. Sur le faisceau $\mathcal{D}_{X/S}$ on a la filtration de Berthelot [B_3] par les faisceaux des opérateurs différentiels " d'échelon " m pour $m \in \mathbf{N}$ qui donne la p-filtration par reduction modulo p en vertu de (1.2.2).

Si on fixe un échelon m on remplace le faisceau des parties, d'ordre l, $\mathcal{P}^{(l)}_{X/S}(\mathcal{O}_X)$ par le faisceau des parties principales, d'ordre l et d'échelon m, $\mathcal{P}^{(l)}_{X/S,m}(\mathcal{O}_X)$. Pour sa définition précise nous renvoyons le lecteur à l'exposé de Berthelot dans ce même volume. On pose alors

$$\mathcal{D}iff^{(l)}_{X/S,m}(\mathcal{O}_X) := hom_{\mathcal{O}_X}(\mathcal{P}^{(l)}_{X/S,m}(\mathcal{O}_X),\mathcal{O}_X).$$

C'est le faisceau des opérateurs différentiels d'ordre l et d'échelon m, voir aussi (1.2.7). C'est un faisceau de $\mathcal{O}_X$-algébres localement libre de type fini. Pour $m \geq m'$ on un morphisme naturel

$$\mathcal{D}iff^{(l)}_{X/S,m'}(\mathcal{O}_X) \to \mathcal{D}iff^{(l)}_{X/S,m}(\mathcal{O}_X)$$

qui fait des $\mathcal{D}iff^{(l)}_{X/S,m}(\mathcal{O}_X)$ (pour l fixé) un systéme inductif essentiellement constant à $\mathcal{D}iff^{(l)}_{X/S}(\mathcal{O}_X)$. On pose alors

$$\mathcal{D}iff_{X/S,m}(\mathcal{O}_X) := \cup_l \mathcal{D}iff^{(l)}_{X/S,m}(\mathcal{O}_X).$$

Le faisceau $\mathcal{D}iff_{X/S,m}(\mathcal{O}_X)$ est un faisceau de $\mathcal{O}_X$-algébres et le faisceau $\mathcal{O}_X$ est $\mathcal{D}iff_{X/S,m}(\mathcal{O}_X)$-module à gauche. La limite inductive des $\mathcal{D}iff_{X/S,m}(\mathcal{O}_X)$ ($m \in \mathbf{N}$) est alors canoniquement isomorphe à $\mathcal{D}iff_{X/S}(\mathcal{O}_X)$. En coordonnées locales l'image de $\mathcal{D}iff_{X/S,m}(\mathcal{O}_X)$ dans $\mathcal{D}iff_{X/S}(\mathcal{O}_X)$ est engendrée en tant que $\mathcal{O}_X$-algébre par les opérateurs Δ^{p^k}, k = 0,..., m et les morphismes de transition ne sont autre que les inclusions canoniques. Le faisceau est $\mathcal{D}iff_{X/S,m}(\mathcal{O}_X)$ cohérent pour tout m. La reduction modulo p de la filtration par les échelons de $\mathcal{D}iff_{X/S}(\mathcal{O}_X)$ n'est autre que la p-filtration.

Remarque **(1.2.6).**— L'extension $\mathcal{D}iff_{X/S,m'}(\mathcal{O}_X) \rightarrow \mathcal{D}iff_{X/S,m}(\mathcal{O}_X)$ pour m > m' *n'est pas* plate bien que sa reduction modulo p est plate!.

Proposition **(1.2.7).**— On peut aussi remarquer que l'anneau des opérateurs d'échelon m est engendré par les opérateurs d'ordre au plus d'ordre p^m aussi bien en caracteristique pure p qu'en inégale caractéristique p. Ceci donne une desciption intrinséque simple de la filtration par les échelons.

Proposition **(1.2.8).**— *Si* S *est le spectre d'un anneau de valuation discrète* W *complet d'inégale caractéristique p pour tout ouvert affine* U *de* X *et pour tout* m *le* complété p-adique *de l'anneau* $\Gamma(U;\mathcal{D}iff_{X/W,m}(\mathcal{O}_X))$ *est de dimension homologique égale à* dim(X/W) + 1.

Preuve. Notons $\widehat{\mathcal{D}iff}_{X/W,m}(\mathcal{O}_X)$ le complété p-adique de $\mathcal{D}iff_{X/W,m}(\mathcal{O}_X)$. La reduction modulo p de l'anneau $\Gamma(U;\widehat{\mathcal{D}iff}_{X/W,m}(\mathcal{O}_X))$ est l'anneau des opérateurs différentiels de p-filtration m+1 sur un corps de caractiristique p qui est de dimension homologique égale à dim(X/W) ([Ch],[Sm]). Un vertu de ([Rt], thm. 9.33) on a l'inégalité $dh(\Gamma(U;\widehat{\mathcal{D}iff}_{X/W,m}(\mathcal{O}_X))) \geq \dim(X/W)+1$. D'autre part le gradué associé à la filtration p-adique de $\Gamma(U;\widehat{\mathcal{D}iff}_{X/W,m}(\mathcal{O}_X))$ est de dimension homologique dim(X/W)+1 car isomorphe à l'algébre des polynômes à une variable sur un anneau de dimension homologique dim(X/W). Donc $dh(\Gamma(U;\widehat{\mathcal{D}iff}_{X/W,m}(\mathcal{O}_X))) \leq \dim(X/W)+1$ en vertu de ([N-V], VII-11, p. 315).
De même la dimension homologique de $\Gamma(U;\mathcal{D}iff_{X/W,m}(\mathcal{O}_X))$ est moins égale à dim(X/W)+1. Il est probable qu'elle est égale à dim(X/W)+1.Si on pose avec Berthelot [B_3] $\mathcal{D}^{\dagger}_{X/W} := \varinjlim_{m} \widehat{\mathcal{D}iff}_{X/W,m}(\mathcal{O}_X)$.

On trouve que la dimension homologique de $\Gamma(U;\mathcal{D}^{\dagger}_{X/W}(\mathcal{O}_X))$ est au plus dim(X/W)+2 car limite inductive d'anneaux de dimension homologique dim(X/X)+1 en vertu de [Ber]. D'autre part sa dimension homologique est au moins dim(X/W)+1 car sa reduction modulo p est de dimension homologique dim(X/W)+1. Il est probable que $dh(\Gamma(U;\mathcal{D}^{\dagger}_{X/W}(\mathcal{O}_X)) = \dim(X/W)+1$.

§ 2. Le théorème des coefficients de de Rham-Grothendieck en caractéristique zéro.

On suppose dans ce § que le schéma de base est un corps k de caractéristique zéro que nous supposerons, pour simplifier, algébriquement clos. Le faisceau des opérateurs $\mathcal{D}_{X/k} := \mathcal{D}iff_{X/k}(\mathcal{O}_X)$ différentiels sur un variété algébrique X non singulière sur k est alors un faisceaux de $\mathcal{O}_X$-algébres de type fini tel que le faisceau gradué $gr(\mathcal{D}_{X/k})$ associé à la filtration $\mathcal{D}^{(m)}_{X/k} := \mathcal{D}iff^{(m)}_{X/k}$ par l'ordre des opérateurs différentiels est un faisceau de $\mathcal{O}_X$-algébres commutatives de type fini. La variété au dessus de X associée à la $\mathcal{O}_X$-algébre $gr(\mathcal{D}_{X/k})$ est le fibré cotangent T*X de X. L'anneau des sections globales au-dessus de tout ouvert affine de coordonnées de X du faisceau $\mathcal{D}_{X/k}$ est næthérien. On en déduit que le faisceau quasi-cohérent $\mathcal{D}_{X/k}$ est en fait un faisceau *cohérent* d'anneaux.

2.1. Le théorème de finitude.

2.1.1. Soit U un ouvert affine de X. On a alors le théorème ([Ro], [Bj], [Ch]) :

Théorème **(2.1.2).**— *La dimension homologique de l'anneau* $\Gamma(U;\mathcal{D}_{X/k})$ *est égale à* la dimension de X.

Tout $\mathcal{D}_{X/k}$-module cohérent $\mathcal{M}$ admet une filtration $\mathcal{M}^{(m)}$ $(m \in \mathbb{N})$ par des $\mathcal{O}_X$-modules cohérents qui est bonne c'est à dire dont le gradué associé $gr(\mathcal{M})$ est un $gr(\mathcal{D}_{X/k})$-module cohérent. La variété réduite $Ch(\mathcal{M})$ du fibré T*X associé à $gr(\mathcal{M})$ ne dépend pas de la bonne filtration choisie. Par définition c'est la variété caractéristique de $\mathcal{M}$. En faite la multiplicité de la variété caractéristique $Ch(\mathcal{M})$ en chacun de ses points ne dépend pas de la bonne filtration choisie. En particulier le cycle $CCh(\mathcal{M})$ associé à $gr(\mathcal{M})$ ne dépend que de $\mathcal{M}$. C'est par définition le cycle caractéristique de $\mathcal{M}$. Voir par exemple ([M_2] ou ([Me_5, I.2.2, 2.4)). On a alors l'inégalité dite de Bernstein cf. [Be] :

Théorème **(2.1.3).**— *La dimension de la variété caractéristique* $CCh(\mathcal{M})$ *d'un* $\mathcal{D}_{X/k}$*-module cohérent non nul* $\mathcal{M}$ *est au moins égale à* dim(X).

Pour une démonstration géométrique du théorème (2.1.3) voir ([M_2] ou ([Me_5], I.2.3)). En fait la théorie des bons anneaux filtrés voir par exemple ([M_2] ou ([Me], I.4.1.3, 4.2.14)) montre que le théorème algébrique (2.1.2) est équivalent au théorème géométrique (2.1.3).

Si on pose pour un $\mathcal{D}_{X/k}$-module cohérent non nul $\mathcal{M}$ $\mathrm{grade}(\mathcal{M}) := \inf\{i;\ \mathcal{E}xt^i_{\mathcal{D}_{X/k}}(\mathcal{M},\mathcal{D}_{X/k}) \neq 0\}$. On a $\mathrm{grade}(\mathcal{M}) = \mathrm{codim}_{T^*X}(Ch(\mathcal{M}))$ cf. ([Me_5], I.4.2.14). Cela améne à poser la définition :

Définition **(2.1.4).**— *On dit qu'un* $\mathcal{D}_{X/k}$*-module cohérent* $\mathcal{M}$ *est holonome s'il est nul ou si la dimension de sa variété caractéristique est égale à* dim(X). *De façon équivalente s'il est nul ou si les faisceaux* $\mathcal{E}xt^i_{\mathcal{D}_{X/k}}(\mathcal{M},\mathcal{D}_{X/k})$ sont nuls pour $i \neq \dim(X)$.

On note par $Mh(\mathcal{D}_{X/k}$-) la catégorie des $\mathcal{D}_{X/k}$-modules à gauche holonome qui est alors une sous-catégorie pleine de la catégorie $\mathcal{D}_{X/k}$-modules cohérents. Dans une suite exacte courte de $\mathcal{D}_{X/k}$-modules cohérents le terme médian est holonome si et seulement si les termes extrêmes sont holonomes. En particulier la catégorie des $\mathcal{D}_{X/k}$-modules holonomes est abélienne. On note $D^b_h(\mathcal{D}_{X/k}\text{-mod})$ la catégorie des complexes $\mathcal{D}_{X/k}$-modules à cohomologie bornée et holonome. C'est alors une sous-catégorie pleine et triangulée de la catégorie $D^b_c(\mathcal{D}_{X/k}\text{-mod})$.

Si $\mathcal{M}$ est un $\mathcal{D}_{X/k}$-module (à gauche) holonome le $\mathcal{D}_{X/k}$-module (à droite) $Ext^n_{\mathcal{D}_{X/k}}(\mathcal{M},\mathcal{D}_{X/k})$ est encore holonome. Autrement dit le foncteur de dualité $\mathcal{M} \to \mathcal{M}^*$ est une anti-équivalence de catégorie de la catégorie des $\mathcal{D}_{X/k}$-modules (à gauche ou à droite) holonomes dans elle même.

2.1.5. Le point clef des propriétés de finitude en caractéristique nulle pour les $\mathcal{D}_{X/k}$-modules holonomes est l'équation fonctionnelle dite de Bernstein-Sato. Motivé par le probléme qui à priori n'a rien à avoir avec la cohomologie des variétés algébriques, posé par I. Gelfand au congrés d'Amsterdam (1954), du prolongement analytique de la distribution $s \to P^s$ où P est un polynôme réel I.N.Bernstein a démontré le théorème suivant. Soit P un polynôme à n variable à coeficients dans k. Notons $D_{k^n/k}$ l'algébre de Weyl c'est à dire $\Gamma(k^n;\mathcal{D}_{k^n/k})$ et $D_{k^n/k}[s]P^s$ le $D_{k^n/k}[s] := D_{k^n/k}\otimes_k k[s]$ module engengré par le symbol P^s. Il faut voir $D_{k^n/k}[s]P^s$ comme sous-module de $A[s,P^{-1}]P^s$ où $A := k[x_1,...x_n]$ et où l'action de $D_{k^n/k}[s]$ sur $A[s,P^{-1}]P^s$ est celle qu'on pense.

Théorème **(2.1.6)** [Be].— *Soit* P *un polynôme non nul de* A *il existe alors un plynôme non nul à une variable* B(s) *à coefficient dans* k *et un opérateur différentiel de* Q(s) *de* $D_{k^n/k}[s]$ *tels que l'on ait l'équation fonctionnelle*

$$B(s)P^s = Q(s)PP^s.$$

Corollaire **(2.1.7)**.— *Si* P *est un polynôme le* $D_{k^n/k}$*-module* $A[P^{-1}]$ *est de* type fini.

En effet pour tout entier m on a un morphisme de spécialisation de $A[s,P^{-1}]P^s$ dans $A[m,P^{-1}]P^m$:

$$\Phi_m : A[s,P^{-1}]P^s \to A[m,P^{-1}]P^m.$$

De l'équation fonctionnelle on déduit que le $D_{k^n/k}$-module $A[P^{-1}]$ est engendré par P^{-m} pour m assez grand.

On peut localiser l'équation fonctionnelle. Soient U un ouvert affine de X où l'on dispose de coordonnées locales, $A := \Gamma(U,\mathcal{O}_X)$, $D_{U/k} := \Gamma(U,\mathcal{D}_{X/k})$ et M l'espace des section globales sur U d'un $\mathcal{D}_{X/k}$-module holonome. Pour tout élément non nul u de M et toute fonction réguliére P sur U on considére le $D_{U/k}[s]$-module $D_{U/k}[s]P^su$. On alors le résultat suivant :

Théorème **(2.1.8)**.— *Soit* P *une fonction non nul de* A *il existe alors un polynôme* B(s) *non nul à une variable à coefficient dans* k *et un opérateur différentiel de* Q(s) *de* $D_{U/k}[s]$ *tels que l'on ait l'équation fonctionnelle*

$$B(s)P^su = Q(s)PP^su.$$

Le théorème (2.1.8) se réduit à (2.1.6) si $U = k^n$, M = A et u = 1. La démonstration de (2.1.8) est *aujourd'hui* élémentaire. A partir de (2.1.2) et de la théorie des bons anneaux filtrés elle repose sur le fait que A est une k-algébre de *type fini* et sur le fait qu'en géométrie algébrique tous les faisceaux cohérents en dehors d'une sous-variété se prolonge en faisceaux cohérents cf. ([Me_5], I.4.2).

Corollaire **(2.1.9).**— *Si* $\mathcal{M}_U$ *est un* $\mathcal{D}_{U/k}$*-module holonome sur* U *complémentaire d'une hypersurface* Z *de* X *l'image directe de* $\mathcal{M}_U$ *par l'inclusion canonique de* U *dans* X *est un* $\mathcal{D}_{X/k}$*-module* holonome.

En effet la question est locale sur X. On peut supposer que X est affine et que Z défini par une équation P. Si i désigne l'inclusion cananique de U dans X, $i_*\mathcal{M}_U$ est un $\mathcal{D}_{X/k}$-modules quasi-cohérent. Pour montrer qu'il est $\mathcal{D}_{X/k}$-cohérent il suffit de montrer que le $D_{X/k}$-module de ses sections globales est de type fini. Mais $\mathcal{M}_U$ admet un plongement holonome cf. ([Me_5], I.4.1.8) à X. L'équation fonctionnelle (2.1.8) appliquée aux sections globales d'un tel prolongement et à ses générateurs montre que $\Gamma(X, i_*\mathcal{M}_U)$ est un $D_{X/k}$-module de type fini. Une fois acquis le fait que $i_*\mathcal{M}_U$ est $\mathcal{D}_{X/k}$-cohérent un argument de spécialisation montre qu'il est holonome cf. ([Me_5], I.8.2, p. 101).

On peut déduire facilement de (2.1.9) que les catégories sont stable par images inverses cf. ([Me_5], I. 8.7). Si $f : X \to Y$ est morphisme de variétés algébriques sur k avec les notations du § 1 :

Corollaire **(2.1.10).**— *Si* $\mathcal{M}$ *est un coefficient de* $D^b_h(\mathcal{D}_{Y/k}\text{-mod})$ alors $f^*\mathcal{M}$ et $f^!\mathcal{M}$ *sont des coefficients de* $D^b_h(\mathcal{D}_{X/k}\text{-mod})$.

Si $\mathcal{M}$ est un complexe de $\mathcal{D}_{X/k}$-modules posons

$$\mathbf{DR}(\mathcal{M}) := \mathbf{R}\mathrm{hom}_{\mathcal{D}_{X/k}}(\mathcal{O}_X, \mathcal{M}).$$

En caractérisque *zéro* on on l'isomorphisme (notations du § 1) :

$$\mathbf{DR}(\mathcal{M})[\dim(X)] \cong \mathcal{M} \overset{\mathbf{L}}{\otimes}_{\mathcal{D}_{X/k}} \mathcal{D}_{\mathrm{speck} \leftarrow X}$$

On déduit de (2.1.8) le théorème de finitude :

Théorème **(2.1.11).**— *Si* $\mathcal{M}$ *est un* $\mathcal{D}_{X/k}$*-module holonome la cohomologie de* $\mathbf{R}\Gamma(X;\mathbf{DR}(\mathcal{M}))$ *sont des espaces vectoriels sur k* de dimension finie.

Preuve. La suite spectrale de Cech-de Rham permet de supposer que X est affine. En Prenant un plongement de X dans un espace numérique, on est réduit à supposer que $X = k^n$. En vertu de (2.1.8) l'image de $\mathcal{M}$ par l'inclusion de k^n dans l'espace projectif P^n de dimension n est un $\mathcal{D}_{pn}$-module cohérent. On est réduit à montrer que le complexe $\mathbf{R}\Gamma(P^n;\mathbf{DR}(\mathcal{M}))$ est à cohomologie de dimension finie sur k pour un $\mathcal{D}_{X/k}$-module cohérent $\mathcal{M}$. Mais tout $\mathcal{D}_{pn}$-module cohérent est quotient d'un module de la forme $\mathcal{D}_{pn} \otimes_{\mathcal{O}_{pn}} \mathcal{F}$ pour un $\mathcal{O}_{pn}$-module cohérent $\mathcal{F}$. Mais $\mathbf{R}\Gamma(P^n; \mathbf{DR}(\mathcal{D}_{pn} \otimes_{\mathcal{O}_{pn}} \mathcal{F}))$ est isomorphisme à $\mathbf{R}\Gamma(P^n; \omega_{X/k} \otimes_{\mathcal{O}_{pn}} \mathcal{F})[-n]$. On est réduit au théorème de finitude de la cohomologie d'un faisceau algébrique cohérent sur un espace projectif. Mais puisque le $\mathcal{D}_{X/k}$-module $\mathcal{O}_X$ est holonome en caractérisque zéro on trouve que la cohomologie de de Rham d'une variété algébrique non singulière sur un corps de caractéristique nulle est de dimension *finie*. Plus généralement si f : X → Y est un morphisme de variétés algébriques non singulières sur k on a, avec les notations du § 1, le théorème :

Théorème **(2.1.12).**— *Si* $\mathcal{M}$ *est un coefficient de* $D_h^b(\mathcal{D}_{X/k}\text{-mod})$ alors $f_*^c\mathcal{M}$ et $f_!^c\mathcal{M}$ *sont des coefficients de* $D_h^b(\mathcal{D}_{Y/k}\text{-mod})$.

Preuve. On factorise f en une immersion fermée suivie d'une projection. Le cas d'une immersion fermée ne pose pas problème. Si f est la projection XxY → Y la suite spectrale Cech-de Rham relatif nous réduit à supposer que X est affine. Un plongement dans un espace numérique réduit à supposer que X est l'espace affine k^n. Le corollaire (2.1.9) réduit à supposer que X est l'espace projectif P^n. Mais tout $\mathcal{D}_{P^n\times Y}$-module cohérent est quotient d'un module de la forme $\mathcal{D}_{P^n\times Y}\otimes_{\mathcal{O}_{P^n\times Y}}\mathcal{F}$ pour un $\mathcal{O}_{P^n\times Y}$-module cohérent $\mathcal{F}$. La $\mathcal{D}_{Y/k}$-cohérence résulte alors de la formule de projection et du théorème d'images directes par un morphisme projectif des faisceaux algébriques cohérents. Une fois acquis la cohérence le théorème de dualité relative pour un morphime projectif qui dit que le foncteur image directe commute à la dualité ([Me$_5$], I. 5.3.13) montre que les faisceaux *Ext* sur $\mathcal{D}_{Y/k}$ et contre $\mathcal{D}_{Y/k}$ des faisceaux de cohomologie des images directes sont concentrés en degré dim(Y) ([Me$_5$], I. 5.4.1, p. 76). Ils sont donc holonomes. En particulier le théorème (2.1.12) établit l'holonomie des modules de Gauss-Manin [G$_2$] qui sont les faisceaux de cohomologie du complexe $f_*^c\mathcal{O}_X$ et cela pour un morphisme f *arbitraire* de variétés algébriques non singulières sur un corps de caractéristique nulle.

Si $\mathcal{M}_1$ et $\mathcal{M}_2$ sont des complexes de $\mathcal{D}_{X/k}$-modules (à gauche) le complexe $\mathcal{M}_1\overset{L}{\otimes}_{\mathcal{O}_X}\mathcal{M}_2$ est un complexe de $\mathcal{D}_{X/k}$-modules (à gauche) et il en résulte facilement de (2.2.4) que le complexe est à cohomologie holonome si $\mathcal{M}_1$ et $\mathcal{M}_2$ le sont.

Autrement dit pour un morphisme f de variétés algébriques non singulières sur k les catégories $D_h^b(\mathcal{D}_{X/k}\text{-mod})$ sont stables par les cinq opérations f^*, $f^!$, f_*^c, $f_!^c$ et $\overset{L}{\otimes}_{\mathcal{O}_X}$ parmi les six opérations de Grothendieck. Les catégories $D_h^b(\mathcal{D}_{X/k}\text{-mod})$ n'étant pas stable par l'opération interne $\mathrm{Rhom}_{\mathcal{O}_X}(\mathcal{M}_1, \mathcal{M}_2)$ qu'il faut remplacer par l'adjoint à gauche du produit tensoriel qui existe ([Me$_5$], II.9.1, p. 185).

On dispose d'une théorie des cycles évanescents, la théorie de la V-filtration, pour les $\mathcal{D}_{X/k}$-modules qui est l'analogue de la théorie des cycles évanescents de SGA 7, cf. par exemple [S-M].

2.1.13. Si on part d'un morphisme f de variétés analytiques complexes ou p-adiques rigides on a encore les catégories $D^b_h(\mathcal{D}_{X/k}\text{-mod})$ qui sont stables par les opérations f^*, $f^!$, $f^!_*$ et $\overset{L}{\otimes}_{\mathcal{O}_X}$. Pour l'image directe il faut supposer le morphisme propre et imposer une condition d'existence de filtrations globales. Mais les démonstrations sont algébriques [M-N$_2$].

2.2. Le théorème de comparaison.

2.2.1. Soit X une variété algébrique complexe non singulière et X^{an} la variété transcendante associée à X. On a un morphisme GAGA :

$$(*)\quad \mathbf{R}\Gamma(X; \mathbf{DR}(\mathcal{O}_X)) \to \mathbf{R}\Gamma(X^{an}; \mathbf{DR}(\mathcal{O}_{X^{an}})).$$

Théorème **(2.2.2)** (Grothendieck).— *Le morphisme* (*) *est un* isomorphisme.

Comme le complexe $\mathbf{DR}(\mathcal{O}_{X^{an}})$ est une résolution du faisceau constant $\mathbf{C}_{X^{an}}$ en vertu du lemme de Poincaré la cohomologie de de Rham algébrique d'une variété algébrique complexe non singulière est isomorphe à sa cohomologie de Betti. Si X est propre le théorème (2.2.2) est une conséquence du théorème GAGA de Serre. Mais si X est affine sa démonstration nécessite la théorie des équations différentielles complexes à points singuliers réguliers.

Plus généralement soit $\mathcal{E}$ un fibré sur X muni d'une connexion intégrable. On a encore un morphisme GAGA :

$$(**)\quad \mathbf{R}\Gamma(X; \mathbf{DR}(\mathcal{E})) \to \mathbf{R}\Gamma(X^{an}; \mathbf{DR}(\mathcal{E}^{an})).$$

Théorème **(2.2.3)** (Deligne).— *Si le fibré image inverse de* $\mathcal{E}$ *sur toute courbe non singulière au-dessus de X n'a que des singularités régulières à l'infini le morphisme* (**) *est un* isomorphisme.

La démonstration du théorème de comparaison de Grothendieck-Deligne ([G$_1$], [D$_1$]) reposait de façon essentielle sur le théorème de résolution des singularités de Hironaka [H]. En ce sens elle restait largement inaccessible vu la difficulté de la démonstration du théorème de Hironaka. Nous allons expliquer que le théorème de Hironaka n'est pas indispensable ici.

2.2.4. Dans les théorèmes de comparaison précédents, comme beaucoup de théorèmes de géométrie algébrique, il s'agit de démontrer que les fibrés à connexion en question ont une ramification *modérée* à l'infinie. En dimension un le théorème de Fuchs affirme qu'un point singulier d'une équation différentielle est régulier si et seulement si le nombre de Fuchs de cette équation attaché à ce point est nul. Le calcul de ce nombre n'offre pas de difficulté. En dimension supérieur il faut faire la même chose. Etant donné une variété algébrique non singulière X sur un corps k de caractéristique nulle et un fibré à connexion $\mathcal{E}_U$ sur le complémentaire U d'une hypersurface Z éventuellement singulière de X nous allons attacher au triplet $(X, Z, \mathcal{E}_U)$ une stratication $\cup Z_i$ de Z et pour chaque strate Z_i un entier m_i

positif ou nul qui mesure parfaitement la ramification de $\mathcal{E}_U$ le long de Z. Si X est une courbe ces nombres m_i ne sont au autre que les nombres de Fuchs du fibré $\mathcal{E}_U$ aux points singuliers Z. Puis il faut donner un critére facile à vérifier assurant que le cycle positif $\sum m_i Z_i$ est nul.

2.2.5. La définiton de ce cycle a pour origine la démonstration de Grothendieck [G_1] de son théorème de comparaison. En effet c'est dans cette démonstration qu'on trouve la manisfestation la plus tangible d'un objet d'une *catégorie dérivée* discrète du type $D^b_c(C_X)$ qui est à la base de la notion de *faisceau pervers*.

Rappelons que si X est variété algébrique complexe un faisceau transcendant $\mathcal{F}$ d'espaces vectoriels complexe de dimension finie sur X^{an} est dit algébriquement constructible s'il existe une stratification algébrique telle la restiction de $\mathcal{F}$ à chaque strate est une systéme local. On note alors $D^b_c(C_X)$ la catégorie des coefficients complexes constructibles c'est à dire des complexes à cohomologie bornée et constructible. On dit qu'un coefficient constructible $\mathcal{F}$ a la propriété de *support* si ses faisceaux de cohomologie $h^i(\mathcal{F})$ sont concentrés en degrés [0, dim(X)] et si la dimension du support du faisceau $h^i(\mathcal{F})$ est au plus égale à dim(X) – i pour tout i dans [0, dim(X)]. On dit qu'un coefficient constructible $\mathcal{F}$ a la propriété de *co-support* si le complexe dual $\mathbf{R}\mathrm{hom}_{C_{X^{an}}}(\mathcal{F}, C_{X^{an}})$ (X étant non singulier) a la propriété de support. On dit qu'un coefficient constructible est un *faisceau pervers* s'il a la propriété de support et de co-support. On note $\mathrm{Perv}(C_X)$ la catégorie des faisceaux pervers. C'est alors une sous-catégorie pleine de $D^b_c(C_X)$ et *abélienne* cf. [B-B-D]. Un coefficient constructible $\mathcal{F}$ sur une sous-variété Z de X est un faisceau pervers sur Z si $\mathcal{F}[-\mathrm{codim}_X(Z)]$ vu comme coefficient constructible sur X est un faisceau pervers sur X. On note $\mathrm{Perv}(C_Z)$ la catégorie des faisceaux pervers sur Z.

2.2.6. Soit $\mathcal{M}$ un $\mathcal{D}_{X/C}$-module holonome sur une variété algébrique complexe non singulière. Alors son complexe de de Rham transcendant $\mathbf{DR}(\mathcal{M}^{an}) := \mathbf{R}\mathrm{hom}_{\mathcal{D}_{X^{an}/C}}(\mathcal{O}_X{}^{an}, \mathcal{M}^{an})$ est un coefficient constructible ayant la propriété de support [K_1], voir aussi [N-M_1]. D'autre part il résulte du théorème de dualité locale qu'il la propriété de co-support [Me_3]. C'est donc un faisceau pervers sur X.

2.2.7. Si Z est une sous-variété fermée de X notons par

$$i : Z \rightarrow X \leftarrow U : j$$

les inclusions canoniques et par

$$i^{an} : Z^{an} \rightarrow X^{an} \leftarrow U^{an} : j^{an}$$

les inclusions transcendantes correspondantes. Rappelons que le foncteur $i^{an!}$ de $D^b_c(C_X)$ dans $D^b_c(C_Z)$

est adjoint à gauche du foncteur $i^{an}{}_*$ de $D^b_c(C_Z)$ dans $D^b_c(C_X)$. Ici $i^{an!}(\mathcal{F})$ est simplement $i^{-1}\mathbf{R}\Gamma_{Z^{an}}(\mathcal{F})$. Soit un $\mathcal{D}_{X/C}$-module holonome $\mathcal{M}$ posons

$$\mathbf{IR}_Z(\mathcal{M}) := i^{an!}(\mathbf{DR}((j_*j^{-1}\mathcal{M})^{an}))[1].$$

En vertu de (2.1.9) $j_*j^{-1}\mathcal{M}$ est un $\mathcal{D}_{X/C}$-module holonome et donc $\mathbf{IR}_Z(\mathcal{M})$ est un coefficient constructible sur Z. On a ainsi défini un foncteur cohomologique de $Mh(\mathcal{D}_{X/C^-})$ dans $D^b_c(C_Z)$. Il transforme suite exacte de $\mathcal{D}_{X/C}$-modules en triangle distingué de $D^b_c(C_Z)$. Pour rassurer le lecteur le complexe $\mathbf{IR}_Z(\mathcal{M})$ n'est rien d'autre que le cône du morphisme naturel

$$\mathbf{DR}((j_*j^{-1}\mathcal{M})^{an}) \to \mathbf{R}j^{an}{}_*j^{-1an}\mathbf{DR}((\mathcal{M})^{an}$$

et sous cette forme qu'il apparaît dans la démonstration de Grothendieck [G_1]. On a alors le théorème suivant [Me_6] :

Théorème **(2.2.8).**— *Si Z est une hypersurface et $\mathcal{M}$ est un $\mathcal{D}_{X/C}$-module holonome le coefficient* $\mathbf{IR}_Z(\mathcal{M})$ *est un* faisceau pervers sur Z.

Comme un triangle distingué formé par des faisceaux est en faite une suite exacte de la catégorie des faisceaux pervers il en résulte que le foncteur

$$\mathbf{IR}_Z : Mh(\mathcal{D}_{X/C^-}) \to Perv(C_Z)$$

est un foncteur *exact* de catégories abéliennes.

*Définition***(2.2.9).**— *Si Z est une hypersurface et $\mathcal{M}$ est un $\mathcal{D}_{X/C}$-module holonome on appelle* faisceau d'irrégularité *de $\mathcal{M}$ le long de Z le faisceau* $\mathbf{IR}_Z(\mathcal{M})$.

Par exemple si X est une courbe et Z un point le faisceau $\mathbf{IR}_Z(\mathcal{M})$ est un espace vectoriel complexe de dimension finie. Sa dimension est égale au nombre de Fuchs en Z de $\mathcal{M}$. Cela résulte du théorème de comparaison de Malgrange [M_1].

Il faut maitenant avoir un critére pour que le faisceau $\mathbf{IR}_Z(\mathcal{M})$ soit nul. On dit qu'un module holonome à support une sous-variété lisse Y est *lisse* si sa variété caractéristique contient au plus que le fibré conormal T^*_YX.

Soient Y une sous-variété d'une variété algébrique complexe non singulière X, Z une hypersurface de X dont la trace sur Y contient le lieu singulier de Y et de codimension un dans Y et $\mathcal{M}$ un $\mathcal{D}_{X/C}$-module holonome à support contenu dans Y et lisse en dehors de Z. On a alors le théorème [Me_7] :

Théorème **(2.2.10).**— *Sous les conditions précédentes le faisceau* $\mathbf{IR}_Z(\mathcal{M})$ *est nul si et seulement si la codimension dans* $Z \cap Y$ *de son support* est au moins égale à un.

Par exemple si on on prend $Y = X$ et $\mathcal{M}$ le fibré $\mathcal{O}_X$ trivial muni de la connexion naturelle on trouve que le faisceau $\mathbf{IR}_Z(\mathcal{O}_X)$ est nul pour toute hypersurface Z puisque de façon naturelle son support est contenu dans le lieu singulier de Z qui est bien de codimension un dans Z. A partir de là on touve par un agument combinatoire que le faiscceau $\mathbf{IR}_Z(H^p_Y(\mathcal{O}_X))$ est nul pour toute sous-variété Y de X, pour toute hypersurface Z et tout p. Si on applique cela à X = l'espace projectif complexe P^m, Y l'adhérance d'une variété algébrique affine non singulière plongée dans l'espace numérique C^m et Z le diviseur à l'infini on trouve que le faisceau $\mathbf{IR}_Z(H^p_Y(\mathcal{O}_X))$ est nul. Mais l'hypercohomologie de ce faisceau est précisément l'obstruction à l'isomorphisme de la cohomologie de de Rham de la variété affine en question avec sa cohomologie de Betti [G_1]. Autrement dit le théorème (2.2.10) implique le théorème de comparaison de Grothendieck [G_1].

Plus généralement soit $\mathcal{E}$ un fibré à connexion intégrable sur une variété algébrique complexe non singulière muni d'un plongement dans l'espace numérique C^m. Notons Y l'adhérance de cette variété dans l'espace projectif P^m et $\overline{\mathcal{E}}$ l'image directe au sens des $\mathcal{D}$-modules de $\mathcal{E}$ par l'immersion dans P^m. $\overline{\mathcal{E}}$ est alors un $\mathcal{D}_{P^m}$-module holonome à support Y et lisse en dehors du diviseur à l'infini Z. Supposons que l'image inverse de $\mathcal{E}$ sur toute courbe non singulière n'a que des singularités régulières à l'infini et Y est normale (pour simplifier car ceci n'est pas une restiction). Alors le faisceau $\mathbf{IR}_Z(H^p_Y(\mathcal{O}_X))$ est à codimension $Y \cap Z$ dans au moins un comme on le voit en faisant passer une courbe générale en un point assez général de Y. Le théorème (2.2.10) implique que le faisceau $\mathbf{IR}_Z(\overline{\mathcal{E}})$ est nul. Mais l'hypercohomologie de ce faisceau est l'obstruction à l'isomorphisme entre cohomologie de de Rham de $\mathcal{E}$ et cohomologie de de Rham du fibré transcendant associé. Autrement le théorème (2.2.10) implique le théorème de comparaison de Deligne [D_1].

2.2.11. Plus généralement soit $\mathcal{M}$ un $\mathcal{D}_X$-module holonome sur une variété algébrique complexe non singulière X. On a alors le théorème suivant [Me_7] :

Théorème **(2.2.12).**— *L'image inverse de* $\mathcal{M}$ *sur toute courbe non singulière au dessus de* X *n'a que des singularités réguliéres (y compris à l'infini)* si et seulement si *son faisceau d'irrégularité le long de tout diviseur (y compris ceux de l'infini) de tout ouvert affine* est nul.

Le théorème (2.2.12) améne à poser la définition :

Définition **(2.2.13).**— *Un* $\mathcal{D}_X$*-module holonome* $\mathcal{M}$ *est dit régulier s'il satisfait au deux conditions équivalentes* (2.2.12).

Le théorème (2.2.10) donne un critére qui *ne dépend pas* de la résolution des singularités pour vérifier si un $\mathcal{D}_X$-module est régulier ou pas.

On note par Mhr($\mathcal{D}_X$-) la catégorie des $\mathcal{D}_X$-modules holonomes réguliers. Dans une suite exacte de $\mathcal{D}_X$-modules holonomes le terme médian est régulier si et seulement si les termes extrêmes sont réguliers. Cela résulte du fait que le foncteur $\mathbf{IR}_Z$ entre catégories abéliennes est *exact* pour toute hypersurface Z. La catégorie Mhr($\mathcal{D}_X$-) est donc abélienne, stable par extension et sous-quotient. On note par $D^b_{hr}(\mathcal{D}_{X/C}\text{-mod})$ la sous-catégorie de $D^b_{hr}(\mathcal{D}_{X/C}\text{-mod})$ des complexes à cohomologie régulière. La catégorie $D^b_{hr}(\mathcal{D}_{X/C}\text{-mod})$ est alors triangulée. On déduit du théorème (2.2.10) que les catégories $D^b_{hr}(\mathcal{D}_{X/C}\text{-mod})$ pour X variables sont stables par les opérations cohomologiques de Grothendieck .

Théorème **(2.2.14).**— *Si* f : X → Y *est un morphisme de variétés algébriques complexe non singulières les foncteurs* f^*, $f^!$, *envoient la catégorie* $D^b_{hr}(\mathcal{D}_{Y/C}\text{-mod})$ *dans la catégorie* $D^b_{hr}(\mathcal{D}_{X/C}\text{-mod})$ *et pour tout complexe* $\mathcal{M}$ *de la catégorie* $D^b_{hr}(\mathcal{D}_{Y/C}\text{-mod})$ *on les isomorphismes entre coefficients constructibles* :

$$f^{an!}\mathbf{DR}(\mathcal{M}^{an}) \cong \mathbf{DR}((f^*\mathcal{M})^{an})$$
$$f^{an-1}\mathbf{DR}(\mathcal{M}^{an}) \cong \mathbf{DR}((f^!\mathcal{M})^{an}).$$

Preuve. [Me$_7$].

Théorème **(2.2.15).**— *Si* f : X → Y *est un morphisme de variétés algébriques complexe non singulières les foncteurs* f^c_*, $f^c_!$, *envoient la catégorie* $D^b_{hr}(\mathcal{D}_{X/C}\text{-mod})$ *dans la catégorie* $D^b_{hr}(\mathcal{D}_{Y/C}\text{-mod})$ *et pour tout complexe* $\mathcal{M}$ *de la catégorie* $D^b_{hr}(\mathcal{D}_{X/C}\text{-mod})$ *on a les isomorphismes entre coefficients constructibles*

$$\mathbf{DR}((f^c_*\ \mathcal{M})^{an}) \cong \mathbf{R}f^{an}{}_*\mathbf{DR}(\mathcal{M}^{an})$$
$$\mathbf{DR}((f^c_!\ \mathcal{M}^{an})) \cong \mathbf{R}f^{an}{}_!\mathbf{DR}(\mathcal{M}^{an}).$$

Preuve. [Me$_7$].

On déduit [Me$_7$] des théorèmes (2.2.14) et (2.2.15) le théorème des coefficients de de Rham-Grothendieck à savoir que les catégories $D^b_{hr}(\mathcal{D}_{X/C}\text{-mod})$ sont stables par les six opérations de Grothendieck et que le foncteur de de Rham transcendant qui est pleinement fidèle respecte les six opérations analogues dans les catégories $D^b_c(C_X)$ de coefficients algébriquement constructibles. En particulier on obtient une démonstration du théorème de la ***régularité*** de la connexion de Gauss-Manin en caractéristique nulle indépendante du théorème de résolution des singularités ([Ka_1], [Ka_2])et ce pour un morphisme arbitraire de variétés algébriques. Ceci remplit parfaitement le programme de Grothendieck en caractéristique nulle ([G_2], p. 312) " It would be convenient to have a more general duality formalisme of type $f_!$, $f^!$, as developed in [R.D], for the De Rham cohomology ". On constate, une fois de plus, que pour avoir une bonne théorie cohomologique des variétés algébriques il faut d'abord dégager une bonne théorie des coefficients qui sont stables par les opérations cohomologiques.

2.2.16. La catégorie $D^b_{hr}(\mathcal{D}_{X/C}\text{-mod})$ est purement algébrique, en effet la condition pour un $\mathcal{D}_{X/C}$-module holonome d'avoir que des singularités régulières sur toute courbe au dessus de X est algébrique. Donc elle garde un sens sur un corps k de base de caractéristique nulle. On déduit alors que les catégries $D^b_{hr}(\mathcal{D}_{X/k}\text{-mod})$ sont stables par les six opérations de Grothendieck. Mais les démonstrations se ramènent par le principe de Leschetz au cas transcendant. Cependant le faisceau transcendant $\mathbf{IR}_Z(\mathcal{M})$ qui est la base des démonstrations a un substitut purement algébrique. Ceci améne à penser qu'on peut démontrer certainement le résultat précédent de façon purement algébrique.

Le faisceau $\mathbf{IR}_Z(\mathcal{M})$, comme tout faisceau pervers, admet un cycle caractéristique $CCh(\mathbf{IR}_Z(\mathcal{M}))$ qui est un cycle lagrangien positif du fibré cotangent T*X cf. [L-M]. Par définition le faisceau $\mathbf{IR}_Z(\mathcal{M})$ s'incère dans une suite exacte de faisceaux pervers sur X :

$$0 \to \mathbf{IR}_Z(\mathcal{M})[-1] \to \mathbf{DR}((j_*j^{-1}\mathcal{M})^{an}) \to \mathbf{R}j^{an}{}_*j^{-1an}\mathbf{DR}(\mathcal{M}^{an}) \to 0.$$

Le cycle $CCh(\mathbf{IR}_Z(\mathcal{M})[-1])$ apparaît comme la différence $CCh(\mathbf{DR}((j_*j^{-1}\mathcal{M})^{an}))$ - $CCh(\mathbf{R}j^{an}{}_*j^{-1an}\mathbf{DR}((\mathcal{M})^{an})$. Mais chaque membre de cette différence est purement algébrique. En effet le cycle $CCh(\mathbf{DR}((j_*j^{-1}\mathcal{M})^{an}))$ est le cycle caractéristique du $\mathcal{D}_{X/C}$-module holonome $j_*j^{-1}\mathcal{M}$ et le cycle $CCh(\mathbf{R}j^{an}{}_*j^{-1an}\mathbf{DR}(\mathcal{M}^{an}))$ est l'image directe au sens des cycles du cycle caractéristique du $\mathcal{D}_{U/C}$-module holonome $j^{-1}\mathcal{M}$ [Me_6].

Nous allons décrire le cycle $CCh(\mathbf{IR}_Z(\mathcal{M})[-1])$ dans le cas important où $\mathcal{M}$ est un fibré vectoriel de rang r au-dessus de U. A ce moment là le cycle $CCh(\mathbf{R}j^{an}{}_*j^{-1an}\mathbf{DR}(\mathcal{M}^{an}))$ est égal à $rCCh(j_*j^{-1}\mathcal{O}_X)$ et se décrit à partir de l'hypersurface Z [L-M]. On a donc

$$CCh(\mathbf{IR}_Z(\mathcal{M}) = CCh(j_*j^{-1}\mathcal{M}) - rCCh(j_*j^{-1}\mathcal{O}_X).$$

Sous cette forme le cycle $CCh(\mathbf{IR}_Z(\mathcal{M})$ garde un sens sur un corps de base de caractéristique nulle. Ce cycle étant positif on obtient une statification de $Z = \cup Z_i$ et pour chaque strate un entier m_i *positif ou nul* et le cycle $\Sigma m_i Z_i$ positif promis en 2.2.4 attaché au triplet (X, Z, $\mathcal{M}$) qui généralise le nombre de Fuchs. Le théorème (2.2.10) assure alors que le cycle $\Sigma m_i Z_i$ est nul *0,,,,,,* si les nombres attachés aux strates de codimension *nulle* sont *nuls*. Le calcul de ces derniers est général accessible. En somme on est dans la même situation en dimension supérieure que celle du théorème de Fuchs en dimension un.

§ 3. Le théorème des coefficients holonomes complexes d'orde infini.

Une fonction g de $(\overline{\mathbf{Q}}_P[x][1/x])^\dagger$ est la somme d'une série $\Sigma a_m/x^{m+1}$ ($a_m \in \overline{\mathbf{Q}}_P$, $m \in \mathbf{N}$) et d'une fonction de $(\overline{\mathbf{Q}}_P[x])^\dagger$ avec la condition de Dwork-Monsky-Washnitzer $\lim_{m \to \infty} |a_m| \varepsilon^{-m} = 0$ pour un réel positif ε assez petit convenable ([Dw], [M-W]). Mais $1/x^{m+1} = (-1)^m\Delta^m(1/x)$ et si on pose $P(\Delta) := \Sigma a_m(-1)^m\Delta^m$ on trouve que $\Sigma a_m/x^{m+1} = P(\Delta)(1/x)$. Autrement dit $(\overline{\mathbf{Q}}_P[x][1/x])^\dagger$ est un module de type fini et même de présentation finie sur un anneau d'opérateurs différentiels d'ordre *infini*.

La situation est pareille à la situation transcendante complexe qu'on a longuement étudiée depuis maintenant presque vingt ans. En effet si g est une fonction holomorphe sur le plan complexe qui a au plus une singularité essentielle à l'origine c'est alors une somme d'une série $\Sigma a_m/x^{m+1}$ ($a_m \in \mathbf{C}$, $m \in \mathbf{N}$) et d'une fonction entière avec la condition pour tout ε positif il existe une constante M_ε telle que $|a_m| \leq M_\varepsilon \varepsilon^m$. On a par le même calcul algébrique que l'anneau des fonctions holomorphes sur le plan complexe ayant au plus une singularité essentielle en zéro est un module de présention finie sur l'anneau des opérateurs différentiels d'ordre infini.

Ceci suggére que si X est une variété affine non singulière sur $\mathbf{F}_p$ et Z une hypersurface de X le faisceau $j^\dagger{}_*\mathcal{O}^\dagger_U[Mr]$, où U := X-Z, est l'analogue du faisceau $j^{an}{}_*\mathcal{O}_{U^{an}}$ de la situation transcendante complexe associée à une situation algébrique complexe. On dispose sur **C** d'un théorème finitude pour ces faisceaux et plus généralement d'une théorie des coefficients holonomes complexe d'ordre infini.

3.1. Le faisceau des opérateurs différentiels complexes d'ordre infini.

Pour une variété analytique complexe X on note $\mathcal{D}_X$ le faisceau $\mathcal{D}iff_{X/C}(\mathcal{O}_X)$.

3.1.1. Soit $(X, \mathcal{O}_X)$ une variété analytique complexe. Le faisceau des fonctions holomorphes est un faisceau de Fréchet-Nucléaire : pour tout ouvert U de X l'espace vectoriel complexe $\Gamma(U;\mathcal{O}_U)$ est un espace vectoriel topologique de Fréchet-Nucléaire et les morphismes de restictions sont continus.

Définition **(3.1.2).**— *Le préfaisceau des opérateurs différentiels d'ordre infini* $\mathcal{D}^\infty_X$ *est définie comme sous préfaisceau du faisceau* $hom_{C_X}(\mathcal{O}_X,\mathcal{O}_X)$ des C_X-endomorphismes continus.

Lemme **(3.1.3).**— *Le préfaisceau* $\mathcal{D}^\infty_X$ *est un* faisceau.

En effet un endomorphisme localement continu est continu.

Proposition **(3.1.4).**— *Si* U *est un ouvert de* X *où sont définies des coordonnées locales* x :=

$(x_1,\ldots,x_n)$ *un endomorphisme au-dessus de* U *est opérateur différentiel d'ordre infini* si et seulement *c'est une somme infinie* $\Sigma\, a_\alpha(x)\Delta^\alpha$ *où* $a_\alpha(x)$ $(\alpha\in N^n)$ *est une suite de fonctions holomorphes au dessus de* U *telle que* $\lim_{|\alpha|\to\infty} |a_\alpha(x)|^{1/|\alpha|}$ uniformément sur tout compact de U.

Preuve. Il résulte des inégalités de Cauchy qu'une somme infini $\Sigma\, a_\alpha(x)\Delta^\alpha$ qui la propriété de (3.1.4) est un opérateur différentiel d'ordre infini. Réciproquement soit P un opérateur d'ordre infini. On définit la suite de fonctions holomorphes $a_\alpha(x)$ en posant

$$a_\alpha(x) := \Sigma_{k\leq\alpha}\, (-1)^k\binom{k}{\alpha}x^{\alpha-k}P(x^k)$$

avec les notations de l'analyse classique. Il faut voir que la suite $a_\alpha(x)$ a la propriété requise. On remarque pour cela que pour tout point y de U on a l'égalité

$$\Sigma_{k\leq\alpha}\, (-1)^k\binom{k}{\alpha}x^{\alpha-k}P(x^k) = \Sigma_{k\leq\alpha}\, (-1)^k\binom{k}{\alpha}(x-y)^{\alpha-k}P((x-y)^k).$$

De cela si pour tout ε assez petit et tout compact K de U on trouve que

$$|\, a_\alpha(x)|_K \leq M_\varepsilon \varepsilon^{|\alpha|}$$

où M_ε est une constante qui ne dépend que K et de ε. Ceci est la condition requise pour que la somme infini $\Sigma_\alpha a_\alpha(x)\Delta^\alpha$ soit un opérateur différentiel d'ordre infini. Par construction P et $\Sigma_\alpha a_\alpha(x)\Delta^\alpha$ coincident sur les polynômes. Par localisation et densité $P = \Sigma_\alpha a_\alpha(x)\Delta^\alpha$.

En particulier on trouve que le faisceau $\mathcal{D}_X$ est le sous faisceau de $\mathcal{D}_X^\infty$ des opérateurs différentiels d'ordre localement fini.

Par construction le faisceau $\mathcal{D}_X^\infty$ est un faisceau d'anneaux et le faisceau $\mathcal{O}_X$ est un $\mathcal{D}_X^\infty$-module à *gauche*.

De même partant du faisceau des n-formes différentielles ω_X qui est aussi un faisceau de Fréchet-Nucléaire muni de sa structure de $\mathcal{D}_X$-module à droite et considérant le faisceau des endomorphismes C_X-linéaires continus on trouve que ce faisceau est canoniquement isomorphe à $\mathcal{D}_X^\infty$ et que la structure de $\mathcal{D}_X$- module à droite de ω_X se prolonge en une structure de $\mathcal{D}_X^\infty$-module *à droite*. On peut aussi remarquer que l'on a $Ext^n_{\mathcal{D}_X^\infty}(\mathcal{O}_X, \mathcal{D}_X^\infty) \cong \omega_X$.

3.1.5. Soient $f : X \to Y$ un morphisme de variétés analytiques complexes lisses et $\Delta_f : X \to X\times Y$ le morphisme graphe de f. Notons par q_1 et q_2 les projections de $X\times Y$ sur X et Y respectivement. Posons

$$\mathcal{D}^\infty_{X\to Y} := q_{1*}\mathcal{H}_X^{\dim(Y)}(q_2^*\omega_Y)$$
$$\mathcal{D}^\infty_{Y\leftarrow X} := q_{1*}\mathcal{H}_X^{\dim(Y)}(q_1^*\omega_X).$$

La structure de $\mathcal{D}^\infty_X$-module *à gauche* de $\mathcal{O}_X$ induit une structure de $\mathcal{D}^\infty_X$-module *à gauche* sur le faisceau $\mathcal{D}^\infty_{X\to Y}$ et la structure de $\mathcal{D}^\infty_Y$-module *à droite* sur ω_Y induit une structure de $f^{-1}\mathcal{D}^\infty_Y$-module *à droite* sur le faisceau $\mathcal{D}^\infty_{X\to Y}$ qui est donc un $(\mathcal{D}^\infty_X, f^{-1}\mathcal{D}^\infty_Y)$-bimodule. De même la structure de $\mathcal{D}^\infty_X$-module *à droite* de ω_X induit une structure de $\mathcal{D}^\infty_X$-module *à droite* sur le faisceau $\mathcal{D}^\infty_{Y\leftarrow X}$ et la structure de $\mathcal{D}^\infty_Y$-module *à gauche* sur $\mathcal{O}_Y$ induit une structure de $f^{-1}\mathcal{D}^\infty_Y$-module *à gauche* sur le faisceau $\mathcal{D}^\infty_{Y\leftarrow X}$ qui est donc un $(f^{-1}\mathcal{D}^\infty_Y, \mathcal{D}^\infty_X)$-bimodule. En particulier si f est le morphisme identique on trouve que le faisceau $p_{1*}\mathcal{H}_X^{\dim(X)}(p_2^*\omega_X)$ est un $(\mathcal{D}^\infty_X, p_{1*}p_2^{-1}\mathcal{D}^\infty_X)$-bimodule et donc un $(\mathcal{D}^\infty_X, \mathcal{D}^\infty_X)$-bimodule.

Proposition **(3.1.6).**— Le $(\mathcal{D}^\infty_X, \mathcal{D}^\infty_X)$*-bimodule* $p_{1*}\mathcal{H}_X^{\dim(X)}(p_2^*\omega_X)$ *est canoniquement* isomorphe *au faisceau d'anneaux* $\mathcal{D}^\infty_X$.

Preuve. Si $K(x,y)dy$ est une classe de cohomologie de $p_{1*}\mathcal{H}_X^{\dim(X)}(p_2^*\omega_X)$ et $g(x)$ une fonction holomorphe la formule de Cauchy montre que $2\pi\sqrt{-1}\int_{|x-y|=\varepsilon} K(x,y)g(y)dy$, pour un cycle $|x-y|=\varepsilon$ convenable est une fonction holomorphe de la forme $P(g)(x) = \Sigma\, a_\alpha(x)\Delta^\alpha g(x)$ qui ne dépend pas du cycle $|x-y|=\varepsilon$. On a alors un homorphisme de $p_{1*}\mathcal{H}_X^{\dim(X)}(p_2^*\omega_X)$ dans $\mathcal{D}^\infty_X$ dont on vérifie par un calcul directe que c'est un isomorphisme de $(\mathcal{D}^\infty_X, \mathcal{D}^\infty_X)$-bimodules.

3.1.7. On ne connait aucune propriété de finitude du faisceau $\mathcal{D}^\infty_X$. En particulier on ignore s'il est cohérent ou pas. Mais on a le résultat suivant ([S.K.K], 3.4.2):

Théorème **(3.1.8).**— *L'extension* $\mathcal{D}_X \to \mathcal{D}^\infty_X$ est fidèlement plate *à droite et à gauche.*

La démonstration de [S.K.K.] est micro-différentielle. Comme nous l'avons dit dans l'introduction de [Me$_5$] nous connaissons une démonstration différentielle.

3.2. Le théorème de bidualité.

Soient X une variété analytique complexe et $\mathcal{M}$ un fibré vectoriel sur X de rang fini à connexion intégrable. Alors son systéme local des sections horizontales $hom_{\mathcal{D}_X}(\mathcal{O}_X,\mathcal{M})$ le déterminent complétement par l'isomorphisme

$$\mathcal{M} \cong hom_{\mathcal{D}_X}(\mathcal{O}_X,\mathcal{M})\otimes_{C_X}\mathcal{O}_X.$$

Il est plus commode de considérer son système des solutions holomorphes $hom_{\mathcal{D}_X}(\mathcal{M},\mathcal{O}_X)$ et l'isomorphisme précédent devient un devient un isomorphisme de $\mathcal{D}_X$-modules à gauche :

$$\mathcal{M} \cong hom_{C_X}(hom_{\mathcal{D}_X}(\mathcal{M},\mathcal{O}_X),\mathcal{O}_X).$$

Mais si $\mathcal{M}$ est un $\mathcal{D}_X$-module holonome son faisceau (pervers) de de Rham $\mathbf{DR}(\mathcal{M}) := \mathbf{R}hom_{\mathcal{D}_X}(\mathcal{O}_X,\mathcal{M})$ ou son faisceau des solutions holomorphes $\mathbf{S}(\mathcal{M}) := \mathbf{R}hom_{\mathcal{D}_X}(\mathcal{M},\mathcal{O}_X)$ ne le determinent pas en général à cause de la ramification *non* modérée. Cependant ils déterminent le $\mathcal{D}_X^\infty$-module $\mathcal{D}_X^\infty\otimes_{\mathcal{D}_X}\mathcal{M}$ obtenu par extension des scalaires. Si $\mathcal{M}^\infty$ est un complexe borné de $\mathcal{D}_X^\infty$-modules (à gauche) on a un morphisme canonique

$$Bd(\mathcal{M}^\infty) : \mathcal{M}^\infty \to \mathbf{R}hom_{C_X}(\mathbf{R}hom_{\mathcal{D}_X^\infty}(\mathcal{M}^\infty,\mathcal{O}_X),\mathcal{O}_X)$$

de complexes de $\mathcal{D}_X^\infty$-modules. Ce morphisme s'explicite en prenant une résolution $\mathcal{D}_X^\infty$-injective $\mathcal{I}$ de $\mathcal{O}_X$ qui reste par platitude $\mathcal{O}_X$-injective et le morphisme $Bd(\mathcal{M}^\infty)$ est simplement le morphisme naturel de bidualité :

$$Bd(\mathcal{M}^\infty) : \mathcal{M}^\infty \to hom_{C_X}(hom_{\mathcal{D}_X^\infty}(\mathcal{M}^\infty,\mathcal{I}),\mathcal{I}).$$

La terminologie peut prêter à confusion avec les théorèmes de bidualité (interne) pour les coefficients cohérents ou discrets constructibles. On a alors le théorème ([Me$_3$], thm. 2.1) :

Théorème **(3.2.1).**— *Si* $\mathcal{M}^\infty$ *est un complexe* parfait *de* $\mathcal{D}_X^\infty$*-modules tel que le complexe* $\mathbf{R}hom_{\mathcal{D}_X^\infty}(\mathcal{M}^\infty,\mathcal{O}_X)$ *est constructible le morphisme* $Bd(\mathcal{M}^\infty)$ *est un* isomorphisme.

Rappelons qu'on appelle parfait un complexe de modules admettant *localement* une résolution finie par des modules libres de type fini [S.G.A. 6].

Soit Z une hypersurface de X et j l'inclusion de U:= X-Z dans X.

Corollaire **(3.2.2).**— *Le morphisme* $\mathcal{D}_X^\infty \otimes_{\mathcal{D}_X} \mathcal{O}_X(*Z) \to j_*j^{-1}\mathcal{O}_X$ est un isomorphisme.

Par dualité ([Me$_3$], Thm. 1.1) le théorème de Grothendieck [G$_1$] est équivalent à l'isomorphisme $S(\mathcal{O}_X(*Z)) \cong j_!j^{-1}\mathbf{C}_X$ et (3.2.2) est conséquence de (3.2.1) parce que $\mathbf{R}\mathrm{hom}_{\mathbf{C}_X}(j_!j^{-1}\mathbf{C}_X, \mathcal{O}_X) \cong \mathbf{R}j_*j^{-1}\mathcal{O}_X \cong j_*j^{-1}\mathcal{O}_X$. Le corollaire (3.2.2) montre en particulier que $j_*j^{-1}\mathcal{O}_X$ est un $\mathcal{D}_X^\infty$-module de présentation finie ce qui est l'analogue transcendant complexe de (2.1.9) pour le fibré trivial. Cependant sa démonstration, bien que maintenant indépendante de la résolution des singularités, nécessite la connaissance des solutions $S(\mathcal{O}_X(*Z))$. Nous ne connaissons pas une démonstration algébrique directe qui puisse nous être utile dans le cas p-adique sauf dans le cas des singularités trés simples (croisements normaux, point quadratique ordinaire) cf. 4.4.

La démonstration de (3.2.1) utilise d'une façon essentiel les propriétés de constructibilité des solutions holomorphes.

La catégorie des complexes $\mathcal{M}^\infty$ parfaits tels que $\mathbf{R}\mathrm{hom}_{\mathcal{D}_X^\infty}(\mathcal{M}^\infty, \mathcal{O}_X)$ est constructible est triangulée.Une sous-catégorie, à priori plus petite, de complexes qui ont les propriétés de (3.2.1) est formée des complexes *localement* de la forme $\mathcal{D}_X^\infty \otimes_{\mathcal{D}_X} \mathcal{M}$ pour un complexe $\mathcal{M}$ de $D_h^b(\mathcal{D}_X\text{-mod})$. Il n'est pas du tout clair que cette derniére soit *triangulée*. Pour élucider cette difficulté il faut un théorème de finitude général

3.3. Le théorème de finitude.

3.3.1. Soient $\mathcal{F}$ un coefficient constructible sur une variété analytique complexe. Alors le complexe $\mathrm{Rhom}_{\mathbf{C}_X}(\mathcal{F}, \mathcal{O}_X)$ est muni d'une structure de $\mathcal{D}_X^\infty$-complexe à gauche induite par la structure de $\mathcal{D}_X^\infty$-module de $\mathcal{O}_X$. On alors le théorème [Me$_4$] :

Théorème **(3.3.2).**— *Le complexe* $\mathrm{Rhom}_{\mathbf{C}_X}(\mathcal{F}, \mathcal{O}_X)$ *est globalement de la forme* $\mathcal{D}_X^\infty \otimes_{\mathcal{D}_X} \mathcal{M}$ *pour un complexe* $\mathcal{M}$ *de la catégorie* $D_{hr}^b(\mathcal{D}_X\text{-mod})$.

La signification de $D_{hr}^b(\mathcal{D}_X\text{-mod})$ est clair. C'est la catégorie des complexes bornés de $\mathcal{D}_X$-modules à cohomologie holonome régulière. Un $\mathcal{D}_X$-module holonome $\mathcal{M}$ est régulier si son faisceau $\mathrm{IR}_Z(\mathcal{M})$ est nul pour toute hypersurface Z.

Soit $j : U \to X$ le complémentaire d'une hypersurface Z de X et $\mathcal{M}_U$ un fibré à connexion intégrable sur U. La structure de $\mathcal{D}_U$-module à gauche de $\mathcal{M}_U$ se prolonge en une structure de $\mathcal{D}_U^\infty$-module à gauche

Un cas particulier de (3.3.2) et le

Corollaire **(3.3.3).**— *Le* $\mathcal{D}_X^\infty$-module $j_*\mathcal{M}_U$ *est de la forme* $\mathcal{D}_X^\infty \otimes_{\mathcal{D}_X} \mathcal{M}_r(*Z)$ *pour un* $\mathcal{D}_X$-*module holonome* $\mathcal{M}_r$ *tel que le faisceau* $\mathbf{IR}_Z(\mathcal{M}_r)$ est nul.

En particulier le $\mathcal{D}_X^\infty$-module $j_*\mathcal{M}_U$ est de présentation finie. La démonstration est hautement transcendante. Bien que indépendante aujourd'hui de la résolution des singularités elle nécessite le théorème d'existence de Riemann : trouver un $\mathcal{D}_X$-module holonome $\mathcal{M}_r$ tel que $\mathbf{IR}_Z(\mathcal{M}_r)$ est nul et qui a même monodromie que $\mathcal{M}_U$. Le prolongement analytique est au fond de la question dans le théorème de finitude (3.3.2) et semble indiquer que pour la démonstration d'un résultat analogue dans le cas p-adique le Frobénius est au fond du problème. Le théorème (3.3.2) utilise toutes les resources de la théories des $\mathcal{D}_X$-modules et en est l'étape ultime. Mais il a de nombreuses conséquences et en particulier clarifie la structure de la catégorie des complexes parfaits de $\mathcal{D}_X^\infty$-modules dont les solutions sont constructibles.

Définition **(3.3.4).**— *On appelle* $\mathcal{D}_X^\infty$*-complexe holonome un* $\mathcal{D}_X^\infty$*-complexe* $\mathcal{M}^\infty$ *tel qu'il existe* localement *sur X un* $\mathcal{D}_X$*-complexe holonome* $\mathcal{M}$ *et un isomorphisme*

$$\mathcal{D}_X^\infty \otimes_{\mathcal{D}_X} \mathcal{M} \cong \mathcal{M}^\infty.$$

Notons Mh($\mathcal{D}_X^\infty$-mod) la catégorie des $\mathcal{D}_X^\infty$-modules holonomes et $D_h^b(\mathcal{D}_X^\infty\text{-mod})$ la sous-catégorie de la catégorie $D^b(\mathcal{D}_X^\infty\text{-mod})$ des complexes holonomes. Les catégories Mh($\mathcal{D}_X^\infty$-mod) et $D_h^b(\mathcal{D}_X^\infty\text{-mod})$ sont définies en terme purement algébriques à partir du faisceau des opérateurs différentiels d'ordre infini. Cependant il n'est pas du tout clair que la catégorie Mh($\mathcal{D}_X^\infty$-mod) est abélienne et stable par extention ni que la catégorie $D_h^b(\mathcal{D}_X^\infty\text{-mod})$ soit triangulée. Le théorème de finitude (3.3.2) va dire que tel est bien le cas.

Notons **T** le foncteur d'extention des scalaires

$$D_{hr}^b(\mathcal{D}_X\text{-mod}) \to D_h^b(\mathcal{D}_X^\infty\text{-mod});\ \mathcal{M} \to \mathbf{T}(\mathcal{M}) := \mathcal{D}_X^\infty \otimes_{\mathcal{D}_X} \mathcal{M};$$

F le foncteur fournit par le théorème de constructibilité

$$D_h^b(\mathcal{D}_X^\infty\text{-mod}) \to D_c^b(C_X);\ \mathcal{M}^\infty \to \mathbf{F}(\mathcal{F}) := \mathrm{Rhom}_{\mathcal{D}_X^\infty}(\mathcal{M}^\infty, \mathcal{O}_X)$$

et G le foncteur fournit par le théorème (3.3.2)

$$D_c^b(C_X) \to D_h^b(\mathcal{D}_X^\infty\text{-mod});\ \mathcal{F} \to G(\mathcal{F}) := \mathrm{Rhom}_{C_X}(\mathcal{F}, \mathcal{O}_X).$$

Théorème **(3.3.5)**[Me_4].— *Les foncteurs* **F** et **G** *sont des* équivalences de catégories triangulées quasi-inverses *l'un de l'autre et le foncteur* **T** *est une* équivalence de catégories.

Comme l'extension $\mathcal{D}_X \to \mathcal{D}_X^\infty$ est fidèlement plate le foncteur **T** induit une équivalence de catégories

$$\mathrm{Mhr}(\mathcal{D}_X\text{-mod}) \to \mathrm{Mh}(\mathcal{D}_X^\infty\text{-mod});\ \mathcal{M} \to T(\mathcal{M}) := \mathcal{D}_X^\infty \otimes_{\mathcal{D}_X} \mathcal{M}.$$

En particulier la catégorie $\mathrm{Mh}(\mathcal{D}_X^\infty\text{-mod})$ est *abélienne* et *stable par extension* et la catégorie $D_h^b(\mathcal{D}_X^\infty\text{-mod})$ est *triangulée.* De plus l'inclusion de la catégorie des $\mathcal{D}_X^\infty$-complexe holonomes dans la catégorie des $\mathcal{D}_X^\infty$-complexes parfaits dont les solutions sont constructibles est une équivalence de catégories. Ceci résulte des théorème (3.2.1) et (3.3.2). Enfin l'inclusion de la catégorie des $\mathcal{D}_X^\infty$-complexes holonomes dans la catégorie des complexes de $\mathcal{D}_X^\infty$-modules à cohomologie bornée et holonomes est une équivalence de catégories par les mêmes théorèmes ce qui justifie la notation $D_h^b(\mathcal{D}_X^\infty\text{-mod})$ pour la catégories des complexes holonomes. On a ainsi clarifié la structure de la catégorie des $\mathcal{D}_X^\infty$-complexes holonomes comme conséquence ultime de la théorie des $\mathcal{D}_X$-modules.

3.3.6. Si X est variété algébrique complexe on définit le faisceau $\mathcal{D}_X^\infty$ comme l'image directe du faisceau des opérateurs différentiels d'ordre infini de la variété transcendente associée par le morphisme canonique $X^{an} \to X$. On définit un $\mathcal{D}_X^\infty$-complexe holonome comme un $\mathcal{D}_X^\infty$-complexe localement pour la toplogie de Zariski isomorphe à un $\mathcal{D}_X^\infty$-complexe obtenu par extention des scalaires à partir d'un $\mathcal{D}_X$-complexe holonome. Alors la catégorie $D_h^b(\mathcal{D}_X^\infty\text{-mod})$ des complexes holonomes est équivalente à la catégorie des coefficients algébriquement constructibles $D_c^b(C_X)$. L'espoir est qu'un analogue de la catégorie $D_h^b(\mathcal{D}_X^\infty\text{-mod})$ garde un sens en caractéristique $p > 0$ et fournit une catégorie de coefficients p-adiques. Aussi nous allons rappeler ([Me_5], II. 9.5.7) le formalisme des six opérations de Grothendieck pour les catégories $D_h^b(\mathcal{D}_X^\infty\text{-mod})$.

Si $f : X \to Y$ est un morphime de variétés algébriques complexes non singulières on pose

$$f^*\mathcal{N} := \mathcal{D}^\infty_{X\to Y} \overset{L}{\otimes}_{f^{-1}\mathcal{D}^\infty_Y} f^{-1}\mathcal{N}$$

$$f^c_* \mathcal{M} :\cong Rf_*\mathcal{D}^\infty_{Y\leftarrow X} \overset{L}{\otimes}_{\mathcal{D}^\infty_X}\mathcal{M}.$$

pour un complexe $\mathcal{N}$ de la catégorie $D^b(\mathcal{D}^\infty_Y$-mod) et pour un complexe $\mathcal{M}$ de la catégorie $D^b(\mathcal{D}^\infty_X$-mod). On obtient ainsi les foncteurs image inverse et image directe entre catégories de complexes de modules d'ordre infini.

Si $\mathcal{M}^\infty$ est un complexe de la catégorie $D^b_h(\mathcal{D}^\infty_X$-mod) il est localement de la forme $\mathcal{D}^\infty_X \otimes_{\mathcal{D}_X} \mathcal{M}$ pour un complexe $\mathcal{M}$ de la catégorie $D^b_h(\mathcal{D}_X$-mod). Le complexe $\mathcal{D}^\infty_X \otimes_{\mathcal{D}_X} \mathcal{M}^*$ est un objet de la catégorie $D^b_h(\mathcal{D}^\infty_X$-mod) donné *localement*. En vertu du théorème (3.3.2) et du théorème de dualité locale c'est en fait un objet *globale* de la catégorie $D^b_h(\mathcal{D}^\infty_X$-mod) qui est le dual $\mathcal{M}^{\infty *}$. On a donc le foncteur de dualité de la catégorie $D^b_h(\mathcal{D}^\infty_X$-mod) qui est un anti-équivalence de catégories. On pose alors pour un complexe $\mathcal{N}$ de la catégorie $D^b_h(\mathcal{D}^\infty_Y$-mod) et pour un complexe $\mathcal{M}$ de la catégorie $D^b_h(\mathcal{D}^\infty_X$-mod)

$$f^!\mathcal{N} := (f^*(\mathcal{N}^*))^*$$
$$f^c_!\mathcal{M} := (f^c_*(\mathcal{M}^*))^*.$$

On alors les foncteurs image inverse et image directe extraordinaires entre catégories des complexes de $\mathcal{D}^\infty_X$-modules.

Le produit tensoriel interne $\otimes$ de la catégorie $D^b_h(\mathcal{D}^\infty_X$-mod) est aussi défini mais c'est le *complété* du produit tensoriel sur $\mathcal{O}_X$ (cf. [Me$_5$], II, § 9, p. 191). On a alors le théorème:

Théorème **(3.3.7).**— *Soit* f : X→Y *un morphisme de variétés algébriques complexe non singulières alors les catégories* $D^b_h(\mathcal{D}^\infty_X$-mod) sont stables par les opérations cohomologiques f^c_* ,$f^c_!$,f^*,$f^!$, $\otimes$ et par la dualité. *Le foncteur contravariant* $Rhom_{\mathcal{D}^\infty_X}(\mathcal{M}^\infty,\mathcal{O}_X)$ *et le foncteur covariant* $Rhom_{\mathcal{D}^\infty_X}(\mathcal{O}_X, \mathcal{M}^\infty)$ *qui s'échange par dualité sont des* équivalences de catégories triangulées *entre la catégorie des coefficients holonomes d'ordre infini* $D^b_h(\mathcal{D}^\infty_X$-mod) *et la catégorie des coefficients constructibles complexes* $D^b_c(C_X)$. *Le foncteur contravariant* $Rhom_{\mathcal{D}^\infty_X}(\mathcal{M}^\infty,\mathcal{O}_X)$ *admet comme* quasi-inverse *le foncteur* $Rhom_{C_X}(\mathcal{F},\mathcal{O}_X)$ *et le foncteur covariant* $Rhom_{\mathcal{D}^\infty_X}(\mathcal{O}_X, \mathcal{M}^\infty)$ *admet comme* quasi-inverse *le foncteur* $Rhom_{C_X}(\mathcal{F}^\vee,\mathcal{O}_X)$. *De plus le foncteur* $Rhom_{\mathcal{D}^\infty_X}(\mathcal{M}^\infty,\mathcal{O}_X)$ *transforme* f^* *en* f^{-1} *mais* *transforme* f_*^c en $f_!$. *Dualement le foncteur* $\mathrm{Rhom}_{\mathcal{D}_X^\infty}(\mathcal{O}_X, \mathcal{M}^\infty)$ *transforme* f^* *en* $f^!$ *et transforme* f_*^c *en* f_*.

La démonstration de (3.3.7) repose de façon essentielle sur le théorème (3.3.2) et donc sur le prolongement analytique complexe. Dans le cas p-adique il s'agit de trouver des substituts au prolongement analytique!.

§ 4. Sur les coefficients p-adiques d'ordre infini.

Soit W un anneau de valuation discrète complet de corps résiduel k de caractéristique $p > 0$ et de corps de fractions K de caractéristique nulle. Soit $X^\dagger := (X, \mathcal{O}_X^\dagger)$ un schéma formel faible sur W [Mr]. Rappelons que l'espace topologique X sous-jacent est une variété algébrique sur k munie de la topologie de Zariski et le faisceau structural $\mathcal{O}_X^\dagger$ est un faisceau de W-algèbres.

4.1.On peut considérer comme en 1.1 le faisceau des opérateurs différentiels :

$$\mathcal{D}iff_{X/W}(\mathcal{O}_X^\dagger) := \cup_l \mathcal{D}iff_{X/W}^{(l)}(\mathcal{O}_X^\dagger).$$

Proposition **(4.1.1).—** *La catégorie des W-schémas formels faibles* admet *des produits fibrés finis.*

Preuve voir [N-M_2]. A partir du produit $X^\dagger \times_W X^\dagger$ on peut construire comme dans ([E.G.A], § 16) les parties principales (séparées), d'ordre l, $\mathcal{P}_{X/W}^{(l)}(\mathcal{O}_X^\dagger)$ et l'on a l'isomorphisme

$$hom_{\mathcal{O}_X}(\mathcal{P}_{X/W}^{(l)}(\mathcal{O}_X^\dagger), \mathcal{O}_X^\dagger) \cong \mathcal{D}iff_{X/W}^{(l)}(\mathcal{O}_X^\dagger).$$

D'où l'on déduit comme dans ([E.G.A.IV]. § 16) que $\mathcal{D}iff_{X/W}(\mathcal{O}_X^\dagger)$ est un faisceau d'anneaux filtré de $\mathcal{O}_X^\dagger$-algèbres. Supposons que la variété X est non singulière et que $\mathcal{O}_X^\dagger$ est W-plat. Notons alors $\mathcal{D}_{X^\dagger/W}^{(l)}$ et $\mathcal{D}_{X^\dagger/W}$ pour $\mathcal{D}iff_{X/W}^{(l)}(\mathcal{O}_X^\dagger)$ et $\mathcal{D}iff_{X/W}(\mathcal{O}_X^\dagger)$. Alors pour tout l les $\mathcal{O}_X^\dagger$-modules $\mathcal{P}_{X/W}^{(l)}(\mathcal{O}_X^\dagger)$, $\mathcal{D}_{X^\dagger/W}^{(l)}$ sont localement libre de type fini. Si $(x_1,\ldots,x_n)$ sont des coordonnées au dessus d'un ouvert affine U de X c'est dire telles que $dx_1,\ldots,dx_n$ forment une base du module des différentielles séparées alors le $\Gamma(U,\mathcal{O}_X^\dagger)$-module $\Gamma(U,\mathcal{D}_{X^\dagger/W}^{(l)})$ est engendré par les opérateurs à puissances divisées Δ^q pour longueur de $q \leq l$ cf.[N-M_2]. Comme on est dans le cas d'un Z_p-module de base W, $\Gamma(U,\mathcal{D}_{X^\dagger/W}^{(l)})$ est filtré par les opérateurs différentiels d'ordre $\leq l$ et d'échelon m pour m variable cf. § 1. En vertu du developpement p-adique d'un entier l'anneau des opérateurs différentiels d'échelon m est engendré par les opérateurs d'ordre au plus p^m On note $\mathcal{D}^{(l)}_{X^\dagger/W,m}$ le faisceau des opérateurs différentiels d'ordre au plus l et d'échelon au plus m et $\mathcal{D}_{X^\dagger/W,m} := \cup_l \mathcal{D}^{(l)}_{X^\dagger/W,m}$ le faisceau des opérateurs différentiels d'échelon m. Pour m variable les faisceaux forment un système inductif et l'on a l'isomorphisme

$$\lim_{\substack{\rightarrow \\ m}} \mathcal{D}_{X^\dagger/W,m} \cong \mathcal{D}_{X^\dagger/W}.$$

Pour tout m le gradué $gr(\mathcal{D}_{X^\dagger/W,m})$ associé à la filtration $\cup_l \mathcal{D}^{(l)}_{X^\dagger/W,m}$ est un faisceau de $\mathcal{O}^\dagger_X$-algébres næthériennes commutatives. On a déduit comme en caractéristique nulle que le faisceau d'anneaux $\mathcal{D}_{X^\dagger/W,m}$ est cohérent et næthérien.

Proposition **(4.1.2).**— *Pour tout* m *et tout ouvert affine* U *de* X *l'anneau* $\Gamma(U;\mathcal{D}_{X^\dagger/W,m})$ *est de dimension homologique égale au moins à* dim(X)+1.

Preuve. C'est la même preuve que (1.2.7).

Comme l'extension $\mathcal{D}_{X^\dagger/W,m'} \rightarrow \mathcal{D}_{X^\dagger/W,m}$ n'est pas plate pour m' < m on ne peut pas déduire que le faisceau $\mathcal{D}_{X^\dagger/W}$ est cohérent. Mais tensorisé avec **Q** ce faisceau à toutes les propriétés du faisceau des opérateurs différentiels sur une variété analytique complexe. Le faisceau d'anneaux $\mathcal{D}_{X^\dagger/W}\otimes_{\mathbf{Z}}\mathbf{Q}$ est cohérent et de dimension homologique dim(X). On définit naturellement la catégorie des $\mathcal{D}_{X^\dagger/W}\otimes_{\mathbf{Z}}\mathbf{Q}$-modules holonomes comme la catégorie des modules cohérents $\mathcal{M}$ tel que $Ext^i_{\mathcal{D}_{X^\dagger/W}\otimes_{\mathbf{Z}}\mathbf{Q}}(\mathcal{M}, \mathcal{D}_{X^\dagger/W}\otimes_{\mathbf{Z}}\mathbf{Q}) = 0$ pour $i \neq \dim(X)$. La catégorie $Mh(\mathcal{D}_{X^\dagger/W}\otimes_{\mathbf{Z}}\mathbf{Q}\text{-})$ des modules holonomes est abélienne et stable par extension et les catégories $D^b_h(\mathcal{D}_{X^\dagger/W}\otimes_{\mathbf{Z}}\mathbf{Q}\text{-mod})$ sont stables par les six opérations de Grothendieck et par la dualité exactement comme dans le cas analytique complexe. Ceci résulte du résultat purement algébrique suivant [N-M_2].

Si A est une K-algébre régulière næthérienne équicodimensionnelle sur un corps K de caractéristique nulle telle qu'il existe $x_1,\ldots,x_n$ élément de A et $\partial_1,\ldots,\partial_n$ des K-dérivations de A tels que $\partial_i(x_j) = \delta_{ij}$ où dim(A) = n posons $D_{A/K} := A[\partial_1,\ldots,\partial_n]$. Si les corps résiduels des points fermés de A sont K-*algébriques* l'anneau $D_{A/K}$ est de dimension homologique égale à dim(A) [Bj]. Sous ces conditions un $D_{A/K}$-module M de type fini tel que $Ext^i_{D_{A/K}}(M, D_{A/K}) = 0$ pour $i \neq \dim(A)$ est dit holonome.

Théorème **(4.1.3).**— *Sous les conditions précédentes soient* P *un élément non nul de* A, M *un* $D_{A/K}$-module holonome et u *un élément non nul de* M. *Alors il existe une équation fonctionnelle non* *nulle*

$$B(s)P^s u = Q(s)PP^s u$$

et le $D_{A/K}$-*module* $M[P^{-1}]$ *est encore* holonome.

Le problème dans (4.1.3) est que l'algébre A n'est pas algébriquement de type fini. Appliquant (4.1.3) à une algébre $A^\dagger$ de DMW (Dwork-Monsky-Washnitzer) on trouve que $A^\dagger[P^{-1}]\otimes_Z Q$ est un $D_{A^\dagger/K}$-module de type fini. Seulement l'algébre $A^\dagger[P^{-1}]$ n'est pas faiblement compléte et est donc distincte de $(A^\dagger[P^{-1}])^\dagger$ qui est l'algébre de coordonnées (dag) du complémentaire de l'hypersurface définie par la reduction modulo p de P. Le cas complexe cf.§ 3 et les exemples évidents suggèrent que ce complété est un module de présentation fini sur le complété faible en un certain sens de l'anneau $D_{A/W}$ une fois tensorisé avec Q bien sûr. Mais c'est là un problème qui semble reléver de la théorie des équations différentielles p-adiques et non un problème purement algébrique.

Par le même théorème on a une bonne théorie des $\mathcal{D}_{X/W}\otimes_Z Q$-modules pour un W-schéma formel toplogiquement de type fini X.

4.2. Soit un schéma faiblement formel $(X,\mathcal{O}_X^\dagger)$ sur W, un anneau de valuation discrète d'inégale caractéristique p, d'idéal maximal *m*, de corps des fractions K et de corps résiduel k.

Définition **(4.2.1).**— *On appelle faisceau des opérateurs différentiels* d'ordre infini *sur le schéma faiblement formel* $(X,\mathcal{O}_X^\dagger)$ *et on note* $\mathcal{D}^\dagger_{X\dagger/W}$ *le sous-faisceau de* $hom_W(\mathcal{O}_X^\dagger, \mathcal{O}_X^\dagger)$ *des endomorphismes* P *tel pour tout* $r\geq 1$ *la réduction modulo* m^r *de* P *est un opérateur* différentiel *sur le schéma* $(X, \mathcal{O}_X^\dagger/m^r\mathcal{O}_X^\dagger)$ *d'ordre inférieur ou égale à* $\lambda(r+1)$ *pour une constante réelle* $\lambda > 0$ *indépendante de* r.

Dans la notation $\mathcal{D}^\dagger_{X\dagger/W}$ l'indice † sur $\mathcal{D}$ rappelle qu'on a complété faiblement en quelque sorte et l'indice † sur X le distingue du faisceau $\mathcal{D}^\dagger_{X/W}$ considéré par Berthelot [B_3] et construit de même à partir d'un W-schéma formel.

Le faisceau $\mathcal{D}^\dagger_{X\dagger/W}$ est l'analogue p-adique du faisceau des opérateurs d'ordre infini dans le cas complexe. Il est tout à fait naturel d'examiner dans quelle mesure les résultats complexes que l'on a décrit dans le § 3 ont des analogues dans le cas p-adique. C'est la piste qui nous semble la plus sérieuse pour démontrer le théorème de finitude de la cohomologie de Dwork-Monsky-Washnitzer d'une variété affine non singulière sur un corps fini.

Cependant il y a des différences importantes. L'extension $\mathcal{D}_{X^\dagger/W} \to \mathcal{D}^\dagger_{X^\dagger/W}$ même tensorisé par $\otimes_Z \mathbf{Q}$ n'est pas fidélement plate. Ceci améne à penser que l'anneau d'une bonne théorie des coefficients dans le cas p-adique est le faisceau $\mathcal{D}^\dagger_{X^\dagger/W} \otimes_Z \mathbf{Q}$.

Par construction le faisceau $\mathcal{O}^\dagger_X$ est un $\mathcal{D}^\dagger_{X^\dagger/W}$-module à gauche. De même en coordonnées locales le faisceau $\omega^\dagger_{X/W}$ est un $\mathcal{D}^\dagger_{X^\dagger/W}$-module à droite. Mais cette structure ne dépend pas des coordonnées parce que l'on a l'isomorphisme $\omega^\dagger_{X/W} \otimes_{\mathcal{D}_{X^\dagger/W}} \mathcal{D}^\dagger_{X^\dagger/W} \cong \omega^\dagger_{X/W}$ et la structure de $\mathcal{D}_{X^\dagger/W}$-module à droite de $\omega^\dagger_{X/W}$ ne dépend pas des coordonnées.

Soient $f : X \to Y$ un morphisme de schémas formels sur W non singuliers et $\Delta_f : X \to X \times_W Y$ le morphisme graphe de f. Notons par q_1 et q_2 les projections de $X \times_W Y$ sur X et Y respectivement. Posons

$$\mathcal{D}^\dagger_{X^\dagger \to Y^\dagger/W} := q_{1*} \mathcal{H}_X^{\dim(Y)}(q_2^* \omega_{Y^\dagger})$$

$$\mathcal{D}^\dagger_{Y^\dagger \leftarrow X^\dagger/W} := q_{1*} \mathcal{H}_X^{\dim(Y)}(q_1^* \omega_{X^\dagger}).$$

La structure de $\mathcal{D}^\dagger_{X^\dagger/W}$-module *à gauche* de $\mathcal{O}_{X^\dagger}$ induit une structure de $\mathcal{D}^\dagger_{X^\dagger/W}$-module *à gauche* sur le faisceau $\mathcal{D}^\dagger_{X^\dagger \to Y^\dagger/W}$ et la structure de $\mathcal{D}^\dagger_{Y^\dagger/W}$-module *à droite* sur $\omega_{Y^\dagger}$ induit une structure de $f^{-1}\mathcal{D}^\dagger_{Y^\dagger/W}$-module *à droite* sur le faisceau $\mathcal{D}^\dagger_{X^\dagger \to Y^\dagger/W}$ qui est donc un $(\mathcal{D}^\dagger_{X^\dagger/W}, f^{-1}\mathcal{D}^\dagger_{Y^\dagger})$-bimodule. De même la structure de $\mathcal{D}^\dagger_{X^\dagger/W}$-module *à droite* de $\omega_{X^\dagger}$ induit une structure de $\mathcal{D}^\dagger_{X^\dagger/W}$-module *à droite* sur le faisceau $\mathcal{D}^\dagger_{Y^\dagger \leftarrow X^\dagger/W}$ et la structure de $\mathcal{D}^\dagger_{Y^\dagger/W}$-module *à gauche* sur $\mathcal{O}_{Y^\dagger}$ induit une structure de $f^{-1}\mathcal{D}^\dagger_{Y^\dagger/W}$-module *à gauche* sur le faisceau $\mathcal{D}^\dagger_{Y^\dagger \leftarrow X^\dagger/W}$ qui est donc un $(f^{-1}\mathcal{D}^\dagger_{Y^\dagger/W}, \mathcal{D}^\dagger_{X^\dagger/W})$-bimodule. En particulier si f est le morphisme identique on trouve que le faisceau $p_{1*}\mathcal{H}_X^{\dim(X)}(p_2^*\omega_{X^\dagger})$ est un $(\mathcal{D}^\dagger_{X^\dagger/W}, p_{1*}p_2^{-1}\mathcal{D}^\dagger_{X^\dagger/W})$-bimodule et donc un $(\mathcal{D}^\dagger_{X^\dagger/W}, \mathcal{D}^\dagger_{X^\dagger/W})$-bimodule.

Proposition **(4.2.2).**— Le $(\mathcal{D}^\dagger_{X^\dagger/W}, \mathcal{D}^\dagger_{X^\dagger/W})$*-bimodule* $p_{1*}\mathcal{H}_X^{\dim(X)}(p_2^*\omega_{X^\dagger})$ *est canoniquement* isomorphe *au faisceau d'anneaux* $\mathcal{D}^\dagger_{X^\dagger/W}$.

Dans un travail ultérieur nous montrerons que le faisceau $\mathcal{D}^\dagger_{X^\dagger/W}$ est filtré par les complétés faibles des faisceaux $\mathcal{D}_{X^\dagger/W,m}$ comme dans la situation de Berthelot. La méthode de Berthelot [B_3] montre alors que le faisceau d'anneaux $\mathcal{D}^\dagger_{X^\dagger/W} \otimes_Z \mathbf{Q}$ est cohérent. De plus nous montrerons que la dimension homologique de $\mathcal{D}^{\dagger}_{X\dagger/W}\otimes_{Z}Q$ est égale à dim(X). Ceci permet de définir les $\mathcal{D}^{\dagger}_{X\dagger/W}\otimes_{Z}Q$-modules holonomes comme les $\mathcal{D}^{\dagger}_{X\dagger/W}\otimes_{Z}Q$-modules cohérents tels que leur dual sur $\mathcal{D}^{\dagger}_{X\dagger/W}\otimes_{Z}Q$ est concentré en degré dim(X).

4.3.Pour f : X→Y un morphisme de schémas formels sur W non singuliers, un complexe de $\mathcal{D}^{\dagger}_{X\dagger/W}$-modules à gauche $\mathcal{M}^{\dagger}$ et $\mathcal{N}^{\dagger}$ un complexe de $\mathcal{D}^{\dagger}_{Y\dagger/W}$-modules à gauche posons :

$$f^{*}\mathcal{N}^{\dagger} := \mathcal{D}^{\dagger}_{X\dagger\to Y\dagger/W} \overset{L}{\otimes}_{f^{-1}\mathcal{D}^{\dagger}_{Y\dagger/W}} f^{-1}\mathcal{N}^{\dagger}$$

$$f_{*}^{c}\, \mathcal{M}^{\dagger} := Rf_{*}\mathcal{D}^{\dagger}_{Y\dagger\leftarrow X\dagger/W} \overset{L}{\otimes}_{\mathcal{D}^{\dagger}_{X\dagger/W}} \mathcal{M}^{\dagger}.$$

On obtient les foncteurs images inverse et directes pour les $\mathcal{D}^{\dagger}_{X\dagger/W}$-modules à gauche. En particulier si Y est un point la cohomologie de Dwork-Monsky-Washnitzer d'un $\mathcal{D}^{\dagger}_{X\dagger/W}$-module à gauche $\mathcal{M}^{\dagger}$ est le complexe de W-modules

$$R\Gamma(X; \mathcal{D}_{\cdot\leftarrow X\dagger/W} \overset{L}{\otimes}_{\mathcal{D}^{\dagger}_{X\dagger/W}} \mathcal{M}^{\dagger})$$

qui tensorisé avec **Q** est isomorphe à

$$R\Gamma(X; \mathbf{DR}(\mathcal{M}^{\dagger}))\otimes_{Z}\mathbf{Q}[\dim(X)] := R\Gamma(X; R\mathrm{hom}_{\mathcal{D}^{\dagger}_{X\dagger/W}}(\mathcal{O}_{X\dagger}, \mathcal{M}^{\dagger}))\otimes_{Z}\mathbf{Q}[\dim(X)].$$

Le point clef sera de dégager la catégorie des $\mathcal{D}^{\dagger}_{X\dagger/W}$-modules stables par les foncteurs images directe et inverse. Comme en caractéristique nulle ceci se raméne à établir la stabilité par immersion ouverte. Pour l'instant on ne peut que se contenter de quelques exemple évidents et d'une conjecture.

4.4.Supposons que le schéma formel faible est l'espace affine Spff(W[X]†) de dimension n. Alors l'anneau des sections globales $D^{\dagger}_{n}$ du faisceau $\mathcal{D}^{\dagger}_{X\dagger/W}\otimes_{Z}\mathbf{Q}$ sont les séries $\Sigma_{\alpha,\beta}a_{\alpha,\beta}x^{\alpha}\Delta^{\beta}$ tel que la série $\Sigma_{\alpha,\beta}a_{\alpha,\beta}x^{\alpha}y^{\beta}$ soit dans l'algébre W[x, y]†. Une question naturel qui se pose est de savoir si l'anneau $D^{\dagger}_{n}\otimes_{Z}\mathbf{Q}$ est næthérien. L'anneau $D^{\dagger}_{n}$ contient l'anneau W[Δ]† des opérateurs à coefficients constants et l'extention W[Δ]† → $D^{\dagger}_{n}$ est fidèlement plate. Donc si W[Δ]$^{\dagger}\otimes_{Z}\mathbf{Q}$ n'était pas næthérien cela entraine que $D^{\dagger}_{n}\otimes_{Z}\mathbf{Q}$ n'est pas næthérien. Seulement :

Proposition **(4.4.1).**— *L'anneau* W[Δ]$^{\dagger}\otimes_{Z}\mathbf{Q}$ *est næthérien.*

Preuve. Ceci est une conséquense du fait que l'anneau $W[\Delta]^{\dagger}\otimes_Z Q$ admet un algorithme de division de Weierstrass-Hironaka. Fixons un ordre total sur N^n on a alors la notion d'exposant pour tout opérateur à coefficients constants défini comme l'exposant de son polynôme initiale pour la valuation *m*-adique.

Proposition **(4.4.2).**— *Soient* $P_1,\ldots,P_r$ *des opérateurs de* $W[\Delta]^{\dagger}\otimes_Z Q$ *alors pour tout opérateur* A *il existe des uniques opérateurs* $Q_1,\ldots Q_r$, et R *de* $W[\Delta]^{\dagger}\otimes_Z Q$ *dont les exposants satisfassent aux propriétés usuelles de la division tels que*

$$A = Q_1P_1 + \ldots + Q_rP_r + R\,.$$

La preuve se fait en considérant l'extension $\widetilde{W}$ de W tel que si t est une uniformisante de W et π une uniformisante de $\widetilde{W}$ alors $t = \pi^{p-1}$. Un opérateur $P(\Delta)$ est dans $W[\Delta]^{\dagger}\otimes_Z Q$ si et seulement si la série $P(\pi^e x)$ appartient à $W[x]^{\dagger}\otimes_Z Q$ où e est l'indice de ramification absolu de W. En fait le morphime de $W[\Delta]^{\dagger}\otimes_Z Q$ dans $W[x]^{\dagger}\otimes_Z Q$ qui à $P(\Delta)$ associe $P(\pi^e x)$ est injectif et fait de la $\widetilde{K}$-algébre DMW une extension libre engendré par $1,\ldots,\pi^{p-2}$ de la K-algébre DMW à puissances divisées. Dans la $\widetilde{K}$-algébre $\widetilde{K}[x]^{\dagger}$ on a un algorithme de division tel que les quotients et le reste proviennent de $W[\Delta]^{\dagger}\otimes_Z Q$ si le dividant et les diviseurs proviennent de $W[\Delta]^{\dagger}\otimes_Z Q$.

On pourait songer à fabriquer un algorithme de division dans l'anneau $D_n^{\dagger}\otimes_Z Q$. C'est possible qu'on se restreint à l'échelon zéro c'est dire à l'algébre de Weyl DMW mais non dans le cas général. En effet considérons à une variable le diviseur $P = \partial - px$. Si on choisi ∂ comme exposant de P pour faire le division et on si divise l'opérateur $\Sigma_m 2^{-m}\partial^{2m}$ qui est dans $W[\Delta]^{\dagger}$ on trouve que le coefficient constant dans le reste, qui une série en x, ne converge pas en caractéristique 2. Si on choisit x comme exposant et si on divise la série $\Sigma_m(-1)^m 2^m x^{2m}$ on trouve que le coefficient constant du reste, qui est une série en ∂, ne converge pas en caractéristique 2. On peut fabriquer des contres exemples en caractéristique différente de 2. Voir en 4.6 l'épilogue pour ce point.

4.5. Une question clef est de montrer que le $D_n^{\dagger}\otimes_Z Q$-module $(W[x][P^{-1}])^{\dagger}\otimes_Z Q$ est de présention fini pour tout polynôme P de W[x].

4.5.1. Si P est le monôme $x_1\ldots x_r$ $(r \le n)$ le même calcul fait dans [Me$_1$] montre que le $D_n^{\dagger}\otimes_Z Q$-module $(W[x][P^{-1}])^{\dagger}\otimes_Z Q$ est engendré par $1/x_1\ldots x_r$ dont l'annulateur est l'idéal $\partial_1 x_1,\ldots,\partial_r x_r, \partial_{r+1},\ldots,\partial_n$.

4.5.2. Si P est la forme quadratique $x_1^2+\ldots x_r^2$ $(r \le n)$ il est encore possible de faire le calcul pour montrer que $1/x_1^2+\ldots+x_r^2$ engendre le $D_n^{\dagger}\otimes_Z Q$-module $(W[x][P^{-1}])^{\dagger}\otimes_Z Q$ parce l'équation fonctionnelle est simple. On a l'équation fonctionnelle $2(s+1)(2s+r)P^s = (\partial_1^2+\ldots+\partial_r^2)P^{s+1}$. Soit g un élément de

$(W[x][P^{-1}])^{\dagger}\otimes_{Z}Q$. C'est une série $\Sigma_{\alpha,\beta}a_{\alpha,\beta}x^{\alpha}P^{-\beta}$ ($\alpha\in N^n,\beta\in N$, $a_{\alpha,\beta}\in K$) avec la condition DMW $v(a_{\alpha,\beta}) \geq \lambda(\beta+\alpha)$ asymptotiquement pour une constante positive λ non nulle. On peut supposer que les $a_{\alpha,\beta}$ sont dans W. Pour $\beta \geq 2$, $P^{-\beta} = (\partial_1^2 +\ldots+ \partial_r^2)^{\beta-1}P^{-1}/4^{\beta-1}(\beta-1)!(\beta-r/2)!$. En caractéristique différente de 2 l'opérateur $\Sigma_{\alpha,\beta}a_{\alpha,\beta}x^{\alpha}(\partial_1^2 +\ldots+ \partial_r^2)^{\beta-1}/4^{\beta-1}(\beta-1)!(\beta-r/2)!$ formel est en fait un opérateur différentiel d'ordre infini. En effet un calcul directe montre qu'il est de la forme $\Sigma_{\alpha,\gamma}a_{\alpha,\gamma}x^{\alpha}\Delta^{\gamma}$ ($\gamma\in N^r$) avec la condition DMW pour la série $\Sigma_{\alpha,\gamma}a_{\alpha,\gamma}x^{\alpha}y^{\gamma}$. En caractéristique 2 on a $x_1^2+\ldots+ x_r^2 = (x_1 +\ldots+ x_r)^2$ mod(2) et l'isomorphisme $(W[x][P^{-1}])^{\dagger} \cong (W[x][(x_1 +\ldots+ x_r)^{-1}])^{\dagger}$. Dans tous les cas ceci montre que le $D_n^{\dagger}\otimes_{Z}Q$-module $(W[x][P^{-1}])^{\dagger}\otimes_{Z}Q$ est de type fini pour la forme quadratique. Cependant comme dans le cas complexe [Me$_2$] cette méthode n'a guère de chance de montrer les propriétés de finitude espérées dans le cas général.

4.5.3. Nous allons formuler une conjecture dans le cas p-adique calquée du cas complexe. Soit X un schéma affine non singulier formel faible sur W d'algébre $A^{\dagger}$ et $\mathcal{M}$ un $\mathcal{D}_{X^{\dagger}/W}$-module tel que $\mathcal{M}\otimes_{Z}Q$ soit holonome qu'on peut prendre algébrique. Si on choisit une section $\sigma : \overline{K}/Z \to \overline{K}$ de la projection $\overline{K} \to \overline{K}/Z$ alors $\mathcal{M}\otimes_{Z}Q$ admet par rapport à toute fonction P de $A^{\dagger}$ un polynôme de Berstein-Sato [N-M$_2$] qui permet de définir une V-filtration canonique (cf. par exemple [S-M]) indéxée par l'image de σ dont le gradué est localement indéxé par un nombre fini d'éléments de $\overline{K}$ que nous appelerons les exposants de $\mathcal{M}$ le long de $V(P)\otimes_{Z}Q$. Faisons l'hypothése que les exposants de $\mathcal{M}$ le long de $V(P)\otimes_{Z}Q$ sont des nombres algébriques en particulier ne sont pas des nombres de Liouville, ceci ne dépend pas du choix de σ. Alors nous pensons que

$$j^{\dagger}{}_{*}j^{\dagger-1}(\mathcal{D}^{\dagger}_{X^{\dagger}/W}\otimes_{\mathcal{D}_{X^{\dagger}/W}}\mathcal{M})\otimes_{Z}Q \text{ est } \mathcal{D}^{\dagger}_{X^{\dagger}/W}\otimes_{Z}Q\text{-}\textit{cohérent et holonome.}$$

où j désigne l'inclusion du complémentaire de l'hypersurface définie par P par reduction modulo p.

Nous avons vérifié cette conjecture en dimension un grâce aux résultats de Robba sur les théorèmes d'indice ([Rb$_1$],[Rb$_2$]) pour un module qui n'a que des singularités régulières complétement soluble ou est de rang un. De façon plus précise :

Proposition (4.5.4).— *Avec les notations précédentes supposons que* Z *est un point d'une courbe affine* X *alors le cohomologie locale de* Z *à valeur dans* $\mathcal{D}^{\dagger}_{X^{\dagger}/W}\otimes_{\mathcal{D}_{X^{\dagger}/W}}\mathcal{M}\otimes_{Z}Q$ *est* $\mathcal{D}^{\dagger}_{X^{\dagger}/W}\otimes_{Z}Q$-*cohérente* si et seulement si *le* $\mathcal{D}_{X^{\dagger}/W}$-*module* $\mathcal{M}\otimes_{Z}Q$ *admet un indice dans l'espace des fonctions analytiques dans le tube de* Z.

4.6. *Epilogue*. Berthelot vient de montrer que l'anneau $D_n^{\dagger}\otimes_{Z}Q$ cf. 4.4.2 *n'est pas* næthérien.

REFERENCES

[B-B-D] BEILINSON, A., BERNSTEIN, I.N., DELIGNE, P., Faisceaux pervers, *Astérisque* **100** (1983).

[Be] BERNSTEIN, I.N., The analytic continuation of generalized functions with respect to parameter, *Funct. Analysis Appl*, **6** (1972), 26-40.

[Ber] BERSTEIN, I., On the dimension of modules and algebras (IX), direct limits, *Nagoya Math J.* 13 (1958), 83-84.

[B_1] BERTHELOT, P., Cohomologie cristalline des schéma de caractéristique p > 0, Lecture Notes in Math **407**, Springer Verlag (1974).

[B_2] —, Géométrie rigide et cohomologie des variétés algébriques de caractéristique p > 0, Journées d'analyse p-adique (1982), in Introduction aux cohomologies p-adiques, Bull. Soc. Math. France, Mémoire **23**, p. 7-32 (1986).

[B_3] —, Cohomologie rigide et théorie des $\mathcal{D}$-modules, dans ce volume.

[B-O] BERTHELOT, P., OGUS, A., Notes on Crystalline Cohomology, Princeton University Press **21**, (1978).

[Bj] BJORK, J.E., The global homological dimension of some algebras of differentiels opérators, *Inv. Math.* **17** (1972), p. 67-78.

[Ch] CHASE, S. U., On the homological dimension of algebras of differential operators, Comm. Algebra **5** (1974), 351-363.

[D_1] DELIGNE, P. Equations différentielles à point singuliers réguliers, Lecture Notes in Math **163**, Springer Verlag (1970).

[D_2] —, Cristaux discontinus, Notes manuscrites non publiées qui datent du début des années 70 (19 pages).

[Dw] DWORK, B., On the Zeta function of hypersurface I, *Publ. Math. I.H.E.S*, **12**, (1962).

[G_1] GROTHENDIECK, A., On the De Rham cohomology of algebraic varieties, *Publ. Math I.H.E.S.*, **29**, (1966), p. 93-103.

[G_2] —, Crystals and the De Rham cohomology of schemes, in Dix exposés sur la cohomologie des schémas, North Holland Company (1968), p. 306-358.

[G_3] —, Groupes de Barsotti-Tate et cristaux de Dieudonné, Les Presses de l'Université de Montréal **45**, (1974).

[Ha] HAASTERT B., On the direct and inverse image of $\mathcal{D}$-modules in prime caracteristic, Manuscripta Math. **62**, (1988), p. 341-354.

[H] HIRONAKA, H., The resolution of the singularities of algebraic varieties over a field of caracteristic zero, *Ann. of Math.* **79** (1964), 109-326.

[K_1] KASHIWARA, M., On the maximally overdetermined systems of differential equations, Publ. RIMS, **10** (1975), 563-579.

[K_2] —, B-function and holonomic systems, *Inv. Math.* 38 (1976) , 33-53.

[Ka_1] KATZ, N., Nilpotent connexions and the monodromy theorem, application of a result of Turittin, *Publ. Math. I.H.E.S.* **39**, (1970), p. 176- 232.

[Ka_2] —, The regularity theorem in algebraic geometry, Actes Congrés intern. Math.,1970, tome I, Gauthier Villars (1971) p. 437-443.

[L-M] LE.D.T., MEBKHOUT Z., Variétés caractéristiques et variétés polaires, C.R.A.S. Acad. Paris, t. **296** (1983), p. 129-132.

[M_1] MALGRANGE B., Sur les points singuliers réguliers des équations différentielles, *Ens. Math.* **20** (1974) , p.147-176.

[M_2] —, Opérateurs différentiels et pseudo-différentiels, Institut Fourier 1976.

[Me_1] MEBKHOUT, Z., Valeur principale et residu des formes à singularité essentielles, In Lecture Notes in Math **482**, Springer Verlag (1975), p. 190-215.

[Me_2] —, Cohomologie locale d'une hypersurface, in Lecture Notes in Math **670**, Springer Verlag, (1977), p. 89-119.

[Me_3] —, Théorèmes de bidualité locale pour les $\mathcal{D}_X$-modules holonomes, *Ark. Mat.* **20** (1982), p. 111-122.

[Me_4] —, Une équivalence de catégories et une autre équivalence de catégories, Comp. Math **51** (1984), p. 51-88.

[Me_5] —, Le formalisme des six opérations de Grothendieck pour les $\mathcal{D}_X$-modules cohérents, Travaux en cours Hermann **35** (1989).

[Me_6] —, Le théorème de positivité de l'irrégularité pour les $\mathcal{D}_X$-modules, in Grothendieck Festschrift, Birkhaüser (1990).

[Me_7] —, Le théorème de comparaison entre cohomologie de de Rham d'une variété algébrique complexe et le théorème d'existence de Riemann, *Publ. Math. I.H.E.S,* **69** (1989), p. 47-89.

[Mr] MEREDITH, D., Weak formel schemes, *Nagoya Math.* J. 45 (1971), 1-38.

[M-W] MONSKY, P., WASHNITZER, G., Formal cohomology I, *Ann. of Math.* **88** (1968), 51-62.

[$N-M_1$] NARVAEZ L., MEBKHOUT Z., Démonstration géométrique du théorème de constructibilité, in Travaux en cours Hermann **35** (1989), p. 248-253.

[$N-M_2$] —, La théorie du polynôme de Bernstein-Sato pour les algébres de Tate et Dwork-Monsky-Washnitzer, *Ann. de l'école Norm. Sup.* (1990) (à paraitre).

[N-O] NASTASESCU, C., VAN OYSTAEYEN, F., Graded rings Theory, North Holland Comp. (1982).

[O] OGUS, A. F-isocrystals and de Rham cohomologie II: Convergent isocristals, Duke Math J. **51** (1984), 765-850.

[S-M] SABBAH, C., MEBKHOUT, Z., $\mathcal{D}_X$-modules et cycles évanescents, in Travaux en cours Hermann (1989) **35** , p. 200-238.

[S] SATO, M., Hyperfunctions and partial diferential equation, Proc. Intern. Conf. on Func. Ana. and Related Topics, 1969, Univ. Tokyo Press, Tokyo,

(1970), p. 91-94.

[S-K-K] SATO, M., KAWAI, T., KASHIWARA, M., Micro-functions ans pseudo-differentiel equations, Lecture Notes in Math **287**, (1973), p. 265-529.

[Sm] SMITH, S.P., The global homological dimension of ring of differential operators on a non singular variety over a field of positive characteristic, *J. of Alg.* **107**, (1987), 98-105.

[Rb_1] ROBBA P., Indice d'un opérateur différentiel p-adique IV, cas des systèmes, *Ann. Inst. Fourier* t. **35**, (1985), p. 13-55.

[Rb_2] —, Livre (à paraitre).

[Ro] ROOS J.E., Determination de la dimension homologique globale des algébres de Weyl, *C.R.A.S., Paris* **274**, (1972), p. 1556-1558.

[Rt] ROTMAN, J.J., An Introduction to Homological Algebra, Academic Press, New York (1979).

Sigles

[E.G.A.IV] Eléments de Géométrie Algébrique, GROTHENDIECK, A., DIEUDONNE, A., Etude locale des schémas, quatrième partie, *Publ. Math. I.H.E.S*, **32** (1967).

[S.G.A.6] Séminaire de Géométrie Algébrique du Bois-Marie (1966-67), GROTHENDIECK, A., BERTHELOT, P., ILLUSIE, L., Théorie des intersections et théorème de Riemann-Roch, Lectures Notes in Math. **225** Springer Verlag.(1971) .

[S.G.A.7] Séminaire de Géométrie Algébrique du Bois-Marie (1967-69), GROTHENDIECK, A., DELIGNE, P., KATZ, N., Groupes de monodromie en géométrie algébrique, Lectures Notes in Math. **280, 340** Springer Verlag (1972-73).

[R-D] Residue and Duality, Séminaire HARTSHORNE, Lectures Notes in Math. **20**, Springer Verlag (1966).

On a Functional Equation of Igusa's Local Zeta Function.*

Diane Meuser
Boston University

Let K be a finite algebraic extension of the field $\mathbf{Q}_p$ of p-adic numbers, R_K the ring of integers of K, P_K the unique maximal ideal of R_K, $\overline{K}$ the residue field and $q = card\ \overline{K}$. Let $|\ |_K$ be the absolute value on K so that the absolute value of an element in $P_K - P_K^2$ is q^{-1}. If $f(x)$ is a polynomial in $R_K[x_1, \ldots, x_n]$ the Igusa local zeta function associated to f is defined for $s \in \mathbf{C}$, $Re(s) > 0$ by

$$Z_K(s) = \int_{R_K^n} |f(x)|_K^s |dx|_K$$

It is a rational function of q^{-s}, and in fact is in $\mathbf{Z}(q^{-1}, q^{-s})$. The denominator can be written as

$$\prod_{i \in I} (1 - q^{-m_i} q^{-N_i s})$$

where (N_i, m_i) are data associated to exceptional divisors E_i, $i \in I$ in a resolution of the singularities of f. The numerator is quite complicated in known examples, and no general result is known.

Many of the known examples of $Z_K(s)$ are for polynomials associated to irreducible regular K-split prehomogeneous vector spaces. These polynomials are invariant polynomials for a group acting transitively on the complement of the hypersurface defined by $f = 0$. They have been classified into 29 cases, given in [4]. with some cases containing infinitely many polynomials.

Igusa observed that there was a functional equation for $Z_K(s)$ in the 20 out of 29 classified cases of the above polynomials for which $Z_K(s)$ has been determined. For these cases one has that if L is any finite algebraic extension of K, q_L the number of elements in the residue field, and $Z_L(s)$ the local zeta function, then $Z_L(s)$ is obtained from $Z_K(s)$ by the replacement $q \mapsto q_L$. This uniquely defines $Z(u, v) \in \mathbf{Q}(u, v)$ such that

$$Z_L(s) = Z(q_L^{-1}, q_L^{-s}) \quad .$$

His result is that for these cases

$$(1) \qquad Z(u^{-1}, v^{-1}) = v^d\ Z(u, v)$$

where d is the degree of f [2].

In order to ask whether a functional equation of this type is true more generally it is necessary to impose a restriction that forces the function $Z \in \mathbf{Q}(u, v)$ satisfying $Z_K(s) = Z(q^{-1}, q^{-s})$ to be well-defined. Hence Igusa introduced the following definition.

Definition. $Z_K(s)$ for $f(x) \in R_K[x_1, \ldots, x_n]$ is universal if there exists a function $Z(u, v) \in \mathbf{Q}(u, v)$ such that

$$Z_L(s) = Z(q_L^{-1}, q_L^{-s})$$

* Supported by National Science Foundation Grant Nos. DMS-8702667 and DMS-8610730.

for all finite extensions L of K.

He then posed two conjectures under the assumption that $f(x)$ has "good reduction" modulo P_K. By "good reduction" it is meant that if $f(x)$ has coefficients in a number field, then the conjecture should hold for all except possibly a finite number of completions.

Conjecture I: For homogeneous polynomials, $Z_K(s)$ universal implies $Z(u,v)$ has the functional equation (1).

Conjecture II: If $f(x)$ is one of the remaining nine types of invariant polynomials associated to a prehomogeneous vector space then $Z_K(s)$ is universal.

Theorem: *Conjecture I is true.*

The general proof will appear in [1]. The proof shows that the functional equation of the Igusa local zeta function is a direct consequence of the functional equation of the Weil zeta function for the smooth, proper varieties given by the intersections of the exceptional varieties in the embedded resolution of the singularities of the projective variety defined by f.

More specifically let $\{E_i\}_{i\in T}$ denote the exceptional divisors of such a resolution, and let $E_I = \bigcap_{i\in I} E_i$. One can use the theory of resolutions to obtain the following expression for $Z_L(s)$ for any finite algebraic extension L of K

$$(2) \qquad Z_L(s) = \frac{1 - q_L^{-1}}{1 - q_L^{-ds-n}} q_L^{-(n-1)} \sum_{I\subseteq T} b_{L,I} \prod_{i\in I}\left(\frac{q_L - 1}{q_L^{N_i s+m_i} - 1} - 1\right)$$

where $b_{L,I}$ is the number of $\bar{L}$ rational points of $\overline{E}_I$ where $\overline{E}_I$ denotes the reduction of E_I modulo P_K. Since $\overline{E}_I$ is smooth and proper the functional equation of the Weil zeta function states that the map $\alpha \mapsto q^{n-1-|I|}/\alpha$ is a bijection of the eigenvalues of Frobenius of $\overline{E}_I$.

Since the definition of universal zeta function has been introduced, the natural question is to ask under what conditions does a given polynomial possess a universal zeta function? In particular the functional equation for the remaining 9 types of polynomials associated to prehomogeneous vector spaces as mentioned above remains a conjecture since one does not know that their zeta functions are universal. However since the first 20 were shown to have universal zeta functions by Igusa's explicit calculations, it is reasonable to conjecture the others do as well.

The following shows that the property of $f(x) \in R_K[x_1,\ldots,x_n]$ satisfying the condition that its Igusa local zeta function is universal implies strong properties for the Weil zeta function for the projective variety X over $\overline{K}$ defined by $\overline{f} = 0$.

Theorem 2: *Suppose $f(x) \in R_K[x_1,\ldots,x_n]$ has the property that its Igusa local zeta function is universal. Let $Z_{red}(t)$ denote the reduced Weil zeta function obtained from the Weil zeta function for X by deleting any common factors in the numerator and denominator. Then the reciprocal zeroes and poles of Z_{red} are all of the form q^i, $0 \leq i \leq n-2$.*

Proof: Let $N_e = Card\{x \in \mathbf{F}_{q^e} \mid \overline{f}(x) = 0\}$. By considering the local zeta function $Z_{K_e}(s)$ for K_e the unique unramified extension of degree e and letting $e \to \infty$ Igusa showed [2] the universal condition implied there existed a polynomial $P(x) \in \mathbf{Q}[x]$ of degree at most $n-1$ such that $N_e = P(q^e)$. The idea is to observe one has

$$\int_{R_{K_e}^{(n)}} |f(x)|^s |dx| = Z(q^{-e}, q^{-es})$$

by universality, and on the other hand

$$\int_{R_{K_e}^{(n)}} |f(x)|^s |dx| = Vol\{x \mid |f(x)| = 1\} + \sum_{i=1}^{\infty} Vol\{x \mid |f(x)| = q^{-ei}\} \cdot q^{-eis} .$$

Letting $Re(s) \to \infty$ one gets $Vol\{x \mid |f(x)| = 1\} = Z(q^{-e}, 0)$ but this volume is $1 - q^{-ne} N_e$. The polynomial expression for N_e is then obtained by considering the power series expansion of $Z(u,0) \in \mathbf{Q}[[u]]$.

Write $N_e = \sum_{i=0}^{n-1} a_i q^{ei}$, $a_i \in \mathbf{Q}$. Choosing $m \in \mathbf{Z}^+$ such that $ma_i \in \mathbf{Z}\ \forall i$, let $N_e^1 = \sum_{i=0}^{n-1} m a_i q^{ei}$. By considering the Weil zeta function for the affine variety defined by $\overline{f} = 0$ we have $\exists\ \alpha_i, \beta_j \in \mathbf{C}$ such that $N_e = \sum \alpha_i^e - \sum \beta_j^e$, so we let $N_e^2 = \Sigma m\alpha_i^e - \Sigma m\beta_j^e$. Since $N_e^1 = N_e^2$, we have

$$\exp \sum_{e=1}^{\infty} \frac{N_e^1 t^e}{e} = \exp \sum_{e=1}^{\infty} \frac{N_e^2 t^e}{e},$$

which in turn gives

$$\frac{\prod_{i|a_i<0}(1-q^i)^{m|a_i|}}{\prod_{i|a_i>0}(1-q^i)^{ma_i}} = \frac{\prod_j (1-\beta_j t)^m}{\prod_i (1-\alpha_i t)^m}.$$

Thus after possible cancellation we have each α_i, β_j is an integral power of q. Letting $\tilde{N}_e$ denote the number of points on the projective variety X and using $N_e = \tilde{N}_e(q-1)+1$ gives the same conclusion for the Weil zeta function of X.

Let $f(x)$ be of degree d in variables. In the case where the variety defined by $\overline{f}(x) = 0$ is projectively nonsingular the Weil zeta function $Z(t)$ is reduced and satisfies

$$Z(t) = \frac{N(t)^{(-1)^{(n-1)}}}{(1-t)(1-qt)\dots(1-q^{(n-2)}t)} .$$

where

$$N(t) = \prod_{i=1}^{b} (1-\alpha_i t) \quad ,$$

b is the betti number of $H^{n-2}(X, \mathbf{Q}_l)$ and $|\alpha_i| = q^{(n-2)/2}$ for all i. Thus n odd implies $b = 0$. But the middle dimensional betti number is given by

$$b = \frac{(d-1)^n + (-1)^n(d-1)}{d} .$$

So if n is odd we must have $d = 2$. Conversely these polynomials all have the property that their Igusa local zeta functions are universal since they are special cases of the invariant polynomials associated to prehomogeneous vector spaces.

If n is even all the middle dimensional eigenvalues of Frobenius α_i must equal $q^{(n-2)/2}$. A further characterization of these varieties can be obtained in terms of properties of the middle dimensional l-adic cohomology group. In particular there is a map from algebraic cycles of codimension $(n-2)/2$ to $H^{n-2}(X, \mathbf{Q}_l)$. A cohomology class is said to be algebraic if it is the image of an algebraic cycle. Frobenius acts trivially on the subspace spanned by algebraic cycles so if $H^{n-2}(X, \mathbf{Q}_l)$ is spanned by algebraic cycles the Igusa local zeta function is universal. Conversely, Tate's conjecture [6] is that the order of the pole of the Weil zeta function at $q^{-(n-2)}/2$ is the rank of the subspace spanned by the algebraic cycles. Thus one has

Theorem 2: *A projectively nonsingular variety X defined by $\overline{f}(x) = 0$ where $\overline{f}(x)$ has an even number of variables has a universal Igusa local zeta function if and only if $H^{n-2}(X, \mathbf{Q}_l)$ is spanned by algebraic cycles, with the only if part assuming Tate's conjecture.*

Any rational variety having a cellular decomposition will satisfy the property that its cohomology is spanned by algebraic cycles. Some other examples are given by Fermat hypersurfaces. The Fermat hypersurface of degree d in n variables where n is even and $d \geq 4$ has $H^{n-2}(X, \mathbf{Q}_l)$ spanned by algebraic cycles if and only if there exists an integer v such that $p^v \equiv -1 \pmod{m}$ where $p = char(\overline{K})$, as was shown by Shioda [5]. These Fermat hypersurfaces are unirational, ie. there exists a rational map of finite degree from a projective space to the variety.

For the case of singular varieties the situation is much less clear since the Weil zeta functions are not as explicit. However this is the primary case of interest since in particular the invariant polynomials associated to prehomogeneous vector spaces define singular varieties. By explicit consideration of the expression of the Igusa local zeta function given in (2) one can easily see

Proposition 1: *If $\{E_i\}_{i\in T}$ denotes the collection of exceptional divisors in a resolution of f, then if every subvariety $E_I = \bigcap_{i\in I} E_i$ for $I \subseteq T$ has the property that its eigenvalues of Frobenius are integral powers of q, then the Igusa local zeta is universal.*

However, whether or not the converse is true remains open. Even for the prehomogeneous cases whose Igusa local zeta functions have been determined it is not known whether their resolutions satisfy this simple property, since in general their resolutions have not been obtained. Igusa's method of determining the zeta functions for these polynomials does not use a resolution. I have shown that in two non-trivial cases the conditions of Proposition 1 are satisfied.

Proposition 2: *If $f(x)$ is the invariant cubic in 27 variables for the group E_6 or the invariant quartic in 56 variables for the group E_7, then the eigenvalues of Frobenius of every subvariety in the resolution are integral powers of q.*

The zeta functions of these are known to be universal by methods of Igusa. On the other hand the numerical data of the exceptional divisors was obtained by Kempf in [3]. For instance for the cubic in 27 variables there are three exceptional divisors E_1, E_2, E_3 with numerical data $(1,1)$, $(2,10)$, $(3,27)$ and the I with $|I| > 1$ satisfying $E_I \neq \emptyset$ are $\{1,2\}$, $\{1,3\}$. By consideration of the formula for the Igusa local zeta function given in (2) it is clear that the only possibly non-universal contribution comes from the $b_{L,I}$. By expanding this formula and $Z(u,v)$ as formal power series in v with coefficients in $\mathbf{Q}(u)$ one can compare coefficients to show each $b_{L,I}$ is given by a function in $\mathbf{Q}(u)$. Then the arguments of Proposition 1 show that the eigenvalues of Frobenius are integral powers of q. A similar but more complicated consideration of the seven exceptional divisors for the polynomial associated to E_7 shows that their intersections also have this property.

Kempf has indicated to me that he had not been aware of the result of the above property nor was he aware of any geometrical property of the exceptional divisors for the above cases that implied this result. Of course the construction of embedded resolutions for the remaining prehomogeneous cases would enable one to settle Igusa's conjecture, but perhaps there are geometrical properties of these varieties that force their local zeta functions to be universal.

References

1. J. Denef and D. Meuser, A functional equation of Igusa's local zeta function, To appear in *Amer. J. Math.*.
2. J. Igusa, On functional equations of local zeta functions, preprint.
3. G. Kempf, The singularities of some invariant hypersurfaces, *Proc. Conference on Algebraic Geometry*, Berlin (1985) Teubner-Texte, 92 (1986), 210-216.
4. T. Kimura and M. Sato, A classification of irreducible prehomogeneous vector spaces and their relative invariants, *Nagoya Math J.*, **65**, (1977) 1-155.
5. T. Shioda and T. Katsura, On Fermat Varieties, *Tôhuku Mat. Journ.*, **31** (1979), 97-115.
6. J. Tate, Algebraic cycles and poles of zeta functions, *Arithmetical Algebraic Geometry* (Proc. Conf. Purdue Univ., 1963), 93-110, Harper and Row (1965).

On Vanishing of Cohomologies of Rigid Analytic Spaces

Yasuo Morita

Mathematical Institute
Tohoku University
Aoba, Sendai 980
Japan

1 Introduction

In our previous paper [Mo] , we have studied relative cohomologies of the rigid analytic space k^n , and, by using a vanishing theorem of relative cohomologies, we have constructed a p-adic theory of hyperfunctions on compact subsets of k^n .
In this paper, we shall review the results of [Mo] , and study relevant vanishing theorems of cohomologies of rigid analytic spaces. In particular, we shall give two open problems concerning vanishing of cohomologies of rigid analytic spaces.

2 Vanishing of the Derived Sheaves $\mathcal{H}^i_S(\mathcal{O}_X)$ $(i \neq n)$

Let $\mathbf{Q}_p$ be the p-adic number field, and let k be a complete nonarchimedean algebraically closed field containing $\mathbf{Q}_p$. For example, the field $\mathbf{C}_p$, which is the completion of the algebraic closure of $\mathbf{Q}_p$, satisfies this condition. We denote by $| \ |$ the standard p-adic valuation of k with $| p | = p^{-1}$.

Let $X = k^n$ be the n-dimensional affine space with the standard rigid analytic structure (cf. e.g. [BGR]) . Let $\mathcal{O}_X$ be the structure sheaf on this rigid analytic space X , let U be an admissible open subset of X , and let $H^i(U, \mathcal{O}_X)$ be the i-th cohomology group. Let K be a compact subset of X , and let $H^i_{K\cap U}(U, \mathcal{O}_X)$ be the i-th relative cohomology group with supports in $K \cap U$, and let $\mathcal{H}^i_K(\mathcal{O}_X)$ be the i-th derived sheaf of $\mathcal{O}_X$ with supports in K (cf. [Mo] for the definition). Then we have proved in [Mo] that the derived sheaf $\mathcal{H}^i_K(\mathcal{O}_X)$ vanishes for any nonnegative integer $i \neq n$. Further, by using the degeneracy of the spectral sequence

$$H^i_{K\cap U}(U, \mathcal{H}^j_K(\mathcal{O}_X)) \Longrightarrow H^h_{K\cap U}(U, \mathcal{O}_X) ,$$

we have obtained the following isomorphism :

$$H^{n+m}_{K\cap U}(U, \mathcal{O}_X) \simeq H^m(U, \mathcal{H}^n_K(\mathcal{O}_X)) \text{ for any integer } m \geq 0.$$

Hence the presheaf $\mathcal{B}_K$ defined by

$$\mathcal{B}_K(K \cap U) = H^n_{K\cap U}(U, \mathcal{O}_X)$$

becomes a (flabby) sheaf on K. Furthermore, we have proved that the n-th relative cohomology group $H^n_K(X, \mathcal{O}_X)$ gives the dual space of the space of locally analytic functions on K (cf. [Mo]).

One of the most essential part of [Mo] is in the following lemma.

Let U be an affinoid subset of $X = k^n$ containing the product of 1-dimensional affinoids $D_1, \ldots, D_n$ and let S be a product of open balls $B_i = \{x \in k \mid |x - a_i| \leq r_i^-\}$ $(i = 1, \ldots, n)$ such that $B_i \subset D_i$ and $r_i \in |k^\times|$. (In [Mo], these conditions on U, D_i, B_i are not correctly stated.)

Then we have proved the following :

LEMMA. Let the notation and assumptions be as above. Then :

(i) The relative cohomology $H^i_S(U, \mathcal{O}_X)$ vanishes for any nonnegative integer i different from n.

(ii) The n-th relative cohomology $H^n_S(U, \mathcal{O}_X)$ can be explicitly calculated.

REMARK. B. Chiarellotto also obtained this result in a more generalized form. He studied the Serre duality in the rigid analytic case (cf. his report in this Proceedings).

Though we have stated in [Mo] only for compact subsets in the affine space $X = k^n$, it is easy to see that, for any compact subset K in the projective space $X = \mathbf{P}^n(k)$, we can prove that the derived sheaf $\mathcal{H}^i_K(\mathcal{O}_X)$ vanishes for any nonnegative integer $i \neq n$. In particular, if we denote by L a locally compact subfield of k containing $\mathbf{Q}_p$, then we can prove that the derived sheaf $\mathcal{H}^i_{\mathbf{P}^n(L)}(\mathcal{O}_X)$ vanishes for any nonnegative integer $i \neq n$. Further, if K is a compact subset of $\mathbf{P}^n(k)$, then we can prove that the cohomology $H^i_{K\cap U}(U, \mathcal{O}_X)$ vanishes for any integer $i \geq n+1$ and for any admissible open subset U of $\mathbf{P}^n(k)$.

In the classical case, it is known that for any n-dimensional real analytic manifold M and for any complexification X of M, the derived sheaf $\mathcal{H}^i_S(\mathcal{O}_X)$ vanishes for any closed subset S of M and for any nonnegative integer $i \neq n$ (cf. e.g. [KKK], Theorem 2.2.1). This is the most essential fact to construct the theory of hyperfunctions on M, and we have used an analogous fact to construct a p-adic analogue of the theory of hyperfunctions (cf. [Mo]). So we have the following problem :

Let L be a locally compact subfield of k containing $\mathbf{Q}_p$, let X be an n-dimensional rigid analytic space defined over L, and let M be the subset of X consisting of all L-valued points of X. Let S be a closed subset of M such that $X - S$ is admissible open in X. Then :

PROBLEM VDS. Does the derived sheaf $\mathcal{H}^i_S(\mathcal{O}_X)$ vanish for any nonnegative integer i different from n?

We have already noted that the answer is yes for the projective space $X = \mathbf{P}^n(k)$.

3 Relation with Vanishing of Cohomologies

Let $(X, \mathcal{O}_X, \mathcal{T})$ be a rigid analytic space, and let $\mathcal{F}$ be a sheaf of $\mathcal{O}_X$-module on $\mathcal{T}$. Let U be an admissible open subset of X, and let S be a subset of U such that $U - S$ is admissible open in U. Let $H^i_S(U, \mathcal{F})$ be the i-th relative cohomology group of $\mathcal{F}$ with supports in S. Then we have a spectral sequence

$$H^i_{S\cap U}(U, \mathcal{H}^j_S(\mathcal{F})) \Longrightarrow H^h_{S\cap U}(U, \mathcal{F}) .$$

Hence we have an isomorphism

$$H^{n+m}_{S\cap U}(U, \mathcal{F}) \simeq H^m(U, \mathcal{H}^n_S(\mathcal{F}))$$

for any integer $m \geq 0$ if the derived sheaf $\mathcal{H}^i_S(\mathcal{F})$ vanishes for any nonnegative integer $i \neq n$ (cf. [G], Théorème 4.4.1).

Since $\mathcal{H}^i_S(\mathcal{F})$ is the sheafication of $H^i_{S\cap U}(U, \mathcal{F})$, the derived sheaf $\mathcal{H}^i_S(\mathcal{F})$ vanishes for any nonnegative integer $i \neq n$ if and only if (1) $H^i_{S\cap U}(U, \mathcal{F})$ vanishes for any integer $i \geq n+1$ and for any admissible open subset U, and (2) $H^i_{S\cap U}(U, \mathcal{F})$ vanishes for any integer $i \leq n-1$ and for any sufficiently small admissible open subset U.

Now we have the following exact sequence (cf. [Mo]) :

$$\begin{aligned} 0 \longrightarrow H^0_S(U, \mathcal{F}) &\longrightarrow H^0(U, \mathcal{F}) \longrightarrow H^0(U - S, \mathcal{F}) \\ \longrightarrow H^1_S(U, \mathcal{F}) &\longrightarrow H^1(U, \mathcal{F}) \longrightarrow H^1(U - S, \mathcal{F}) \longrightarrow \\ \ldots \longrightarrow H^i_S(U, \mathcal{F}) &\longrightarrow H^i(U, \mathcal{F}) \longrightarrow H^i(U - S, \mathcal{F}) \longrightarrow \ldots \quad \text{(exact)} . \end{aligned}$$

Hence, $H^i_S(U, \mathcal{F})$ vanishes for any nonnegative integer $i \neq n$ if and only if the following three conditions are satisfied :

$$H^i(U, \mathcal{F}) \simeq H^i(U - S, \mathcal{F}) \quad \text{for any integer } i \neq n-1,\ n ,$$

$$H^{n-1}(U, \mathcal{F}) \longrightarrow H^{n-1}(U - S, \mathcal{F}) \quad \text{is injective} ,$$

$$H^n(U, \mathcal{F}) \longrightarrow H^n(U - S, \mathcal{F}) \quad \text{is surjective} .$$

Hence $\mathcal{H}^i_S(\mathcal{F})$ vanishes for any nonnegative integer $i \neq n$ if and only if

$$(1') \quad \begin{cases} H^n(U, \mathcal{F}) \longrightarrow H^n(U - S, \mathcal{F}) & \text{is surjective, and} \\ H^i(U, \mathcal{F}) \simeq H^i(U - S, \mathcal{F}) & \text{for any integer } i \geq n+1 \end{cases}$$

holds for any admissible open subset U of X, and

$$(2') \quad \begin{cases} H^{n-1}(U, \mathcal{F}) \longrightarrow H^{n-1}(U - S, \mathcal{F}) & \text{is injective, and} \\ H^i(U, \mathcal{F}) \simeq H^i(U - S, \mathcal{F}) & \text{for any integer } i \leq n-2 \end{cases}$$

holds for any sufficiently small admissible open subset U of X.

Therefore, if the derived sheaf $\mathcal{H}^i_S(\mathcal{O}_X)$ vanishes for any nonnegative integer $i \neq n$, then we have

$$H^i(U, \mathcal{O}_X) \simeq H^i(U - S, \mathcal{O}_X)$$

for any integer $i \geq n+1$ and for any admissible open subset U of X .

In the classical case, it is known that for any coherent $\mathcal{O}_X$-module $\mathcal{F}$ on an n-dimensional complex manifold X , the cohomology group $H^i(X, \mathcal{O}_X)$ vanishes for any integer $i \geq n+1$ (cf. [Mal] , Théorèm 3) . Further, it is known that the cohomology $H^n(X, \mathcal{F})$ also vanishes if X is not compact and if $\mathcal{F}$ is a locally free coherent $\mathcal{O}_X$-module (cf. [Mal] , Problèmes 1) .

Malgrange used the Dolbeault complex to prove these facts, so it is not easy to translate the proof into the p-adic case.

We note that the cohomologies $H^i(X, \mathcal{F})$ $(i \geq n+1)$ may not vanish if $\mathcal{F}$ is not coherent.

On the other hand, by a result of A. Grothendieck, the cohomology $H^i(X, \mathcal{F})$ vanishes for any integer $i \geq n+1$ and for any sheaf of abelian groups $\mathcal{F}$ on X if X is a noetherian topological space of dimension n .

In our case, M. van der Put obtained the following (cf. [P] , p.174, Lemma) :

THEOREM VC . If X is an n-dimensional affinoid space, then the cohomology group $H^i(X, \mathcal{F})$ vanishes for any integer $i \geq n+1$ and for any sheaf $\mathcal{F}$ of abelian groups on X.

Van der Put expressed the cohomology $H^i(X, \mathcal{F})$ in terms of the Čech cohomology group $\check{H}^i(X, \mathcal{F})$, and expressed it by the Čech cohomology $\check{H}^i(\tilde{X}, \tilde{\mathcal{F}})$ of the reduced affine variety $\tilde{X}$. Then he proved the vanishing of the last cohomology.

It seems that his result holds not only for affinoid spaces but also for rigid analytic spaces satisfying a countable condition (cf. [P] , p.172, Proposition 1.4.4) , but the author could not check it because only a sketch of the proof is given in [P] .

REMARK. Though the cohomological dimension of a paracompact topological space can be calculated locally, its proof seems to have a difficulty to translate into the rigid analytic case (cf. [G] , Théorème 4.14.1) .

In the classical case, we can prove that the cohomology $H^i(X, \mathcal{F})$ vanishes for any integer $i \geq 1$ if $\mathcal{F}$ is a soft sheaf. But soft sheaves seem to be not so nice in the rigid analytic case.

Related with this question, we are also interested in the following problem :

PROBLEM VAC. Does the cohomology group $H^i(X, \mathcal{F})$ vanish for any integer $i \geq n$ and for any locally free coherent $\mathcal{O}_X$-module $\mathcal{F}$ if X is an admissible open subset of a quasi Stein space ?

REFERENCES

[A] M. Artin, Grothendieck topologies, Lecture Notes of Harvard University, 1962.

[BGR] S. Bosch, U. Günter and R. Remmert, Non-Archimedean Analysis, Springer-Verlag, 1984.

[G] R. Godement, Topologie algébrique et théorie des faisceaux, Hermann, 1958.

[KKK] M. Kashiwara, T. Kawai and T. Kimura, Foundations of algebraic analysis, Princeton Univ. Press, 1986.

[Mal] B. Malgrange, Fasceaux sur variétés analytiques réelles, Bull. Soc. math. France, **83**(1955), 231-237.

[Mo] Y. Morita, A p-adic theory of hyperfunctions, II. In : Algebraic Analysis, Vol. I, 457-472, Academic Press Inc., 1989.

[SM] W. H. Schikhof and Y. Morita, Duality of projective limit spaces and inductive limit spaces over a nonspherically complete nonarchimedean field, Tohoku Math. J., **38**(1986), 387-397.

[P] M. van der Put, Cohomology on affinoid spaces, Composito Mathematica, **45** (1982), 165-198.

A p-adic Analogue of the Chowla-Selberg Formula

Arthur Ogus *
University of California, Berkeley

Let $X/\mathbf{C}$ be an elliptic curve, with complex multiplication by a quadratic imaginary field E. Then $X/\mathbf{C}$ automatically and uniquely descends to the algebraic closure $\overline{\mathbf{Q}}$ of $\mathbf{Q}$ in $\mathbf{C}$ and has good reduction at every prime $\wp$ of $\overline{\mathbf{Q}}$. Let $K_\wp$ denote the algebraic closure of $\mathbf{Q}_p$ in the completion of $\overline{\mathbf{Q}}$ at $\wp$, and let $W_\wp$ denote the set of automorphisms of $\overline{\mathbf{Q}}$ which preserve $\wp$ and which act as some integral power of the Frobenius automorphism on the residue field $k_\wp$ at $\wp$. These automorphisms extend to automorphisms of $K_\wp$ by continuity. As explained in [3], there is a canonical semilinear crystalline action of this Weil group on $H^1_{DR}(X/\overline{\mathbf{Q}}) \otimes K_\wp$, compatible with the action of E. The principle goal of this paper is to give an "explicit formula" for a matrix representation of this action, with respect to a basis for $H^1_{DR}(X/\overline{\mathbf{Q}})$.

As we shall see, our formula, 3.15 below, bears a striking resemblance to the classical Chowla-Selberg formula [15] for the classical period matrix of $X/\overline{\mathbf{Q}}$, with the classical gamma function replaced by Morita's p-adic gamma function. As a matter of fact, our proof is inspired by and patterned after Gross's proof of the classical formula in *op. cit.*, in which the formula for elliptic curves is deduced from a similar formula for the periods of Fermat curves. In our case, we use the formalism of absolute Hodge cycles to reduce to the computation of the Frobenius matrix of Fermat curves. When p does not divide the degree, this computation goes back a long way, but in fact no convenient reference seemed to exist in the literature until Coleman's article [6], which was written in response to my queries. More recently, Coleman has been able to carry out his computations for all odd primes. In fact, in the case of bad reduction he was only able to calculate part of the Frobenius matrix, but, remarkably, this partial calculation is sufficient for the determination of the calculation of the periods of the "semiversal ulterior cyclotomic motive" constructed here (*c.f.* Theorem 3.1). Consequently, using the results of this paper, Coleman was finally able to bootstrap the calculations and obtain an explicit formula for the entire Frobenius matrix of the Fermat curve. We refer to his paper for the explicit formulas and the detailed story of the relationship between his results and ours [5].

I should point out that our formula is a rather special consequence of the general philosophy that Hodge cycles should be compatible with the action of Frobenius on crystalline cohomology, as conjectured in [17]. In fact, recent work of Fontaine-Messing [14] and Faltings [12] has made it possible to strengthen this conjecture, and, as Don Blasius has pointed out [4], to prove it for abelian varieties. The strengthened conjecture, which includes a p-adic analogue of Shimura's monomial period relations, has been considered by other authors as well [7]. I only emphasize the Chowla-Selberg formula itself to try

*Partially supported by NSF Grant No. DMS-8502783

to render crystalline cohomology more "explicit" and "concrete," at least in the setting of the theory of complex multiplication.

Our paper is organized as follows. The first sections review the terminology and formalism of motives of CM-type and their classical and crystalline periods. In particular we introduce the concept of a "marked motive," which is a motive endowed with a choice of basis for its cohomology. We construct an abelian group out of the category of marked motives and state the main theorem, which asserts that the periods of a marked motive depend only on the corresponding marked CM-type. This more delicate formulation is essential for Coleman's bootstrap in the determination of the Frobenius matrix for Fermat curves in the case of bad reduction. Section 2 reviews the classical and crystalline cohomology of Fermat curves, being of necessity rather careful about specific choice of bases for cohomology and about rational factors. Section 3 includes the construction of the semiversal ulterior motive M alluded to above, which can probably be viewed as a geometric incarnation of Anderson's constructions in [1]. Finally, section 4 contains a proof of the general result on absolute Hodge cycles on which our formula relies.

I would like to thank Don Blasius for sharing a preliminary version of his proof of the "De Rham conjecture" for absolute Hodge cycles upon which our results are based, as well as Robert Coleman for the dedication with which he pursued the precise forms of the formulas I needed. Thanks also go to Hendrik Lenstra and Ken Ribet for useful conversations and hints about Galois cohomology and CM-types.

1 Complex Multiplication

We begin by recalling the basic facts about complex multiplication and motives of CM-type.

Let $E/\mathbf{Q}$ be a CM-algebra—that is, a product of CM-fields—and for any $\mathbf{Q}$-algebra A, let $S(A) =: Mor(E, A)$ for any. Then $E \otimes K \cong K^{S(K)}$ for any algebraically closed field K of characteristic zero. Let $X/\overline{\mathbf{Q}}$ be an abelian variety with complex multiplication by E, so that (by definition) E operates on $X/\overline{\mathbf{Q}}$ and the corresponding action of E on De Rham cohomology makes $H^1_{DR}(X/\overline{\mathbf{Q}})$ a free $E \otimes \overline{\mathbf{Q}}$-module of rank one. For each $s \in S(\overline{\mathbf{Q}})$, the fiber $H^1_{DR}(s)$ of $H^1_{DR}(X/\overline{\mathbf{Q}})$ at s has dimension one over $\overline{\mathbf{Q}}$, and there is a canonical isomorphism: $H^1_{DR}(X/\overline{\mathbf{Q}}) \cong \Pi\{H^1_{DR}(s) : s \in S\}$.

We shall find it extremely convenient to consider motives of higher weight constructed from abelian varieties of CM-type by linear algebra operations. We refer to [10] for the formal definition, contenting ourselves here with the examples and properties we need. Associated with an E-motive $X/\overline{\mathbf{Q}}$ of rank r are its various cohomological realizations:

- $H_B(X/\overline{\mathbf{Q}})$—a free E-module of rank r.
- $H_{DR}(X/\overline{\mathbf{Q}})$—a free $\overline{\mathbf{Q}} \otimes E$-module of rank r.
- $H_{\acute{e}t}(X/\overline{\mathbf{Q}})$—a free $\mathbf{Q}_p \otimes E$-module of rank r.

These realizations are functorial and come equipped with other functorial data as well. For example, the De Rham cohomology comes with the Hodge filtration, and there are canonical isomorphisms between the De Rham and Betti cohomologies (after tensoring with $\mathbf{C}$). We shall be more explicit about these and other data later.

Here are some examples of constructions of E-motives:

- If $A/\overline{\mathbf{Q}}$ is an abelian variety with complex multiplication by E, then the degree one part of its cohomology is an E-motive of rank one.
- If $a: E \to E'$ is a homomorphism of CM-algebras and $X/\overline{\mathbf{Q}}$ is an E-motive of rank r, then $a^*X =: X \otimes_E E'$ is an E'-motive of rank r.
- $E \otimes H^i(\mathbf{P}^1)$ is an E-motive of rank one and weight i if $i = 0$ or 2. We denote this motive by E if $i = 0$ and by $E(-1)$ if $i = 2$.
- If X and Y are E-motives, we can define $X \otimes_E Y$ and $Hom_E(X, Y)$ so that

$$H(X \otimes_E Y) \cong H(X) \otimes H(Y) \quad \text{and}$$
$$H(Hom(X,Y)) \cong Hom(H(X), H(Y))$$

  Here the "H" stands for any cohomological realization, and the tensor product and Hom are taken over the appropriate coefficient ring.

There are several ways to define morphisms, and in particular isomorphisms, of E-motives. For the sake of definiteness, let us choose here to use the strictest definition, originally proposed by Grothendieck—morphisms of motives are induced by algebraic correspondences. The point is that any such morphism is necessarily compatible with the all the structures on and compatibilities between the various cohomological realizations.

The Hodge filtration on the De Rham cohomology of an E-motive $X/\overline{\mathbf{Q}}$ is especially important. If $X/\overline{\mathbf{Q}}$ has rank one, this filtration can be described purely combinatorially. We identify $S(\overline{\mathbf{Q}})$ with the set of maximal ideals of $\overline{\mathbf{Q}} \otimes E$, and the fiber $H_{DR}(s)$ of $H_{DR}(X/\overline{\mathbf{Q}})$ at s can be identified as

$$H_{DR}(s) \cong \{x \in H_{DR}(X/\overline{\mathbf{Q}}) : ex = s(e)x \text{ for all } e \in E\}$$

Then for each $s \in S(\overline{\mathbf{Q}})$, we define

$$\tau(X)(s) =: \sup\{i : H_{DR}(s) \in F^i H_{DR}(X/\overline{\mathbf{Q}})\}$$

Since each stalk is one dimensional, it follows immediately that

$$F^i H_{DR}(X/\overline{\mathbf{Q}}) = \sum_s \{H_{DR}(s) : \tau(X)(s) \geq i\}$$

Attached to each motive is a weight, *i.e.* the weight n of the underlying Hodge structure, which can be expressed in terms of τ and the natural action of complex conjugation on $S(\overline{\mathbf{Q}})$. Namely, one finds easily that $\tau(s) + \tau(\overline{s}) = n$ for every $s \in S(\overline{\mathbf{Q}})$. If E is a field, n is just an integer, and in general it is a function on $S(\overline{\mathbf{Q}})$ which is constant on the orbits of $Gal(\overline{\mathbf{Q}}/\mathbf{Q})$ acting on $S(\overline{\mathbf{Q}})$. In general, a function $\tau: S(\overline{\mathbf{Q}}) \to \mathbf{Z}$ such that $\tau(s) + \tau(\overline{s}) = n$ is called a "CM-type of weight n." Evidently the sum of a CM-type of weight n and a CM-type of weight m is a CM-type of weight $n + m$, and the set of all CM-types forms a subgroup of the group of all functions $S(\overline{\mathbf{Q}}) \to \mathbf{Z}$. We denote this subgroup by $CM(E)$.

Let $Mot(E/\overline{\mathbf{Q}})$ denote the set of isomorphisms classes of E-motives of rank one. Using the isomorphisms:

$$X \otimes_E Y \cong Y \otimes_E X \quad \text{and} \quad (X \otimes_E Y) \otimes_E Z \cong X \otimes_E (Y \otimes_E Z),$$

we find that $Mot(E/\overline{\mathbf{Q}})$ becomes an abelian group. Since the Hodge filtration of $X \otimes_E Y$ is the tensor product filtration, we have $\tau(X \otimes_E Y) = \tau(X) + \tau(Y)$. Similarly, one finds that if $a: E \to E'$ is a homomorphism with corresponding map $a^*: S'(\overline{\mathbf{Q}}) \to S(\overline{\mathbf{Q}})$, then $\tau(a^* X) = \tau(X) \circ a^*$. We summarize:

Proposition 1.1 *Formation of the CM-type induces a surjective homomorphism of abelian groups:*

$$\tau: Mot(E/\overline{\mathbf{Q}}) \to CM(E)$$

If $a: E \to E'$ is homomorphism, let a^* denote the corresponding maps

$$S'(\overline{\mathbf{Q}}) \to S(\overline{\mathbf{Q}}), \quad CM(E) \to CM(E'), \quad \text{and} \quad Mot(E/\overline{\mathbf{Q}}) \to Mot(E'/\overline{\mathbf{Q}})$$

Then there is a commutative diagram:

$$\begin{array}{ccc} Mot(E/\overline{\mathbf{Q}}) & \xrightarrow{\tau} & CM(E) \\ \downarrow{\scriptstyle a^*} & & \downarrow{\scriptstyle a^*} \\ Mot(E'/\overline{\mathbf{Q}}) & \xrightarrow{\tau} & CM(E') \end{array}$$

Proof: The only part of this proposition that is not yet clear is the surjectivity of τ. This is classical; we recall the outline of the argument. Notice that if $\tau \in CM(E)$ takes its values in $\{0,1\}$, then the weight w of τ is either 1 or 2. In the latter case, τ is constant and equal to $\tau(E(-1))$. If the weight is one, we can regard τ as defining a partition of $S(\overline{\mathbf{Q}})$ into two subsets Σ and $\overline{\Sigma}$. Then there is a well-known complex analytic construction of a complex torus on which the ring of integers $\mathcal{O}$ of E operates: $\mathcal{O}$ is a lattice in $\mathbf{C}^{\overline{\Sigma}}$, and the quotient is the desired torus. One finds that the action of E on cohomology is such that one recovers τ as the CM-type. Then one proves that the complex torus is in fact an abelian variety, and that this variety, together with the action of $\mathcal{O}$, descends uniquely to $\overline{\mathbf{Q}}$. To prove the proposition, one now is reduced to checking that the group $CM(E)$ is generated by elements τ which take their values in $\{0,1\}$, which is easy. ∎

It is sometimes possible and useful to make more precise statements, with respect to a specific basis. To this end, we define a "bimarking" of an E-motive X to be a pair (ξ, η), where ξ is an $E \otimes \overline{\mathbf{Q}}$ basis of $H_{DR}(X/\overline{\mathbf{Q}})$ and η is a $\mathbf{Q}$-basis for $H_B(X)$. Similarly, we define a "De Rham marking" of X to be an $E \otimes \overline{\mathbf{Q}}$-basis ξ of $H_{DR}(X/\overline{\mathbf{Q}})$. For example, $H_{DR}(E)$ is just $E \otimes \overline{\mathbf{Q}}$, which has a canonical basis 1, and $H_B(E)$ is just E, which also has a canonical basis 1. Similarly, $H_{DR}(E(-1))$ is $E \otimes \overline{\mathbf{Q}}$, with basis 1, and $H_B(E(-1)$ is $(2\pi i)^{-1}E$, with basis $(2\pi i)^{-1}$. Then we have the obvious notions of the categories of bimarked and De Rham marked motives, and the tensor product operation induces a group structure on the corresponding sets of isomorphism classes, which we denote by $BMot(E)$ and $DMot(E)$, respectively. Notice that an element e of $\mathbf{T}_E(\mathbf{Q})$ induces an isomorphism of bimarked motives

$$(X, \xi, \eta) \cong (X, e\xi, e\eta),$$

so that these two define the same element of $BMot(E)$.

If A is any $\mathbf{Q}$-algebra, we let $\mathbf{T}_E(A)$ denote the group of units of $E \otimes A$; we have a canonical isomorphism $\mathbf{T}_E(K) \cong K^{*S(K)}$ if K is an algebraically closed field. There are then evident exact sequences:

$$0 \to \mathbf{T}_E(\mathbf{Q}) \to BMot(E) \to DMot(E) \to 0$$
$$0 \to \mathbf{T}_E(\overline{\mathbf{Q}})/\mathbf{T}_E(\mathbf{Q}) \to DMot(E) \to Mot(E) \to 0$$
$$0 \to \mathbf{T}_E(\overline{\mathbf{Q}}) \to BMot(E) \to Mot(E) \to 0$$

Let us next review the formalism of periods of E-motives. The theory of integration defines a canonical isomorphism:

$$\sigma_B: H_{DR}(X/\overline{\mathbf{Q}}) \otimes_{\overline{\mathbf{Q}}} \mathbf{C} \to H_B(X/\overline{\mathbf{Q}}) \otimes_{\mathbf{Q}} \mathbf{C}$$

If we choose a basis ξ for $H_{DR}(X/\overline{\mathbf{Q}})$ over $E \otimes \overline{\mathbf{Q}}$ and a basis η for $H_B(X/\overline{\mathbf{Q}})$ over E, then the matrix for σ_B with respect to these bases is a unit of $E \otimes \mathbf{C}$, which we denote by $\gamma(X, \xi, \eta)$. Its image in $\mathbf{T}_E(\mathbf{C})/\mathbf{T}_E(\mathbf{Q})$ depends only on X and ξ, and its image in $\mathbf{T}_E(\mathbf{C})/\mathbf{T}_E(\overline{\mathbf{Q}})$ depends only on X; we denote these by $\gamma(X, \xi)$ and $\gamma(X)$, respectively. One checks easily that in fact γ defines homomorphisms:

$$BMot(E, \overline{\mathbf{Q}}) \to \mathbf{T}_E(\mathbf{C})$$
$$DMot(E, \overline{\mathbf{Q}}) \to \mathbf{T}_E(\mathbf{C})/\mathbf{T}_E(\mathbf{Q})$$
$$Mot(E, \overline{\mathbf{Q}}) \to \mathbf{T}_E(\mathbf{C})/\mathbf{T}_E(\overline{\mathbf{Q}}),$$

all of which we denote simply by γ. For example,

$$\gamma(E, 1, 1) = 1 \quad \text{and} \quad \gamma(E(-1), 1, (2\pi i)^{-1}) = 2\pi i$$

For the crystalline version, we let $\wp$ be a prime of $\overline{\mathbf{Q}}$ and let ψ be a member of the Weil group $W_\wp$. Since we are working over $\mathbf{Q}_p$, we can identify $W_\wp$ with the "crystalline Weil group" constructed in [3]. Since abelian varieties of CM-type have potentially good reduction everywhere, the crystalline Weil group acts on the completion of their De Rham cohomology, and hence also on the completion of the cohomology of any motive constructed from abelian varieties of CM-type. If again ξ is an $E \otimes \overline{\mathbf{Q}}$-basis for $H_{DR}(X/\overline{\mathbf{Q}})$, we let $\gamma_\wp(X, \xi, \psi) \in \mathbf{T}_E(K_\wp)$ denote the matrix for the action of ψ with respect to ξ. (This also makes sense for any $E \otimes K_\wp$-basis.)

The group $W_\wp$ also acts on $\mathbf{T}_E(K_\wp)$ through its action on $K_\wp$, and if $\lambda \in \mathbf{T}_E(K_\wp)$ and $\xi \in H_{DR}(X/K_\wp)$ is a basis, we have the following formulas:

$$\begin{aligned} \psi(\lambda\xi) &= \psi(\lambda)\psi(\xi) \\ \gamma_\wp(X, \xi, \psi'\psi) &= \gamma_\wp(X, \xi, \psi')\psi'(\gamma_\wp(X, \xi, \psi)) \\ \gamma_\wp(X, \lambda\xi, \psi) &= \psi(\lambda)\lambda^{-1}\gamma_\wp(X, \xi, \psi). \end{aligned} \tag{1}$$

The second of these says that $\gamma_\wp(X, \xi)$ defines a crossed homomorphism

$$W_\wp(K_\wp) \to \mathbf{T}_E(K_\wp)$$

Lemma 1.2 *The crossed homomorphism $\gamma_\wp(X, \xi): W_\wp(K_\wp) \to \mathbf{T}_E(K_\wp)$ is continuous, i.e., it vanishes on a subgroup of finite index of the inertia group $I_\wp$. Specifically, if $K \subseteq K_\wp$ is a finite extension of $\mathbf{Q}_p$ such that X is defined and has good reduction over K and such that $\xi \in H_{DR}(X/K)$, then $\gamma_\wp(X, \xi)$ vanishes on $Gal(K_\wp/K) \cap I_\wp$.*

Proof: This follows immediately from the definitions in [3]. Namely, if $W(\wp)$ is the Witt ring of the residue field k of K and X_k is the special fiber of X/K, there is a canonical isomorphism:

$$H_{DR}(X/K) \otimes K_\wp \cong H_{cris}(X_\wp, W(\wp)) \otimes_{W(\wp)} K_\wp$$

Then if $\psi \in W_\wp(K_\wp)$ has degree n, ψ is defined to act on $H_{cris}(X_\wp, W_\wp) \otimes K_\wp$ as $H_{cris}(F_X^n) \otimes \psi$. In particular, if $n = 0$ and $\psi \in Gal(K_\wp/K)$, then $\psi(\xi) = \xi$ and $\gamma_\wp(X, \xi, \psi) = 1$. ∎

We find that calculation of periods induces a a homomorphism of abelian groups:

$$\gamma_\wp : DMot(E/\overline{\mathbf{Q}}) \to Z^1(W_\wp, \mathbf{T}_E(K_\wp))$$

The formulas 1 above imply that the image of $\gamma_\wp(X, \xi)$ modulo coboundaries from $\mathbf{T}_E(\overline{\mathbf{Q}})$ is independent of the choice of basis ξ. We will therefore find it convenient to make the following definition.

Definition 1.3 *Let G be a group and M and N are G-modules with $N \subseteq M$. Then $H^1(G, M; N)$ is the quotient of the module $Z^1(G, M)$ of crossed homomorphisms $G \to M$ by the coboundaries $B^1(G, N)$ coming from N.*

Perhaps it is worth remarking that there are natural exact sequences

$$0 \to N/N^G \to Z^1(G, M) \to H^1(G, M; N) \to 0$$

$$0 \to M/(N + M^G) \to H^1(G, M; N) \to H^1(G, M) \to 0$$

$$0 \to H^1(G, N) \to H^1(G, M; N) \to Z^1(G, M/N)$$

Thus, $\gamma_\wp(X)$ should be viewed as an element of $H^1(W_\wp, \mathbf{T}_E(K_\wp); \mathbf{T}_E(\overline{\mathbf{Q}}))$. For typographical reasons, we shall abbreviate this by $H^1(W_\wp, \mathbf{T}_E(K_\wp; \overline{\mathbf{Q}}))$.

Lemma 1.4 *The groups $H^1(W_\wp, \mathbf{T}_E(K_\wp; \overline{\mathbf{Q}}))$ and $H^1(W_\wp, \mathbf{T}_E(\overline{\mathbf{Q}}))$ are torsion free.*

Proof: The group $Z^1(\mathbf{T}_E(K_\wp)/\mathbf{T}_E(\overline{\mathbf{Q}})$ is evidently torsion free, so we see from the exact sequences above that it suffices to show that $H^1(W_\wp, \mathbf{T}_E(\overline{\mathbf{Q}}))$ is torsion free. To this end, let $\mathbf{T}_E(\overline{\mathbf{Q}})_n$ denote the kernel of the surjective map multiplication by n on $\mathbf{T}_E(\overline{\mathbf{Q}})$. Let $D_\wp \subseteq Gal(\overline{\mathbf{Q}}/\mathbf{Q})$ be the decomposition group at $\wp$. We have a commutative diagram:

$$\begin{array}{ccccc}
H^1(D_\wp, \mathbf{T}_E(\overline{\mathbf{Q}})_n) & \to & H^1(D_\wp, \mathbf{T}_E(\overline{\mathbf{Q}})) & \xrightarrow{n} & H^1(D_\wp, \mathbf{T}_E(\overline{\mathbf{Q}})) \\
\downarrow{\scriptstyle res_n} & & \downarrow{\scriptstyle res} & & \downarrow{\scriptstyle res} \\
H^1(W_\wp, \mathbf{T}_E(\overline{\mathbf{Q}})_n) & \to & H^1(W_\wp, \mathbf{T}_E(\overline{\mathbf{Q}})) & \xrightarrow{n} & H^1(W_\wp, \mathbf{T}_E(\overline{\mathbf{Q}}))
\end{array}$$

Since $\mathbf{T}_E(\overline{\mathbf{Q}})_n$ is a finite set, any cocycle $z \in Z^1(W_\wp, \mathbf{T}_E(\overline{\mathbf{Q}})_n)$ automatically factors through a finite quotient, and hence prolongs uniquely to a cocycle in $Z^1(D_\wp, \mathbf{T}_E(\overline{\mathbf{Q}})_n)$. Thus, the map res_n in the diagram above is surjective (and even an isomorphism). On the other hand, Hilbert's theorem 90 implies that $H^1(D_\wp, \mathbf{T}_E(\overline{\mathbf{Q}}))$ vanishes. A diagram chase now shows that multiplication by n on $H^1(W_\wp, \mathbf{T}_E(\overline{\mathbf{Q}}))$ is injective. ∎

We can summarize the situation so far as follows:

Corollary 1.5 *Let $K_p =: \overline{\mathbf{Q}} \cap \mathbf{Q}_p$. Then there is a commutative diagram with exact rows:*

$$\begin{array}{ccccccccc}
0 & \to & \mathbf{T}_E(\overline{\mathbf{Q}})/\mathbf{T}_E(K_p) & \to & Z^1(W_p, \mathbf{T}_E(K_p)) & \to & H^1(W_p, \mathbf{T}_E(K_p;\overline{\mathbf{Q}})) & \to & 0 \\
 & & \uparrow & & \uparrow{\scriptstyle \gamma_p} & & \uparrow{\scriptstyle \gamma_p} & & \\
0 & \to & \mathbf{T}_E(\overline{\mathbf{Q}})/\mathbf{T}_E(\mathbf{Q}) & \to & DMot(E) & \to & Mot(E) & \to & 0 \\
 & & \| & & \downarrow{\scriptstyle \gamma} & & \downarrow{\scriptstyle \gamma} & & \\
0 & \to & \mathbf{T}_E(\overline{\mathbf{Q}})/\mathbf{T}_E(\mathbf{Q}) & \to & \mathbf{T}_E(\mathbf{C})/\mathbf{T}_E(\mathbf{Q}) & \to & \mathbf{T}_E(\mathbf{C})/\mathbf{T}_E(\overline{\mathbf{Q}}) & \to & 0.
\end{array}$$

Remark 1.6 If $E/\mathbf{Q}$ is Galois (as we may as well assume), then $Gal(E/\mathbf{Q})$ operates in a rather trivial way on the right on the category of bimarked E-motives, via the base change operation 1.1. $Gal(E/\mathbf{Q})$ also acts on $CM(E)$, $\mathbf{T}_E(\mathbf{C})$ and on $\mathbf{T}_E(K_p)$ on the right, and it is clear that the maps τ, γ, and γ_p are compatible with these actions.

The main result which underlies our proofs of Chowla-Selberg states that both the period homomorphisms of Corollary 1.5 depend only on the CM-type. For the classical period γ, this result is essentially equivalent to Shimura's monomial period relations and is due to Shimura and Deligne [11].

Theorem 1.7 *The homomorphisms γ and γ_p above factor through τ. That is, there is a commutative diagram:*

$$\begin{array}{ccc}
 & & \mathbf{T}_E(\mathbf{C})/\mathbf{T}_E(\overline{\mathbf{Q}}) \\
 & \nearrow{\scriptstyle \gamma} & \uparrow{\scriptstyle \gamma} \\
Mot(E) & \xrightarrow{\tau} & CM(E) \\
 & \searrow{\scriptstyle \gamma_p} & \downarrow{\scriptstyle \gamma_p} \\
 & & H^1(W_p, \mathbf{T}_E(K_p;\overline{\mathbf{Q}}))
\end{array} \quad \blacksquare$$

To express a convenient analog for marked motives, we let $BCM(E)$ and $DCM(E)$ denote the fiber products in the diagram below:

$$\begin{array}{ccccccc}
DCM(E) & \to & CM(E) & & BCM(E) & \to & CM(E) \\
\downarrow & & \downarrow{\scriptstyle \gamma} & & \downarrow & & \downarrow{\scriptstyle \gamma} \\
\mathbf{T}_E(\mathbf{C})/\mathbf{T}_E(\mathbf{Q}) & \to & \mathbf{T}_E(\mathbf{C})/\mathbf{T}_E(\overline{\mathbf{Q}}) & & \mathbf{T}_E(\mathbf{C}) & \to & \mathbf{T}_E(\mathbf{C})/\mathbf{T}_E(\overline{\mathbf{Q}})
\end{array} \qquad (2)$$

It is clear that we have natural maps:

$$\tau_{DM}: DMot(E, \overline{\mathbf{Q}}) \to DCM(E) \quad \text{and} \quad \tau_{BM}: BMot(E, \overline{\mathbf{Q}}) \to BCM(E).$$

We can now formulate the following slightly stronger version of Theorem 1.7.

Theorem 1.8 *The homomorphism γ_p factors through τ_{DM}, and the induced map γ_p fits into an exact ladder:*

$$\begin{array}{ccccccccc} 0 & \rightarrow & \mathbf{T}_E(\overline{\mathbf{Q}})/\mathbf{T}_E(\mathbf{Q}) & \rightarrow & DCM(E) & \rightarrow & CM(E) & \rightarrow & 0 \\ & & \downarrow & & \downarrow{\scriptstyle \gamma_p} & & \downarrow{\scriptstyle \gamma_p} & & \\ 0 & \rightarrow & \mathbf{T}_E(\overline{\mathbf{Q}})/\mathbf{T}_E(K_p) & \rightarrow & Z^1(W_p, \mathbf{T}_E(K_p)) & \rightarrow & H^1(W_p, \mathbf{T}_E(K_p; \overline{\mathbf{Q}})) & \rightarrow & 0 \end{array}$$

We shall explain the proofs of 1.7 and 1.8 in section 4. Let us close this section by mentioning the following natural question.

Problem 1.9 *Are the maps*

$$\gamma{:}\, CM(E) \rightarrow \mathbf{T}_E(\mathbf{C})/\mathbf{T}_E(\overline{\mathbf{Q}}) \quad \text{and} \quad \gamma_p{:}\, CM(E) \rightarrow H^1(W_p, \mathbf{T}_E(K_p; \mathbf{T}_E(\overline{\mathbf{Q}}))$$

injective?

2 Periods of Fermat Curves

In this section we calculate the periods described in the previous section for the motives constructed from the Jacobians of Fermat curves. These calculations are due to other authors; it is simply a matter of licking the formulas into shape. We have to be a little careful to keep track of the rational factors neglected in [11] and to be specific about the choice of bases. It seems best to recall in detail the situation, following the notation of [11] as closely as possible.

Let X_m denote the Fermat curve of degree m, with homogeneous equation $x_0^m + x_1^m + x_2^m = 0$. Let μ_m denote the group of roots of unity in $\overline{\mathbf{Q}}$ and let Λ_m denote the group μ_m^3 modulo the diagonal, acting on the right on X_m by

$$(x_0 : x_1 : x_2)(\zeta_0 : \zeta_1 : \zeta_2) = (x_0\zeta_0^{-1} : x_1\zeta_1^{-1} : x_2\zeta_2^{-1})$$

Then Λ_m acts on the left on coordinates (and cohomology), and in particular if $\zeta \in \Lambda$, $\zeta x_i = \zeta_i^{-1} x_i$ The group Λ_m^* of $\overline{\mathbf{Q}}$-valued characters of Λ_m can be identified in an obvious way with the subgroup of $(m^{-1}\mathbf{Z}/\mathbf{Z})^3$ whose coordinates sum to zero. If $\chi =: (\chi_0, \chi_1, \chi_2)$ is such a character, we let H_χ denote the χ-eigensubspace of $H_{DR}(X_m, \overline{\mathbf{Q}})$. Recall that H_χ has dimension one if no $\chi_i = 0$ and is zero otherwise; in the former case we say that χ is a "Fermat character."

To describe the Hodge level of H_χ, let $\langle a \rangle$ denote the representative of a in $\mathbf{Q} \cap (0, 1]$. (Note that the endpoint convention is that of [11] rather than [6]; it makes $\langle a \rangle + \langle -a \rangle$ equal to 1 if a is a nonzero element of $\mathbf{Q}/\mathbf{Z}$ and to 2 if a is zero.) For any Fermat character χ, we let $\langle \chi \rangle$ denote $\sum_i \langle \chi_i \rangle$.

Lemma 2.1 **([11](7.6))** *If χ is a Fermat character, the Hodge level of H_χ is $\langle \chi \rangle - 1$. That is, $Gr_F^j H_\chi$ is zero unless $j = \langle \chi \rangle - 1$, and $\tau(X)(\chi) = \langle \chi \rangle - 1$.*

The group algebra $\mathbf{Q}[\Lambda_m]$ acts on the Jacobian of X_m and its various realizations; in fact, even $\overline{\mathbf{Q}}[\Lambda]$ acts on $H_{DR}(X_m/\overline{\mathbf{Q}})$. Each $\overline{\mathbf{Q}}$-valued character $\chi{:}\, \Lambda_m \rightarrow \overline{\mathbf{Q}}$ defines a homomorphism $s_\chi{:}\, \mathbf{Q}[\Lambda_m] \rightarrow \overline{\mathbf{Q}}$, and there is a corresponding idempotent e_χ of $\overline{\mathbf{Q}}[\Lambda_m]$: $e_\chi =: |\Lambda_m|^{-1} \sum_\lambda \chi(\lambda)\lambda$. We will allow ourselves to identify χ and s_χ, and we let F_m

denote the group of all Fermat characters. The Galois group $Gal(\overline{\mathbf{Q}}/\mathbf{Q})$ operates on the set of characters Λ_m^* and on the group algebra; this action factors through the cyclotomic character $Gal(\overline{\mathbf{Q}}/\mathbf{Q}) \to \mathbf{Z}/m\mathbf{Z}^*$, and $ge_\chi = e_{g\chi}$. Thus, the $\mathbf{Z}/m\mathbf{Z}^*$-orbits of Λ_m^* induce Galois equivariant idempotents of $\overline{\mathbf{Q}}[\Lambda_m]$ *i.e.* idempotents of $\mathbf{Q}[\Lambda_m]$, and we can decompose the Jacobian of X_m (and its corresponding motive) according to these idempotents. The sum $\sum e_\chi \mathbf{Q}[\Lambda_m]$ over all the Fermat characters is $Gal(\overline{\mathbf{Q}}/\mathbf{Q})$-invariant, and hence descends to a $\mathbf{Q}$-algebra Φ_m; the characters of Φ_m are precisely the Fermat characters. Thus, $H_B(X)$ is a free Φ_m-module of rank one.

To describe the markings on X_m that we need, let Y_m denote the complement in X_m of the hyperplane $x_0 = 0$ and let $y_i =: x_i/x_0$. Then Y_m is the affine variety with equation $y_1^m + y_2^m + 1 = 0$, and $\lambda \in \Lambda_m$ operates on these coordinates by by $\lambda y_i = \lambda_0 \lambda_i^{-1} y_i$. For any pair of positive integers a_1 and a_2, let

$$\omega_{\mathbf{a}} = y_1^{a_1} y_2^{a_2} \frac{dy_1}{y_1}$$

Let $a_0 =: -a_1 - a_2$. Then if no a_i is divisible by m, $\omega_{\mathbf{a}}$ is a differential of the second kind on X_m, and hence defines a cohomology class $\xi_{\mathbf{a}}$ in $H^1_{DR}(X_m/\mathbf{Q})$. One sees immediately that $\xi_{\mathbf{a}} \in H\chi$, where χ_i is the image of $-a_i/m$ mod $\mathbf{Z}$. Following Coleman's observations in [6], we shall find it slightly more convenient to consider

$$\omega_\chi =: -m a_0 a_2^{-1} \langle \chi_0 \rangle^{2-\langle\chi\rangle} \omega_{\mathbf{a}} \tag{3}$$

and its corresponding class $\xi_\chi \in H_{DR}$.

Let Δ denote the standard one-simplex in $\mathbf{R}^2$, let $\epsilon_m = e^{\pi i/m}$, and let

$$\sigma : \Delta \to Y_m \quad \text{by} \quad (t_1, t_2) \mapsto (\epsilon_m t_1^{1/m}, \epsilon_m t_2^{1/m})$$

The group algebra $\mathbf{Q}[\Lambda_m]$ operates on the vector space of $\mathbf{Q}$-chains of $Y_m(\mathbf{C})$, and we let δ_i be the element of Λ_m which is ϵ_m^2 in the ith place and zero elsewhere. Then $\eta' =: (1-\delta_1)(1-\delta_2)\sigma$ is a *cycle* of $Y_m(\mathbf{C})$, and hence defines an element η of $H_B(X_m)$, dual to the homology basis η'.

If $m = dn$, there is an obvious map $f_{m,n}: X_m \to X_n$, defined by its action on points: $f_{m,n}(x_0 : x_1 : x_2) = (x_0^d : x_1^d : x_2^d)$, and on coordinates by $f_{m,n}^*(y_i) = y_i^d$. This map is compatible with the actions of Λ_m and Λ_n, via the homomorphism $\Lambda_m \to \Lambda_n$ sending an element to its dth power. One finds easily that $f_{m,n}^*(\omega_\chi) = \omega_\chi$ and $f_{m,n*}(\eta') = \eta'$. We can conclude:

Corollary 2.2 *If n divides m, the natural map $\Phi_m \to \Phi_n$ induces an isomorphism of bimarked Φ_n-motives*

$$(X_n, \xi, \eta) \cong (X_m, \xi, \eta) \otimes_{\Phi_m} \Phi_n. \quad \blacksquare$$

To describe the classical periods, we shall find it convenient to define, for any $z \in \mathbf{C}$, $\Gamma_\pi(z) =: (2\pi i)^{-1}\Gamma(z)$. If $z \in \mathbf{Q}/\mathbf{Z}$, we let $\Gamma_\pi\langle z\rangle =: \Gamma_\pi(\langle z\rangle)$.

Proposition 2.3 *The cohomology class η forms a Φ_m-basis of $H_B(X_m)$, and the class $\xi =: \sum_\chi \xi_\chi$ forms a Φ_m-basis of $H_{DR}(X_m/\mathbf{Q})$. The classical periods of X_m are given by*

$$\int_{\eta'} \omega_\chi = \Gamma_\pi\langle 0\rangle \prod_i \Gamma_\pi\langle\chi_i\rangle^{-1}.$$

Proof: This formula is essentially classical; we follow the treatment of [11]. It can probably also be found, at least implicitly in [1]. The following formula is contained in the proof of Lemma 7.12 of [11]; it appears on the bottom of page 89. Recall that a_1 and a_2 are positive integers; we let $a_0 =: -a_1 - a_2$ and assume that no a_i is divisible by m. Then:

$$\Gamma(1-\frac{a_0}{m})\int_\sigma \omega_{\mathbf{a}} = a_2\epsilon_m^{a_1+a_2}m^{-2}\Gamma(\frac{a_1}{m})\Gamma(\frac{a_2}{m}) \tag{4}$$

Recall the formula:

$$\Gamma(s)\Gamma(1-s) = \pi(\sin\pi s)^{-1} = (2\pi i)e^{-\pi i s}(1-e^{-2\pi i s})^{-1} \tag{5}$$

Hence,

$$\Gamma(\frac{a_i}{m}) = 2\pi i\Gamma(1-\frac{a_i}{m})^{-1}\epsilon_m^{-a_i}(1-\epsilon_m^{-2a_i})^{-1} \tag{6}$$

Now suppose that χ is a Fermat character and $a_i = m\langle -\chi_i\rangle$ for $i = 1,2$. Then also $\langle\chi_i\rangle = 1 - a_i/m$ for $i = 1,2$, and hence

$$\Gamma(\frac{a_i}{m}) = \Gamma_\pi\langle\chi_i\rangle^{-1}\epsilon_m^{-a_i}(1-\epsilon_m^{-2a_i})^{-1} \tag{7}$$

Substituting this into (4), we find

$$\Gamma(1-\frac{a_0}{m})\int_\sigma \omega_{\mathbf{a}} = a_2 m^{-2}\Gamma_\pi\langle\chi_1\rangle^{-1}\Gamma_\pi\langle\chi_2\rangle^{-1}(1-\epsilon_m^{-2a_1})^{-1}(1-\epsilon_m^{-2a_2})^{-1} \tag{8}$$

Continuing to follow [11], we note that if $\lambda \in \Lambda_m$, $\lambda^*\omega_{\mathbf{a}} = \chi(\lambda)\omega_{\mathbf{a}}$. Hence the group algebra element δ_i operates by $\delta_i^*\omega_{\mathbf{a}} = \epsilon_m^{-2a_i}\omega_{\mathbf{a}}$. Let $\delta =: (1-\delta_1)(1-\delta_2)$; then we can write:

$$\begin{aligned}
\int_{\eta'}\omega_{\mathbf{a}} &= \int_{\delta\sigma}\omega_{\mathbf{a}} \\
&= \int_\sigma (1-\delta_1)^*(1-\delta_2)^*\omega_{\mathbf{a}} \\
&= \int_\sigma (1-\epsilon_m^{-2a_1})(1-\epsilon_m^{-2a_2})\omega_{\mathbf{a}} \\
&= a_2 m^{-2}\Gamma(1-\frac{a_0}{m})^{-1}\Gamma_\pi\langle\chi_1\rangle^{-1}\Gamma_\pi\langle\chi_2\rangle^{-1}
\end{aligned} \tag{9}$$

Note that

$$1-\frac{a_0}{m} = \langle\chi_0\rangle + 3 - \langle\chi\rangle$$

The Hodge level $\langle\chi\rangle - 1$ of H_χ is either zero or one, and using the fact that $\Gamma(s+1) = s\Gamma(s)$ one checks that in either case,

$$\Gamma(1-\frac{a_0}{m}) = 2\pi i\frac{-a_0}{m}\langle\chi_0\rangle^{2-\langle\chi\rangle}\Gamma_\pi\langle\chi_0\rangle \tag{10}$$

Substituting 10 into 9 gives

$$\int_{\eta'}\omega_{\mathbf{a}} = -a_2 m^{-1}a_0^{-1}\langle\chi_0\rangle^{\langle\chi\rangle-2}(2\pi i)^{-1}\prod_{i=0}^{2}\Gamma_\pi\langle\chi_i\rangle^{-1} \tag{11}$$

Recalling that $\Gamma_\pi\langle 0\rangle = (2\pi i)^{-1}$ and the definition (3) of ω_χ, we find that

$$\int_{\eta'} \omega_\chi = \Gamma_\pi\langle 0\rangle \prod_i \Gamma_\pi\langle \chi_i\rangle^{-1},$$

as claimed. ∎

For the crystalline analogue, we define, for $a \in \mathbf{Q}/\mathbf{Z}$ with denominator prime to p, $\Gamma_p\langle a\rangle =: \Gamma_p(\langle a\rangle)$, where Γ_p is Morita's p-adic Γ-function.

Proposition 2.4 *If $\psi \in W_{cris}$ has degree one and p does not divide m,*

$$\psi\xi_\chi = (-p)^{\langle\chi\rangle-1} \prod_i \Gamma_p\langle -p\chi_i\rangle^{-1}\xi_{p\chi}$$

Proof: This follows from Coleman's formulas in [5]; we simply need a dictionary comparing his notation to ours. His (r,s) is $(-\chi_1, -\chi_2)$, so that $r+s = \chi_0$, and his $\epsilon(r,s)$ is $2 - \langle\chi\rangle$. Furthermore, let us note that

$$d(y_1^{a_1}y_2^{a_2}) = [(a_1+a_2)y_1^{a_1}y_2^{a_2} + a_2 y_1^{a_1}y_2^{a_2-m}]dy_1/y_1,$$

and hence

$$a_2\omega_{(a_1,a_2-m)} = a_0\omega_{\mathbf{a}} + d(y_1^{a_1}y_2^{a_2})$$

Using also the fact that $dy_1/y_1 = -y_2^m(y_2/y_1)d(y_1/y_2)$, we find that ω_χ is cohomologous to

$$-m\langle\chi_0\rangle^{2-\langle\chi\rangle}\omega_{(a_1,a_2-m)} = m\langle\chi_0\rangle^{2-\langle\chi\rangle} y_1^{a_1}y_2^{a_2}y_2/y_1 d(y_1/y_2)$$

This is just $m\eta_{r,s}$ in Coleman's notation, and as $\epsilon(-r,-s) = 2 - \langle -\chi\rangle = \langle\chi\rangle - 1$, we find that his Corollary (1.9) of [5] is precisely our formula. ∎

Let $\xi = \sum_\chi \xi_\chi$, which forms a Φ_m-basis for $H_{DR}(X/\mathbf{Q})$ as a Φ_m-algebra. Recall that $CM(\Phi_m)$ is a subset of the set of functions $F_m \to \mathbf{Q}$ and that $\mathbf{T}_{\Phi_m}(\mathbf{C})$ can be viewed as the set of functions $F_m \to \mathbf{C}^*$ (and analogously for K_p). It is clear that the cocycle $\gamma_p(X,\xi,\eta)$ in $Z^1(W_p, \mathbf{T}_{\Phi_m}(K_p))$ is determined by its values on elements of degree one. Then in the language of the previous section, we can state:

Theorem 2.5 *The CM-type and periods of the marked Φ_m-motive (X_m,ξ,η) are given as follows:*

$$\begin{aligned}\tau(X_m)(\chi) &= \sum_i \langle\chi_i\rangle - 1\\ \gamma(X_m,\xi,\eta) &= \Gamma_\pi\langle 0\rangle \prod_i \Gamma_\pi\langle\chi_i\rangle^{-1}\end{aligned}$$

and, if $\deg\psi = 1$ and p does not divide m,

$$\gamma_p(X_m,\xi,\psi)(\chi) = (-p)^{\langle\psi^{-1}\chi\rangle-1}\prod_i \Gamma_p\langle -\chi_i\rangle^{-1} \quad ∎$$

We shall discuss the case in which p divides m in this next section.

The symmetric group S_3 also acts on X_m, and it is clear that $\sigma^* H_\chi = H_{\sigma\chi}$ for $\sigma \in S_3$ and χ a Fermat character. However, we shall need the following more delicate remark:

Lemma 2.6 *If $\sigma \in S_3$ and χ is a Fermat character, $\sigma^*\xi_\chi = sgn(\sigma)\xi_{\sigma\chi}$*

Proof: It suffices to check this for the two transpositions $\tau =: (1,2)$ and $\sigma =: (0,2)$, as these generate S_3. Let $z =: y_1/y_2$, and $a_i =: \langle -\chi_i \rangle$, so that $\omega_\chi = m\langle\chi_0\rangle^{2-\langle\chi\rangle} y_1^{a_1} y_2^{a_2} dz/z$. Since $\tau^* z = z^{-1}$, we have $\tau^*(dz/z) = -dz/z$ and

$$\tau^*\omega_\chi = -m\langle\chi_0\rangle^{2-\langle\chi\rangle} y_2^{a_1} y_1^{a_2} dz/z = -\omega_{\tau\chi}$$

We have $\sigma^* y_1 = y_1/y_2$ and $\sigma^* y_2 = y_2^{-1}$, so $\sigma^* z = y_1$ and $\sigma^*(dz/z) = dy_1/y_1 = -y_2^m dz/z$. Let $\chi' =: \sigma\chi$ and $a_i' =: \langle -\chi_i' \rangle$. Then

$$\sigma^*\omega_\chi = -m\langle\chi_0\rangle^{2-\langle\chi\rangle} y_1^{a_1} y_2^{m-a_1-a_2} dz/z \tag{12}$$

We consider separately the two possibilities $\langle\chi\rangle = 1, 2$, *i.e.* $a_1 + a_2 > m$ and $a_1 + a_2 < m$, respectively. In the second case,

$$m - a_1 - a_2 = m\langle -\chi_0 \rangle = m\langle -\chi_2' \rangle = a_2',$$

and we have

$$\begin{aligned} \sigma^*\omega_\chi &= -m y_1^{a_1} y_2^{m-a_1-a_2} dz/z \\ &= -m y_1^{a_1'} y_2^{a_2'} dz/z \\ &= -\omega_{\chi'} \end{aligned}$$

If $\langle\chi\rangle = 1$, we have

$$\begin{aligned} a_2' = m\langle -\chi_0 \rangle &= 2m - a_1 - a_2 \\ m\langle\chi_0\rangle &= a_1 + a_2 - m = m - a_2' \\ m\langle\chi_0'\rangle &= a_1' + a_2' - m \end{aligned}$$

Substituting these expressions into equation 12 and using also

$$d(y^{a_1'} y^{a_2'-m}) = (a_1' + a_2' - m) y_1^{a_1'} y_2^{a_2'} + (a_2' - m) y_1^{a_1'} y_2^{a_2'-m},$$

we find that

$$\begin{aligned} \sigma^*\omega_\chi &= (a_2' - m) y_1^{a_1'} y_2^{a_2'-m} \\ &= -(a_1' + a_2' - m) y_1^{a_1'} y_2^{a_2'} + d(y^{a_1'} y^{a_2'-m}) \\ &= -m\langle\chi_0'\rangle y_1^{a_1'} y_2^{a_2'} + d(y^{a_1'} y^{a_2'-m}). \end{aligned}$$

Thus, we see that $\sigma^*\xi_\chi = -\xi_{\sigma\chi}$ in this case also. ∎

Each element of S_3 induces an automorphism σ of Φ_m, and hence we can pull back any Φ_m-motive by σ, obtaining an action of S_3 on the group of bimarked Φ_m-motives.

Corollary 2.7 *If $\sigma \in S_3$, $\sigma^*\xi = sgn(\sigma)\xi$ and $\sigma^*\eta = sgn(\eta)\eta$. Consequently there is an isomorphism of bimarked Φ_m-motives $\sigma^*(X, \xi, \eta) \cong (X, \xi, \eta)$.*

Proof: It is clear from Lemma 2.6 that $\sigma^*(\xi) = sgn(\sigma)\xi$. Since $\int_{\sigma_*\eta'} \xi = \int_{\eta'} \sigma^*\xi = sgn(\sigma) \int_{\eta'} \xi$, we must also have $\sigma_*\eta' = sgn(\sigma)\eta'$ and hence $\sigma^*\eta = sgn(\sigma)\eta$. ∎

3 The Abelian Case

In this section we give the "explicit" formula for the periods described in section 1 when E is abelian and $\wp$ is unramified in E. It suffices to treat the case of a cyclotomic field $E_m =: \mathbf{Q}(\mu_m) \subset \overline{\mathbf{Q}}$. Actually we shall find it more convenient to work with the algebra $C_m =: \mathbf{Q}[T]/(T^m - 1) \cong \prod_{d|m} E_d$. If $m = en$, the map $T \mapsto T^e$ induces a map $C_m \to C_n$. This map can be identified with the obvious projection $\prod_{d|m} E_d \to \prod_{d|n} E_d$. For any $\mathbf{Q}$-algebra R, $Mor(C_m, R)$ is canonically isomorphic to the set $\mu_m(R)$ of mth roots of unity of R. In particular, if we fix an orientation on $\mathbf{C}$, there is a natural bijection:

$$m^{-1}\mathbf{Z}/\mathbf{Z} \to \mu_m : a \mapsto e^{2\pi i a}$$

and if K is a field containing $\overline{\mathbf{Q}}$, we have a natural isomorphism

$$\mathbf{T}_{C_m}(K) \cong K^{\mu_m} \cong K^{m^{-1}\mathbf{Z}/\mathbf{Z}}$$

¿From now on we make these identifications without further comment. Notice that U_m acts naturally on $\mu_m \cong m^{-1}\mathbf{Z}/\mathbf{Z}$. Furthermore, if n divides m, we have natural maps

$$\pi: U_m \to U_n \quad \text{and} \quad i: \mu_n \to \mu_m$$

and $i(\pi(u)a) = ui(a)$ if $u \in U_m$ and $a \in \mu_n$. Thus we can let $U =: \varprojlim U_m$ act on $\mathbf{Q}/\mathbf{Z} = \varinjlim m^{-1}\mathbf{Z}/\mathbf{Z} \cong \mu_\infty$, compatibly with our actions at finite levels. Furthermore, we let C stand for the inverse system $\{C_m : m \in \mathbf{Z}^+\}$, regarded as a pro-object, so that $Mor(C, R) =: \varinjlim Mor(C_n, R) = \mu_\infty(R)$. If K is a field containing $\overline{\mathbf{Q}}$, we have an identification

$$\mathbf{T}_C(K) \cong K^{*\mathbf{Q}/\mathbf{Z}} \cong K^{*\mu_\infty}$$

The character group of the torus $\mathbf{T}_C$ over $\overline{\mathbf{Q}}$ is the free abelian group on $\mathbf{Q}/\mathbf{Z}$, which we denote by $B_{\mathbf{Q}/\mathbf{Z}}$. Similarly, the character group of $\mathbf{T}_{C_m}$ is the free abelian group on $m^{-1}\mathbf{Z}/\mathbf{Z}$, which we denote by B_m

Our first goal is to factor the bimarked motive (X_m, ξ, η) in a way which reflects the factorization 2.5 of its periods into a product of values of the Γ-function. This step can be viewed as a geometric incarnation of Anderson's factorization in [1]. Strictly speaking, the factors M_m will not be motives, but "ulterior motives," *i.e.* elements of the abelian group $BMot(C_m) \otimes \mathbf{Q}$. If n divides m, the map $C_m \to C_n$ will induce an isomorphism $M_m \otimes_{C_m} C_n \cong M_n$. We then say, by abuse of language, that we have constructed an "ulterior motive M on the pro-algebra C." For example, the system $\{(X_m, \xi, \eta) : m \in \mathbf{Z}^+\}$ forms an ulterior motive on the pro-algebra $\Phi =: \text{"}\varprojlim\text{"}\Phi_m$. Because the inverse system $\{C_m : m \in \mathbf{Z}^+\}$ can be identified with the product $\prod_m E_m$, to give such a motive M on C is the same as giving a system of motives $A_m =: M \otimes_C E_m$ on each E_m (with no compatibilities).

For each $i \in \{0, 1, 2\}$, let $j_i: \mu_m \to \Lambda_m$ be the map inserting ζ into the ith place, and also use j_i for the corresponding map $C \to C_m \to \mathbf{Q}[\Lambda_m] \to \Phi_m$. Recall that $X_m(-1)$ is the Tate twist of the Fermat curve of degree m; it has as a bimarking the pair $(\xi, \eta/2\pi i)$.

Theorem 3.1 *There is a bimarked ulterior C-motive (M, ξ, η) of rank one with the following properties:*

1. *The bimarked CM-type $\tau_{BM}(X(-1),\xi,\eta/2\pi i)$ of $X(-1)$ is isomorphic to the product $\prod_i j_i^* \tau_{BM}(M,\xi,\eta)$ in $BCM(\Phi)$.*
2. *The CM-type $\tau_M \in \mathbf{Q}^{\mathbf{Q}/\mathbf{Z}}$ of M is the function $\langle\ \rangle$.*
3. *The complex period $\gamma \in \mathbf{T}_C(\mathbf{C}) \otimes \mathbf{Q}$ of (M,ξ,η) is the function $\Gamma_\pi\langle\ \rangle^{-1}$.*

Proof: Let E by the cyclotomic field $\mathbf{Q}(\zeta_m)$. If $a: C \to E$ is any homomorphism, the ulterior CM-type $\langle\ \rangle$ on C induces by pullback an ulterior CM-type $a^*\langle\ \rangle$ on E. As $\mathbf{T}_C(E)$ is the free abelian group on the set of all such a, we obtain in this way a homomorphism

$$t^E: \mathbf{T}_C(E) \to CM(E) \otimes \mathbf{Q} : \alpha_a \mapsto a^*\langle\ \rangle$$

We construct a homomorphism $\Gamma_\pi^E\langle\ \rangle: \mathbf{T}_C(E) \to \mathbf{T}_C(\mathbf{C}) \otimes \mathbf{Q}$ in the same way. Specifically, if we view $\mathbf{T}_C(E)$ as the free abelian group on $\mathbf{Q}/\mathbf{Z}$, and if $\alpha =: \sum_a \alpha_a$ is an element of $\mathbf{T}_C(E)$, then we have:

$$\begin{aligned} t^E\alpha(u) &= \sum_a \alpha_a\langle ua\rangle \\ \Gamma_\pi^E(\alpha)(u) &= \prod_a (\Gamma_\pi\langle ua\rangle)^{\alpha_a} \end{aligned}$$

If $\chi: \Phi \to E$ is a Fermat character, let

$$X_\chi(-1) =: \chi^*(X(-1),\xi,\eta/2\pi i) \in BMot(E),$$

and let $\alpha_\chi =: \sum_i \chi_i$. It follows from Proposition 2.7 that X_χ is independent of the ordering of (χ_0,χ_1,χ_2), *i.e.* depends only on α_χ. We shall prove Theorem 3.1 simultaneously with the following result.

Theorem 3.2 *Let E be the cyclotomic field $\cup_m \mathbf{Q}(\zeta_m)$. There is a homomorphism*

$$F: B_{\mathbf{Q}/\mathbf{Z}} \otimes \mathbf{Q} \to BMot(E) \otimes \mathbf{Q}$$

with the following properties:

1. *If $a \in \mathbf{Q}/\mathbf{Z} \cong Mor(C,E)$, $F(\alpha_a) = a^*M$.*
2. *The CM-type $\tau \in \mathbf{Q}^U$ of $F(\alpha)$ is the function $t^E(\alpha)$.*
3. *The complex period $\gamma \in \mathbf{T}_E(\mathbf{C}) \otimes \mathbf{Q}$ of $F(\alpha)$ is the function $\Gamma_\pi^E(\alpha)^{-1}$.*
4. *F is $Gal(E/\mathbf{Q})$ equivariant and maps $B_m \otimes \mathbf{Q}$ to $BMot(E_m) \otimes \mathbf{Q}$.*
5. *For any Fermat character χ,*

$$\tau_{BM}(X_\chi(-1)) = \tau_{BM}(F(\alpha_\chi))$$

Proofs: If we have defined a Galois equivariant homomorphism F satisfying 3.2.2–3.2.5, then in particular for each $m \in \mathbf{Z}^+$ we have an ulterior E_m-motive $F(\alpha_{1/m})$. As explained above, the set of all these defines an ulterior C-motive M such that $1/m^*M = F(\alpha_{1/m})$. Then for any $u \in U_m$, $u/m^*M = u^*(1/m^*M) = u^*F(\alpha_{1/m}) = F(\alpha_{u/m})$, by the Galois equivariance, and hence 3.2.1 holds. Then 3.1.2 and 3.1.3 follow from the corresponding statements 3.2.2 and 3.2.3; 3.1.1 follows immediately also, *e.g.* from Theorem 2.5.

To prove 3.2, it will suffice to define the map F on any basis for $B_{\mathbf{Q}/\mathbf{Z}} \otimes \mathbf{Q}$. The natural basis $\mathbf{Q}/\mathbf{Z}$ is not as convenient as the one provided by the following lemma.

Lemma 3.3 *For each $b \in \mathbf{Q}/\mathbf{Z}$, let $f_b =: 2\alpha_b + \alpha_{-2b}$. Then the set of all f_b, for all $b \in \mathbf{Q}/\mathbf{Z}$, is a basis for the vector space $B_{\mathbf{Q}/\mathbf{Z}} \otimes \mathbf{Q}$.*

Proof: It suffices to show that the set of all f_b's for $b \in m^{-1}\mathbf{Z}/\mathbf{Z}$ spans B_m. Let B'_m be the space thus spanned. It suffices to show that for each for each $a \in m^{-1}\mathbf{Z}/\mathbf{Z}$, α_a belongs to B'_m. It is clear that $-2\alpha_a \cong \alpha_{-2a} \bmod B'_m$, and by induction $(-2)^j\alpha_a \cong \alpha_{(-2)^j a}$ for all j. Since $m^{-1}\mathbf{Z}/\mathbf{Z}$ is a finite set, there exist two distinct integers j and k such that $(-2)^j a = (-2)^k a$, whence $(-2)^j\alpha_a \cong (-2)^k\alpha_a \bmod B'_m$ and $\alpha_a \in B'_m$. ∎

Returning to the proof of 3.2, we note first that $E_0 = E_1 = E_2 = \mathbf{Q}$. We define

$$F(f_0) =: (3\mathbf{Q}(-1), 1, (2\pi i)^{-3}) \quad \text{and} \quad F(f_{1/2}) =: (2\mathbf{Q}(-1), 2i, (2\pi i)^{-2}). \tag{13}$$

Because $f_0 = 3\alpha_0$, $F(\alpha_0)$ is $(\mathbf{Q}(-1), 1, (2\pi i)^{-1})$, and we can check 3.2.1 and 3.2.2 for α_0 instead of f_0.

$$\begin{aligned} t^{\mathbf{Q}}(\alpha_0) &= \langle 0\rangle \\ &= 1 \\ &= \tau(\mathbf{Q}(-1)) \\ \Gamma_\pi^{\mathbf{Q}}(\alpha_0)^{-1} &= (\Gamma_\pi\langle 0\rangle)^{-1} \\ &= 2\pi i \\ &= \gamma(E(-1), 1, (2\pi i)^{-1}) \end{aligned}$$

Checking $f_{1/2} = 2\alpha_{1/2} + \alpha_0$, we note that

$$\begin{aligned} t^{\mathbf{Q}}(f_{1/2}) &= 2\langle 1/2\rangle + 1 = 2 \\ &= 2\tau(\mathbf{Q}(-1)) \\ \Gamma_\pi^{\mathbf{Q}}(f_{1/2})^{-1} &= (2\pi i)^3\Gamma(1/2)^{-2}\Gamma(1)^{-1} \\ &= (2\pi i)^2(2i) \\ &= \gamma(2\mathbf{Q}(-1), 2i, (2\pi i)^{-2}) \end{aligned}$$

Now if $a \notin \{0, 1/2\}$ and we choose an ordering of the set of elements $\{a, a, -2a\}$ we find a Fermat character χ. Notice that the isomorphism class of the bimarked E-motive $X_\chi =: \chi^* X_m$ is in fact independent of the ordering, by Corollary 2.7. It is also independent of the choice of m, by 2.2 We define $F(f_a) =: X_\chi(-1)$. Then Theorem 2.5 shows that the CM-type and periods of $F(f_a)$ are as predicted by the statement of the theorem. The $Gal(E/\mathbf{Q})$-equivariance of F follows from Remark 1.6. Statement 3.2.5 is just a restatement of the first two, using the definition of BCM and the formulas 2.5. ∎

Remark 3.4 Although we deduced Theorem 3.1 from Theorem 3.2, let us note that it is also true that Theorem 3.2 is an easy consequence of Theorem 3.1. Indeed, we can use 3.2.1 above to define the homomorphism F. Then properties 3.2.2 and 3.2.3 follow from their counterparts 3.1.2 and 3.1.3, and 3.2.5 follows from theorem 2.5. The Galois equivariance follows from Remark 1.6.

Remark 3.5 The Hodge conjecture for all products of Fermat varieties would imply that the factorization 3.1.1 above takes place in $BMot(C_m) \otimes \mathbf{Q}$, not just in $BCM(C_m) \otimes \mathbf{Q}$. In fact, it seems likely that one could in fact verify the algebraicity of the specific Hodge cycles needed for this statement directly. This would free Coleman's calculation of the Fermat period matrix in the bad reduction case from its dependence on Theorem 1.8, if not entirely from the formalism developed here. Unfortunately we have not had time to complete this task, and we will simply refer to the paper [2] in which a similar calculation was carried out based on Shioda's "inductive structure" of Fermat varieties [18]. Let us also note that our construction of F can be refined in an obvious way to produce a homomorphism from the subgroup B_F of $B_{\mathbf{Q}/\mathbf{Z}}$ generated by the set of all α_χ for all Fermat characters χ to $BCM(E)$. One simply defines $F(\alpha_\chi)$ to be the bimarked E-motive of X_χ, and verifies that F prolongs to a homomorphism by inspection from Theorem 2.5. Conjecturely, one would like to say that F factors through $BMot(E)$.

We are now ready to look for our p-adic analogue of the Chowla-Selberg formula. Attached to the marked ulterior C-motive (M, ξ) is the cocycle

$$\gamma_\mathfrak{p}(M,\xi) \in Z^1(W_\mathfrak{p}, \mathbf{T}_C(K_\mathfrak{p})) \otimes \mathbf{Q} \cong Z^1(W_\mathfrak{p}, \mathbf{T}_C(K_\mathfrak{p})/\mu_\infty)$$

Our aim is to give an explicit formula for this cocycle in terms of the Morita p-adic gamma-function. We shall need the following proposition.

Proposition 3.6 *Suppose* $G \in Z^1(W_\mathfrak{p}, \mathbf{T}_C(K_\mathfrak{p})/\mu_\infty)$, *and let*

$$B \in Z^1(W_\mathfrak{p}, \mathbf{T}_{\Phi_m}(K_\mathfrak{p})/\mu_\infty) =: \prod_i j_i^* G$$

Suppose that for every $\psi \in W_\mathfrak{p}$,

$$\gamma_\mathfrak{p}(X(-1), \xi, \psi)(a, a, -2a) = B(\psi)(a, a, -2a), \quad \textit{if} \quad a \in \mathbf{Q}/\mathbf{Z}\backslash\{0, 1/2\} \quad \textit{and}$$

$$G(\psi)(0) = p^{\deg \psi} \quad , \quad G(\psi)(1/2) = p^{\deg \psi/2}$$

Then $\gamma_\mathfrak{p}(M, \xi) = G$ *and* $\gamma_\mathfrak{p}(X_m(-1), \xi) = B$.

Proof: Fix $\psi \in W_\mathfrak{p}$ of degree one, and define homomorphisms

$$\gamma_\psi \, , \, G_\psi : B_{\mathbf{Q}/\mathbf{Z}} \to K_\mathfrak{p}/\mu_\infty \quad \text{by :}$$

$$\gamma_\psi(\alpha) =: \prod_a [\gamma_\mathfrak{p}(M, \xi, \psi)(a)]^{\alpha_a} \quad \text{and} \quad G_\psi(\alpha) =: \prod_a [G(\psi)a]^{\alpha_a}$$

For any Fermat character χ, let

$$\beta_\psi(\chi) =: \gamma_\mathfrak{p}(X(-1), \xi, \psi)(\chi) \quad \text{and} \quad B_\psi(\chi) =: B(\psi)(\chi) = \prod_i G_\psi(\chi_i)$$

We have to prove that $\beta_\psi = B_\psi$ and that $\gamma_\psi = G_\psi$. By the constructions in Theorems 3.2 and 3.1, we also have the bimarked motive E_m-motive $F(\alpha_\chi) = \prod_i \chi_i^* M$.

Suppose first that $a \in \mathbf{Q}/\mathbf{Z}\backslash\{0, 1/2\}$. Then $\chi =: (a, a, -2a)$ is a Fermat character, and by hypothesis, $B_\psi(\chi) = \beta_\psi(\chi)$. On the other hand, F was constructed so that $F(f_a) = \chi^*(X_m(-1), \xi)$. It follows that

$$\begin{aligned} \gamma_\psi(f_a) &= \gamma_p(F(f_a), \psi)(1) \\ &= \gamma_p(\chi^*(X(-1), \xi, \psi)(1) \\ &= \gamma_p(X(-1), \psi)(\chi) \\ &= \beta_\psi(\chi) \\ &= B(\psi)(\chi) \\ &= G_\psi(f_a). \end{aligned}$$

Furthermore,

$$G_\psi(f_0) = G_\psi(\alpha_0)^3 = p^3 \quad \text{and} \quad \gamma_\psi(f_0) = [\gamma_p(M, \psi)(0)]^3 = \gamma_p(\mathbf{Q}(-1)^3, 1) = p^3$$

Similarly,

$$G_\psi(f_{1/2}) = p^2 \quad \text{and} \quad \gamma_\psi(f_{1/2}) = \gamma_p(\mathbf{Q}(-1)^2, 2i) = -p^2\psi(i)i$$

(Recall that we are working modulo roots of unity, so $-psi(i)i \equiv 1$.) As the set of all f_a's forms a basis for $B_{\mathbf{Q}/\mathbf{Z}}$, it follows that $\gamma_\psi = G_\psi$.

Now 3.1.1 tells us that the marked CM-type of $(X_m(-1), \xi)$ is $\prod_i \tau_{DM} j_i^*(M)$, and hence by Theorem 1.8 we can conclude that $\beta_\psi = \prod_i j_i^* \gamma_\psi = B_\psi$. ∎

Remark 3.7 Coleman has asked me to point out that Proposition 3.6 can be refined in the following way. Let $S \subset \mathbf{Q}/\mathbf{Z}$ be a subset stable under $Gal(\overline{\mathbf{Q}}/\mathbf{Q})$, corresponding to quotient $C \to C_S$ of the pro-algebra C. If S is stable by multiplication by 2, then Lemma 3.3 still applies, and we find that the $B_S \otimes \mathbf{Q}$ has as a basis the set of elements $f_a : a \in S$. Let Φ_S be the quotient of Φ whose characters consist of the Fermat characters χ such that each χ_i belongs to S. Finally, let M_S and X_S be the bimarked ulterior motives on C_S and Φ_S respectively obtained from M and X by base change. Then Proposition 3.6 applies to M_S and X_S, with the obvious modification.

Proposition 3.6 shows that a determination of the action of W_p on the part of the cohomology of X_m corresponding to the Fermat characters of the form $(a, a, -2a)$ is enough to determine the entire action. Remarkably, this is precisely the portion of the action that Coleman is able to compute in the case in which p divides m, and so he is able (with the aid of an additional congruence to pin down the roots of unity) to obtain an exact formula for the entire matrix. We refer to his paper [5] for the details and formulas. In anticipation of the aesthetics of our eventual Theorem 3.13, we make the following definition (in the general case).

Definition 3.8 *If $\psi \in W_p$ and $a \in \mathbf{Q}/\mathbf{Z}$, $\Gamma_p(\psi)\langle a\rangle =: \gamma_p(M, \xi, \psi)(a) \in K_p/\mu_\infty$.*

We shall justify this definition by giving an explicit formula for Γ_p in terms of the Morita p-adic Γ-function in the unramified case. Thus, we let C' denote the limit of the C_m's over the multiplicative set of m's which are prime to p, and let M' be $M \otimes_C C'$ Then $\mathbf{T}_{C'}(K_p) \cong K_p^{*\mathbf{Q}/\mathbf{Z}'}$, where $\mathbf{Q}/\mathbf{Z}'$ is the subgroup of $\mathbf{Q}/\mathbf{Z}$ consisting of the prime-to-p torsion. Note that W_p operates on $\mathbf{Q}/\mathbf{Z}'$ by the rule $\psi(a) = p^{\deg\psi}a$. Since we are working modulo torsion, we can neglect roots of unity—in particular, signs. For any rational number r, we have a well-defined element p^r of K_p/μ_∞.

Theorem 3.9 *With the above notation,* $\Gamma_p \in Z^1(W_p, \mathbf{T}_{C'}(K_p)/\mu_\infty)$ *is the unique cocycle* $\tilde{\Gamma}_p$ *such that*

$$\tilde{\Gamma}_p(\psi)(a) = p^{\langle\psi^{-1}a\rangle}\Gamma_p\langle a\rangle \quad \text{whenever } \deg\psi = 1$$

Explicitly, if $d =: \deg\psi$,

$$\Gamma_p(\psi)\langle a\rangle = \prod_{i=0}^{d-1} p^{\langle\psi^{-i-1}a\rangle}\Gamma_p\langle\psi^i a\rangle$$

Proof: Since the elements $p^r \in K_p/\mu_\infty$ and the values of the p-adic gamma-function are fixed by W_p, it is clear that the formulas above do indeed define a unique cocycle $\tilde{\Gamma}_p$ with the prescribed values on elements of degree one. To prove that the formula is correct, it suffices to apply Proposition 3.6 and Remark 3.7. Indeed, if χ is *any* Fermat character, Theorem 2.5 tells us that for all ψ of degree one and all $a \in \mathbf{Q}/\mathbf{Z}'$

$$\gamma_p(X(-1), \xi, \psi)(\chi) = p^{\langle\psi^{-1}\chi\rangle}\prod_i \Gamma_p\langle-\chi_i\rangle^{-1}$$

Since $\Gamma_p(1-x)\Gamma_p(x) = \pm 1$ and we are working modulo roots of unity, we find

$$\begin{aligned}\gamma_p(X_m(-1), \xi, \psi)(\chi) &= p^{\langle\psi^{-1}\chi\rangle}\prod_i \Gamma_p\langle\chi_i\rangle \\ &= \prod_i p^{\langle\psi^{-1}\chi_i\rangle}\Gamma_p\langle\chi_i\rangle \\ &= \prod_i \tilde{\Gamma}_p(\psi)\langle\chi_i\rangle\end{aligned}$$

Finally, we note that

$$\begin{aligned}\tilde{\Gamma}_p(\psi)(0) &= p^{\langle 0\rangle}\Gamma_p\langle 0\rangle \\ &= p\end{aligned}$$

and

$$\begin{aligned}\tilde{\Gamma}_p(\psi)(1/2) &= p^{\langle 1/2\rangle}\Gamma_p\langle 1/2\rangle \\ &= p^{1/2}\end{aligned}$$

Thus Proposition 3.6 applies, with $G =: \tilde{\Gamma}_p$. ∎

Knowledge of the periods of the ulterior motive C determines the periods of all ulterior E-motives (with E abelian), because of the following fact.

Proposition 3.10 *If* E *is a cyclotomic field, the map*

$$t^E : \mathbf{T}_C(E) \otimes \mathbf{Q} \to CM(E) \otimes \mathbf{Q}$$

described above is surjective.

Proof: This proposition is well-known, but we give the proof in a form which is computationally useful, for example, for our proof of the Chowla-Selberg formulas for elliptic curves. The argument is essentially taken from [16].

We use the action of the group U_m on the set $m^{-1}\mathbf{Z}/\mathbf{Z}$. For each divisor d of m, the stabilizer subgroup of the image of $1/d$ in $m^{-1}\mathbf{Z}/\mathbf{Z}$ is the subgroup $I_d \subseteq U_m$ of elements congruent to 1 mod d, and the orbit is naturally isomorphic to $U_d \cong U_m/I_d$. The orbits are thus in bijection with the divisors of m, and hence we obtain a canonical direct sum decomposition

$$B_m \cong \bigoplus_{d:d|m} A_d^*,$$

where A_d^* is the free abelian group generated by U_d. If $\alpha \in B_m$, we write $\alpha = \sum_d \alpha^d$ under this decomposition. Thus, $\alpha_v^d = \alpha_{v/d}$ for $v \in U_d$.

To show that our map t^E is surjective, we may first complexify it to obtain a map $B_m \otimes \mathbf{C} \to CM(E_m) \otimes \mathbf{C}$, also denoted by t. Regard $CM(E_m) \otimes \mathbf{C}$ as a linear subspace of the space of all functions from the abelian group $U_m \to \mathbf{C}$. This space has a natural inner product structure, corresponding to the Haar measure on the group U_m in which U_m has measure 1, and the set of complex characters of U_m forms an orthonormal basis. Observe first that $CM(E_m)\otimes\mathbf{C}$ is precisely the span of the trivial character and the set of odd characters. Indeed, it is clear that each of these characters belongs to $CM(E_m)\otimes\mathbf{C}$. On the other hand, if χ is an even character and $f \in CM(E_m) \otimes \mathbf{C}$ has weight w, then one finds easily that that $< f, \chi >$ vanishes if χ is nontrivial and is $w/2$ if χ is trivial.

Thus, to show that t is surjective, it suffices to show that its image contains the trivial character and every odd character. As we have seen, t_{α_0} is twice the trivial character. Suppose that ϵ is nontrivial and odd, and suppose that e is the smallest divisor of m such that ϵ factors through $U_m \to U_e$; let ϵ_e be the corresponding character of U_e.

Claim 3.11

$$t(\alpha_\epsilon) = -L(0,\epsilon_e)\epsilon, \quad \text{where} \quad \alpha_\epsilon =: \sum_{u\in U_e} \epsilon_e(u)\alpha_{u/e} \in B_m \otimes \mathbf{C}$$

Proof: Suppose α is any member of $B_m \otimes \mathbf{C}$ and χ is a character of U_m. We recall from [16] that

$$< t(\alpha), \chi >= -\sum_d L(0,\chi_d) < \alpha^d, \chi_d >,$$

where the sum is taken over all the divisors d of m such that χ is pulled back from a character χ_d of U_d. But α_ϵ lives on the orbit of $1/e$, and so $\alpha_\epsilon^d = 0$ if $d \neq e$, while $\alpha_\epsilon^e = \epsilon_e$. Thus, $< t_{\bar\epsilon}, \chi >$ is zero unless $\chi = \epsilon$, and $< t_{\bar\epsilon}, \epsilon >= -L(0,\epsilon_e)$. As the set of all such χ's form an orthonormal basis, the claim is proved.

Since ϵ_e is an odd primitive character, the factor $L(0,\epsilon_e)$ does not vanish, and thus 3.11 shows that ϵ lies in the image of t. ∎

Remark 3.12 An explicit set of generators for the kernel of t^E was given in [16].

Theorem 3.13 *Let E be the cyclotomic field $\mathbf{Q}(\zeta_m)$. Suppose that $X \in Mot(E) \otimes \mathbf{Q}$ and $\alpha \in B_m \otimes \mathbf{Q}$ are such that $\tau(X) = t^E(\alpha)$. Then*

1. $\gamma(X)$ *is the class of* $\Gamma_\pi^E(\alpha)^{-1}$ *in* $\mathbf{T}_E(\mathbf{C})/\mathbf{T}_E(\overline{\mathbf{Q}})$
2. $\gamma_p(X)$ *is the class of* $\Gamma_p^E(\alpha)$ *in* $H^1(W_p, \mathbf{T}_E(K_p;\overline{\mathbf{Q}}))$

Proof: Consider the bimarked ulterior E-motive $F(\alpha)$ constructed in 3.2, and let X' be $F(\alpha)$ stripped of its markings. It is clear that $\tau(X) = \tau(X')$, and hence Theorem 1.7 implies that $\gamma(X) = \gamma(X')$ and that $\gamma_p(X) = \gamma_p(X')$. Thus the formulas 1 and 2 follow from 3.2.3 and 3.8, respectively. ∎

Here is a slight refinement:

Corollary 3.14 *Suppose* X, E, *and* α *are as above. Then there exist bases* ξ *for* $H_{DR}(X/\overline{\mathbf{Q}})$ *and* η *for* $H_B(X)$ *such that*

1. $\gamma(X,\xi,\eta) = \Gamma_\pi^E(\alpha)^{-1}$ *in* $\mathbf{T}_E(\mathbf{C})\otimes\mathbf{Q}$
2. $\gamma_p(X,\xi) = \Gamma_p^E(\alpha)$ *in* $Z^1(W_p, \mathbf{T}_E(K_p))\otimes\mathbf{Q}$

Proof: The first statement of the previous theorem tells us precisely that there exist ξ and η such that 3.14.1 holds. It follows that $\tau_{BM}(F(\alpha)) = \tau_{BM}(X,\xi,\eta)$ in $BM(E)\otimes\mathbf{Q}$, and hence Theorem 1.8 implies that $\gamma_p(F(\alpha)) = \gamma_p(X,\xi)$. ∎

Let us make this explicit for elliptic curves. Let E be the quadratic imaginary field with discriminant $-m$ and let $X/\overline{\mathbf{Q}}$ be an elliptic curve with complex multiplication by E. Let τ be the CM-type of X, which in this case amounts to just an embedding of E in $\mathbf{C}$. If we use s to obtain an identification ι of the set of embeddings of E in $\mathbf{C}$ with $Gal(E/\mathbf{Q}) \cong \{\pm 1\}$, then $\tau = (\iota+1)/2$. The image of E under s is contained in the cyclotomic field E_m, and the corresponding mapping on Galois groups $U_m \to \mu_2$ can be regarded as a primitive character ϵ of U_m. The pullback of X to E_m then has as CM-type $(\epsilon+1)/2$. According to 3.11 we can write $-L(0,\epsilon)\epsilon = t_{\alpha_\epsilon}$, where $\alpha_\epsilon =: \sum_u \epsilon(u)\alpha_{u/m}$. Write L for the integer $L(0,\epsilon)$, so that $\tau(X\otimes_E E_m) = t^E(1/2 - \alpha_\epsilon/L)$.

Theorem 3.15 *With the above notation, there exist bases*

$$\xi_s \in F^1 H^1_{DR}(X/\overline{\mathbf{Q}}), \quad \xi_{\bar{s}} \in H^1_{DR}(X/\overline{\mathbf{Q}})_{\bar{s}}$$

and a basis η' *for the singular homology of* X *such that*

$$\begin{aligned}
\int_{\eta'} \xi_s &= (2\pi i)^{1/2} \prod_{u\in U_m} \Gamma_\pi\langle u/m\rangle^{\epsilon(u)/2L} && \text{in } \mathbf{C}^*/\mu_\infty \\
\int_{\eta'} \xi_{\bar{s}} &= (2\pi i)^{1/2} \prod_{u\in U_m} \Gamma_\pi\langle -u/m\rangle^{\epsilon(u)/2L} && \text{in } \mathbf{C}^*/\mu_\infty \\
\psi\xi_s &= p^{\deg\psi/2} \prod_u \Gamma_p(\psi)\langle \psi u/m\rangle^{-\epsilon(u)/2L} \xi_{\psi s} && \text{in } K_p/\mu_\infty \\
\psi\xi_{\bar{s}} &= p^{\deg\psi/2} \prod_u \Gamma_p(\psi)\langle -\psi u/m\rangle^{-\epsilon(u)/2L} \xi_{\psi\bar{s}} && \text{in } K_p/\mu_\infty
\end{aligned}$$

Proof: If we view s as a map $E \to E_m$, $s * X$ is an E_m motive, and its CM-type is $t^E(\alpha_0 - \alpha_\epsilon/L)$. Then our formulas follow immediately from Theorem 3.13 and the definitions. ∎

4 Absolute Hodge Cycles

In this section we explain the proof of Theorems 1.7 and 1.8 from the point of view of the theory of absolute Hodge cycles, following the theme of [11].

Theorem 4.1 *Let $X =: (X_1, X_2 \dots X_n)$ be a family of abelian varieties and projective spaces over $K_\wp$ with good reduction and T a tensor construction as explained in [17]. Suppose that $\xi_{DR} \in T_{DR}(X)$ is a Hodge class,* i.e. *$\xi_{DR} \in F^0 T_{DR}(X)$ and $\sigma_B(\xi_{DR}) \in T_B(X)$. Then ξ_{DR} is fixed by $W_\wp(K_\wp)$.*

Proof of 1.7 and 1.8: If (X, ξ) is a marked E_m-motive of rank one, its cohomology groups can be expressed as a tensor construction involving abelian varieties and projective spaces having good reduction everywhere. If $\tau(X) = 0$, $H_{DR}(X)$ is purely of type $(0,0)$ and so $H_{DR}(X) \otimes \mathbf{C}$ is trivially generated by Hodge classes. In particular, if $\tau_{DM}(X, \xi) = 0$ in $\mathbf{T}_E(\mathbf{C})/\mathbf{T}_E(\mathbf{Q})$, $\sigma_B(\xi) \in H_B(X)$ is a Hodge class, so 4.1 implies that $W_\wp$ fixes ξ and hence $\gamma_\wp(X, \xi) = 1$. This proves 1.8. To prove 1.7, we note that Deligne's theory of absolute Hodge cycles [11] tells us that the Hodge classes of $H_{DR}(X) \otimes \mathbf{C}$ lie in fact in $H_{DR}(X/\overline{\mathbf{Q}})$. Hence we may multiply ξ by an element of $\mathbf{T}_E(\overline{\mathbf{Q}})$ then and apply the previous argument. ∎

Theorem 4.1 follows from a stronger result, conjectured by Deligne and proved by Blasius. In order to state it, we have to use the additional structure on motives provided by the Fontaine-Messing-Faltings machinery comparing De Rham and étale cohomologies [14,12]. Recall that there is a field $B_{DR} \supseteq K_\wp$ endowed with a canonical filtration F and an action of the Galois group $G := Gal(K_\wp/\mathbf{Q}_p)$. Furthermore [12], for each smooth projective scheme $X/K_\wp$, there is a canonical isomorphism:

$$\sigma_{\acute{e}t}: H_{DR}(X/\overline{\mathbf{Q}}) \otimes B_{DR} \to H_{\acute{e}t}(X/K_\wp) \otimes B_{DR}$$

If $X/K_\wp$ is obtained by base extension from $X/\overline{\mathbf{Q}}$, we also have natural isomorphisms (which we regard as identifications):

$$\begin{aligned} H_{\acute{e}t}(X/\overline{\mathbf{Q}}, \mathbf{Q}_p) &\cong H_{\acute{e}t}(X/K_\wp, \mathbf{Q}_p) \\ H_{\acute{e}t}(X/\overline{\mathbf{Q}}, \mathbf{Q}_p) &\cong H_{\acute{e}t}(X/\mathbf{C}, \mathbf{Q}_p) \\ H_{\acute{e}t}(X/\mathbf{C}, \mathbf{Q}_p) &\cong H_B(X) \otimes \mathbf{Q}_p \end{aligned}$$

Theorem 4.2 (Blasius) *In the notation of Theorem 4.1, suppose that $\xi \in T_{DR}(X)$ is a Hodge class. Then $\sigma_{\acute{e}t}(\xi) = \sigma_B(\xi)$.*

Proof: For the reader's convenience we include a simplified and (somewhat generalized) version of Blasius's proof. Following Deligne's strategy in [11], we have to establish principles A and B in the above context. Principle A is quite trivial: Suppose that $\eta_i \in T_{i,B}$ for $i = 1 \dots n-1$, and let $G \subseteq Gl(H_B(X))$ be the subgroup consisting of those elements g such that $g\eta_i = \eta_i$ for all i; regarded as an algebraic group over $\mathbf{Q}$. Suppose that each η_i is a Hodge class and let ξ_i denote the corresponding element in $T_{i,DR}(X/\mathbf{C})$; recall that in fact ξ_i necessarily lies in $T_{i,DR}(X/\overline{\mathbf{Q}})$ by the main result of [11]. Now let (ξ_n, η_n) be the DeRham and Betti components of another Hodge class, and let $G' \subseteq G$ be group fixing $(\eta_1, \dots, \eta_{n-1}, \eta_n)$. Suppose that $\sigma_{\acute{e}t}(\xi_i) = \eta_i$ for all $i < n$ and that $G' = G$. Then I claim that $\sigma_{\acute{e}t}(\xi_n) = \eta_n$ also. To see this, we adapt the argument of [11] and [17].

For any $\overline{\mathbf{Q}}$-algebra A, let $P(A)$ denote the set of isomorphisms $\phi: H_{DR}(X/\overline{\mathbf{Q}}) \otimes_{\overline{\mathbf{Q}}} A \to H_B(X) \otimes_{\mathbf{Q}} A$ such that $\phi(\xi) = \eta$. It is clear that P is representable by a scheme of finite type over $\overline{\mathbf{Q}}$, and that $P(A)$ is a $G(A)$-pseudotorsor, *i.e.* is either empty or a $G(A)$-torsor. The similarly defined P' is representable by a closed subscheme of P, and is a G' pseudotorsor. By hypothesis, $G' = G$, and it follows that $P'(A) = P(A)$ provided that $P'(A)$ is not empty. Since the classical period isomorphism σ_B belongs to $P'(\mathbf{C})$, we see that $P(\mathbf{C})$ is not empty, and by Hilbert's Nullstellensatz $P(\overline{\mathbf{Q}})$ is also not empty, so in fact $P' = P$. By hypothesis, $\sigma_{ét}$ belongs to $P(B_{DR})$, and hence also to $P'(B_{DR})$; this says exactly that $\sigma_{ét}(\xi_n) = \eta_n$.

We next must prove Principle B. The following is a generalization and simplification of Blasius's proof.

Proposition 4.3 *Let $f: X \to S$ be a smooth proper morphism of $\overline{\mathbf{Q}}$-schemes of finite type, with S connected, and let η be a global section of $F^0Tf_*^{an}\mathbf{Q}$, where T a tensor construction as above. Suppose that for some $s_0 \in S(\overline{\mathbf{Q}})$, $\sigma_{ét}(\eta(s_0)_{DR}) = \eta(s_0)_{ét}$. Then the same is true at every point.*

Proof: Let $\overline{X}$ be a smooth compactification of X; one knows by the theory of weights [9] and the degeneration of the Leray spectral sequence [8] that the image of

$$i_s^*: T(\overline{X}) \to T(X(s))$$

can be identified with $H^0(S, Tf_*^{an}\mathbf{Q})$. In particular, the kernel $K(s)$ of i_s^* is independent of s. Moreover, as i_s^* is strictly compatible with the Hodge filtrations, one can choose an element $\overline{\eta}$ of $F^0T(\overline{X})$ which lifts η. It suffices to prove that $\sigma_{ét}(\eta_{DR}) - \eta_{ét} \in K(s)$ for every s. As this is true for s_0 by hypothesis, the proposition is proved. ∎

The relationship between Theorems 4.2 and 4.1 uses also the existence of the filtered subring B_{cris} of B_{DR} with its Frobenius endomorphism F. We obtain a semilinear action ρ_{cris} of $W_\wp$ on $B_{cris} \otimes K_\wp$ in the obvious way, and there is a natural injection $B_{cris} \otimes K_\wp \to B_{DR}$. The following is an expression of Fontaine's conjecture on potentially crystalline representations [13], proved by Faltings in [12]. Messing has pointed out that for abelian varieties, (the only case we need), this result can in fact be deduced from results in the literature.

Theorem 4.4 (Faltings) *Suppose $X/K_\wp$ is a smooth projective variety with good reduction. Then there exists a natural isomorphism*

$$H_{DR}(X/K_\wp) \otimes B_{cris} \otimes K_\wp \cong H_{ét}(X/K_\wp, \mathbf{Q}_p) \otimes B_{cris} \otimes K_\wp$$

This isomorphism is compatible with the isomorphism $\sigma_{ét}$ above, and under it the action of $W_\wp$ on De Rham cohomology corresponds to $id_{H_{ét}} \otimes \rho_{cris}$. ∎

It is clear that this result and Theorem 4.2 together imply Theorem 4.1.

References

[1] Greg Anderson. Cyclotomy and a covering of the Taniyama group. *Compositio Mathematicae*, 57(153–217), 1985.

[2] Greg Anderson. Torsion points on Fermat varieties, roots of circular units, and relative singular homology. *Duke Mathematics Journal*, 54:501–561, 1987.

[3] Pierre Berthelot and Arthur Ogus. F-isocrystals and de Rham cohomology I. *Inventiones Mathematicae*, 72:159–199, 1983.

[4] Don Blasius. P-adic Hodge cycle on abelian varieties. in preparation.

[5] Robert Coleman. On the Frobenius matrices of Fermat curves. This volume.

[6] Robert Coleman. The Gross-Koblitz formula. *Advanced Studies in Pure Mathematics*, 12:21–52, 1987.

[7] Ehud de Shalit. On monomial relations between p-adic periods. *Journal für die Reine und Angewandte Mathematik (Crelle)*, 374:193–207, 1987.

[8] Pierre Deligne. Théorème de Lefschetz et critères de dégénéscence de suites spectrales. *Publications Mathématiques de l'I.H.E.S.*, 35:107–126, 1968.

[9] Pierre Deligne. Théorie de Hodge II. *Publications Mathématiques de l'I.H.E.S.*, 40:5–57, 1972.

[10] Pierre Deligne. Valeurs de fonctions L et périodes d'intégrales. *Proceedings of Symposia in Pure Mathematics*, 33(part 2):313–346, 1979.

[11] Pierre Deligne. Hodge cycles on Abelian varieties. In *Hodge Cycles, Motives, and Shimura Varieties*, volume 900 of *Lecture Notes in Mathematics*. Springer Verlag, 1982.

[12] Gerd Faltings. Crystalline cohomology and p-adic galois representations. To appear in the *American Journal of Mathematics*.

[13] Jean-Marc Fontaine. Sur certains types de représentations p-adiques du groupe de Galois d'un corps local, construction d'un anneau de Barsotti-Tate. *Annals of Mathematics*, 115:529–577, 1982.

[14] Jean-Marc Fontaine and William Messing. P-adic periods and p-adic étale cohomology. In *Current Trends in Arithmetical Algebraic Geometry*, volume 67 of *Contemporary Mathematics*. American Mathematical Society, 1985.

[15] Benedict Gross. On the periods of Abelian integrals and a formula of Chowla and Selberg. *Inventiones Mathematicae*, 45:193–211, 1978.

[16] Neal Koblitz and Arthur Ogus. Algebraicity of some products of values of the γ-function. *Proceedings of Symposia in Pure Mathematics*, 33:343–346, 1979. (Appendix to "Valeurs de Fonctions L et Périodes d'Intégrales" by P. Deligne).

[17] Arthur Ogus. Hodge cycles and crystalline cohomology. In *Hodge Cycles, Motives, and Shimura Varieties*, volume 900 of *Lecture Notes in Mathematics*. Springer Verlag, 1982.

[18] Tetsuji Shioda and Toshiyuki Katsura. On Fermat varieties. *Tohoku Mathematical Journal*, 31(1):97–115, 1979.

THE COMPLEMENTATION PROPERTY OF ℓ^∞ IN p-ADIC BANACH SPACES

Wim H. Schikhof
Department of Mathematics, K.U. Nijmegen
Toernooiveld, 6525 ED Nijmegen, The Netherlands

ABSTRACT. Let K be a non-archimedean valued field whose valuation is complete and non-trivial. It is shown that ℓ^∞ is complemented in any polar K-Banach space (Theorem 1.2) which opens the way for various descriptions of complemented subspaces of ℓ^∞ (Theorem 2.3).

INTRODUCTION. (For unexplained terms see below.) Let K be as above. If K is spherically complete and E is a K-Banach space then, like in the 'classical' theory, each closed subspace is weakly closed, each subspace D of E has the weak extension property (WEP) i.e. each $f \in D'$ can be extended to an $\widetilde{f} \in E'$. It is also well known that these statements are false if K is not spherically complete. In fact, the following questions are posed in [1].

Q.1. Does every weakly closed subspace have the WEP ?

Q.2. Is every closed subspace having the WEP weakly closed?

It was a close look at these questions (for the (negative) answers, see the Remark following Proposition 1.5 and Example 3.3) that revealed the above-mentioned complementation property of ℓ^∞. Again, for spherically complete K this is nothing new as it follows directly from the spherical completeness of ℓ^∞. However, for nonspherically complete K the results came as a surprise.

PRELIMINARIES. Let K be as above. From now on until §5 we suppose that K is NOT spherically complete. We use the terminology of [2].

P.1. For an absolutely convex subset A of a K-vector space we set $A^e := \cap\{\lambda A : \lambda \in K, |\lambda| > 1\}$. A is *edged* if $A = A^e$.

P.2. Let E, F be K-Banach spaces. $\mathcal{L}(E,F)$ is the Banach space of all continuous linear maps $E \to F$ with the operator norm, $E' := \mathcal{L}(E,K)$. E and F are *isomorphic* if there exists a surjective linear isometry $E \to F$. We write $E \sim F$ if E, F are linearly homeomorphic. We shall say that E and F are *almost isomorphic* if for each $\varepsilon > 0$ there exists a linear homeomorphism $T : E \to F$ with $\|T\| \leq 1 + \varepsilon$, $\|T^{-1}\| \leq 1 + \varepsilon$.

$T \in \mathcal{L}(E,F)$ is a *quotient map* if T maps $\{x \in E : \|x\| < 1\}$ onto $\{x \in F : \|x\| < 1\}$. The closed unit ball $\{x \in E : \|x\| \le 1\}$ is denoted B_E. The norm closure of a set $X \subset E$ is $\overline{X}$. Its weak closure, i.e. the closure with respect to the weak topology (w-topology) $\sigma(E,E')$, is $\overline{X}^w$. Similarly, if $Y \subset E'$ then $\overline{Y}^{w'}$ is the w'-closure of Y i.e. the closure with respect to the w'-topology $\sigma(E',E)$.

For $X \subset E$ we set $X^0 := \{f \in E' : |f(x)| \le 1 \text{ for all } x \in X\}$
For $Z \subset E'$ we set $X_0 := \{x \in E : |f(x)| \le 1 \text{ for all } f \in Z\}$

X is a *polar set* if $(X^0)_0 = X$.
Let $A \in \mathcal{L}(E,F)$. If $\overline{AB_E}$ is a neighbourhood of 0 in F then $AB_E = \overline{AB_E}$. (The 'classical' proof of (one half) of Banach's Open Mapping Theorem yields that AB_E is a zero neighbourhood. By absolute convexity AB_E is open, hence closed.)

P.3. Let E, F be K-Banach spaces. The natural map $j_E : E \to E''$ satisfies $\|j_E\| \le 1$. We call E a *polar Banach space* (and the norm a *polar norm*) if j_E is an isometry into E''. Following [2], E is *topologically pseudoreflexive* if j_E is a linear homeomorphism into E''. E is *reflexive* if j_E is an isomorphism. Recall ([2], 4.17) that c_0 and ℓ^∞ are reflexive and duals of one another via the pairing $\langle x, y\rangle = \sum x_n y_n$ ($x \in c_0, y \in \ell^\infty$). In the same spirit, $c_0(I)$ and $\ell^\infty(I)$ are reflexive if the cardinality of I is nonmeasurable ([2], 4.21).

For $A \in \mathcal{L}(E,F)$ we define its *adjoint* $A' \in \mathcal{L}(F',E')$ by $A'(f) = f \circ A$ ($f \in F'$). We have $\|A'\| \le \|A\|$. The diagram

$$\begin{array}{ccc} E & \xrightarrow{A} & F \\ \downarrow{\scriptstyle j_E} & & \downarrow{\scriptstyle j_F} \\ E'' & \xrightarrow[A'']{} & F'' \end{array}$$

commutes. A subspace D of E has the WEP in E if the adjoint $E' \to D'$ of the inclusion map $D \to E$ is surjective.

P.4. The following two statements are not hard to prove.

1. Let $(E, \|\ \|)$ be a K-Banach space. Let $\varepsilon_1 > \varepsilon_2 > \ldots, \lim_{n\to\infty} \varepsilon_n = 0$. If $p_1, p_2, \ldots$ are polar norms on E such that $\|x\| \le p_n(x) \le (1+\varepsilon_n)\|x\|$ ($n \in \mathbb{N}, x \in E$) then $\|\ \|$ is a polar norm.
2. If E is a polar K-Banach space and $E \sim \ell^\infty$ then E is almost isomorphic to ℓ^∞. (By reflexivity, $E' \sim c_0$. By choosing t-orthogonal bases of E', where t is close to 1 one proves that c_0 and E' are almost isomorphic, hence so are E and ℓ^∞.)

P.5. Let A be a complete metrizable absolutely convex compactoid in a Hausdorff locally convex space E over K. Let $\lambda \in K$, $|\lambda| > 1$. Then there exist $e_1, e_2, \ldots$ in λA with $\lim_{n\to\infty} e_n = 0$ such that, with B the closed absolutely convex hull of $\{e_1, e_2, \ldots\}$,

(i) $A \subset B \subset \lambda A$.
(ii) Each $x \in B$ has a unique representation $x = \sum_{n=1}^\infty \lambda_n e_n$, where $\lambda_n \in K$, $|\lambda_n| \le 1$ for each n.
(iii) If $\lambda_n \in K$, $|\lambda_n| \le 1$ for each n then $\sum_{n=1}^\infty \lambda_n e_n \in B$.

(Proof. By [4], 6.1 A, as a topological module over $\{\lambda \in K : |\lambda| \le 1\}$, is isomorphic to a closed compactoid of c_0. Then apply [2], 4.37 (α) $\Rightarrow$ (ε).)

P.6. (p-adic Alaoglu Theorem) Let E be a K-Banach space. Then $B_{E'}$ is, for the w'-topology, a complete compactoid. (The proof is standard.)

§1 THE COMPLEMENTATION PROPERTY OF ℓ^∞.

Proposition 1.1. *Let E be a polar K-Banach space, let D be a closed subspace with inclusion map $i : D \to E$. Then $(\overline{i'(B_{E'})}^{w'})^e = B_{D'}$.*

Proof. Since $\|i'\| \le 1$ and $B_{D'}$ is w'-closed and edged we have $S := (\overline{i'(B_{E'})}^{w'})^e \subset B_{D'}$. Suppose we had an $f \in B_{D'}$, $f \notin S$. As the w'-topology on D' is of countable type and S is w'-closed and edged, S and $\{f\}$ can be separated by a w'-continuous linear function $D' \to K$ ([3], 4.4 and 4.7). But such functions are evaluations ([5], Th. 4.10) so there is an $x \in D$ with $|f(x)| > 1$ and $|h(x)| \le 1$ for all $h \in S$. We have $1 < |f(x)| \le \|f\|\,\|x\| \le \|x\|$ so $\|x\| > 1$. On the other hand, by polarity,

$$\begin{aligned}\|x\| = \|i(x)\| &= \sup\{|g \circ i(x)| : g \in B_{E'}\}\\ &= \sup\{|i'(g)(x)| : g \in B_{E'}\}\\ &\le \sup\{|h(x)| : h \in S\} \le 1,\end{aligned}$$

which is a contradiction.

Theorem 1.2. *Let E be a subspace of some polar K-Banach space X. Let $E \sim \ell^\infty$. Then, for each $t \in (0,1)$, E has a t-orthogonal complement in X.*

Proof. Let $i : E \to X$ be the inclusion map.

I. We show that the adjoint $i' : X' \to E'$ is a quotient map. By reflexivity the w'-topology on E' equals the weak topology. The norm topology on E' is strongly polar ([3], 4.4) so every norm-closed absolutely convex edged set is weakly closed. ([3], 4.9). Together with Proposition 1.1 this leads to

$$B_{E'} = (\overline{i'(B_{X'})}^{w'})^e = (\overline{i'(B_{X'})}^{w})^e = (\overline{i'(B_{X'})})^e$$

implying that $\overline{i'(B_{X'})}$ is a norm neighbourhood of 0 in E'. Then (see P.2) $i'(B_{X'})$ is closed so we find $B_{E'} = i'(B_{X'})^e$, which is what we wanted to prove.

II. Let $t \in (0,1)$. The space E' is of countable type so it has a $\sqrt{t}$-orthogonal base $e_1, e_2, \ldots$. Now $i' : X' \to E'$ is a quotient map so we can choose $z_1, z_2, \ldots \in X'$ such that $i'(z_n) = e_n$, $\|z_n\| \le t^{-\frac{1}{2}}\|e_n\|$ for each n. The formula

$$T\Big(\sum_{n=1}^{\infty} \lambda_n e_n\Big) = \sum_{n=1}^{\infty} \lambda_n z_n \quad (\lambda_n \in K,\ \lim_{n\to\infty} \|\lambda_n e_n\| = 0)$$

defines a $T \in \mathcal{L}(E', X')$ for which $\|T\| \le t^{-1}$ and $i' \circ T$ is the identity on E'. Then $T' \circ i''$ is the identity on E'' and it is easily seen from the diagram

$$\begin{array}{ccccc} E'' & \xrightarrow{i''} & X'' & \xrightarrow{T'} & E'' \\ \uparrow{\scriptstyle j_E} & & \uparrow{\scriptstyle j_X} & & \downarrow{\scriptstyle j_E^{-1}} \\ E & \xrightarrow[i]{} & X & & E \end{array}$$

that $P := j_E^{-1} \circ T' \circ j_X$ is in $\mathcal{L}(X, E)$, has norm $\leq t^{-1}$ and is a projection onto E.

Remark. The space E of Theorem 1.2 is almost isomorphic to ℓ^∞ (see P.4.2).

Corollary 1.3. *Let E be a subspace of some topologically pseudoreflexive K-Banach space X. Let $E \sim \ell^\infty$. Then E is complemented in X.*

Corollary 1.4.

(i) *Let E, X be as in 1.2. Then, for each $\varepsilon > 0$, each $f \in E'$ can be extended to an $\tilde{f} \in X'$ with $\|\tilde{f}\| \leq (1+\varepsilon)\|f\|$.*

(ii) *Let E, X be as in 1.3. Then E has the WEP in X.*

Corollary 1.4 yields the following result that may look bizarre at first sight.

Proposition 1.5. *Suppose $\#K$ is nonmeasurable. Let E be a K-Banach space such that E' separates the points of E. Suppose $E' \sim \ell^\infty$. Then $E \sim c_0$.*

Proof. $j_E : E \to E''$ is injective, so $\#E \leq \#E'' = \#c_0 = \#K$ and $\#E$ is nonmeasurable. Thus we can, in a standard way, construct a quotient map $\pi : c_0(I) \to E$ where $\#I$ is nonmeasurable. Then $\pi' : E' \to c_0(I)'$ is an isometry. By Corollary 1.4 $\pi'(E')$ has the WEP in $c_0(I)'$ so π'' : $c_0(I)'' \to E''$ is surjective. From the reflexivity of $c_0(I)$ and the commutative diagram

$$\begin{array}{ccc} c_0(I) & \xrightarrow{\pi} & E \\ \Big\downarrow{\scriptstyle j_{c_0(I)}} & & \Big\downarrow{\scriptstyle j_E} \\ c_0(I)'' & \xrightarrow[\pi'']{} & E'' \end{array}$$

it follows that j_E is surjective. So, $E \sim E'' \sim c_0$.

Remark. It may be worth noticing that the 'dual' statement of Proposition 1.5 is false! In fact, in [2], 4.J a closed subspace D of ℓ^∞ is constructed for which $D' \sim c_0$ but $D \not\sim \ell^\infty$. This D furnishes also a negative answer to the question 2 in the Introduction: it is easily seen that each $f \in D'$ has a unique extension $\tilde{f} \in (\ell^\infty)'$, so D has the WEP in ℓ^∞, but also D is weakly dense so that D is not weakly closed.

§2 COMPLEMENTED SUBSPACES OF ℓ^∞.

We first prove two useful lemmas. Let E, F, G be K-Banach spaces, let $i \in \mathcal{L}(E, F)$, $\pi \in \mathcal{L}(F, G)$. Suppose Im i = Ker π, π is surjective. A standard application of the Open Mapping Theorem shows that for the adjoint sequence $G' \xrightarrow{\pi'} F' \xrightarrow{i'} E'$ we have Im π' = Ker i'. If i' is surjective we can apply the same argument to $G' \xrightarrow{\pi'} F' \xrightarrow{i'} E'$ yielding the following.

Lemma 2.1. *Let D be a closed subspace of a K-Banach space E with inclusion map $i : D \to E$ and quotient map $\pi : E \to E/D$. Suppose D has the WEP in E. Then in the commutative* *diagram*

$$\begin{array}{ccccc} D & \xrightarrow{i} & E & \xrightarrow{\pi} & E/D \\ \big\downarrow{\scriptstyle j_D} & & \big\downarrow{\scriptstyle j_E} & & \big\downarrow{\scriptstyle j_{E/D}} \\ D'' & \xrightarrow[i'']{} & E'' & \xrightarrow[\pi'']{} & (E/D)'' \end{array}$$

we have $\operatorname{Im} i'' = \operatorname{Ker} \pi''$, *and* i'' *is injective.*

Lemma 2.2. (Compare also Example 3.3) *Let D be a closed subspace of a polar K-Banach space E, and let D have the* WEP *in E.*

(i) *If D is reflexive then D is weakly closed.*

(ii) *If E is reflexive, D is weakly closed then D is reflexive.*

Proof. Straightforward from Lemma 2.1 and 'diagram chasing'. (Observe that D is weakly closed if and only if $j_{E/D}$ is injective.)

Remark. The following corollary is a partial positive answer to question 2 (see also the remark following Proposition 1.5). *Let E be a K-Banach space with a base whose cardinality is nonmeasurable. Then every closed subspace with the WEP in E is weakly closed.* (**Proof.** Without loss, assume $E = c_0(I)$ where $\#I$ is nonmeasurable. By Gruson's Theorem [2], 5.9 any closed subspace is isomorphic to $c_0(J)$ for some set J. Then $\#J \leq \#I$ so $c_0(J)$ is reflexive. Now apply Lemma 2.2.(i).)

Theorem 2.3. *For a closed subspace D of ℓ^∞ the following are equivalent.*

(a) *For each $t \in (0,1)$, D has a t-orthogonal complement.*

(b) *D is complemented.*

(c) *D is weakly closed.*

(d) *Either D is finite dimensional or $D \sim \ell^\infty$.*

(e) *Either ℓ^∞/D is finite dimensional or $\ell^\infty/D \sim \ell^\infty$.*

(f) *ℓ^∞/D is a polar K-Banach space.*

(g) *$(\ell^\infty/D)'$ separates the points of ℓ^∞/D.*

(h) *D is reflexive and has the WEP in ℓ^∞.*

(i) *B_D is weakly closed in ℓ^∞.*

Proof. We first prove (a) $\Rightarrow$ (b) $\Rightarrow$ (c) $\Rightarrow$ (d) $\Rightarrow$ (a). Trivially (a) $\Rightarrow$ (b) $\Rightarrow$ (c); (d) $\Rightarrow$ (a) is Theorem 1.2. To establish (c) $\Rightarrow$ (d), let D be weakly closed, D infinite dimensional. By reflexivity, $D = A^0$ where A is a closed subspace of c_0. Then A has an infinite dimensional complement B in c_0 ([2], 3.16(v)) so $B \sim c_0$ and $D \sim B' \sim \ell^\infty$.

Next, we prove (a) $\Rightarrow$ (f) $\Rightarrow$ (g) $\Rightarrow$ (c) of which only (a) $\Rightarrow$ (f) needs some attention. Let $0 < t_1 < t_2 < \dots < 1$, $\lim_{n\to\infty} t_n = 1$. For each $n \in \mathbb{N}$ there is a projection $P_n : \ell^\infty \to D$ with $\|P_n\| \leq t_n^{-1}$. Then $Q_n := I - P_n$ is a projection with norm $\leq t_n^{-1}$ and kernel D. So there is a bijection $A_n \in \mathcal{L}(\ell^\infty/D, Q_n\ell^\infty)$ making

$$\begin{array}{ccc} \ell^\infty & \xrightarrow{\pi} & \ell^\infty/D \\ Q_n \searrow & & \swarrow A_n \\ & Q_n\ell^\infty & \end{array}$$

commute. For each n the norm $z \mapsto ||A_n z||$ $(z \in \ell^\infty/D)$ is polar. Also we have

$$||z|| \leq ||A_n z|| \leq t_n^{-1}||z|| \qquad (z \in \ell^\infty/D)$$

so that (see P.4.1) the quotient norm on ℓ^∞/D is polar. We have also (b) $\Rightarrow$ (e) (E/D is linearly homeomorphic to a complemented subspace of ℓ^∞) and (e) $\Rightarrow$ (g). So at this stage we have proved that (a) - (g) are equivalent. The implication (a) $\Rightarrow$ (h) is easy, (h) $\Rightarrow$ (c) is Lemma 2.2 (i) and (c) $\Rightarrow$ (i) is obvious. So we shall complete the proof by showing (i) $\Rightarrow$ (d). Let D be infinite dimensional. The 'closed' unit ball of ℓ^∞ is for the w'-topology $\sigma(\ell^\infty, c_0)$ (which equals the weak topology) a metrizable ([3],8.3), complete (P.6), edged compactoid, hence so is B_D. Let $\lambda \in K, |\lambda| > 1$. By P.5 there exist $f_1, f_2, ... in \lambda B_D$ with $\lim_{n\to\infty} f_n = 0$ weakly such that

$$B_D \subset \overline{co}^w\{f_1, f_2, ...\} \subset \lambda B_D$$

and such that each element of $\overline{co}^w\{f_1, f_2, ...\}$ has a unique representation $\sum_{n=1}^{\infty} \lambda_n f_n$ where $\lambda_n \in K, |\lambda_n| \leq 1$ (the summation is with respect to the weak topology of ℓ^∞). The formula

$$(\lambda_1, \lambda_2, ...) \overset{T}{\mapsto} \sum_{n=1}^{\infty} \lambda_n f_n$$

defines therefore a linear bijection $T : \ell^\infty \to D$. It is easily seen that $||x|| \leq ||Tx|| \leq |\lambda|\ ||x||$ $(x \in \ell^\infty)$. Thus, D is linearly homeomorphic to ℓ^∞.

Remark. The implication (i) $\Rightarrow$ (c) is an ultrametric version of the classical Banach-Dieudonné Theorem which states that a subspace D of the dual of a complex Banach space is w'-closed as soon as the closed unit ball of D is w'-closed. In the Appendix we shall prove a stronger version (Krein-Šmulian Theorem) for Banach spaces over a spherically complete ground field.

§3 HOW ABOUT $\ell^\infty(I)$?

It is natural to ask to what extent the previous results can be generalized to $\ell^\infty(I)$. The only positive result we have is in fact Proposition 3.1. The examples 3.2 and 3.3 show that several implications in Theorem 2.3 fail if we replace ℓ^∞ by $\ell^\infty(I)$.

Proposition 3.1. *Let I be a set whose cardinality is nonmeasurable. For a closed subspace D of $\ell^\infty(I)$ the following are equivalent.*
(a) D is complemented.
(b) $D \sim \ell^\infty(J)$ for some set J where $\#J$ is nonmeasurable. D has the WEP in $\ell^\infty(I)$.
Proof.
(a) $\Rightarrow$ (b). Clearly D has the WEP, is weakly closed, so D is reflexive by Lemma 2.2 (ii). ($\ell^\infty(I)$ is reflexive). Let $P : \ell^\infty(I) \to D$ be a linear continuous surjection. Then $P' : D' \to c_0(I)$ is a norm homeomorphism into $c_0(I)$ so by Gruson's Theorem $D' \sim c_0(J)$ where $\#J \leq \#I$ is nonmeasurable. So $D \sim D'' \sim \ell^\infty(J)$.

(b) $\Rightarrow$ (a). Let $i : D \hookrightarrow \ell^\infty(I)$ be the inclusion map. By (b) the adjoint $c_0(I) \xrightarrow{i'} D'$ is surjective. Now $D \sim \ell^\infty(J)$, where $\#J$ is nonmeasurable so $D' \sim c_0(J)$ and there is a map $T \in \mathcal{L}(D', c_0(I))$ such that $i' \circ T$ is the identity on D'. Then $T' \circ i''$ is the identity on D'' and $j_D^{-1} \circ T' \circ j_{\ell^\infty(I)}$ is a projection of $\ell^\infty(I)$ onto D.

Example 3.2. *Let $\#I = \#K$ be nonmeasurable. Then there exists an infinite dimensional closed subspace A_1 of $\ell^\infty(I)$ that has the WEP in $\ell^\infty(I)$ and is of countable type.* (Hence, D is weakly closed, reflexive, but not complemented (Lemma 2.2 (i) and Proposition 3.1).)

Proof. We can make, in a standard way, a quotient map $c_0(I) \xrightarrow{\pi} \ell^\infty$. By reflexivity π'' is surjective, so $A_1 := \pi'((\ell^\infty)')$ has the WEP in $c_0(I)'$ and is of countable type.

Example 3.3. (Negative answer to question 1) *Let I, K be as above. Then there exists a weakly closed subspace A_2 of $\ell^\infty(I)$ such that A_2 is of countable type, but A_2 does not have the WEP in $\ell^\infty(I)$.*

Proof. Let D be as in the Remark following Proposition 1.5. Again, make a quotient map $\pi : c_0(I) \to D$. It is easily seen that $A_2 := \pi'(D')$ is weakly closed, of countable type. If A_2 had the WEP then π'' would be surjective. Then, by reflexivity of $c_0(I)$, j_D would be surjective conflicting the nonreflexivity of D.

§4 SOME CONSEQUENCES FOR STRONGLY POLAR SPACES.

Recall that a K-Banach space E is *strongly polar* ([3], 3.5) if $\sup\{|f| : f \in E', |f| \leq p\} = p$ for each continuous seminorm p on E. It is proved in [3], 4.2 that E is strongly polar if and only if for each continuous seminorm p, for each subspace $D \subset E$, for each $f \in D'$ with $|f| \leq p$, for each $\varepsilon > 0$, there exists an extension $\tilde{f} \in E'$ such that $|\tilde{f}| \leq (1+\varepsilon)p$. It is still an intruiging open problem whether each strongly polar K-Banach space is of countable type. The previous theory yields the following results.

Proposition 4.1. *If E is reflexive and strongly polar then each closed subspace is reflexive.*

Proof. Any closed subspace is weakly closed. Now apply Lemma 2.2(ii).

Proposition 4.2. *Let E be reflexive and strongly polar. Let E' be a subspace of some polar K-Banach space X. Then E' has the* WEP *in X and E' is weakly closed in X.*

Proof. The first statement follows from Part I of the proof of Theorem 1.2. For the second statement apply Lemma 2.2(i).

Proposition 4.3. *Let E be a reflexive strongly polar space. If E' is linearly homeomorphic to a subspace of ℓ^∞ then E is of countable type.*

Proof. Assume $\dim E = \infty$. By Proposition 4.2 and Theorem 2.3 (h) $\Rightarrow$ (d), $E' \sim \ell^\infty$. Then $E \sim E'' \sim c_0$.

§5 APPENDIX: THE p-ADIC KREIN-ŠMULIAN THEOREM.

Throughout §5 K is spherically complete. By modifying 'classical' techniques we shall prove:

Theorem 5.1. *Let E be a K-Banach space. A convex subset C of E' is w'-closed if and only if for each $n \in \mathbf{N}$ the set $C \cap \{f \in E' : \|f\| \le n\}$ is w'-closed.*

Proof. We only need to prove the 'if' part.

(1) From the assumption on C one easily deduces that $C \cap B$ is w'-closed in B for every bounded set $B \subset E'$.

(2) Let bw' be the topology on E' of uniform convergence on compact subsets of E. Then bw' is locally convex, stronger than w', but coincides with w' on bounded subsets of E'. As $j_E(E) = (E', w')'$ and bw' is admissible we have $(E', bw')' = j_E(E)$ so a convex subset of E' is w'-closed if and only if it is bw'-closed. (See [5] for details)

(3) Let $a \in E' \backslash C$; it suffices by (2) to find a bw'-neighbourhood U of a for which $U \subset E' \backslash C$. We may assume $a = 0$, see (1). For each $r > 0$ set $B_r := \{x \in E : \|x\| \le r\}$, $B'_r := \{f \in E' : \|f\| \le r\}$. We shall find finite subsets $F_0, F_1, \ldots$ of E such that $F_n \subset B_{1/n}$ for each $n \in \{1,2,3,\ldots\}$ and $F_0^0 \cap F_1^0 \cap \ldots \cap F_n^0 \cap B'_{n+1} \subset E' \backslash C$ for each $n \in \{0,1,2,\ldots\}$. (Then $X := \bigcup_n F_n \cup \{0\}$ is compact so $U := X^0$ is a bw'-zero neighbourhood, $U \subset E' \backslash C$.) As $C \cap B'_1$ is w'-closed there is a finite set $F_0 \subset E$ for which $F_0^0 \cap B'_1 \subset E' \backslash C$. Suppose we have chosen $F_0, F_1, \ldots, F_{n-1}$ with the required properties, in particular

$$(*) \qquad F_0^0 \cap F_1^0 \cap \ldots \cap F_{n-1}^0 \cap B'_n \subset E' \backslash C$$

and suppose there is no F_n that meets the requirements. Then, for each finite subset F of $B_{1/n}$ we have

$$A_F := F^0 \cap F_0^0 \cap F_1^0 \cap \ldots \cap F_{n-1}^0 \cap B'_{n+1} \cap C \neq \emptyset$$

The sets A_F, where F is a finite subset of $B_{1/n}$, are c-compact in the w'-topology and have the finite intersection property. So there is an $f \in \bigcap_F A_F$. Then $f \in C$ and $|f| \le 1$ on each finite subset of $B_{1/n}$, so $\|f\| \le n$ i.e. $f \in B'_n$. Then, by (*),

$$f \in F_0^0 \cap F_1^0 \cap \ldots \cap F_{n-1}^0 \cap B'_n \subset E' \backslash C$$

contradicting $f \in C$.

Corollary 5.2. *A subspace D of E' is w'-closed if and only if B_D is w'-closed.*

Proof. Suppose B_D is w'-closed. Let $\lambda \in K, |\lambda| > 1$. For each $n \in \mathbf{N}$ the set $D \cap \{f \in E' : \|f\| \le |\lambda|^n\} = \lambda^n B_D$ is w'-closed. Let $r > 0$. For large n we have $|\lambda|^n \ge r$ so that $D \cap \{f \in E' : \|f\| \le r\} = \lambda^n B_D \cap \{f \in E' : \|f\| \le r\}$ is w'-closed. Now apply Theorem 5.1.

REFERENCES

[1] N. de Grande-de Kimpe and C. Perez-Garcia: Weakly closed subspaces and the Hahn-Banach extension property in p-adic analysis. Proc. Kon. Ned. Akad. Wet. 91, 253-261 (1988).

[2] A.C.M. van Rooij: Non-archimedean functional analysis. Marcel Dekker, New York (1978).

[3] W.H. Schikhof: Locally convex spaces over nonspherically complete valued fields. Groupe d'étude d'analyse ultramétrique 12 no. 24, 1-33 (1984/85).

[4] W.H. Schikhof: A connection between p-adic Banach spaces and locally convex compactoids. Report 8736, Department of Mathematics, Catholic University, Nijmegen, 1-16 (1987).

[5] J. van Tiel: Espaces localement K-convexes. Indag. Math. 27, 249-289 (1965).

Gross-Koblitz formula for function fields

Dinesh S. Thakur
School of Mathematics
Institute for Advanced Study
Princeton, NJ 08540, USA

The Gross-Koblitz formula, based on crucial earlier work by Honda, Dwork and Katz, expresses Gauss sums lying above a prime p in terms of values of Morita's p-adic gamma function at appropriate fractions (see [GK],[K]). Now various analogies between the global fields have been quite useful, so I will discuss and sketch a proof of an analogue of the Gross-Koblitz formula in the theory of function fields over finite fields.

Let K be a function field of one variable over finite field $\mathbf{F}_q$. Fix any place ∞ of K and let A be the ring of elements of K integral outside ∞. Basic analogies are,

$$\mathbf{Q} \leftrightarrow K, \;\; \mathbf{Z} \leftrightarrow A, \;\; \mathbf{C} \leftrightarrow \Omega =: \hat{\bar{K}}_\infty \tag{1}$$

Gauss sums and gamma functions are both closely related to the cyclotomic theory. Over $\mathbf{Q}$, one has basic cyclotomic extensions $\mathbf{Q}(\mu_n)$'s and the Kronecker-Weber theorem says that any finite abelian extension of $\mathbf{Q}$ is contained in one of these. Over K, usual cyclotomic extensions $K(\mu_n)$'s are just constant field extensions and there are many more abelian extensions eg. Kummer and Artin-Schreier extensions. Carlitz [C2] in 1930's and Drinfeld and Hayes [D],[H1],[H2] in 1970's produced other 'Cyclotomic families' $K(\Lambda_a)$'s ($a \in A$) where Λ_a is the set of a-torsion points of suitable rank one Drinfeld module $A \to \mathrm{End}G_a$, in analogy with μ_n, which is the set of n-torsion points of $\mathbf{Z} \to \mathrm{End}G_m$ (where integer n gives n'th power endomorphism). In Gekeler's talk, we have learnt the basics about Drinfeld modules (see [Ge] for more details). So I will just present a simple example, due essentially to Carlitz [C2], in detail.

Let $A = \mathbf{F}_q[T]$. It is easy to see that $\mathrm{End}G_a$ is the (non-commutative) ring of polynomials in Frobenius. Consider the ring homomorphism $A \to \mathrm{End}G_a (a \longmapsto C_a)$ given by

$$C_T(u) =: Tu + u^q, \;\; C_\theta(u) =: \theta u, \;\; (u \in \Omega, \theta \in \mathbf{F}_q) \tag{2}$$

For $a \in A$, let $\Lambda_a =: \{u \in \Omega : C_a(u) = 0\}$ For example, T-torsion points are just solutions of 'T-th cyclotomic equation' $u^q + Tu = 0$. For nonzero $a \in A$, $K(\Lambda_a)$ is an abelian extension of K with Galois group $(A/a)^*$. Let me mention in passing, that for general K, the maximal abelian extension of K is the compositum of all such $K(\Lambda_a)$'s over a's and A's (i.e. all possible choices of ∞'s). In our case of $A = \mathbf{F}_q[T]$, it can also be described [H1] as the compositum of constant field extensions, $K(\Lambda_a)$'s and '$K(\Lambda_a)$'s for $A = \mathbf{F}_q[1/T]$'. Gauss sums that we will now consider arise in the mixture of cyclotomic families $K(\mu_n)$'s and $K(\Lambda_a)$'s.

Classically, a Gauss sum is defined to be

$$-\sum_{x \in \mathbf{F}_{p^m}^*} \chi(x)\psi(Trx)$$

where χ is a non-trivial multiplicative character $\chi : \mathbf{F}_{p^m}^* \to \mathbf{C}^*$, ψ is a non-trivial additive character $\psi : \mathbf{F}_p \to \mathbf{C}^*$. We view ψ rather as an isomorphism of $\mathbf{Z}$-modules $\mathbf{Z}/p \to \mu_p$ and replace it by an isomorphism of A-modules $\psi : A/\wp \to \Lambda_\wp$ (Here $\wp$ is a monic irreducible polynomial of positive degree h of A and hence is a prime of A). Notice that ψ is no longer a character in usual sense. Let k be a finite field of 'characteristic $\wp$' i.e. a finite extension of $A/\wp$. To obtain non-trivial Gauss sums, we restrict the class of multiplicative characters to those giving $\mathbf{F}_q$ homomorphisms $\phi : k \to L$, where L is a field containing $K(\Lambda_\wp)$. Then Gauss sum [T1],[T2] is defined as

$$g(\phi) = -\sum_{x \in k^*} \phi(x^{-1})\psi(Trx) \tag{3}$$

It's easy to see that one only has to consider $k = A/\wp$ and that there are h basic Gauss sums, say g_j ($j$mod$h$) with $\phi = \chi_j$ being $\mathbf{F}_q$-homomorphisms $A/\wp \to L$, indexed so that $\chi_j^q = \chi_{j+1}$($j$mod$h$).

One can prove [T1],[T2] analogues of various results on classical Gauss sums. We just state here a weak form (without congruences) of an analogue of Stickelberger's theorem.

Let $K_h =: K(\mu_{q^h-1})$ and $L =: K(\Lambda_\wp)K_h$. Now $\wp$ splits in K_h completly into $\wp_j =: T - \chi_{1-j}(T)$($jmodh$) and $\wp_j$ totally ramify to power $q^h - 1$ in L, let $\overline{\wp_j}$ be the unique prime above $\wp_j$ in the integral closure of A in L.

Theorem 1 *('Stickelberger factorization'): With the notation described above, we have, in L,*

$$(g_j) = \overline{\wp}_{1-j}\overline{\wp}_{2-j}^{q} \cdots \overline{\wp}_{h-j}^{q^{h-1}} \tag{4}$$

Even though the proof is quite different than the classical case, we omit the proof, as the classical version is easy to prove and well-known.

Now we turn to the gamma side. Classically, the exponential function e^z is nothing but the entire function (normalized) satisfying the functional equation $e^{nz} = (e^z)^n$ corresponding to $\mathbf{Z} \to \mathrm{End}G_m$. Similarly, in this game, the exponential $e(z)$ is defined to be the entire function (normalized to be tangent to identity at Lie algebra level i.e. linear term is z) satisfying the functional equation $C_a(e(z)) = e(az)$ corresponding to $A \to \mathrm{End}G_a$ of (2). Classically, $e^z = \sum z^n/n!$, here $e(z)$ being linear, one can write $e(z) = \sum z^{q^n}/D_n$ (normalization corresponds to $D_0 = 1$) and hence one can regard D_n's as factorials of q^n by analogy. One gets the recursion relation

$$D_i = [i]D_{i-1}^q, \quad [i] = T^{q^i} - T \tag{5}$$

by equating the coefficients of z^{q^i} in the functional equation for $e(z)$, for $a = T$. For $n \in \mathbf{N}$, define the factorial of n to be (due to Carlitz)

$$\Pi(n) =: \prod D_j^{n_j} \tag{6}$$

where $n = \sum n_j q^j$ is the base q expansion.

Why is this a good notion of factorial? For one thing, classically $n! = \prod p^{n_p}, n_p = \sum_{e \geq 1}[n/\mathbf{N}p^e]$, where $\mathbf{N}$ is the norm and the product is over positive primes. In our case also, as Sinnott noticed (see [Go1]), the same formula holds, if p's are replaced by the monic primes of $\mathbf{F}_q[T]$.

Now from (5), it's easy to see that $[i]$ is the product of all monic irreducible polynomials of degree dividing i and D_i is the product of all monic polynomials of degree i. Hence 'removing $\wp$-factors', Goss [Go2] made Morita-style $\wp$-adic factorial $\Pi_\wp$ as follows. Define $\tilde{D}_i$ to be the product of all monic polynomials, prime to $\wp$, which are of degree i. Goss showed that $-\tilde{D}_i \to 1$ in $\wp$-adic topology and hence defined, for $n \in \mathbf{Z}_p, n = \sum n_j q^j, 0 \le n_j < q$,

$$\Pi_\wp(n) =: \prod(-\tilde{D}_j)^{n_j} \in A_\wp \tag{7}$$

Now we can state our main result:

Theorem 2 *('Analogue of the Gross-Koblitz formula'):*

$$g_j = -(-\wp^{q^j})^{\frac{1}{q^h-1}} / \Pi_\wp\left(\frac{q^j}{1-q^h}\right), \quad 0 \le j < h \tag{8}$$

Proof: (For simplicity, we will not describe which q^h-1-th root we have chosen and how we choose the embeddings). Clearly, $\tilde{D}_a = D_a / D_{a-h}\wp^l$, where l is such that $\tilde{D}_a$ is unit at $\wp$. Hence, using the base q expansion $q^j/(1-q^h) = \sum q^{j+ih}$ we get,

$$\Pi_\wp\left(\frac{q^j}{1-q^h}\right) = \lim(-1)^{m+1}\tilde{D}_j \cdots \tilde{D}_{j+mh} = \lim(-1)^{m+1} D_{j+mh}/\wp^w$$

where $w = \mathrm{ord}_\wp D_{j+mh}$. Applying the recursion formula (5), h times, we have

$$D_{j+mh} = [j+mh][j-1+mh]^q \cdots [j+1+(m-1)h]^{q^{h-1}} D_{j+(m-1)h}^{q^h}$$

Let $T = au$ be the decomposition in $K_\wp$ of T, a unit at $\wp$ (without loss of generality, $\wp \ne T$), as the product of its 'Teichmuller representative' a and its one unit part u. As $a^{q^{mh}} = a$ and $u^{q^t} \to 1$ as $t \to \infty$, we have, as $m \to \infty$, $[l+mh] = ((au)^{q^{mh+l}} - T) \to (a^{q^l} - T)$, which is just (negative of) one of the $\wp_j$'s above. Using this in the limit above and counting powers of $\wp$, using the description of $[i]$ given above, we see that

$$\Pi_\wp\left(\frac{q^j}{1-q^h}\right)^{1-q^h} = (-\wp_{1-j})(-\wp_{2-j})^q \cdots (-\wp_{-j})^{q^{h-1}} / \wp^{q^j}$$

Comparing with the Stickelberger factorization (4) (and using more precise information of Stickelberger congruences, which we have not stated, to fix units), one immediately deduces (8)

Q.E.D.

Note that this proof is quite direct and does not need a lot of machinery, unlike the proof in the classical case.

Gauss sums described above in detail for $\mathbf{F}_q[T]$ can be similarly defined for any A using $\wp$-torsion points for suitable Drinfeld modules, but in general they might also have prime divisors not above $\wp$. For more peculiarities of the general theory, see [T4].

Now, if we drop the insistence on the analogy described above, then 'Gauss sums' i.e. elements providing correct Stickelberger elements', were shown to exist in the general situation by Tate and Deligne; Hayes provided an explicit construction. (See [H3],[H4] for the details). Interesting feature of his construction is that 'Gauss sum for a prime $\wp$' occurs as a torsion point for some rank one Drinfeld-A-module, where the infinite place for A lies above $\wp$. Hence an infinity-adic formula for a torsion point is a '$\wp$-adic' formula for Gauss sums. Again, for simplicity, we assume $A = \mathbf{F}_q[T]$, though it is not really necessary.

Classically, n-th roots of unity are special values $e^{2\pi i m/n}$ of the exponential function. Now $2\pi i\mathbf{Z}$ can be defined to be the solution set of $e^z = 1$. Since we are dealing with the additive rather than the multiplicative group, and since the kernel of $e(z)$ is of the form $\tilde{\pi}A$, for some $\tilde{\pi} \in \Omega$, we consider $\tilde{\pi}$ to be an analogue of $2\pi i$. Then a-th torsion points for C are given by $e(\tilde{\pi}b/a)$, as is easy to verify using the functional equation $C_a(e(z)) = e(az)$.

In our non-archimedean situation, meromorphic functions are determined, up to multiplication by a nonzero constant, by their divisors, so that

$$e(z) = z \prod_{\lambda \in \tilde{\pi}A}{}' (1 - \frac{z}{\lambda}) \tag{9}$$

But from the point of view of the divisors, classically, gamma function is a meromorphic function with no zeros and simple poles at zero and negative integers. So we define (using analogy between positive and monic)

$$\Gamma(z) =: \frac{1}{z} \prod_{n \text{ monic}} (1 + \frac{z}{n})^{-1}, \;\; \Pi(z) =: z\Gamma(z), \;\; (z \in \Omega) \tag{10}$$

With these definitions, we have

Theorem 3 *('Reflection formula'):*

$$\prod_{\theta \in A^*} \Pi(\theta z) = \frac{\tilde{\pi} z}{e(\tilde{\pi} z)} \tag{11}$$

Proof: If one compares (9) and (10), apart from the scaling factor $\tilde{\pi}$, the main difference between the two is that in (9) all the signs are allowed, whereas in (10) we restrict to the monic elements. But the product over θ's takes care of this difference and the formula follows. Q.E.D.

Compare this with its classical counterpart

$$\prod_{\theta \in \mathbf{Z}^*} (\theta z)! = \frac{\pi z}{\sin \pi z}$$

which is nothing but $\Gamma(z)\Gamma(1-z) = \pi/\sin \pi z$ in disguise. (Note that from the point of view of the divisors, $e(z)$ is an analogue of the classical sine function).

On the other hand, one can also interpolate [Go2],[T1],[T3] the factorial function (6) ∞-adically to get $\Pi_\infty : \mathbf{Z}_p \to K_\infty$, and show [T1],[T3] that essentially

$$\Gamma_\infty(\frac{1}{2}) = \sqrt{\tilde{\pi}}, \;\; \Gamma_\infty(0) = \tilde{\pi} \tag{12}$$

where $\Gamma_\infty(z) =: \Pi_\infty(z-1)$ and the first equation holds only for odd characteristic. This also generalizes to general A [T1],[T3].

Observations above together with (11) and (12), express 'Gauss sums' in terms of values of '$\wp$-adic gamma functions' at fractions. In this sense, this is an analogue of the Gross-Koblitz formula, the main difference being that in this case the gamma function is constructed by uniform procedure at all places, instead of being interpolated from a fixed gamma function. This gamma function can also be interpolated [T3] and one might ask for Gross-Koblitz phenomenon using the interpolated gamma function.

For more information on a two variable gamma function of Goss, functional equations, algebraicity and transcendence results, Chowla-Selberg phenomenon, more analogies and some striking differences, see the references below.

References

[C1] L. Carlitz, On certain functions connected with polynomials in a Galois field, Duke Math.J. (1935),pp.137-168.

[C2] L.Carlitz, A class of polynomials, Trans. A.M.S. 43 (1938), pp.167-182.

[D] V. Drinfeld, Elliptic modules (English translation) Math. Sbornik, vol 23,(1974),561-592.

[Ge] E. U. Gekeler, Drinfeld modular curves, Lecture notes in math. 1231.

[Go1] D. Goss, von-Staudt for $\mathbf{F}_q[T]$, Duke Math.J. 45 (1978),pp.885-910.

[Go2] D. Goss, Modular forms for $\mathbf{F}_q[T]$. J. reine angew. Math. vol.317, (1980),163-191.

[Go3] D. Goss, The Γ-function in the arithmetic of function fields. Duke Math J. vol. 56 (1988), 163-191.

[G-K] B.H.Gross, N. Koblitz, Gauss sums and the p-adic Γ-function. Ann. Math. 109, 569-581 (1979).

[H1] D. Hayes, Explicit class field theory for rational function fields. Trans. Am. Math. Soc. 189, 77-91(1974).

[H2] D. Hayes, Explicit class field theory in global function fields. G.C. Rota (ed), Studies in Algebra and Number theory. Academic press 173-217 (1979)

[H3] D. Hayes, Stickelberger elements in function fields. Comp. Math. 55, 209-235 (1985).

[H4] D. Hayes, The refined $\wp$-adic abelian Stark conjecture in function fields. Inv. Math. 94, 505-527 (1988).

[K] N. Katz, Crystalline cohomology, Dieudonne modules and Jacobi sums. In Automorphic forms, representation theory and arithmetic. (TIFR, Bombay, 1979), 165-245.

[T1] D. Thakur, Gamma functions and Gauss sums for function fields and periods of Drinfeld modules. Thesis, Harvard University(1987).

[T2] D. Thakur, Gauss sums for $\mathbf{F}_q[T]$. Inv. Math. 94,105-112 (1988).

[T3] D. Thakur, Gamma functions for function fields and Drinfeld modules, To appear in Annals of Math.

[T4] D. Thakur, Gauss sums for function fields, To appear in Journal of Number Theory.

Three generalizations of Mahler's expansion for continuous functions on $\mathbb{Z}_p$.

Lucien VAN HAMME

Vrije Universiteit Brussel, Faculty of Applied Sciences
Pleinlaan 2, B-1050 Brussels, Belgium

1. Introduction - Notations.

Let p be a prime number. $\mathbb{Z}_p$ and $\mathbb{C}_p$ denote, respectively, the ring of p-adic integers and the completion of the algebraic closure of the field of p-adic numbers. The valuation on $\mathbb{C}_p$ will be denoted $|\ |$.
Consider a continuous function $f : \mathbb{Z}_p \to \mathbb{C}_p$. Let Δ be the difference operator defined by
$(\Delta f)(x) = f(x+1) - f(x)$. Applying Δ n-times we get

$$(\Delta^n f)(x) = \Delta^n f(x) = \sum_{k=0}^{n} \binom{n}{k} (-1)^{n-k} f(x+k)$$

The continuity of f implies that $\lim_{n\to\infty} \Delta^n f(x) = 0$ uniformly. From this it is easy to get Mahler's expansion (see e.g. [2]) :

$$f(x) = \sum_{n=0}^{\infty} \Delta^n f(0) \binom{x}{n}$$

The purpose of this paper is to prove three results which include Mahler's expansion as a special or limiting case. In the next section we state our results and give a few examples. In sections 3, 4 and 5 complete proofs are given.

2. The Theorems.

<u>Theorem 1</u> : If $f : \mathbb{Z}_p \to \mathbb{C}_p$ is a continuous function and $g : \mathbb{N} \to \mathbb{C}_p$ is a bounded sequence then

$$\sum_{n=0}^{\infty} \binom{x}{n} g(n)\, \Delta^n f(0) = \sum_{n=0}^{\infty} \binom{x}{n} \Delta^n g(0)\ \Delta^n f(t)\Big|_{t=x-n}$$

$\Delta^n f(t)\,|_{t=x-n}$ is the value of $\Delta^n f$ at the point x–n.
This is a generalization of Mahler's expansion since for $g(n) = 1$, $n = 0, 1, 2, \ldots$,
we get $\Delta^n g(0) = 0$ for $n \geq 1$ and the formula reduces to Mahler's expansion.
The following special cases are worth mentioning.
Take $f(u) = (1+t)^u$, $|t| < 1$. This gives

$$\sum_{n=0}^{\infty} \binom{x}{n} g(n)\, t^n = (1+t)^x \sum_{n=0}^{\infty} \binom{x}{n} \Delta^n g(0) \left(\frac{t}{1+t}\right)^n \qquad (1)$$

Putting x = – 1 and replacing t by –t we get the well-known expansion

$$\sum_{n=o}^{\infty} g(n)\, t^n = \sum_{n=o}^{\infty} \Delta^n g(0)\, \frac{t^n}{(1-t)^{n+1}} \qquad (2)$$

Take $g(n) = (-1)^n \binom{y}{n}$ in (1) and replace t by –t. This gives

$$\sum_{n=o}^{\infty} \binom{x}{n}\binom{y}{n} t^n = (1-t)^x \sum_{n=o}^{\infty} \binom{x}{n}\binom{-1-y}{n}\left(\frac{t}{t-1}\right)^n \quad ; \ |t|<1$$

which can be written as a (known) relation between hypergeometric series

$${}_2F_1(-x,-y;1;t) = (1-t)^{-x}\, {}_2F_1(-x, 1+y; 1; \tfrac{t}{t-1})$$

Finally, take $f(y) = \binom{y+z}{m}$ then $\Delta^n f(0) = \binom{z}{m-n}$, $\Delta^n f(t)\big|_{t=x-n} = \binom{x+z-n}{m-n}$

and the theorem becomes

$$\sum_{n=o}^{m} \binom{x}{n}\binom{z}{m-n} g(n) = \sum_{n=o}^{m} \binom{x}{n}\binom{x+z-n}{m-n} \Delta^n g(0)$$

For x = z= m this gives

$$\sum_{n=o}^{\infty} \binom{m}{n}^2 g(n) = \sum_{n=o}^{m} \binom{m}{n}\binom{m+n}{n} \Delta^{m-n} g(0)$$

Theorem 2 : If $f : \mathbb{Z}_p \to \mathbb{C}_p$ is a continuous function then

$$f(x) = f(0) + \sum_{n=1}^{\infty} \frac{x}{x+yn}\binom{x+yn}{n} \Delta^n f(t)\big|_{t=-yn}$$

For y = 0 we again get Mahler's expansion.

The third generalization gives an expression for the remainder in Mahler's expansion. We need the notion of convolution of two continuous functions.

The convolution of two sequences f(n) and g(n) is, by definition, the sequence given by

$$(f*g)(n) = \sum_{k=o}^{n} f(k)\, g(n-k)$$

In order to extend this definition to continuous functions defined on $\mathbb{Z}_p$, we need the following lemma.

Lemma 1 : If the sequences f(n) and g(n) are the restriction to N of two continuous functions from $\mathbb{Z}_p$ to $\mathbb{C}_p$ then $\lim_{n\to\infty} \Delta^n(f*g)(0) = 0$.

Corollary : The sequence $n \to (f*g)(n)$ is the restriction of a continuous function from $\mathbb{Z}_p$ to $\mathbb{C}_p$, denoted by (f∗g).

This corollary is well-known (see e.g. [2] p. 106) but the published proofs are somewhat complicated. An easier and more direct proof can be based on (2).

Proof of lemma 1 :

$$\sum_{n=0}^{\infty} (f*g)(n)\, T^n = \left\{\sum_{n=0}^{\infty} f(n)\, T^n\right\}\left\{\sum_{n=0}^{\infty} g(n)\, T^n\right\}$$

$$= \frac{1}{(1-T)^2}\left\{\sum_{n=0}^{\infty} \Delta^n f(0)\frac{T^n}{(1-T)^n}\right\}\left\{\sum_{n=0}^{\infty} \Delta^n g(0)\frac{T^n}{(1-T)^n}\right\}$$

Define numbers a_n by

$$\sum_{n=0}^{\infty} (f*g)(n)\, T^n =$$

$$\frac{1}{(1-T)^2}\sum_{n=0}^{\infty} a_n \frac{T^n}{(1-T)^n} = \frac{1}{1-T}\left(1+\frac{T}{1-T}\right)\sum_{n=0}^{\infty} a_n \frac{T^n}{(1-T)^n}$$

$$= \frac{a_0}{(1-T)} + \sum_{n=1}^{\infty} (a_n + a_{n-1})\frac{T^n}{(1-T)^{n+1}}$$

Then $\Delta^n(f*g)(0) = a_n + a_{n-1}$ for $n \geq 1$.
Since $\lim_{n\to\infty} \Delta^n f(0) = \lim_{n\to\infty} \Delta^n g(0) = 0$ we see that $\lim_{n\to\infty} a_n = 0$ which proves the lemma.

We now introduce another convolution which is, in a sense, more natural.
Definition : $(f©g)(x) = (f*g)(x-1) \qquad x \in \mathbb{Z}_p$
If U is the function defined by $U(x) = 1$ for all $x \in \mathbb{Z}_p$ then

$$(f©U)(n) = f(0) + f(1) + \dots + f(n-1)$$

and f©U is the "indefinite sum" of f i.e. $\Delta(f©U) = f$
We can now state our third result.

Theorem 3 : Mahler's expansion with remainder.
If $f : \mathbb{Z}_p \to \mathbb{C}_p$ is continuous then

$$f(x) = f(0) + \Delta f(0)\binom{x}{1} + \dots + \Delta^n f(0)\binom{x}{n} + \binom{x}{n} © (\Delta^{n+1} f)(x)$$

Corollary : If n tends to infinity we get Mahler's expansion.
To see this, observe that

$$\| f©g \| \leq \| f \| . \| g \| \text{ where } \| f \| = \sup_{x\in\mathbb{Z}_p} | f(x) |.$$

Hence

$$\left\| \binom{x}{n} © \Delta^{n+1} f \right\| \leq \left\| \binom{x}{n} \right\| . \| \Delta^{n+1} f \| \leq \| \Delta^{n+1} f \| \to 0$$

3. Proof of Theorem 1

Since $\lim_{n\to\infty} \Delta^n f(x) = 0$ uniformly, the series on the l.h.s. and on the r.h.s. are uniformly convergent which means that their sums are continuous functions of x.
Hence it is sufficient to prove the theorem for $x = n \in \mathbb{N}$.
Start from the identity $(1+vu-v)^n = ((v-1)(u-1)+u)^n$

$$\sum_{k=o}^{n} \binom{n}{k} v^k (u-1)^k = \sum_{k=o}^{n} \binom{n}{k} (v-1)^k u^{n-k} (u-1)^k \qquad (3)$$

The linear map $L_1 : \mathbb{C}_p[u] \to \mathbb{C}_p : u^k \to f(k)$ maps $(u-1)^k$ on $\Delta^k f(0)$.
If we apply first L_1 to (3) and then the map $L_2 : \mathbb{C}_p[v] \to \mathbb{C}_p : v^k \to g(k)$ we get

$$\sum_{k=o}^{n} \binom{n}{k} g(k)\, \Delta^k f(0) = \sum_{k=o}^{n} \binom{n}{k} \Delta^k g(0)\, \Delta^k f(t)\Big|_{t=n-k}$$

which proves the theorem for $x = n$.

Remark : By imposing more severe conditions on the function f one can deduce a formula for f'(0) from theorem 1. Write the theorem in the form

$$\sum_{n=1}^{\infty} \binom{x-1}{n-1} \frac{g(n)}{n} \Delta^n f(0) = g(0) \cdot \frac{f(x)-f(0)}{x} + \sum_{n=1}^{\infty} \binom{x-1}{n-1} \frac{\Delta^n g(0)}{n} \Delta^n f(t)\Big|_{x-n}$$

If $\lim_{n\to\infty} \frac{\Delta^n f(x)}{n} = 0$, uniformly in $\mathbb{Z}_p$, one can take the limit for $x \to 0$.

This gives $$g(0)f'(0) = \sum_{n=1}^{\infty} \frac{(-1)^{n-1}}{n} \left[g(n)\, \Delta^n f(0) - \Delta^n g(0)\, \Delta^n f(t)\Big|_{t=-n} \right]$$

If, moreover, $f(x) = f(-x)$ then $\Delta^n f(t)\big|_{t=-n} = (-1)^n \Delta^n f(0)$, $f'(0) = 0$ and we get a formula involving only differences at O :

$$\sum_{n=1}^{\infty} \frac{(-1)^n}{n} g(n)\, \Delta^n f(0) = \sum_{n=1}^{\infty} \frac{\Delta^n g(0)}{n} \Delta^n f(0)$$

4. Proof of Theorem 2

Put $A_n(x,y) = \frac{x}{x+yn} \binom{x+yn}{n} = \frac{x}{n} \binom{x+y\,n-1}{n-1}$ for $n \geq 1$. $A_o(x,y) = 1$

We first need a lemma.
Lemma 2 : $|A_n(x,y)| \leq 1 \qquad x,y \in \mathbb{Z}_p$
Proof : There are three cases to consider.
1^e case : $|x| > |ny|$.
Then $|x+yn| = |x|$ and $|A_n(x,y)| = \left|\frac{x}{x+yn}\right| \left|\binom{x+yn}{n}\right| \leq 1.$

2^e case : $|x| < |ny|$.

Then $|x+yn| = |ny|$ and $\left|\frac{x}{x+yn}\right| = \left|\frac{x}{ny}\right| < 1$. Hence $|A_n(x,y)| \le 1$.

3^e case : $|x| = |ny|$.

Then $|\frac{x}{n}| = |y| \le 1$ and $|A_n(x,y)| = |\frac{x}{n}| \cdot \left|\binom{x+n\,y-1}{n-1}\right| \le 1$.

We can now prove theorem 2.

We first consider the case where $f(x) = \binom{x}{n}$ $n \ge 1$.

The theorem then reduces to

$$\binom{x}{n} = \sum_{k=o}^{n} \frac{x}{x+yk}\binom{x+yk}{k}\binom{-yk}{n-k} = \sum_{k=o}^{n} A_k(x,y)\binom{-yk}{n-k} \tag{4}$$

This identity follows from the more general identity

$$\binom{x+z+ny}{n} = \sum_{k=o}^{n} A_k(x,y)\binom{x+(n-k)y}{n-k} \tag{5}$$

by taking $z+ny = 0$. Formula (5) can be found in [1] p. 172.

We now use (4) to prove the theorem.

$$\sum_{n=1}^{N} \Delta^n f(0)\binom{x}{n} = \sum_{n=1}^{N}\sum_{k=1}^{n} A_k(x,y)\,\Delta^n f(0)\binom{-yk}{n-k}$$

$$= \sum_{k=1}^{N} A_k(x,y)\sum_{n=1}^{N}\Delta^n f(0)\binom{-yk}{n-k}$$

From $f(x) = \sum_{n=o}^{\infty}\Delta^n f(0)\binom{x}{n}$ we get $\Delta^k f(x) = \sum_{n=k}^{\infty}\Delta^n f(0)\binom{x}{n-k}$

This means that we can write

$$\sum_{n=k}^{N}\Delta^n f(0)\binom{x}{n-k} = \Delta^k f(x) + r_{k,N} \text{ with } \lim_{N\to\infty} r_{k,N} = 0$$

Hence

$$\sum_{n=1}^{\infty}\Delta^n f(0)\binom{x}{n} = \sum_{k=1}^{N} A_k(x,y)\,\Delta^k f(-yk) + \sum_{k=1}^{N} A_k(x,y)\, r_{k,N}$$

Noting that $|r_{k,N}| \le |r_{0,N}|$ and using lemma 2 we see that the last sum tends to zero if $N\to\infty$.

This proves the theorem.

5. Proof of Theorem 3

The proof will follow from two lemmas. Recall that U is the function with $U(x) = 1$ for all $x \in \mathbb{Z}_p$.

Lemma 3 :

(i) $\binom{x}{n} © U = \binom{x}{n+1}$, $n \in \mathbb{N}$, $x \in \mathbb{Z}_p$

(ii) $\underbrace{(U © U © U © \ldots © U)}_{(n+1) \text{ factors}}(x) = \binom{x}{n}$

Proof : Since all functions involved are continuous it is sufficient to verify the statements for $x = k \in \mathbb{N}$.

For $n = 0$ the assertion (i) reduces to $(U © U)(k) = k$ and this is equivalent with $\sum_{i=0}^{k-1} 1 = k$. (ii) is trivial for n=0

For $n > 0$ we use induction on n. The lemma follows from

$$\binom{k}{n} © U(k) = \sum_{i=0}^{k-1} \binom{i}{n} = \binom{k}{n+1} \text{ since } \binom{i+1}{n+1} - \binom{i}{n+1} = \binom{i}{n}$$

Lemma 4 : $f = f(0). U + U © \Delta f$

Proof : Since f is continuous it is sufficient to verify this when the variable $x = n \in \mathbb{N}$. But in that case the lemma reduces to

$$f(n) = f(0) + \sum_{k=0}^{n-1} \Delta f(k)$$

which is obvious.

Theorem 3 now follows by using lemma 4 repeatedly.

$$f = f(0)\, U + U © \Delta f$$

$$\Delta f = \Delta f(0)\, U + U © \Delta^2 f$$

Hence $f = f(0)\, U + \Delta f(0)\, .\, U © U + U © U © \Delta^2 f$

Continuing in this way we get

$$f = f(0)\, U + \Delta f(0)\, .\, U © U + \ldots + \Delta^n f(0)\, .\, \underbrace{U © U © \ldots © U}_{(n+1) \text{ factors}} + \underbrace{U © U © \ldots © U}_{(n+1) \text{ factors } U} © \Delta^{n+1} f$$

Evaluating this at the point x we obtain (from lemma 3) theorem 3.

REFERENCES

[1] J. RIORDAN : Combinatorial Identities.
Robert Krieger Publishing Company, Huntington, New York 1979.

[2] W. SCHIKHOF : Ultrametric Calculus.
Cambridge University Press, 1984.

P-ADIC SYMMETRIC DOMAINS

Harm Voskuil
Department of Mathematics, University of Groningen
P.O. Box 800, 9700 AV Groningen, The Netherlands

§0 INTRODUCTION

We will study p-adic symmetric domains. The complex symmetric domains are well-known (See [C] and [H]). We will briefly recall their construction.

Let G be a real non-compact semi-simple and connected Lie group. The group G contains maximal compact subgroups K, they are all conjugated. Now $X=G/K$ is the symmetric space associated to G. An arithmetic subgroup $\Gamma\subset G$ acts properly on G/K. Now X/Γ need not to be compact, but there exists a space $\bar{X}\supset X$ such that $\bar{X}/\Gamma$ is compact (See [BB], [BS], [N] and [AMRT]).

The construction above does not work for p-adic Lie groups, since the maximal compact subgroups are not all conjugated in this case. Our construction is based on the work of Kurihara (See [Ku]).

Let G be a split simply-connected linear algebraic group defined over a non-archimedean local field K. Let $\Gamma\subset G$ be a discrete co-compact subgroup. Now we call X a symmetric space for the pair (G,Γ) if X/Γ is a proper rigid analytic variety. To construct such a space X, we start with a projective variety $Y=G/P$, here $P\subset G$ is a maximal parabolic subgroup defined over K. We then try to construct a space $X\subset Y$, which is a symmetric domain. In fact the construction is such that we have:

$$Y-(H_B\cap Y)\subset X\subset Y$$

Here H_B is defined as follows. Let Y be imbedded in a projective space $\mathbb{P}^n_K$. Let $T\subset G$ be a K-rational torus of maximal rank. Let x_i, $i=1..n+1$ be the coördinates associated to a basis of $\mathbb{P}^n_K$ such that T acts diagonally on $\mathbb{P}^n_K$. Now the union of all these hyperplanes $x_i=0$, $i=1..n+1$ for all maximal K-rational tori $T\subset G$ is H_B.

Many examples of discrete co-compact subgroups Γ of p-adic Lie groups G are known (See [Kan]). Furthermore it is known that the groups Γ are S-arithmetic if rank$(G)>1$ (See [Mar], [T.1] and [V]). For more information about S-arithmetic groups we refer to [S] and the literature given there.

This article is divided into three paragraphs.

In §1 we briefly recall the construction of the Bruhat-Tits building B for a reductive linear algebraic group G. Our construction of a symmetric domain X uses the building B.

In §2 we give a simplification of the construction of X in the case $G=PGL(n,K)$ acting on $Y=\mathbb{P}^{n-1}_K$. In this case X is very well-known, $X=\mathbb{P}^{n-1}_K-\{K\text{-rational hyperplanes}\}$ (See [D], [Mus] and [Ku]).

In §3 we study the split orthogonal groups $G=PSO(f,K)$, where f is a quadratic form in n variables, acting on the projective space $Y\subset\mathbb{P}_K^{n-1}$ defined by $f=0$. In this case we can not construct a symmetric domain for $\Gamma\subset G$ discrete and co-compact, but we can construct a "symmetric" space for $\Gamma\subset H$ discrete and co-compact. Here $H\subset G$ is a group isomorphic to $GL(l,K)$ with $l=\left[\frac{n}{2}\right]$.

In [F] there is indicated another approach to p-adic symmetric spaces. Last but not least I would like to thank Marius van der Put for his help while I was (and am) studying this subject.

§1 THE BRUHAT-TITS BUILDING

Here we will briefly recall the construction of the Bruhat-Tits building (or affine building) of a reductive linear algebraic group G.
For more precise statements and proofs see [BT] and [T.2]. We also will give some extra information about the buildings needed in §2 and §3.

Notation:

K	a finite extension of $\mathbb{Q}_p$ or $\mathbb{F}_q((t))$
v	the additive valuation on K, normalised such that $v(K^*)=\mathbb{Z}$
G	a reductive linear algebraic group defined over K
$T\subset G$	a torus defined over K with maximal rank, so $T(K)\cong(K^*)^n$ with n maximal
rank (G)	the rank of G which is the maximal rank of a K-rational torus $T\subset G$
$N\subset G$	the normaliser of T
Φ	a finite root system
W	a finite Weyl group
$\Phi_{af},\check{\Phi}_{af}$	an affine root system
W_{af}	an affine Weyl group

A group G is called *split* if $\text{rank}_K(G)=\text{rank}_L(G)$ for every finite extension $K\subset L$.
We will always assume that G is simple and split.

Let us fix T and G. To $T\subset G$ belongs a *finite root system* $\Phi\subset\mathbb{R}^n$, where $n=\text{rank}(G)$. On this root system acts a finite *Weyl group* W. Furthermore we have $W\subset N/T$, $(\# N/T<\infty)$. Now N acts by conjugation on the set of T-invariant additive subgroups $U\subset G$. We have a bijection:
$j:\{\alpha|\alpha\in\Phi\}\xrightarrow{\sim}\{U|U\subset G$ a T-invariant additive subgroup$\}$, such that $U_\alpha:=j(\alpha)$ and $w(U_\alpha)=U_{w(\alpha)}$ for all $w\in W\subset N/T$.

Starting with the root system Φ we can construct an *affine root system* Φ_{af} and the dual affine root system $\check{\Phi}_{af}$. The affine root system Φ_{af} consists of the affine-linear functions $k+\alpha$: $\mathbb{R}^n\to\mathbb{R}$, with $\alpha\in\Phi$, $k\in\mathbb{Z}$ and $k\equiv 1 \bmod 2$ if $\frac{1}{2}\alpha\in\Phi$, defined by: $k+\alpha(x)=(\alpha,x)+k$ for $x\in\mathbb{R}^n$, $(.,.)$ is the inner product. For the definition of $\check{\Phi}_{af}$ we refer to [Mac]. We can identify

the affine root $k+\alpha\in\Phi_{af}$ with the halfspace $\{x\in\mathbb{R}^n \mid k+(\alpha,x)\geq 0\}$.
The boundary $\{x\in\mathbb{R}^n \mid k+(\alpha,x)=0\}$ of an affine root is called a *wall*. The drawing of all the walls gives a simplicial decomposition of $\mathbb{R}^n$.

Example

Φ, type A_2

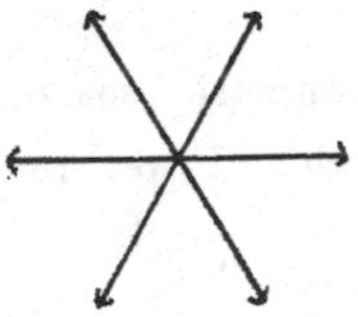

$\Phi_{af}\simeq\check{\Phi}_{af}$, type $\tilde{A}_2$

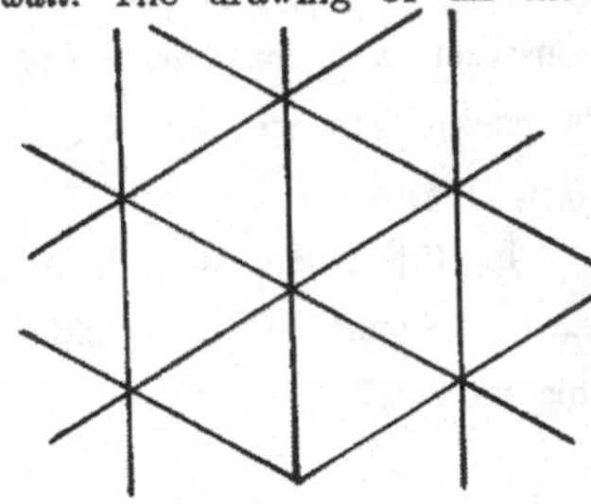

The maximal simplices are called *chambers* (In the example above, these are the triangles). A chamber C is the intersection of exactly $n+1$ affine roots (i.e. halfspaces). These $n+1$ affine roots form a *simple basis* Δ_{af} of the affine root system Φ_{af} (See [Mac]). On the affine root system acts an *affine Weyl group* W_{af}, which is generated by the reflections in the walls.

Now we return to our group G. Let T be as before and let $T_0\subset T$ be defined by:

$$T_0:=\{t\in T \mid v(t_{ii})=v(t_{jj}) \quad \forall\, i,j\}$$

(Here we assume that T is represented by diagonal matrices). The group N/T_0 contains an affine Weyl group W_{af}. We can fix a set of isomorphisms $\varphi_\alpha\colon U_\alpha\xrightarrow{\sim}K$, $\alpha\in\Phi$, satisfying certain conditions (See [BT]). The additive valuation v of K gives us a valuation $v\circ\varphi_\alpha$ on U_α. We can use these valuations $v\circ\varphi_\alpha$ to refine the set $\{U_\alpha \mid \alpha\in\Phi\}$ into a set of T_0-invariant additive subgroups $\{U_{k+\alpha} \mid k+\alpha\in\Phi_{af}^{(\vee)}\}$. Here $U_{k+\alpha}$ is defined by:

$$U_{k+\alpha}:=\{u\in U_\alpha \mid k+v\circ\varphi_\alpha(u)\geq 0\}$$

Again we have: $w(U_{k+\alpha})=U_{w(k+\alpha)} \quad \forall w\in W_{af}\subset N/T_0$.

All K-rational tori $T\subset G$ of maximal rank are conjugated. Therefore we can associate to each torus $T\subset G$ as above a space $\mathbb{R}^n$ with its simplicial decomposition given by the affine root system. Such a space is called an *appartment* A.

Let A be the appartment belonging to T. The *affine building* B of G is defined by:

$$B:=\bigcup_{g\in G} g\cdot A/\sim$$

Here $g\cdot A$ is the appartment belonging to gTg^{-1}. The equivalence-relation $\sim$ is given by identifying all the affine roots (half-appartments) belonging to the same T_0-invariant additive subgroup.

The stabiliser of a simplex $S\in B$ is called a *parahoric subgroup*. In [BT] it is proved that the parahoric subgroups of G are compact and that every compact subgroup of G is included in some parahoric subgroup of G. Now the following two statements are clear:

a) *$\Gamma\subset G(K)$ discrete $\Leftrightarrow$ Γ acts with finite stabilisers of simplices on B.*

b) *$\Gamma\subset G(K)$ co-compact $\Leftrightarrow$ B/Γ is finite.*

For the next paragraphs we need to know the structure of the *Iwahori-groups* P_C, the stabiliser of the chamber C.

Proposition: *Let $C\in A\subset B$ be a chamber. Let T_0 be as above and let Δ_{af} be the simple basis of Φ_{af} belonging to C. Now we have:*

$$P_C=<T_0,U_\alpha|\alpha\in\Delta_{af}>$$

For later use we will now give some more detailed information about the buildings of the group $SL(n,K)$ and the split orthogonal groups $SO(f,K)$, where f is a quadratic form.

The building of $SL(n,K)$: (See [BT] §10.2 and [T.2] §1.14)

This is a building of type $\tilde{A}_{n-1}$ i.e. the affine root system is of type $\tilde{A}_{n-1}$. Let $T\subset G$ be a fixed torus and let $G=SL(n,K)$ act on K^n. We choose a basis e_i, $i=1..n$ of K^n such that T acts diagonally. Now the T_0-invariant additive subgroups $U_{k+\alpha_{ij}}$ are:

$$U_{k+\alpha_{ji}}:=\{g\in G|g(e_i)=e_i+ce_j,\ g(e_s)=e_s,\ s\neq i,\ v(c)\geq k\}\text{ for } i\neq j.$$

On the affine roots we have a relation: $\alpha_{ij}=-\alpha_{ji}$

For a standard chamber $C\in A$ the simple basis Δ_{af} of Φ_{af} is:

$$\alpha_{i,i+1},\ i=1..n-1,\quad 1+\alpha_{n,1}$$

The roots in Δ_{af} satisfy the following relation:

$$\sum_{i=1}^{n-1}\alpha_{i,i+1}+(1+\alpha_{n,1})=1$$

The building of $SO(f,K)$, $f=\sum_{i=1}^{l}x_ix_{2l+1-i}$, $l>1$: (See [BT] §10.1)

This is a building of type $\tilde{A}_1\times\tilde{A}_1$, $l=2$, $\tilde{A}_3$, $l=3$ and $\tilde{D}_l$ for $l>3$. Let $T\subset G=SO(f,K)$ be a fixed torus and let G act on K^{2l}. We choose coördinates e_i, $i=1..2l$ of K^{2l} such that T acts diagonally. The T_0-invariant additive subgroups are:

$$U_{k+\alpha_{ji}}:=\{g\in G|g(e_i)=e_i+ce_j,\ g(e_{2l+1-j})=e_{2l+1-j}-ce_{2l+1-i},\ g(e_s)=e_s,\ s\neq i,\ 2l+1-j,\ v(c)\geq k\}$$

for $i\neq j$, $2l+1-j$.

On the affine roots we have the following relations:

$$\alpha_{ij}=-\alpha_{ji},\quad \alpha_{ij}=\alpha_{2l+1-j,\ 2l+1-i}$$

For a standard chamber $C\in A$ the simple basis Δ_{af} of Φ_{af} is:

$$\alpha_{i,i+1},\ i=1..l-1,\ \alpha_{l,l+2},\ 1+\alpha_{2l,2}$$

The roots in Δ_{af} satisfy the following relation:

$$\alpha_{1,2}+(1+\alpha_{2l,2})+2\sum_{i=2}^{l-2}\alpha_{i,i+1}+\alpha_{l-1,l}+\alpha_{l,l+2}=1, \qquad l\geq 4$$

$$\alpha_{1,2}+(1+\alpha_{6,2})+\alpha_{2,3}+\alpha_{3,5}=1 \qquad l=3$$

The building of $SO(f,K)$, $f=x_0^2+\sum_{i=1}^{l}x_ix_{2l+1-i}$, $l>1$: (See [BT] §10.1)

This is a building of type $\tilde{B}_l$. Let $T\subset G$ be a fixed torus and let $G=SO(f,K)$ act on K^{2l+1}. We choose a basis e_i, $i=0..2l$ of K^{2l+1} such that T acts diagonally. The T_0-invariant additive subgroups are:

$$U_{k+\alpha_{ji}}:=\{g\in G|\ g(e_i)=e_i+ce_j,\ g(e_{2l+1-j})=e_{2l+1-j}-ce_{2l+1-i},\ g(e_s)=e_s,\ s\neq i,2l+1-j,\ v(c)\geq k\}$$

for $i\neq j,2l-j$ and $i,j\neq 0$ and:

$$U_{k+\alpha_{0,j}}:=\{g\in G|\ g(e_j)=e_j+2cx_0+dx_{2l+1-j},\ g(e_0)=e_0-ce_{2l+1-j},$$
$$g(e_s)=e_s,\ s\neq j,0,\ c^2+d=0,\ v(d)\geq 2k\}$$

On the affine roots we have the following relations:

$$\alpha_{ij}=-\alpha_{ji},\quad \alpha_{ij}=\alpha_{2l+1-j,\ 2l+1-i},\quad i,j\neq 0$$

$\alpha_{0,j}=-\alpha_{0,2l+1-j}$ (and we define $\alpha_{j,0}=-\alpha_{0,j}$)

For a standard chamber $C\in A$ the simple basis Δ_{af} of Φ_{af} is:

$$1+\alpha_{2l,2},\ \alpha_{i,i+1},\ i=1..l-1,\ \alpha_{0,l+1}$$

The roots in Δ_{af} satisfy the following relation:

$$(1+\alpha_{4,2})+2\alpha_{1,2}+\alpha_{0,3}=1,\quad l=2$$

$$(1+\alpha_{2l,2})+2\sum_{i=1}^{l-1}\alpha_{i,i+1}+\alpha_{0,l+1}=1,\quad l>2$$

§2 THE GROUP $PSL(n,K)$ ACTING ON $\mathbb{P}_K^{n-1}$

We will give the construction of a symmetric domain for $PSL(n,K)$ starting with its action on $Y=\mathbb{P}_K^{n-1}$. This symmetric domain is also constructed in [D], [Mus] and [Ku]. Our construction is a simplification of the one given by Kurihara. We describe the construction in terms of rigid analytic geometry (See [BGR] or [FP]). Let us first define what we mean by a symmetric domain.

Definition: Let G be a reductive linear algebraic group defined over a non-archimedean local field K. Let $\Gamma\subset G$ be a discrete co-compact subgroup. A symmetric domain X for the pair (G,Γ) is an analytic space X such that X/Γ is a proper rigid analytic variety.

Notation: Let B be the building of $G=PSL(n,K)$. Let $A\subset B$ be a fixed appartment and let $T_A\subset G$ be the K-rational torus belonging to A. We fix a basis $\{e_1,\dots,e_n\}$ of $\mathbb{P}_K^{n-1}$, such that T_A acts diagonally on $\mathbb{P}_K^{n-1}$ with respect to this basis. Let $\{x_1,\dots,x_n\}$ be the associated basis of $\mathcal{O}(1)$.

To each root $\alpha\in\Phi$ belongs a character χ_α of T_A with which T_A acts on the additive subgroup U_α. To each affine root $k+\alpha\in\Phi_{af}$ we associate the meromorphic function $\pi^k\frac{x_i}{x_j}$, where T acts with character χ_α on $\frac{x_i}{x_j}$. In the notation of §1 we associate to the affine root $k+\alpha_{ij}$ the subgroup $U_{k+\alpha_{ij}}$ and the function $\pi^k\frac{x_i}{x_j}$.

Proposition 2.1: *Let $C\in A$ be a chamber, Δ_{af} the simple basis defined by C and $P_C\subset PSL(n,K)$ the Iwahori-group stabilising C. Now we have:*

a) *The set $X_{C,A}:=\{p\in\mathbb{P}_K^{n-1}\mid |\frac{g^*x_i}{x_i}(p)|\leq 1,\ i=1..n,\ \forall g\in P_C\}$ is an affinoid subspace of $Y=\mathbb{P}_K^{n-1}$. (Here $g^*x_i=x_i\circ g^{-1}$).*

b) *The affinoid algebra belonging to $X_{C,A}$ is:*

$$K<\frac{g^*x_i}{x_i}\mid i=1..n,\ g\in P_C> \;=K<\pi^k\frac{x_i}{x_j}\mid k+\alpha_{ij}\in\Delta_{af}>$$

(So for our standard chamber C of §1 this is $K<\frac{x_1}{x_2},\frac{x_2}{x_3},\dots\frac{x_{n-1}}{x_n},\frac{\pi x_n}{x_1}>$.)

Proof: This follows directly from the description of P_C and the T_0-invariant additive subgroups given in §1.

Definitions: Let H_A be the union of the hyperplanes defined by $x_i=0$, $i=1..n$. Let $f\colon \mathbb{P}_K^{n-1}-H_A\to A\ (=\mathbb{R}^{n-1})$ be the map defined by:

$$f(p)=q \text{ if and only if } v\left(\left(\frac{x_i}{x_j}\right)(p)\right)=\alpha_{ij}(q),\ \forall i,j=1..n,\ i\neq j\ .$$

Of course here the affine roots are interpreted as affine-linear functions on $\mathbb{R}^{n-1}$.

Proposition 2.2: *Let f as above.*

a) $f(p)\in C \Leftrightarrow p\in X_{C,A}$

b) $X_A:=\bigcup_{C\in A}X_{C,A}=\mathbb{P}_K^{n-1}-H_A$

c) *The covering $\mathcal{C}:=\{X_{C,A}\mid C\in A\}$ of X_A is pure.*

d) *The reduction of X_A with respect to the covering $\mathcal{C}$ has for every 0-simplex $S\in A$ exactly one proper component.*

Proof: a and b follow directly from the definitions.

c) This is proved by using the function f and statements a and b of the proposition.

d) This is proved by using a,b,c and the theory of toroidal embeddings (See [KKMS], [O.1] or [O.2]). Note that one needs, in order to apply the theory of toroidal embeddings, to change the innerproduct of $A \simeq \mathbb{R}^{n-1}$ such that the lattice of vertices of A becomes a sublattice of $\mathbb{Z}^{n-1}$ with the standard orthogonal inner product.

Definitions: Let us define now:

$$X_C := \bigcap_{A \ni C} X_{C,A}, \quad X := \bigcap_{A \subset B} X_A, \quad H_B := \bigcup_{A \subset B} H_A.$$

Let us fix an appartment A and a basis $\{x_1,\dots,x_n\}$ of $\mathcal{O}(1)$ belonging to T_A. Now we define a function $r\colon \mathbb{P}_K^{n-1} \to \mathbb{R}$ by:

$$r(p) = \begin{cases} 0 & \text{if } \prod_{i=1}^{n} x_i(p) = 0 \\ \inf \left\{ \prod_{i=1}^{n} \left|\frac{g^* x_i}{x_i}(p)\right| \;\middle|\; g \in PSL(n,K) \right\} & \text{otherwise.} \end{cases}$$

Lemma 2.1:

a) $X_{C,A} \cap X_{C,A'} = X_{C,A} - R^{-1}(\bar{H}_{A'} \backslash \bar{H}_A) = X_{C,A'} - R^{-1}(\bar{H}_A \backslash \bar{H}_{A'})$

b) $X_C = X_{C,A} - R^{-1}(\bar{H}_B \backslash \bar{H}_A)$. *(Here $R\colon X_{C,A} \longrightarrow \overline{X_{C,A}}$ is the reduction-map.)*

Proof: a) This is proved by using the description of $X_{C,A}$ given in proposition 2.1.

b) This is proved by using statement a repeatedly. Note that even though $C \in A$ for an infinite number of appartments, we only have a finite number of different affinoid spaces $X_{C,A}$ for a fixed chamber C.

Lemma 2.2: *Let r be as above. The function r has the following two properties.*

a) $r(p) = 0 \Leftrightarrow p \in H_B$

b) $r(p) \neq 0 \Rightarrow \exists (g \in PSL(n,K)) \prod_{i=1}^{n} \left|\frac{g^* x_i}{x_i}(p)\right| = r(p)$.

Proof: This is clear.

Definition: We call an analytic space Z *locally proper* if Z has two admissible affinoid coverings $\{A_i \mid i \in I\}$ and $\{A'_i \mid i \in I\}$ such that:

$$A_i \Subset A'_i, \ \forall i \in I.$$

If I is finite then Z is proper in the sense of Kiehl. If the two coverings are invariant under the action of a group Γ which acts discontinuously on Z and has a finite number of orbits on the A_i, $i \in I$ then Z/Γ is proper.

Theorem 2.1: a) $X = \bigcup_{C \in B} X_C = \bigcap_{A \subset B} X_A = \mathbb{P}_K^{n-1} - \{K\text{-}rational\ hyperplanes\}$.

b) *The covering $\mathcal{C} := \{X_C \mid C \in B\}$ of X is pure and invariant under the action of $PSL(n,K)$.*

c) *X is locally proper.*

Proof: a) We use the function r defined above. If $r(p)=0$ then $p\in H_B$ and $p\notin X_C$ $\forall C\in B$ since $X_C\subset \mathbb{P}_K^{n-1}-H_B$.

If $r(p)\neq 0$ then $p\in X=\mathbb{P}_K^{n-1}-H_B$ and we can find an element $g\in PSL(n,K)$ such that $r(p)=\prod_{i=1}^{n}|\frac{g^*x_i}{x_i}(p)|$. There exists a chamber $C\in g(A)$ such that $p\in X_{C,A}$. Using lemma 2.1 one sees that in fact $p\in X_C$.

b) The fact that the covering C is pure follows from the fact that for $C,C'\in B$ there exists an appartment $A\subset B$ such that $C,C'\in A$, and proposition 2.2.c. The invariance under the action of $PSL(n,K)$ is clear.

c) This is proved by using the fact that we have:

$$X_C=\{p\in \mathbb{P}_K^{n-1}|\quad |\frac{g^*x_i}{x_i}(p)|=1,\ i=1..n,\ \forall g\in P_C\}$$

Clearly the affinoid space $X_{C,\varepsilon}:=\{p\in \mathbb{P}_K^{n-1}|\varepsilon^{-1}\le |\frac{g^*x_i}{x_i}(p)|\le \varepsilon,\quad i=1..n,\ \forall g\in P_C\}$ with $\varepsilon>1$ is contained in X. Now $X_C \Subset X_{C,\varepsilon}$ and $\{X_{C,\varepsilon}|C\in B\}$ is an admissible covering of X. This proves c.

Remark: We can now construct a map λ: $X\rightarrow B$ as follows:

For $p\in X$ we take an appartment $g(A)\subset B$ such that $\prod_{i=1}^{n}|\frac{g^*x_i}{x_i}(p)|=r(p)$ and we use the function f: $\mathbb{P}_K^{n-1}-H_{gA}\rightarrow g(A)$ to define $\lambda(p)=f(p)$. The proof of theorem 2.1.a shows that this is well defined.

The map λ is the same map as is used in [D] §6 and [Be] §4.3. Now $X_C=\lambda^{-1}(C)$.

Remarks: 1) It is known that the space X has a reduction consisting of a projective space $\mathbb{P}_{\bar{K}}^{n-1}$ with all $\bar{K}$-linear subspaces blown-up for every vertex $S\in B$ (See [Mus] and [Ku]).

Our covering C gives this reduction. This can be calculated by using the theory of toroidal embeddings. First one determines the reduction of X_A (See proposition 2.2.d). Now $X_C\subset X_{C,A}$ is open, so the reduction of X_C is an open subspace of the reduction of $X_{C,A}$. One can determine the reduction of X by using the fact that the parahoric subgroup $P_S\subset G$ is transitive on the chambers C containing S and also on the appartments A containing S.

2) If $\Gamma\subset G$ is not co-compact then in some cases one can construct a subcomplex $B_\Gamma\subset B$ such that Γ acts discontinuously on $X_\Gamma:=\bigcap_{A\subset B_\Gamma} X_A$ and X_Γ/Γ is a proper rigid analytic variety (See [Mus].).

3) The following examples of spaces X/Γ are known:

a) Mumford-curves (See [Mum.1] and [GP].)

b) Mumfords fake projective plane (See [Mum.2] and [I].)

(I will give some other examples of surfaces X/Γ in my thesis.)

§3 THE SPLIT ORTHOGONAL GROUPS

We take $G=PSO(f,K)$, here f is a quadratic form $f=\sum_{i=1}^{l} x_i x_{2l+1-i}$ or $f=x_0^2+\sum_{i=1}^{l} x_i x_{2l+1-i}$. Now G acts on the projective variety Y defined by $f=0$ in $\mathbb{P}_K^{2l-1}$ (resp. $\mathbb{P}_K^{2l}$).

We exclude the case $f=x_1x_4+x_2x_3$. Then $Y\simeq\mathbb{P}_K^1\times\mathbb{P}_K^1$ and G is isogenous with $PSL(2,K)\times PSL(2,K)$. In this case Kurihara's original construction does work (See [Mus]).

First we will indicate why the construction given in §2 does not work in this case. Then we will improve the construction and get a space X_A with properties similar to those stated in proposition 2.2.

Now the torus T_A acts on X_A, but the normalizer N_A of T_A does not. This makes it impossible to construct a symmetric domain for G itself. But we can construct a "symmetric" space for a subgroup $H\subset G$, which stabilises two disjoint maximal K-rational linear isotropic subspaces of Y. Clearly $H\simeq GL(l,K)$, where l is as above.

Now we will show why the construction described in §2 does not work in this situation. We only treat the case $f=\sum_{i=1}^{l} x_i x_{2l+1-i}$. The other case is similar.

Let x_i, $i=1..2l$ be as above and let $T_A\subset G$ be the maximal K-rational torus, which acts diagonally on $\mathbb{P}_K^{2l-1}$ with respect to these coördinates. As in §2 we can now for a chamber C in the appartment $A\subset B$ belonging to T_A define an affinoid subspace $X_{C,A}\subset Y$. We have:

$$X_{C,A}:=Sp\ K<\frac{g^*x_i}{x_i}\mid \forall g\in P_C,\ i=1..2l>\cap Y$$

$$=Sp\ K<\pi^n\frac{x_i}{x_j}\mid n+\alpha_{ij}\in\Delta_{af}>\cap Y$$

The differences with the situation in §2 are the following:

1) For each $n+\alpha_{ij}\in\Delta_{af}$ we have exactly two meromorphic functions on $\mathbb{P}_K^{2l-1}$. They are $\pi^n\frac{x_i}{x_j}$ and $\pi^n\frac{x_{2l+1-j}}{x_{2l+1-i}}$.

2) The components of the reduction of $\bigcup_{C\in A} X_{C,A}$ are not in accordance with the vertices (0-simplices) of A. This is a consequence of the fact that the two meromorphic functions mentioned in 1 need not have the same absolute value for $p\in\bigcup_{C\in A} X_{C,A}$. There do exist

$$p\in\bigcup_{C\in A} X_{C,A}\text{ such that }\left|\pi^n\frac{x_i}{x_j}(p)\right|\neq\left|\pi^n\frac{x_{2l+1-j}}{x_{2l+1-i}}(p)\right|.$$

3) The components of the reduction of $\bigcup_{C\in A} X_{C,A}$ are not proper.

The second problem can be solved by replacing $X_{C,A}$ by another affinoid subspace $\tilde{X}_{C,A}$ defined by:

$$\tilde{X}_{C,A} := \{p\in X_{C,A} \mid |p_i p_{2l+1-i}| = |p_j p_{2l+1-j}| \ \forall i,j=1..2l\}$$

The two meromorphic functions mentioned in 1) have the same absolute value for $p\in \bigcup_{C\in A} \tilde{X}_{C,A}$.

Still the components of the reduction of $\bigcup_{C\in A} \tilde{X}_{C,A}$ are not proper. This is because $\bigcup_{C\in A} \tilde{X}_{C,A}$ and $\bigcup_{C\in A} X_{C,A}$ are rather unnatural subspaces of $Y-H_A$, where H_A is the union of the hyperplanes $x_i=0$ for $i=1..2l$.

In order to make a space X_A with properties as in §2 we need a way to associate to each affine root $n+\alpha_{ij}$ an unique meromorphic function. This function turns out to be $\pi^n\frac{y_i}{y_j}$, where the y_i are defined by:

$$y_i = \begin{cases} x_i & i=1..l \\ \dfrac{x_s x_{2l+1-s}}{x_{2l+1-i}} & i=l+1..2l \end{cases}$$

Here x_s is choosen such that $|x_s x_{2l+1-s}| = \max_{k=1..l} |x_k x_{2l+1-k}|$. So the function is not uniquely defined, but its absolute value is. Note that we have made a choice here. The coördinates x_i, $i=1..l$ define a maximal isotropic subspace $x_1=..=x_l=0$. We could in the definition of the y_i above replace the x_i, $i=1..l$, by some other set of x_j's defining a maximal isotropic subspace.

Definitions: For $i=1..l$ we define for $C\in A$ the following affinoid subspace of Y:

$$X^i_{C,A} := Sp\,K< \frac{x_j x_{2l+1-j}}{x_i x_{2l+1-i}},\ \pi^n\frac{y_r}{y_s} \mid j=1..l,\ n+\alpha_{rs}\in\Delta_{af} > \cap Y$$

Here of course we have: $y_r = \begin{cases} x_r & r=1..l \\ \dfrac{x_i x_{2l+1-i}}{x_{2l+1-r}} & r=l+1..2l \end{cases}$

Furthermore we define:

$$X^i_A := \bigcup_{C\in A} X^i_{C,A}, \qquad X_A := \bigcup_{i=1}^{l} X^i_A = \bigcup_{C\in A}\bigcup_{i=1}^{l} X^i_{C,A}$$

We now can define the function $f_i\colon X^i_A \to A(=\mathbb{R}^l)$ by:

$$f_i(p)=q \Leftrightarrow v\left(\frac{y_r}{y_s}(p)\right) = \alpha_{rs}(q) \quad r,s=1..2l,\ r\neq s$$

Since $f_i \equiv f_j$ on $X^i_A \cap X^j_A$ we can glue the functions f_i together into a function $f\colon X_A \to A$. The function f is well defined on $Y-Z_A$, where Z_A is the union of the hyperplanes $x_i=0$ for $i=1..l$ and the maximal isotropic subspace given by $x_{l+1}=x_{l+2}=\ldots=x_{2l}=0$.

Proposition 3.1: a) *$f_i(p)\in C \Leftrightarrow p\in X^i_{C,A}$ for $p\in X^i_A$.*

b) *$f(p)\in C \Leftrightarrow p\in \bigcup_{i=1}^{l} X^i_{C,A}$ for $p\in Y-Z_A$.*

Therefore $X_A = Y-Z_A$.

c) *The covering $C=\{X^i_{C,A} \mid i=1..l,\ C\in A\}$ of X_A is pure.*

d) *X_A is locally proper.*

Proof: a) and b) follow directly from the definitions of f_i and f.

c) Using the definition of f_i one sees that the covering $\{X^i_{C,A}|C\in A\}$ of X^i_A is pure. So we only have to look at affinoids $X^i_{C,A}$ and $X^j_{C',A}$ $i\neq j$. In this case $X^i_{C,A}\cap X^j_{C',A}$ is contained in $X^i_A\cap X^j_A$. Now using f_i or f_j one sees that the intersection is non-empty if and only if $C\cap C'\neq\phi$. Furthermore the intersection is open.

d) This can be proved by using the admissable covering $\{X^{i,\varepsilon}_{C,A}|i=1..l,\ C\in A\}$ of X_A. Here $X^{i,\varepsilon}_{C,A}:=\{p\in Y|\ \left|\frac{x_jx_{2l+1-j}}{x_ix_{2l+1-i}}(p)\right|<\varepsilon,\ \left|\pi^n\frac{y_r}{y_s}(p)\right|<\varepsilon,\ j=1..l,\ n+\alpha_{rs}\in\Delta_{af}\}$. Of course we have for $\varepsilon>1$:

$$X^i_{C,A}\Subset X^{i,\varepsilon}_{C,A}$$

Remark: Instead of $X^i_{C,A}$ we can look at the affinoid subspaces

$$V^i_{C,A}:=Sp\ K<\pi^n\frac{y_r}{y_s},\ \frac{x_jx_{2l+1-j}}{x_ix_{2l+1-i}}\ |\ j=1..l,\ n+\alpha_{rs}\in\Delta_{af}>\ \text{of}\ \mathbb{P}^{2l-1}_K-Z_A.$$

This gives us a pure T-invariant covering $\{V^i_{C,A}|i=1..l,\ C\in A\}$ of $\mathbb{P}^{2l-1}_K-Z_A$. We have $X^i_{C,A}=V^i_{C,A}\cap Y$. As in proposition 3.1 one now proves that $\mathbb{P}^{2l-1}_K-Z_A$ is locally proper.

Remark: For the standard chamber C of §1 we can give $X^i_{C,A}$ explicitly as follows:

$$X^i_{C,A}=Sp\ K<\frac{x_1}{x_2},\ \frac{x_2}{x_3},\dots,\frac{x_{l-1}}{x_l},\ \frac{x_lx_{l-1}}{x_ix_{2l+1-i}}\ \frac{\pi x_ix_{2l+1-i}}{x_1x_2},\ \frac{x_jx_{2l+1-j}}{x_ix_{2l+1-i}}\ |\ j=1..l,\ j\neq i>\cap Y$$

The generators of the affinoid algebra of $X^i_{C,A}$ satisfy the following two relations:

a) $$\frac{x_1}{x_2}\cdot\left(\frac{x_2}{x_3}\right)^2\cdot\dots\cdot\left(\frac{x_{l-2}}{x_{l-1}}\right)^2\left(\frac{x_{l-1}}{x_l}\right)\cdot\frac{x_{l-1}x_l}{x_ix_{2l+1-i}}\cdot\frac{\pi x_ix_{2l+1-i}}{x_1x_2}=\pi,\ l>3$$

$$\frac{x_1}{x_2}\cdot\frac{x_2}{x_3}\cdot\frac{x_3x_2}{x_ix_{2l+1-i}}\cdot\frac{\pi x_ix_{2l+1-i}}{x_1x_2}=\pi,\ l=3$$

b) $$\sum_{j=1}^{l}\frac{x_jx_{2l+1-j}}{x_ix_{2l+1-i}}=0$$

Relation a corresponds with the relation satisfied by the affine roots forming a simple basis Δ_{af} of Φ_{af}. Relation b corresponds with the equation $f=0$. This shows that we have:

$$X^i_{C,A}=R^i_{C,A}\times S^i_{C,A},\ \text{with}\ R^i_{C,A}:=Sp\,K<\pi^n\frac{y_r}{y_s}\ |\ n+\alpha_{rs}\in\Delta_{af}>$$

and $S^i_{C,A}:=Sp\ K<\frac{x_jx_{2l+1-j}}{x_ix_{2l+1-i}}\ |\ j=1..l>/(f=0)$.

Proposition 3.2: a) *The reduction of X^i_A with respect to the covering $\{X^i_{C,A}|C\in A\}$ consists of one component D_S for every vertex $S\in A$.*

b) *$D_S\simeq E_S\times\mathbb{A}^{l-2}_{\bar{k}}$ and E_S is proper and non-singular.*

c) *The reduction of $X_A=Y-Z_A$ with respect to the covering $\{X^i_{C,A}|i=1..l,\ C\in A\}$ consists of one proper component F_S for every vertex $S\in A$.*

These components F_S are such that we have a surjective map

$$\varphi\colon F_S\longrightarrow\mathbb{P}^{l-2}_{\bar{k}}\ \textit{with}\ \varphi^{-1}(p)\simeq E_S\ \textit{for all}\ p\in\mathbb{P}^{l-2}_{\bar{k}}.$$

Proof: a) Relation a in the remark above shows that the reduction of $X^i_{C,A}$ has exactly one component for every vertex $S \in C$. Glueing these components together for all $C \in A$ containing S we get the component D_S of the reduction of X^i_A.
b) The function $f_i: X^i_A \to A$ maps $X^i_{C,A}$ into $C \in A$. The image $f_i(p)$ of $p \in X^i_{C,A} = R^i_{C,A} \times S^i_{C,A}$ depends only on the coördinates of the projection of p into $R^i_{C,A}$. Now the appartment A is, after having made the lattice of its vertices a sublattice of $\mathbb{Z}^l$, the picture of the torus embedding belonging to $\{R^i_{C,A} | C \in A\}$. The theory of torus embeddings tells us that we have for every $S \in A$ a component E_S in the reduction of $\{R^i_{C,A} | C \in A\}$ which is proper and non-singular.

The reduction $\overline{S^i_{C,A}}$ of $S^i_{C,A}$ is of course an $\mathbb{A}^{l-2}_{\bar{k}}$. The reduction of $X^i_{C,A}$ is $\overline{X^i_{C,A}} = \overline{R^i_{C,A}} \times \overline{S^i_{C,A}} = \overline{R^i_{C,A}} \times \mathbb{A}^{l-2}_{\bar{k}}$. The reduction of X^i_A is now easily determined, since we only have to glue along open subspaces $U \times \mathbb{A}^{l-2}_{\bar{k}}$ where $U \subset \overline{R^i_{C,A}}$. So the reduction of X^i_A consists of one component $E_S \times \mathbb{A}^{l-2}_{\bar{k}}$ for every vertex $S \in A$.
c) We already know the reduction of the X^i_A for $i = 1..l$. To determine the reduction of X_A we only have to glue these together. The intersection $X^i_{C,A} \cap X^j_{C,A}$ is given by $\left|\frac{x_i x_{2l+1-i}}{x_j x_{2l+1-j}}\right| = 1$ in both $X^i_{C,A}$ and $X^j_{C,A}$.

The reductions $\overline{S^i_{C,A}}$ of $S^i_{C,A}$ for a fixed chamber C glue together into an $\mathbb{P}^{l-2}_{\bar{k}}$ with coördinates $x_i x_{2l+1-i}$, $i = 1..l$ satisfying the relation $\sum_{i=1}^{l} x_i x_{2l+1-i} = 0$. For a fixed $\bar{p} \in \mathbb{P}^{l-2}_{\bar{k}}$ we only have to look at one $\overline{S^i_{C,A}}$ with $\bar{p} \in \overline{S^i_{C,A}}$. Now part b of the proposition shows that the reductions of the $X^i_{C,A}$, $C \in A$ and i fixed glue together into components $E_S \times \mathbb{A}^{l-2}$. Now it is clear that the reduction of X_A consists of exactly one component F_S for every vertex $S \in A$, such that we have a surjective map $\varphi: F_S \to \mathbb{P}^{l-2}_{\bar{k}}$ with $\varphi^{-1}(p) \simeq E_S$ for all $p \in \mathbb{P}^{l-2}_{\bar{k}}$.

Remark: Looking more precisely at the appartment A and the associated torus-embedding one sees that the reduction of X_A has the properties:
1) $F_S \cap F_{S'} \neq \phi \iff \{S, S'\}$ is a simplex in A
2) $F_S \cap F_{S'} \neq \phi \iff F_S \cap F_{S'}$ is of codimension one in F_S and $F_{S'}$

Remark: Now we are going to construct for the subgroup $H \subset G$, stabilising the pair of maximal isotropic subspaces $x_1 = x_2 = \ldots = x_l = 0$ and $x_{l+1} = x_{l+2} = \ldots = x_{2l} = 0$ a space $X_H \subset Y$ such that X_H / Γ is a proper rigid analytic variety for $\Gamma \subset H$ discrete and co-compact.

First we define a subcomplex $B_H \subset B$ on which H acts. Then $X_H := \bigcap_{A \subset B_H} X_A$ is a locally proper space which has a H-invariant affinoid covering.

Definition: Let H be as above. Let $A \subset B$ be a fixed appartment such that the maximal K-rational torus T_A belonging to A stabilises the two maximal isotropic subspaces mentioned above, so $T_A \subset H$. Now we define the subcomplex $B_H \subset B$ as follows:

$$B_H := \bigcup_{g \in H} g \cdot A \subset B$$

Lemma 3.1: *$B_H \simeq \tilde{B} \times \mathbb{R}$, where $\tilde{B}$ is the building of $SL(l,K)$.*

Proof: It is clear that $H \simeq GL(l,K)$. Therefore all maximal K-rational tori $T \subset H$ are conjugated and correspond with an appartment $A \subset B_H$.
To every maximal torus $T \subset H$ corresponds exactly one maximal torus $\tilde{T} \subset SL(l,K)$. We have $\mathrm{rank}(T) = \mathrm{rank}(\tilde{T}) + 1$ and T is isogenous with $\tilde{T} \times T_C$, where T_C is the torus of rank one which is in the center of $H = GL(l,K)$.

Now T_C acts on $\mathbb{R}$ and $\tilde{T}$ on $\tilde{B}$. This proves the lemma. Note that we don't care about the simplicial structure.

Lemma 3.2: *Let $A, A' \subset B_H$ be appartments such that $A \cap A'$ contains a chamber C. If $g \in P_C$ such that $g(A) = A'$ then $g \in H$.*

Proof: Let m and m' denote the two maximal isotropic subspaces of Y stabilised by H. Since $T_A, T_{A'} \subset H$, they stabilise m and m'.

Furthermore there exist $g \in P_C$ such that $g(A) = A'$ and $T_{A'} = gT_Ag^{-1}$. Now $gT_Ag^{-1}(m) = m$ implies that $T_Ag^{-1}(m) = g^{-1}(m)$. So the maximal isotropic subspace $g^{-1}(m)$ is stabilised by T_A. This proves that $g^{-1}(m) = w(m)$ for some $w \in W$, where W is the finite Weylgroup N_A/T_A. This cannot be, unless $w = id$. So $g(m) = m$. Of course also $g(m') = m'$ and therefore $g \in H$.

Definitions: Let $T_A \subset H$ be a torus and let x_i, $i = 1..2l$ be the associated coördinates of $\mathbb{P}_K^{2l-1}$ such that T_A acts diagonally. Let m be the maximal isotropic subspace $x_1 = x_2 = \ldots = x_l = 0$ and let m' be the maximal isotropic subspace $x_{l+1} = x_{l+2} = \ldots = x_{2l} = 0$.
Let $H \subset G$ be the stabiliser of m and m'.
Let Z_A be the union of the hyperplanes $x_i = 0$, $i = 1 \ldots l$ and m'.
Now $X_A = Y - Z_A$. For every $A' \subset B_H$ there is a $g \in H$ such that $g(A) = A'$. Our choice of Z_A is such that $X_{A'} = g(X_A)$ and $Z_{A'} = g(Z_A)$ are uniquely defined. Therefore the following definitions are allowed:

$$X_H := \bigcap_{A' \subset B_H} X_{A'}, \qquad Z_H := \bigcup_{A' \subset B_H} Z_{A'}$$

We will now construct a pure H-invariant covering of X_H.

Lemma 3.3:

a) $p \in X^i_{C,A} \Rightarrow \left| \frac{g^* x_j}{y_j}(p) \right| \le 1 \quad \forall g \in P_C,\ \forall j = 1..2l$

b) $p \in X^i_{C,A} \Rightarrow \left| \frac{g^* x_j g^* x_{2l+1-i}}{x_i x_{2l+1-i}}(p) \right| \le 1 \ \forall g \in P_C,\ \forall j = 1..l$

Proof: a) The definition of the y_j is such that $\left| \frac{x_j}{y_j}(p) \right| \le 1 \ \forall j = 1..2l$ for $p \in X^i_{C,A}$. We can now associate to $p \in X^i_{C,A}$ a point $\tilde{p} \in \mathbb{P}_K^{2l-1}$ defined by $\tilde{p}_j := y_j(p)$, $j = 1..2l$. The definition of the y_j now shows that $\tilde{p} \in V_{C,A} := Sp\, K < \pi^n \frac{x_r}{x_s} | n + \alpha_{rs} \in \Delta_{af} >$.

As in lemma 2.1.b one can show that $V_{C,A} = Sp\, K < \frac{g^* x_j}{x_j} | g \in P_C,\ j = 1..2l >$.

Therefore we have: $\left| \frac{g^* x_j(\tilde{p})}{x_j(\tilde{p})} \right| = \left| \frac{g^* x_j(\tilde{p})}{y_j(\tilde{p})} \right| \le 1.$

For $g\in P_C$ we have: $g^*x_j(\tilde{p})=\sum_{k=1}^{2l}a_{jk}x_k(\tilde{p})$ with $\left|\frac{a_{jk}x_k}{x_j}(\tilde{p})\right|\le 1$, since $\beta=v(a_{jk})+\alpha_{kj}\in\Phi^+_{af}$, i.e. $\beta=\sum_{\beta_i\in\Delta_{af}}n_i\beta_i$ with $n_i\in\mathbb{Z}_{\ge 0}$.

So we have: $\left|\frac{g^*x_j}{x_j}(\tilde{p})\right|\le \max_{k=1..2l}\left|a_{jk}\frac{x_k}{x_j}(\tilde{p})\right|$

$$\Rightarrow \left|\frac{g^*x_i}{y_j}(p)\right|=\left|\frac{g^*x_i(p)}{x_j(\tilde{p})}\right|\le \max_{k=1..2l}\left|a_{jk}\frac{x_k(p)}{x_j(\tilde{p})}\right|\le \max_{k=1..2l}\left|a_{jk}\frac{x_k(\tilde{p})}{x_j(\tilde{p})}\right|\le 1$$

This proves *a*.

b) This follows from *a* since we have for $p\in X^i_{C,A}$:

$$|g^*x_jg^*x_{2l+1-j}(p)|\le|y_jy_{2l+1-j}(p)|=|x_ix_{2l+1-i}(p)|$$

Remark: A chamber $C\in B_H$ is contained in infinitely many appartments $A\subset B_H$. But there exists a finite number N of appartments $A_1,\dots,A_N$ such that every $X^i_{C,A}$ is equal to X^i_{C,A_j} for some $j\in\{1,\dots,N\}$. The appartments A_j, $j\in\{1,\dots,N\}$ depend on C of course.

Definition: Let $C\in B_H$ and N be as above.
For $s\in I=\{1,\dots,l\}^N$ we now define:

$$X^s_C:=\bigcap_{j=1}^{N}X^{s_j}_{C,A_j}$$

Lemma 3.4:

a) $\quad X^i_{C,A}\cap X^j_{C,A'}\subset X^i_{C,A'},\ X^j_{C,A'}$ *is open.*

b) $\quad X^s_C\subset X^{s_j}_{C,A_j}$ *is open for all* $j=1..N$.

Proof: a) Let $g\in P_C$ be such that $gA=A'$. Now lemma 3.3.a gives us:

$$p\in X^i_{C,A}\cap X^j_{C,gA}\Rightarrow\left|\frac{g^*x_k}{x_k}(p)\right|=\left|\frac{g^*y_k}{y_k}(p)\right|=1,\ k=1..l \qquad (*)$$

From lemma 3.3.b we conclude:

$$p\in X^i_{C,A}\cap X^j_{C,gA}\Rightarrow\left|\frac{g^*x_ig^*x_{2l+1-i}}{x_ix_{2l+1-i}}(p)\right|=1 \qquad (**)$$

The equations (*) and (**) together define an open subspace of $X^i_{C,A}$ and $X^j_{C,A'}$ which is in fact $C^i_{C,A}\cap X^j_{C,A'}$.

b) This follows from the repeated use of *a*.

Definition: Let $r: Y \to \mathbb{R}$ be defined by

$$r(p)=\begin{cases}0 & \text{if } \prod_{i=1}^{l}|x_i(p)|=0\\ \inf\left\{\frac{1}{|\det(g)|}\prod_{i=1}^{l}\left|\frac{g^*x_i}{x_i}(p)\right| \;|\; g\in H\right\} & \text{otherwise}\end{cases}$$

Here $\det(g)$ is the determinant of g as an element of $GL(l,K)$.

Lemma 3.5:

a) $r(p)=0 \Rightarrow \exists(g\in H)\ \prod_{i=1}^{l}|g^*x_i(p)|=0$

b) $r(p)\neq 0 \Rightarrow \exists(g\in H)\ \frac{1}{|\det(g)|}\prod_{i=1}^{l}\left|\frac{g^*x_i}{x_i}(p)\right|=r(p)$

c) $r(p)\neq 0 \wedge p\notin m' \Leftrightarrow p\in X_H$

Proof: a) This follows from the completeness of K.

b) The coördinates of $p\in\mathbb{P}_K^{2l-1}$ are defined over some finite extension $L\supset K$. The set $\{|s| \;|s\in L,\ |s|\geq r>0\}\subset\mathbb{R}$ is discrete. From this b follows

c) This follows from a and b.

Theorem 3.1: a) $X_H=\bigcup_{C\in B_H}\bigcup_{s\in I}X_C^s$

b) *The covering* $\mathcal{C}=\{X_C^s|C\in B_H,\ s\in I\}$ *of* X_H *is pure and* H*-invariant.*

c) X_H *is locally proper.*

Proof: a) We will show that for every $p\in X_H$ there exists a $C\in B_H$ such that $p\in X_C^s$ for some $s\in I$.

Since $p\in X_H$ the set $F=\{g\in H|\ \frac{1}{|\det(g)|}\prod_{i=1}^{l}\left|\frac{g^*x_i}{x_i}(p)\right|=r(p)\}$ exists and is non-empty. Now $t=\inf\{\max_{i=1..l}|\frac{g^*x_i g^*x_{2l+1-i}}{x_k x_{2l+1-k}}(p)|\ |g\in F\}$ exists for a fixed k such that $x_kx_{2l+1-k}(p)\neq 0$ and $t\neq 0$, since $p\notin m'$. Let F' be defined by $F'=\{g\in F|\max_{i=1..l}|g^*x_ig^*x_{2l+1-i}(p)|=t\cdot|x_kx_{2l+1-k}(p)|\}$.

Now F' is non-empty and in fact $F'=F$ (we will not prove this).

Take an element $g\in F'$. There exists a chamber $C\in gA$ such that $p\in X_{C,gA}^i$ for some $i\in\{1..l\}$. In fact $p\in X_C^s$ for some $s\in I$. This follows from lemma 3.2 and lemma 3.4 and the choice of g.

b) The fact that $\mathcal{C}$ is H-invariant is clear. The pureness of the covering follows from proposition 3.1 and the fact that for $C,C'\in B_H$ there exists an appartment $A\subset B_H$ such that $C,C'\in A$.

c) This is proved as in proposition 3.1.d and theorem 2.1.c.

Remark: Let T_C denote the torus of rank one in the center of H. Let $\Gamma_0\subset T_C$ be a discrete co-compact subgroup. Let $\Gamma_1\subset SL(l,K)$ be a discrete co-compact subgroup. Then $\Gamma=\Gamma_0\times\Gamma_1$ is a discrete co-compact subgroup of $H\cong GL(l,K)$.

Proposition 3.3: *Let* $\Gamma\subset H$ *be a discrete co-compact subgroup without torsion of the form* $\Gamma=\Gamma_0\times\Gamma_1$, *where* Γ_0 *and* Γ_1 *are as in the remark above. Now the proper rigid analytic variety* X_H/Γ *is non-projective.*

Proof: We have a surjective map $\varphi: X_H \longrightarrow \mathbb{P}_K^{l-1} - \{K\text{-rational hyperplanes}\} := \Omega^{l-1}$, defined by $\varphi((x_1,..,x_{2l})) = (x_1,..,x_l)$. The fibre of a point $p \in \Omega^{l-1}$ is isomorphic to $\mathbb{A}_K^{l-1} - \{0\}$.

Since $\Gamma = \Gamma_0 \times \Gamma_1$ the morphism φ induces a map $\psi: X_H/\Gamma \longrightarrow \Omega^{l-1}/\Gamma_1$ with fibre $\mathbb{A}_K^{l-1} - \{0\}/\Gamma_0$. The algebraic dimension of the Hopf-manifold $\mathbb{A}_K^{l-1} - \{0\}/\Gamma_0$ is strictly less than l-1 (See [Mus]). Furthermore Ω^{l-1}/Γ_1 is an projective variety of dimension l-1.

Therefore the algebraic dimension of X_H/Γ is strictly less than $2l$-2. So X_H/Γ can not be a projective variety.

Remarks: 1) In the proofs above we never needed the fact that $p \in X_H$ satisfies the equation $f=0$. We can delete this equation in the definition of the X_C^s, $C \in B_H$, $s \in I$. This gives us a pure H-invariant covering of $\mathbb{P}_K^{2l-1} - Z_H$. The space $\mathbb{P}_K^{2l-1} - Z_H/\Gamma$ for $\Gamma \subset H$ discrete and co-compact is proper.

2) All we did in §3 is also true for Y defined by $f = x_0^2 + \sum_{i=1}^{l} x_i x_{2l+1-i}$ in $\mathbb{P}_K^{2l}$. In this case we have for $A \subset B$, where B is the building of $G = PSO(f,K)$ again a space X_A. Now $X_A := Y - Z_A$, where Z_A is the union of the hyperplanes $x_i = 0$, $i = 1..l$ and the space m defined by $x_0 = x_{l+1} = \ldots = x_{2l} = 0$. Note that $m \cap Y$ is the maximal isotropic space $x_{l+1} = \ldots = x_{2l} = 0$.

We can cover X_A with the affinoids $X_{C,A}^i$, $C \in A$, $i = 0..l$.

Now $X_{C,A}^i := Sp\, K < \pi^n \frac{y_r}{y_s}, \frac{x_j x_{2l+1-j}}{x_i x_{2l+1-i}}, \frac{x_0^2}{x_i x_{2l+1-i}} \mid n + \alpha_{rs} \in \Delta_{af},\ j = 1..l >$ for $i = 1..l$.
Here y_j is defined by:

$$y_j = \begin{cases} x_j, & j = 1..l \\ \dfrac{x_i x_{2l+1-i}}{x_{2l+1-j}}, & j = l+1..2l \\ \sqrt{x_i x_{2l+1-j}}, & j = 0 \end{cases}$$

We define $X_{C,A}^0$ by $X_{C,A}^0 = Sp\, K < \pi^n \frac{y_r}{y_s}, \frac{x_j x_{2l+1-j}}{x_0^2} \mid n + \alpha_{rs} \in \Delta_{af},\ j = 1..l > \cap Y$ and here the y_j are given by:

$$y_j = \begin{cases} x_j & j = 0..l \\ \dfrac{x_0^2}{x_{2l+1-j}}, & j = l+1..2l \end{cases}$$

The subgroup H is now the stabiliser of the maximal isotropic subspaces defined by $x_1 = x_2 = \ldots = x_l = 0$ and $x_{l+1} = \ldots = x_{2l} = 0$.

3) For a discrete co-compact subgroup $\Gamma \subset PSO(f,K)$ it is not clear yet how to construct a space $X \subset Y$ such that X/Γ proper. If we assume that Γ is neat (See [B]) then $\Gamma \cap N_A \subset T_A$, here N_A is the normaliser of T_A. Now $\Gamma \cap N_A$ acts properly on X_A. The group Γ has infinitely many orbits on appartments $A \subset B$. It is not clear wether there exists a Γ-invariant choice of the X_A such that $\bigcap_{A \subset B} X_A/\Gamma$ is proper.

If such a choice exists, this would give the p-adic analogon of the bounded symmetric domain which has $SO(n)/SO(n-2) \times SO(2)$, $(n = 2l(+1))$ as its compact dual (See [Ku] §0.4). This symmetric space is denoted by BDI $(q=2)$ in [H] p.354 and by IV_n^* in [N] p.114.

REFERENCES

[AMRT] A. Ash, D. Mumford, M. Rapoport and Y. Tai, *Smooth compactification of locally symmetric varieties*, Math. Sci. Press, 1975.

[BB] W.L. Baily and A. Borel, Compactification of arithmetic quotients of bounded symmetric domains, *Ann. of Math.* **84** (1966), 442–528.

[Be] V. Berkovich, Non-archimedean analytic spaces and buildings of semi-simple groups, *Preprint I.H.E.S.*, february 1989.

[B] A. Borel, *Introduction aux groupes arithmétiques*, Hermann, 1969.

[BS] A. Borel and J.P. Serre, Corners and arithmetic groups, *Comm. Math. Helv.* **48** (1973), 436–491.

[BGR] S. Bosch, U. Güntzer and R. Remmert, *Non-archimedean analysis*, Springer Verlag, 1984.

[BT] F. Bruhat and J. Tits, Groupes réductifs sur un corps local I: Données radicielles valuées, *Publ. Math. I.H.E.S.* **41** (1972), 5–251.

[C] É. Cartan, Sur les domaines bornes homogènes de l'espace de n variables complexes, *Abh. Math. Sem. Univ. Hamburg* **11** (1935), 116–162; Also, *Oeuvres Complètes* part I, 1259–1308.

[D] V.G. Drinfeld, Elliptic Modules, *Math. USSR-Sb.* **23** (1974), 561–592

[F] G. Faltings, F-isocrystals on open varieties: Results and conjectures, *Preprint Princeton University*, 1989.

[FP] J. Fresnel et M. van der Put, Geométrie analytique rigide et applications, *Progress in Math.* **18**, Birkhäuser, 1981.

[GP] L. Gerritzen and M. van der Put, Schottky groups and Mumford curves, *Lect. Notes in Math.* **817**, Springer Verlag, 1980.

[H] S. Helgason, *Differential geometry and symmetric spaces*, Academic Press, 1962.

[I] M.N. Ishida, An elliptic surface covered by Mumford's fake projective plane, *Tôhoku Math. J.* **40** (1988), 367–396.

[Kan] W.M. Kantor, Reflections on concrete buildings, *Geometricae dedicata* **25** (1988), 121–145.

[KKMS] G. Kempf, F. Knudsen, D. Mumford and B. Saint-Donat, Toroïdal embeddings I, *Lect. Notes in Math.* **339**, Springer Verlag, 1973.

[Ku] A. Kurihara, Construction of p-adic unit balls and the Hirzebruch proportionality, *Amer. J. Math.* **102** (1980), 565–648.

[Mac] I.G. Macdonald, Affine root systems and Dedekind's η function, *Inv. Math.* **15** (1972), 91–143.

[Mar] G.A. Margulis, Arithmeticity of irreducible lattices in the semisimple groups of rank greater than one, *Inv. Math.* **76** (1984), 93–120.

[Mum.1] D. Mumford, An analytic construction of degenerating curves over complete local rings, *Compositio Math.* **24** (1972), 129–174.

[Mum.2] D. Mumford, An algebraic surface with K ample, $(K^2)=9$, $p_g=q=0$, *Amer. J. Math.* **101** (1979), 233–244.

[Mus] G.A. Mustafin, Nonarchimedean uniformization, *Math. USSR-Sb.* **34** (1978), 187–214.

[N] Y. Namikawa, Toroidal compactification of Siegel spaces, *Lect. Notes in Math.* **812**, Springer Verlag, 1980.

[O.1] T. Oda, Lectures on torus embeddings and applications, *Tata Inst. of Fund. Research* **58**, Springer Verlag, 1978.

[O.2] T. Oda, *Convex bodies and algebraic geometry*, Springer Verlag, 1988.

[S] J.P. Serre, Arithmetic groups, in *"Homological group theory"* edited by C.T.C. Wall, *London Math. Soc. Lect. Notes Series* **36** (1979), 105–136.

[T.1] J. Tits, Travaux de Margulis sur les sous-groupes discrets de groupes de Lie, Sem. Bourbaki 1975/76, *Lect. Notes in Math.* **567**, Springer Verlag, 1977, 174–190.

[T.2] J. Tits, Reductive groups over local fields, *Proc. A.M.S. Symp. Pure Math.* **33** (1979), 29–69.

[V] T.N. Venkataramana, On superrigidity and arithmeticity of lattices in semisimple groups over local fields of arbitrary characteristic, *Inv. Math.* **92** (1988), 255–306.

LIST OF PARTICIPANTS

Alan Adolphson, Department of Mathematics, Oklahoma State University, Stillwater, 74078 Oklahoma, U.S.A.

Yves André, Intitut H. Poincaré, 11 rue P. et M. Curie, F-75231 Paris 5, France

Jesùs Araujo, Departamento de Matemàticas, E.T.S.I. Industriales, Castiello de Bernueces, Universidad de Oviedo, 33204 Gijon, Spain

Francesco Baldassarri, Dipartimento di Matematica, Università di Padova, Via Belzoni 7, 35131 Padova, Italy

Edoardo Ballico, Dipartimento di Matematica, Università di Trento, 38050 Povo (TN), Italy

Luca Barbieri Viale, Dipartimento di Matematica, Università di Genova, Via L.B. Alberti 4, 16132 Genova, Italy

Jose M. Bayod, Departamento de Teoria de Funciones, Facultad de Ciencias, Avda. de los Castros s/n, 39071 Santander, Spain

Vladimir Berkovich, Department of Theoretical Mathematics, The Weizmann Institute of Science, P.O.B. 26, 76100 Rehovot, Israel

Pierre Berthelot, IRMAR, Université de Rennes 1, Campus de Beaulieu, F-35042 Rennes Cedex, France

Daniel Bertrand, Université de Paris VI, UER de Mathématiques, Tour 46 5ème Etage 45-46, F-75252 Paris Cedex 05, France

Frits Beukers, Mathematics Institute, State University of Utrecht, Budapestlaan 6, P.O. Box 80.010, 3508TA Utrecht, The Netherlands

Siegfried Bosch, Mathematisches Institut, Westfälische Wilhelms-Universität, Roxeler Straße 64, 4400 Münster, West Germany

Abdelbaki Boutabaâ, G 511, R.U. "Jean Zay", F-92160 Antony, France

Stefaan Caenepeel, Vrije Universiteit Brussel, Faculty of Applied Sciences, Pleinlaan 2, B-1050 Brussel, Belgium

Maurizio Candilera, Dipartimento di Matematica, Università di Padova, Via Belzoni 7, 35131 Padova, Italy

Michel Carpentier, Mathématiques, Université Paris VI, 4 Place Jussieu, F-75230 Paris Cedex 05, France

Pierrette Cassou Nogues, Mathématiques et Informatique, Univ. Bordeaux 1, 351 cours de la Libération, 33405 Talence Cedex, France

Bruno Chiarellotto, Dipartimento di Matematica, Università di Padova, Via Belzoni 7, 35131 Padova, Italy

Gilles Christol, Université de Paris VI, UER de Mathématiques, Tour 47 5ème Etage 45-46, 75230 Paris Cedex 05, France

Robert Coleman, Department of Mathematics, University of California, Berkeley Ca. 94720, USA

Matthijs Coster, C.W.I., Kruislaan 413, 1098 SJ Amsterdam, The Netherlands

Valentino Cristante, Dipartimento di Matematica, Università di Padova, Via Belzoni 7, 35131 Padova, Italy

Nicole de Grande-de Kimpe, Vrije Univ. Brussel, Faculteit Wetenschappen, Plenlann 2, 10F7, B-1050 Brussel, Belgium

Jan Denef, Department of Mathematics, University of Leuven, Celestijnenlaan 200B, B-3030 Leuven, Belgium

Bernard Dwork, Department of Mathematics, Fine Hall, Princeton University, Princeton NJ 08540, USA

Alain Escassut, Départment de Mathématiques Pures, Université Blaise Pascal, Complex des Cézaux BP45, F-63170 Aubière, France

J. Y. Etesse, IRMAR, Univ. de Rennes 1, Campus de Beaulieu, F-35042 Rennes Cedex, France

Jean Fresnel, UER Mathématiques et Informatique, Université Bordeaux 1, 351 cours de la Libération, F-33405 Talence Cedex, France

Ernst Ulrich Gekeler, Institute for Advanced Study, School of Mathematics, Princeton NJ 08540, U.S.A.

Giovanni Gerotto, Dipartimento di Matematica, Università di Padova, Via Belzoni 7, 35100 Padova, Italy

Hiroshi Gunji, Department of Mathematics, University of Wisconsin, Madison, WI 53703, U.S.A.

Frank Herrlich, Mathematisches Institut, Ruhr Universität, Postfach 102148, D-4630 Bochum 1, West Germany

Ernst Kani, Department of Mathematics, Queen's University, Kingston K71 3N6, Ontario, Canada

Ha Huy Khoai, Institute of Mathematics, P.O. Box 631, Buu dien Bo Ho, 10000 Hanoi, Vietnam

Pierre Jarraud, 5 Avenue de la Porte de Villiers, F-75017 Paris, France

Bernard Le Stum, 12 rue de Brest, 35000 Rennes, France

Quing Liu, UER Mathématiques et Informatique, Université Bordeaux 1, 351 cours de la Libération, 33405 Talence Cedex, France

Fraçois Loeser, Centre de Mathématiques, Ecole Polytechnique, F-91128 Palaiseau Cedex, France

Werner Lütkebohmert, Mathematisches Institut, Universität Münster, Einsteinstr. 62, D-4400 Münster, West Germany

J. Martinez-Maurica, Departamento de Teoria de Funciones, Facultad de Ciencias, Avda. de los Castros, s/n, 3907, Santander, Spain

M. Matignon, UER Mathématiques et Informatique, Université Bordeaux 1, 351 cours de la Libération, F-33405 Talence Cedex, France

Zoghman Mebkhout, UER de Mathématiques L.A. n. 212, Université Paris VII, 2 Place Jussieu, F-25175 Paris Cedex 05, France

Diane Meuser, Department of Mathematics, Boston University, Boston, Ma 02215, U.S.A.

Yasuo Morita, Department of Mathematics, Faculty of Sciences, Tohoku University, Aoba, Sendai 980, Japan

Elhan Motzkin, 196 rue du château des rentiers, F-75013 Paris, France

Samuel Navarro H., Dpto. de Matemática, Universidad Santiago, Casilla 5659 Correo 2, Santiago, Chile

Peter Norman, Department of Mathematics, University of Massachusetts, Amherst, 01002 Massachusetts, U.S.A.

Arthur Ogus, Department of Mathematics, University of California at Berkeley, Berkeley, Ca. 94720, USA

Pier Ivan Pastro, Dipartimento di Matematica, Università di Padova, Via Belzoni 7, 35131 Padova, Italy

Meinolf Piwek, Ruhr-Universität Bochum, Fakultät für Mathematik, Postfach 102 148, D-4630 Bochum, West Germany

Marc Polzin, UER Mathématique et Informatique, Université de Bordeaux 1, 351 cours de la Libèration, F-33405 Talence, France

Marc Reversat, Laboratoire D'Analyse sur les Varietès, Université Paul Sabatier, 118 route de Narbonne, F-31062 Toulouse Cedex, France

Alain Salinier, 6, Avenue Montjovis, 87100 Limoges, France

Paul J. Sally, Department of Mathematics, University of Chicago, 5734 South University Avenue, Chicago, 60637 Illinois, U.S.A.

Roberto Sanchez-Peregrino, Dipartimento di Matematica, Università di Padova, Via Belzoni 7, 35100 Padova, Italy

Roberto Scaramuzzi, Department of Mathematics, Louisiana State University, Baton Rouge, LA 70803, USA

Wim H. Schikhof, Math. Institut, Katholike Universiteit, Toernooiveld, 6525 ED Nijmegen, The Netherlands

Claus Schmidt, Max-Planck-Institut für Mathematik, Gottfried-Claren-Straße 26, D-5300 Bonn 3, West Germany

Steven Sperber, Department of Mathematics, University of Minnesota, Minneapolis, 55455 Minnesota,U.S.A.

Harvey Stein, Department of Mathematics, University of California, Berkeley, CA 94770, U.S.A.

Francis J. Sullivan, Dipartimento di Matematica, Università di Padova, Via Belzoni 7, 35100 Padova, Italy

Marko Tadič, Department of Mathematics, University of Zagreb, P.O. Box 187, 41001 Zagreb, Yugoslavia

Dinesh S. Thakur, School of Mathematics, Institute for Advanced Study, Princeton, 08540 New Jersey, U.S.A.

Peter Ullrich, Mathematisches Institut, Universität Münster, Roxeler Str. 64, D-4400 Münster, West Germany

Bert van der Marel, Mathematisch Instituut, Rijksuniversiteit Groningen, Postbus 800, 9700 AV Groningen, The Netherlands

Marius van der Put, Mathematisch Instituut, Rijksuniversiteit Groningen, Postbus 800, 9700 AV Groningen, The Netherlands

Lucien van Hamme, Faculty of Applied Sciences, Vrije Universiteit Brussel, Pleinlann 2, B-1050 Brussel, Belgium

Guido van Steen, Dept. of Math., Rijksuniv. Centrum Antwerpen, 171 Groenenborgerlaan, B-2020 Antwerpen, Belgium

Harm Voskuil, Mat. Instituut, Rijksuniversiteit Groningen, Postbus 800, 9700 AV Groningen, The Netherlands

Taoufik Youssefi, Mathématiques et Informatique, Université de Bordeaux I, 351 Cours de la Libération, F-33405 Talence Cedex, France